physica status solidi (b)

asic research

Volumes 151 to 156

Subject and Author Index 1989

Volumes 111 to 116

physica status solidi (a)

pplied research

ADEMIE-VERLAG BERLIN

N 0370-1972/0031-8965 phys. stat. sol. (b)/(a), Berlin **R22** (1989) 1 – 570

International Classification System for Physics*)

60. Condensed matter: structure, mechanical and thermal properties
    61. Structure of liquids and solids; crystallography
        (see also 68.20. Solid surface structure, 71. Electron states)
    62. Mechanical and acoustical properties of condensed matter
        (see also 61.70. Defects in crystals, 68.30. Surfaces and interfaces)
    63. Lattice dynamics and crystal statistics
        (see also 65. Thermal properties, 66.70. Thermal conduction, 68.30. Dynamics of surface and interface vibrations, 78.30. Infrared and Raman spectra)
    64. Equations of state, phase equilibria, and phase transitions
    65. Thermal properties of condensed matter
        (see also 63. Lattice dynamics; for thermodynamic properties of quantum fluids, see 67.40; for thermal properties of solid helium, see 67.80)
    66. Transport properties of condensed matter (nonelectronic)
    67. Quantum fluids and solids; liquid and solid helium
    68. Surfaces and interfaces; thin films and whiskers
        (for impact phenomena, see 79; for crystal growth, see 61.50)

70. Condensed matter: electronic structure; electrical, magnetic, and optical properties
    71. Electronic states
        (see also 63. Lattice dynamics, 73. Electronic structure and electrical properties of surfaces, interfaces, and thin films)
    72. Electronic transport in condensed matter
        (for surfaces, interfaces, and thin films, see 73)
    73. Electronic and electrical properties of surfaces, interfaces, and thin films
    74. Superconductivity
    75. Magnetic properties and materials
    76. Magnetic resonances and relaxation in condensed matter; Mössbauer effect
    77. Dielectric properties and materials
        (for conductivity phenomena, see 72.20 and 72.80)
    78. Optical properties and condensed-matter spectroscopy and other interactions matter with particles and radiation
        (for phonon spectra, see 63)
    79. Electron and ion emission by liquids and solids; impact phenomena

85. Devices**)

---

*) Excerpt; reproduced with permission of International Council for Scientific and Technical Information (ICSTI).

**) Outside the ICSTI Classification for Physics.

(The Substance Classification is given on cover three)

physica status solidi (b)

basic research

physica status solidi (a)

applied research

SUBJECT AND AUTHOR INDEX

Volumes 151 to 156, 1989 Volumes 111 to 116, 1989

ISSN 0370-1972/0031-8965

THEORIA CUM PRAXI

AKADEMIE-VERLAG · BERLIN

Subscriptions, and orders for this Subject and Author Index
or single copies of physica status solidi (b) and physica status solidi (a) should be directed:

to the Akademie-Verlag Berlin, Leipziger Str. 3–4, O-1086 Berlin,
to a book-shop or
to the wholesale distributing agency Kunst und Wissen, Erich Bieber, a Faxon Company, Postfach 102844, W-7000 Stuttgart 70;

In USA and CANADA:

to VCH Publishers, Inc., 303 NW 12th Avenue, Deerfield Beach, FL 33442–1788, USA;

in other countries:

to the international book and periodical trading,
to the distributing centre Kunst und Wissen, Erich Bieber GmbH, a Faxon Company, Stampfenbachstr. 73, Postfach, CH-8035 Zürich,
to the competent newspaper sales agency of the country, or
to Akademie-Verlag Berlin, Leipziger Str. 3–4, O-1086 Berlin.

Schriftleiter und verantwortlich für den Inhalt der Zeitschriften:

Professor Dr. E. Gutsche, Leipziger Straße 3–4, Postfach 1233, O-1086 Berlin.

Verantwortlich für den Inhalt der Register:

Dipl.-Phys. K. Müller, Leipziger Str. 3–4, Postfach 1233, O-1086 Berlin.

Verlag:

Akademie-Verlag Berlin, Leipziger Str. 3–4, O-1086 Berlin; Fernruf: 2236221 und 2236229;
Telex-Nr.: 114420; Bank: Berliner Stadtbank AG, Kto.-Nr.: 6651-25-130512.

Veröffentlicht unter der Registriernummer 1310 und 1620.

Gesamtherstellung:

Druckhaus „Thomas Müntzer“ GmbH, O-5820 Bad Langensalza.

Bestellnummer dieses Bandes: 1068/85 R 22.

AN (EDV) 20831/20735

Preface

The twenty second index volume contains volumes 151 to 156 of "physica status solidi (b) – basic research" as well as volumes 111 to 116 of "physica status solidi (a) – applied research", i.e. all volumes having appeared in 1989 containing 1492 papers.

Beginning from January 1986 the "International Classification System for Physics" (Condensed Matter, items 60 and 70) of the International Council for Scientific and Technical Information, Abstracting Board completed by item 85 and a Substance Classification (both outside the ICSTI Classification for Physics) will be used for quoting papers in physica status solidi. Nevertheless, before taking over these subject classifications into the index volume they were once more carefully revised and in some cases improved and completed in order to get a most uniform and clear classification for all papers. The average number of quotations per paper in this index volume is about 2.7.

I have to express my thanks to the readers for their interest, and I will always readily take up any critical remarks and proposals for improvements.

Berlin, August 1990

Karin Müller

Contents

International Classification System for Physics 6
Substance Classification (according to physica status solidi) 11
Some Comments on the Use of the Subject and Author Index 13
60. Condensed Matter: Structure, Mechanical, and Thermal Properties 17
61. Structure of Liquids and Solids; Crystallography 17
62. Mechanical and Acoustical Properties of Condensed Matter 70
63. Lattice Dynamics and Crystal Statistics 82
64. Equations of State, Phase Equilibria, and Phase Transitions 91
65. Thermal Properties of Condensed Matter 106
66. Transport Properties of Condensed Matter (Nonelectronic) 111
68. Surfaces and Interfaces; Thin Films and Whiskers 118
70. Condensed Matter; Electronic Structure; Electrical, Magnetic, and Optical Properties 137
71. Electronic States 137
72. Electronic Transport in Condensed Matter 174
73. Electronic Structure and Electrical Properties of Surfaces, Interfaces, and Thin Films 198
74. Superconductivity 217
75. Magnetic Properties and Materials 228
76. Magnetic Resonances and Relaxation in Condensed Matter; Moessbauer Effect 255
77. Dielectric Properties and Materials 269
78. Optical Properties and Condensed-Matter Spectroscopy and Other Interactions of Matter with Particles and Radiation 279
79. Electron and Ion Emission by Liquids and Solids; Impact Phenomena 322
85. Devices 328
S. Substances 333
Author Index 493

International Classification System for Physics*)

60. Condensed matter: structure, mechanical and thermal properties
61. Structure of liquids and solids; crystallography
    (see also 68.20. Solid surface structures, 71. Electron states)
    61.10. X-ray determination of structures
        (for specific determinations, see 61.55 to 61.80)
    61.12. Neutron determination of structures
        (for specific determinations, see 61.55 to 61.80)
    61.14. Electron determination of structures
        (for specific determinations, see 61.55 to 61.80)
    61.16. Other determination of structures
        (for specific determination, see 61.55 to 61.80)
    61.20. Classical, semiclassical, and quantum theories of liquid structure
        (for electronic states, see 71; for liquid helium, see 67)
    61.25. Studies of specific liquid structures
    61.30. Liquid crystals
    61.40. Amorphous and polymeric materials
    61.50. Crystalline state (including molecular motions in solids)
        (for magnetic structure and spin systems, see 75.25)
    61.55. Specific structures of elements and alloys
    61.60. Specific structures of inorganic compounds
    61.65. Specific structures of organic compounds
    61.70. Defects in crystals
        (see also 61.80. Radiation damage, 62. Mechanical and acoustical properties,
        71.55. Impurities and defect levels, 76.30. EPR of color centers and other defects,
        78.50. Impurity and defect absorption in solids)
    61.80. Radiation damage and other irradition effects
        (for techniques of structure determination, see 61.10 to 61.16; for electron and ion impact phenomena, see 79.20)
    61.90. Other topics in structure of liquids and solids

62. Mechanical and acoustical properties of condensed matter
    (see also 61.70. Defects in crystals, 68.30. Surfaces and interfaces)
    62.10. Mechanical properties of liquids
        (for viscosity of liquids, see also 66.20)
    62.20. Mechanical properties of solids (related to microscopic structure)
    62.30. Mechanical and elastic waves
    62.40. Anelasticity, internal friction, and mechanical resonances
        For thermomechanical effects, see 65.70
        For magnetomechanical effects, see 75.80
        For piezoelectric effects, see 77.60
        For elasto-optical effects, see 78.20
    62.50. High-pressure and shock-wave effects in solids
    62.60. Acoustical properties of liquids
        For lattice dynamics, phonons, see 63
        For second sound in quantum fluids, see 67.40
    62.65. Acoustical properties of solids
        For magnetoacoustic effects in metals, see 75.80
        For acoustoelectric effects, see 72.50
        For acousto-optical effects, see 78.20
    62.80. Ultrasonic relaxation
        (see also 74.30. Ultrasonic attenuation in superconductors)
    62.90. Other topics in mechanical and acoustical properties of condensed matter

*) Excerpt, reproduced with permission of International Council for Scientific and Technical Information (ICSTI).

63. Lattice dynamics and crystal statistics
(see also 65. Thermal properties, 66.70. Thermal conduction, 68.30. Dynamics of surface and interface vibrations, 78.30. Infrared and Raman spectra)
63.10. General theory
63.20. Phonons and vibrations in crystal lattices
63.50. Vibrational states in disordered systems
63.70. Statistical mechanics of lattice vibrations
(see also 65. Thermal properties of condensed matter, and 66.70. Thermal conduction)
63.75. Statistical mechanics of displacive phase transitions
For order–disorder and statistical mechanics of model systems, see 64.60
For crystallographic aspects of polymorphic and order–disorder transformations, see 61.50
63.90. Other topics in lattice dynamics and crystal statistics

64. Equations of state, phase equilibria, and phase transitions
64.10. General theory of equations of state and phase equilibria
64.30. Equations of state of specific substances
(see also 65.70. Thermal expansion)
64.60. General studies of phase transitions
(for magnetic, superconducting, and quantum fluid critical phenomena, see 75, 74, and 67, respectively)
64.70. Phase equilibria, phase transitions, and critical points of specific substances
64.75. Solubility, segregation, and mixing
64.80. Other phase properties of systems
64.90. Other topics in equations of state, phase equilibria, and phase transitions

65. Thermal properties of condensed matter
(see also 63. Lattice dynamics; for thermodynamic properties of quantum fluids, see 67.40; for thermal properties of solid helium, see 67.80)
65.20. Heat capacities of liquids
65.40. Heat capacities of solids
(for specific heat of superconductors, see 74.30; for specific heat of magnetic systems, see 75.40)
65.50. Thermodynamic properties and entropy
65.70. Thermal expansion and thermomechanical effects
(see also 64.30. Equations of state)
For thermal conduction in nonmetallic liquids, see 66.60; for nonmetallic solids, see 66.70
For electronic thermal conduction, see 72.10 and 72.20
For thermal conductivity of superconductors, see 74.30
For pyroelectric and electrocaloric effects, see 77.70
65.90. Other topics in thermal properties of condensed matter

66. Transport properties of condensed matter (nonelectronic)
66.10. Diffusion and ionic conduction in liquids
66.20. Diffusive momentum transport
(for viscosity of liquids, see also 62.10)
66.30. Diffusion in solids
66.60. Thermal conduction in nonmetallic liquids
(for thermal conduction in liquid metals, see 72.15)
66.70. Nonelectronic thermal conduction and heat-pulse propagation in nonmetallic solids
(for thermal conduction in solid metals, see 72.15; for statistical mechanics of lattice vibrations, see 63.70)
66.90. Other topics in nonelectronic transport properties

67. Quantum fluids and solids; liquid and solid helium
    67.20. Quantum effects on the structure and dynamics of nondegenerate fluids
    67.40. Boson degeneracy and superfluidity of helium-4
    67.50. Fermi fluids; liquid helium-3
    67.60. Mixed systems; liquid helium-3, -4 mixtures
    67.70. Films (including physical adsorption)
    67.80. Solid helium and related quantum crystals
    67.90. Other topics in quantum fluids
    (e.g., neutron-star matter)

68. Surfaces and interfaces; thin films and whiskers
    (for impact phenomena, see 79; for crystals growth, see 61.50)
    68.10. Fluid surfaces and fluid–fluid interfaces
    68.15. Liquid thin films
    68.20. Solid surface structures
    68.25. Mechanical and acoustical properties of solid surfaces and interfaces
    68.30. Dynamics of solid surfaces and interface vibrations
    68.40. Surface energy of solids; thermodynamic properties
    68.45. Solid–fluid interface processes
    68.48. Solid–solid interfaces (including bicrystals)
    (for grain boundaries, see 61.70)
    68.55. Thin film growth, structure, and epitaxy
    68.60. Physical properties of thin films: nonelectronic
    68.70. Whiskers and dendrites: growth, structure, and nonelectronic properties
    68.90. Other topics in the structure and nonelectronic properties of surfaces and thin films

70. Condensed matter: electronic structure; electrical, magnetic, and optical properties

71. Electronic states
    (see also 63. Lattice dynamics, 73. Electronic structure and electrical properties of surfaces, interfaces, and thin films)
    71.10. General theories and computational techniques
    71.20. Electronic density of states determinations
    (including energy states of liquid semiconductors)
    (see also 65.40. Electronic heat capacity)
    71.25. Nonlocalized single-particle electronic states
    71.30. Metal–insulator transitions
    71.35. Excitons and related phenomena (including electron–hole drops)
    71.36. Polaritons
    (including photon–phonon and photon–magnon interactions)
    71.38. Polarons and electron–phonon interactions
    (see also 63.20, phonon–electron interactions in lattices)
    71.45. Collective effects
    71.50. Localized single-particle electronic states (excluding impurities)
    71.55. Impurity and defect levels
    71.60. Positron states
    (see also 78.70. Positron annihilation)
    71.70. Level splitting and interactions
    (see also 75.10. — in magnetic phenomena, 75.30. Exchange and superexchange interactions, 73.20. Electronic surface states)
    71.90. Other topics in electron states

72. Electronic transport in condensed matter
    (for surfaces, interfaces, and thin films, see 73)
    72.10. Theory of electronic transport; scattering mechanism
    72.15. Electronic conduction in metals and alloys
    72.20. Conductivity phenomena in semiconductors and insulators
        (for nonelectronic conduction, see 66.70)
    72.30. High-frequency effects; plasma effects
    72.40. Photoconduction and photovoltaic effects; photodielectric effects
    72.50. Acoustoelectric effects
    72.55. Magnetoacoustic effects
    72.60. Mixed conductivity and conductivity transitions
    72.70. Noise processes and phenoma
    72.80. Conductivity of specific semiconductors and insulators
    72.90. Other topics in electronic transport in condensed matter

73. Electronic structure and electrical properties of surfaces, interfaces, and thin films
    73.20. Electronic surface states
        (for emission and impact phenomena, see 79)
    73.25. Surface conductivity
    73.30. Surface double layers, Schottky barriers, and work functions
    73.40. Interfaces
    73.60. Electronic properties of thin films
    73.90. Other topics in electrical properties of surfaces, interfaces, and thin films

74. Superconductivity
    74.10. Occurrence, critical temperature
    74.20. Theory
    74.30. General properties
    74.40. Fluctuations and critical effects
    74.50. Tunneling phenomena, Josephson effect, and proximity effects
    74.55. Type-I superconductivity
    74.60. Type-II superconductivity
    74.70. Superconducting materials
    74.90. Other topics in superconductivity

75. Magnetic properties and materials
    75.10. General theory and models of magnetic ordering
        (see also 71.25. Nonlocalized single-particle electronic states, 71.70. Level splitting and interactions)
    75.20. Diamagnetism and paramagnetism
    75.25. Spin arrangements in magnetically ordered materials
        (neutron studies, etc.)
    75.30. Magnetically ordered materials, other intrinsic properties
        (for critical-point effects, see 75.40)
    75.40. Critical-point effects, specific heats, short-range order
        (see also 65.40. Heat capacities)
    75.50. Studies of specific magnetic materials
    75.60. Domain effects, magnetization curves, and hysteresis
    75.70. Magnetic films and plates
    75.80. Magnetomechanical and magnetoelectric effects, magnetostriction
    For galvanomagnetic effects, see 72.15 and 72.20
    For magneto-optical effects, see 78.20
    75.90. Other topics in magnetic properties and materials

76. Magnetic resonances and relaxation in condensed matter; Mössbauer effect
- 76.20. General theory of resonances and relaxation
- 76.30. Electron paramagnetic resonance and relaxation
- 76.40. Diamagnetic and cyclotron resonances
- 76.50. Ferromagnetic, antiferromagnetic, and ferrimagnetic resonances; spin-wave resonance (see also 75.30. Spin waves)
- 76.60. Nuclear magnetic resonance and relaxation
- 76.70. Magnetic double resonances and cross effects
- 76.80. Mössbauer effect; other γ-ray spectroscopy
- 76.90. Other topics in magnetic resonances and relaxation

77. Dielectric properties and materials
(for conductivity phenomena, see 72.20 and 72.80)
- 77.20. Permittivity
- 77.30. Polarization and depolarization effects
- 77.40. Dielectric loss and relaxation
- 77.50. Dielectric breakdown and space-charge effects
- 77.55. Dielectric thin films
- 77.60. Piezoelectricity and electrostriction (for piezo-optical effects, see 78.20)
- 77.70. Pyroelectric and electrocaloric effects
- 77.80. Ferroelectricity and antiferroelectricity
- 77.85. Electrical resonances
- 77.90. Other topics in dielectric properties and materials

78. Optical properties and condensed-matter spectroscopy and other interactions of matter with particles and radiation
(for phonon spectra, see 63)
- 78.20. Optical properties and materials
- 78.30. Infrared and Raman spectra and scattering
- 78.35. Brillouin and Rayleigh scattering
- 78.40. Visible and ultraviolet spectra
- 78.45. Stimulated emission
- 78.50. Impurity and defect absorption in solids
- 78.55. Photoluminescence
- 78.60. Other luminescence and radiative recombination

For photoconduction and photovoltaic effects, see 72.40
- 78.65. Optical properties of thin films
- 78.70. Other interactions of matter with particles and radiation
- 78.90. Other topics in optical properties of condensed matter and other interactions of matter with particles and radiation

79. Electron and ion emission by liquids and solids; impact phenomena
- 79.20. Impact phenomena, including electron spectra and sputtering
- 79.40. Thermoionic emission
- 79.60. Photoemission and photoelectron spectra
- 79.70. Field emission and field ionization
- 79.75. Exoelectron emission
- 79.80. Resonance tunneling
- 79.90. Other topics in emission and impact phenomena in condensed matter

85. Devices*)

*) Outside the ICSTI Classification for Physics.

Substance Classification (according to physica status solidi)*)

S1. Transition metals and their alloys and compounds
(for oxides, see S10)
- S1.1. Fe
- S1.2. Other 3d metals
- S1.3. Noble metals
- S1.4. Lanthanides
- S1.5. Actinides
- S1.61. Carbides, silicides, germanides
- S1.62. Nitrides, phosphides, arsenides
- S1.63. Sulfides, selenides, tellurides

S2. Alkali metals and their alloys and compounds
(for alkali halides, see S9.11)

S3. Alkaline earth metals and their alloys and compounds
(for alkaline earth halides, see S9.12)

S4. Other metals and their alloys

S5. Nonmetallic elements
- S5.11. Silicon
- S5.12. Germanium
- S5.2. Solidified Gases

S6. Group IV compounds
(for transition metal compounds, see S1.61)

S7. Group V compounds
(for transition metal compounds, see S1.62)
- S7.1. III–V compounds
- S7.11. Phosphides
- S7.12. Arsenides
- S7.13. Antimonides
- S7.15. Mixed crystals
- S7.16. Higher compounds

S8. Chalcogenides
(for transition metal compounds, see S1.63)
- S8.1. II–VI compounds
- S8.11. Sulfides
- S8.12. Selenides
- S8.13. Tellurides
- S8.15. Mixed crystals
- S8.16. Higher compounds

S9. Halides
- S9.11. Alkali halides
- S9.12. Alkaline earth halides
- S9.15. Mixed crystals
- S9.16. Higher compounds

S10. Oxides
- S10.1. Simple oxides
- S10.15. Mixed crystals

S11. Other higher compounds
- S11.1. Salts of oxy-acids
- S11.2. Spinels, garnets, ferrites

S12. Organic compounds
- S12.1 Quasi-one-dimensional systems

*) Outside the ICSTI classification for Physics.

Some Comments on the Use of the Subject and Author Index

The "International Classification System for Physics" (the condensed matter items 60 and 70) completed by item 85 and a Substance Classification (according to physica status solidi) is used. On cover two the second-level headings of the condensed matter items and on pages 6 to 10 the third-level headings are reproduced. As a help for using the subject index on pages 14 and 15 in addition also a series of important fourth-level headings is given. The substance classification is published on cover three and on page 11.

All papers will be quoted under each relevant section. There is made a distinction between one or two principal entries (refering to the main subject matter of the paper) and secondary entries (refering to subsidiary subject matter and substance, respectively).

On the line following the title the subject classification is given in the same manner as in the volumes. Each entry gives the volume, the number of the issue, the initial and final page numbers. A paper of physica status solidi (b) or physica status solidi (a) is marked by (b) or (a) in front of the line, respectively. The letter "K" before the page numbers indicates a short note published in the offset part of the journal.

The entries for review articles, original papers, short notes, and errata are made together in a single list for each subject. These are ordered according to data of publication, irrespective of being a paper of physica status solidi (b) or physica status solidi (a). Original papers are sometimes published in several parts under the same principal title (in general with special subtitles). The various parts of papers of this type are treated separately and each part is classified according to its particular contents.

The author index gives the names of the authors in alphabetical order. Each name is followed by the bibliographic details of the paper. A paper of physica status solidi (b) or physica status solidi (a) is again marked by (b) or (a) in front of the number of the volume, respectively. In the last column the subject classification of the paper is presented in the same manner as in the volumes. When several papers have been published by the same author they are arranged according to data of publication, irrespective of being a paper of physica status solidi (b) or physica status solidi (a). To find a paper of your interest please choose first the special entry of the subject index and then use the volume number and page.

In using the subject index attention should be given to the following details:

61.10. to 61.16. Theory and experimental techniques of determination of structures
— for specific determinations see 61.55 to 61.80
61.16. Other determination of structures
— electron microscopy and field-ion microscopy determinations
— EPR and NMR determinations
61.25. Studies of specific liquid structures
— macromolecular and polymer solutions
— molten salts
— liquid metals
61.40. Amorphous and polymeric materials
— glasses, polymers, elastomers, plastics
61.50. Crystalline state (including molecular motions in solids)
— physics of crystal growth
— crystal symmetry: models, space groups, crystalline systems and classes
— crystal morphology and orientation
— crystallographic aspects of polymorphic and order–disorder transformations
61.70. Defects in crystals
— grain and twin boundaries
— stacking faults and other planar or extended defects
— doping and implantation of impurities
— impurity concentration, distribution, and gradients
61.80. Radiation damage and other irradiation effects
— channeling, blocking, energy loss of particles
— X-rays, γ-rays, electrons and positrons, neutrons, atoms and molecules, ions
— for ion implantation see 61.70
63.20. Phonons and vibrations in crystal lattices
— phonon–phonon, phonon–electron, and phonon–defect interactions
64.60. General studies of phase transitions
— order–disorder and statistical mechanics of model systems
— equilibrium properties near single critical points, critical exponents
— dynamic critical phenomena
— multicritical points
— metastable phases
64.80. Other phase properties of systems
— stoichiometry and homogeneity
— microstructure
66.30. Diffusion in solids
— theory of diffusion and ionic conduction in solids
— self-diffusion and ionic conduction in non-metals
— diffusion, migration, and displacement of impurities and other defects
71.25. Nonlocalized single-particle electronic states
— techniques of band-structure calculation
— measurements of Fermi surface parameters (including dHvA, magnetoacoustic, positron annihilation, and cyclotron resonance studies, etc.)
— effective mass and *g*-factors
— band structure of crystalline metals, semiconductors, and insulators
71.45. Collective effects
— exchange, correlation, dielectric and magnetic functions, plasmons
— Fermi-Thomas models
— calculations of total electronic binding energy
71.70. Level splitting and interactions
— crystal and ligand fields
— spin–orbit coupling, Zeeman, Stark, and strain splitting
— exchange interactions
— nuclear states and interactions

72.10. Theory of electronic transport; scattering mechanism
- general formulation of transport theory
- scattering by phonons, magnons, and other nonlocalized excitations
- scattering by point defects, dislocations, surfaces, and other imperfections (including Kondo effect)

72.15. Electronic conduction in metals and alloys
- galvanomagnetic and other magnetotransport effects
- thermomagnetic and thermoelectric effects
- collective modes
- scattering mechanism and Kondo effect

72.20. Conductivity phenomena in semiconductors and insulators
- scattering mechanism
- piezoresistance
- galvanomagnetic and other magnetotransport effects
- thermomagnetic and thermoelectric effects

73.40. Interfaces
- contact resistance, contact potential, rectification
- tunneling: general
- p–n junctions
- MIS, MIM, MSM, SIS, and SMS structures

75.30. Magnetically ordered materials, other intrinsic properties
- saturation moments and magnetic susceptibility
- spin waves
- exchange and superexchange interactions
- anisotropy
- magnetic impurity interactions
- magnetic phase boundaries
- magnetocaloric effect

75.50. Studies of specific magnetic materials
- ferromagnetism
- antiferromagnetics, ferrimagnetics
- amorphous magnetic materials
- magnetic liquids

75.60. Domain effects, magnetization curves, and hysteresis
- high coercitivity materials
- fine-particle systems
- magnetic aftereffects
- magnetic annealing and temperature-hysteresis effects

78.20. Optical properties and materials
- general theory
- optical constants and parameters
- optical rotatory power
- birefringence
- piezo-, elasto-, acousto-, electro-, magneto-, and thermo-optical effects

78.60. Other luminescence and radiative recombination
- electroluminescence
- cathodo- and iono-luminescence
- thermoluminescence
- sono- and tribo-luminescence
- chemiluminescence

78.70. Other interactions of matter with particles and radiation
- positron annihilation
- X-ray scattering, absorption, and emission
- microwave and radiofrequency (including resonances)

79.20. Impact phenomena, including electron spectra and sputtering
- electron impact: Auger and secondary emission
- atom, molecule, and ion impact
- atomic and molecular beam interactions

Subject Index

60. Condensed Matter: Structure, Mechanical, and Thermal Properties

61. Structure of Liquids and Solids; Crystallography
(see also 68.20. Solid surface structures, 71. Electron states)

61.10. X-ray determination of structures
(for specific determinations, see 61.55 to 61.80)

S.A. MAKSIMENKO
Covariant Theory of the Multiwave Bragg-Laue X-Ray Diffraction in Crystals
61.10
(b) Vol.151,No.1, 17-22

V. HOLÝ and J. KUBĚNA
On the Integrated Intensity of X-Ray Diffraction in Crystals with Randomly Distributed Defects
61.10 and 61.70; S5.11
(b) Vol.151,No.1, 23-28

V.S. HAROUTYUNYAN and P.H. BEZIRGANYAN
Method of Determing Supersmall Dilatations in a Crystal Lattice
61.10
(a) Vol.111,No.1, 41-46

A. SCHUBERT and B. MICHEL
Simultaneous Line Broadening and Texture Analysis
61.10 and 68.20; S1.1
(a) Vol.111,No.2, K137-K140

V. GEIST and D. STEPHAN
Investigation of Lattice Strain in Proton-Irradiated Semiconductor Materials
61.80; 61.10; S5.11; S5.12; S7.1
(a) Vol.112,No.1, 79-88

J. AULEYTNER
Structural Changes in Solids Induced by Heavy Ions at Energies of MeV/n. Results of X-Ray and Electron Microscopic Studies
61.80; 61.10; 61.16; S5.11; S11.2
(a) Vol.112,No.2, 679-688

C.N.W. DARLINGTON
On the Changes in Structure of PLZT(8.7/65/35) between 80 and 750K
64.70; 61.10; 77.80
(a) Vol.113,No.1, 63-70

T. FUKUMORI and K. FUTAGAMI
Nonuniformity on GaAs Wafers Revealed by an X-Ray Double-Crystal Method
61.10 and 61.70; S7.12
(a) Vol.113,No.1, K19-K22

N.N. FALEEV, L.I. FLAKS, and S.G. KONNIKOV
Double Crystal X-Ray Diffraction Characterization of Multilayer Heteroepitaxial Structures with Submicron Layers
68.55; 61.10; S7.15
(a) Vol.113,No.2, 431-438

A.M. AFANASEV, R.M. IMAMOV, A.V. MASLOV, and E.KH. MUKHAMEDZHANOV
On the Yield of Photoelectrons in X-Ray Diffraction under the Conditions of Total External Reflection
61.10
(a) Vol.113,No.2, K153-K156

D.Y. LI, X.F. WU, and T. KO
The Effect of Stress on the Lattice Instability of a TiNi Alloy
64.70; 61.10; 62.20; 72.15; S1.2
(b) Vol.154,No.1, 85-96

D.V. NOVIKOV, E.A. KONDRASHKINA, and S.A. STEPANOV
Study of X-Ray Surface Backscattering
61.10; 68.20; S5.12
(a) Vol.114,No.1, K7-K10

E. WOLSKA and U. SCHWERTMANN
Selective X-Ray Line Broadening in the Goethite-Derived Hematite Phase
61.10 and 64.70; 61.60; S10.1
(a) Vol.114,No.1, K11-K16

V. HOLÝ and J. KUBĚNA
Characterization of Microdefects in Silicon by Means of X-Ray Reflection Curves
61.10 and 61.70; S5.11
(b) Vol.155,No.2, 339-348

J. DEXPERT-GHYS, Y. CHARREIRE, P. ESTEBAN-PUGES. L. ALBERT, and H. DEXPERT
Optical and EXAFS Investigations of ZnS:Tm^{3+} Powders
71.55; 61.10; 78.55; S8.11
(b) Vol.155,No.2, 571-580

J. HAERTWIG and S. GROSSWIG
Measurement of X-Ray Diffraction Angles of Perfect Monocrystals with High Accuracy Using a Single Crystal Diffractometer
61.10; S5.11
(a) Vol.115,No.2, 369-382

T. KITANO, T. ISHIKAWA, and J. MATSUI
Contrast Formation Mechanism around the Cell Walls in Equi-Lattice-Spacing Mapping X-Ray Topographs for an Undoped...
61.10; S7.12
(a) Vol.115,No.2, 383-388

R. FABBRI, M. SERVIDORI, A. ZANI, and F. CEMBALI
Two-Stage Recovery of Lattice Damage Induced in Silicon by Ion Implantation below Amorphization Threshold. Analysis by...
61.80; 61.10; S5.11
(a) Vol.115,No.2, 437-444

R.M. IMAMOV, A.A. LOMOV, and D.V. NOVIKOV
Grazing Incidence Diffraction X-Ray Topography
61.10; 68.20; S5.11
(a) Vol.115,No.2, K133-K134

S.N. MAITI, S. SENGUPTA, and D. ROY
Theoretical Interpretation of EXAFS Results for the Relaxation in Mixed Crystals
64.75; 61.10; S9.11
(b) Vol.156,No.1, 83-92

M.YA. GAMARNIK and YU.YU. SIDORIN
Change of the Unit Cell Parameters in Highly Dispersed Gold, Silver, and Copper Powders
61.10 and 68.90; S1.3
(b) Vol.156,No.1, K1-K4

A.M. AFANASEV, R.M. IMAMOV, E.KH. MUKHAMEDZHANOV, and A.A. BZHEUMIKHOV.
Double-Channel X-Ray Standing Wave Technique
61.10; 78.70; S7.12
(a) Vol.116,No.1, 197-202

WENHUI SU, CHANGQING JIN, LANYING HE, YIFENG WANG, HONGJIAN LIU, and XIANG LONG
Study on Oxygen Hole and Superconductivity in $Y_1Ba_2Cu_3O_{7-\delta}$ Material
64.70 and 74.70; 61.10; S10.15
(a) Vol.116,No.1, 267-274

H. HERMANN and M. ERMRICH
Absorption of Scattered X-Rays in Stochastic Multi-Phase Systems
61.10
(a) Vol.116,No.1, K19-K24

P.A. BEZIRGANYAN, A.P. BEZIRGANYAN, and S.E. BEZIRGANYAN
X-Ray Diffraction by a Dielectric with Cosine-Like Polarizability ($0 \lesssim B \lesssim \pi/2$ Symmetrical Bragg Case)
61.10
(a) Vol.116,No.2, 469-482

A.H. ROSTOMYAN, P.H. BEZIRGANYAN, and A.M. ROSTOMYAN
X-Ray Resonators. I.Theory
61.10 and 78.70
(a) Vol.116,No.2, 483-492

A.M. ROSTOMYAN, A.H. ROSTOMYAN, and P.H. BEZIRGANYAN
X-Ray Resonators. II. Calculation and Experiment
61.10 and 78.70
(a) Vol.116,No.2, 493-502

V.I. KHRUPA, V.V. NIKOLAEV, and M.YA. SKOROKHOD
On the Transition between Dynamical and Kinematical X-Ray Diffraction in Thin Crystals with Randomly Distributed Dislocations
61.10; 61.70; S5.11
(a) Vol.116,No.2, K141-K146

61.12. Neutron determination of structures
(for specific determinations, see 61.55 to 61.80)

R. MICHALEC, B. CHALUPA, and J. VÁVRA
The Influence of a High-Magnetic Field on the Neutron Diffraction by Perfect Crystals
61.12
(a) Vol.111,No.2, K133-K136

R. MICHALEC, B. CHALUPA, and J. VÁVRA
Neutron Acoustic Resonance in Perfect InSb Single Crystals
61.12 and 62.65; S7.13
(a) Vol.112,No.1, 29-34

O. UEMURA, T. AKAI, Y. KAMEDA, and T. SATOW
The Molten Structure of $CuTlSe_2$ and $CuTlTe_2$
61.40; 61.12; S8.16
(a) Vol.112,No.2, 467-472

J. KULDA and P. LUKÁŠ
Plane Wave Diffraction by a Deformed Crystal
61.12
(b) Vol.153,No.2, 435-442

R. MICHALEC
Vibrating Phase Echo Concept for Small Angle Inelastic Neutron Scattering
61.12
(a) Vol.113,No.2, K157-K160

A.I. GUSEV and A.A. REMPEL
Local Static and Dynamic Atomic Displacements in Disordered Niobium Carbide
61.50; 61.12; S1.61
(b) Vol.154,No.2, 453-460

J. WĄSICKI, W. NAWROCIK, Z. PAJĄK, I. NATKANIEC, and A.V. BELUSHKIN
Neutron Scattering Study of Some Pyridinium Salts
64.70; 61.12; S9
(a) Vol.114,No.2, 497-504

P. LUKÁŠ and J. KULDA
Quasi-Classical Trajectory of Thermal Neutrons in Elastically Deformed Crystals
61.12
(b) Vol.156,No.1, 41-48

W. ŁUŻNY, S. NIZIOŁ, J. MAYER, and I. NATKANIEC
Inelastic Neutron Scattering on Solid and Liquid Gallium
61.12; S4
(a) Vol.116,No.1, K25-K28

G. ANDRÉ, A. OLEŚ, and A. SZYTUŁA
Neutron Diffraction Study of NdCoSi
61.12; S1.61
(a) Vol.116,No.1, K29-K32

V. PIERRON-BOHNES, C. LEROUX, J.P. AMBROISE, A. MENELLE, and P. BASTIE
Measurement of the Mean Static Displacement and of the Debye Waller Factor between 5 and 1464 K in Fe-19.5at%Al
61.55; 61.12; 76.80; S1.1
(a) Vol.116,No.2, 529-536

61.14. Electron determination of structures
(for specific determinations, see 61.55 to 61.80)

YIFAN CHENG and RENHUI WANG
Dynamical Theory of Electron Diffraction for Quasicrystals
61.14; 61.70; S1.2; S4
(b) Vol.152,No.1, 33-38

D. HOEHL and H. BARTSCH
Results of In-Situ Annealing of Ion-Implanted Silicon in a High Voltage Electron Microsope
61.70 and 61.80; 61.14; 68.48; S5.11
(a) Vol.112,No.1, 419-424

A.V. VINOGRADOV, O.A. ZHAROV, V.V. OVCHAROV, and A.V. SUDARKIN
Computer Simulation of Void Parameter Reconstruction under the Conditions of the Two-Beam Dynamical Theory
61.14
(a) Vol.113,No.2, K149-K152

H.S. KIM and S.S. SHEININ
The Many Beam Dynamical Theory of Contrast in Electron Microscope Images of Microtwins for the Non-Symmetrical Laue ...
61.14; 61.70
(a) Vol.115,No.1, 51-60

H. ZANGAR, J.L. MIANE, C. COURSEILLE, N.B. CHANH, M. COUZI, and Y. MLIK
Structural Phase Transition in the Perovskite-Type Layer Compound $(C_3H_7NH_3)_2PbCl_4$
64.70; 61.14; 77.20; S11
(a) Vol.115,No.1, 107-118

N. VOUROUTZIS and C. MANOLIKAS
Phase Transformations and Discommensurations in Hexagonal Cuprous Telluride
61.14 and 64.70; S1.63
(a) Vol.115,No.2, 399-412

G. MOELLENSTEDT
My Early Work on Convergent-Beam Electron Diffraction
61.14
(a) Vol.116,No.1, 13-22

TH. LEUTHNER, H. LICHTE, and K.-H. HERRMANN
STEM-Holography Using the Electron Biprism
61.14
(a) Vol.116,No.1, 113-122

L. TERZIEV, J. LECOMTE-BECKERS, P. RASHEV, and T. GREDAY
Study of Transition Phases during Isothermal Ageing at 250 °C of Zn-27%Al Alloy
61.55; 61.14; 64.70; S1.2
(a) Vol.116,No.2, 521-528

61.16. Other determination of structures
(for specific determinations, see 61.55 to 61.80)

H.S. KIM and S.S. SHEININ
An Electron Microscope Method for Determing the Nature of Stacking Faults in F.C.C. Materials which Does not Rely on ...
61.16 and 61.70; S1.2
(a) Vol.111,No.1, 47-52

J. HEROLD, H. BARTSCH, W. WESCH, and G. GOETZ
Investigation of Low-Temperature Epitaxial Regrowth of Ion-Implanted Amorphous GaAs
68.48; 61.16; 61.80; S7.12
(a) Vol.111,No.1, 59-70

F. HAIDER
Calculation of TEM Contrast of Small Coherent Precipitates
61.16; 61.70; S1.2; S1.3
(a) Vol.111,No.1, 71-80

B.C. DE COOMAN and C.B. CARTER
The Formation of Faulted Dipoles in Plastically Deformed GaAs
61.70; 61.16; 62.20; S7.12
(a) Vol.112,No.1, 41-54

D.M. VANDERWALKER
Hydride Formation on Dislocations in Titanium
61.70; 61.16; S1.2
(a) Vol.112,No.1, 73-78

M. VERWERFT, D. VAN DYCK, V.A.M. BRABERS, J. VAN LANDUYT, and S. AMELINCKX
Electron Microscopic Study of the Phase Transformations in $AgNbO_3$
61.16 and 64.60; S11.1
(a) Vol.112,No.2, 451-466

B.C. DE COOMAN and C.B. CARTER
High-Resolution Microscopy of Dissociated Screw Dislocations in GaAs
61.70; 61.16; S7.12
(a) Vol.112,No.2, 473-492

J. AULEYTNER
Structural Changes in Solids Induced by Heavy Ions at Energies of MeV/n. Results of X-Ray and Electron Microscopic Studies
61.80; 61.10; 61.16; S5.11; S11.2
(a) Vol.112,No.2, 679-688

J. MITTENBACHER and R. MUEHLE
A Scanning Ion Microbeam System with a Liquid Metal Ion Source
61.80 and 79.70; 61.16; S1.3; S4
(a) Vol.112,No.2, 825-828

Z.C. KANG, D. MORIN, L. PIERRE, O. MONNEREAU, and C. BOULESTEIX
AEM Study of the High Temperature Superconductor $Bi_2Sr_2CaCu_2O_8$
61.60 and 74.70; 61.16; S10
(a) Vol.113,No.1, 23-34

M. GUYMONT, A. TOMAS, M.-P. PARDO, and M. GUITTARD
Electron Microscopy Study of gamma-Ga_2S_3
61.16 and 61.60; S8
(a) Vol.113,No.1, K5-K8

N.I. BORGARDT, S.K. MAKSIMOV, and V.V. OVCHAROV
Defocused Electron Microscopy Images of Small Defects with Black-and-White Contrast
61.16 and 61.70
(a) Vol.113,No.1, K23-K26

N.I. BORGARDT, V.N. KUKIN, S.K. MAKSIMOV, and D.I. PISKUNOV
Defocused Images of Stacking Faults
61.16 and 61.70
(a) Vol.113,No.2, 251-268

A.V. LATYSHEV, A.L. ASEEV, A.B. KRASILNIKOV, and S.I. STENIN
Initial Stages of Silicon Homoepitaxy Studied by in situ Reflection Electron Microscopy
68.55; 61.16; S5.11
(a) Vol.113,No.2, 421-430

Z.C. KANG, C. BOULESTEIX, G. NIHOUL, O. MONNEREAU, F. REMY, and A. CASALOT
High Resolution Image of a New Mixed Valence Compound: $Al_2Ti_7O_{15}$
61.60; 61.16; S10
(a) Vol.114,No.1, 73-82

K.Z. BOTROS and A.S. TITI
On Assessing Deviations from Symmetrical Laue Diffraction Conditions Using TEM Images of Stacking Faults
61.16 and 61.70; S1.2
(a) Vol.114,No.1, 83-90

Z.C. KANG, F. REMY, O. MONNEREAU, A. CASALOT, and C. BOULESTEIX
Plane Defects in a New Mixed Valence Compound $Al_2Ti_7O_{15}$
61.60; 61.16; 61.70; S10
(a) Vol.114,No.2, 441-450

K. SATO, P.E. DONOVAN, and W.M. STOBBS
Optimisation of the Fourier Component of Potential in Bloch Wave TEM Image Contrast Calculations
61.16; S1.2; S4
(a) Vol.115,No.1, 61-74

N.Y. JIN, F. PHILLIPP, and A. SEEGER
Investigation of Defect Ordering in Heavily Irradiated Metals by High-Voltage Electron Microscopy
61.70; 61.16; S1.2; S1.3
(a) Vol.116,No.1, 91-112

K. SCHEERSCHMIDT, R. HILLEBRAND, and J. HEYDENREICH
Computer Simulation of Diffraction Contrast Images and Lattice Fringe Patterns of Small Spherical Inclusions
61.16; 61.70
(a) Vol.116,No.1, 123-144

F. HEINRICHT
Reflection Electron Microscopy of Ultra Fast Surface Processes
61.16; 68.20
(a) Vol.116,No.1, 145-152

E. PIPPEL and J. WOLTERSDORF
Electron Microscopy of TiC Containing Alumina Ceramics
61.70; 61.16; S1.61; S10
(a) Vol.116,No.1, 165-178

A. SCHAPER, D. ZENKE, E. SCHULZ, R. HIRTE, and M. TAEGE
Structure-Property Relationships of High-Performance Polyethylene Fibres
61.40; 61.16; 62.20; S12
(a) Vol.116,No.1, 179-196

A.L. TÓTH
Monte-Carlo Simulation of Diffusion Length Measurements by a Wedge Absorber in the Scanning Electron Microscope (SEM)
61.16 and 72.20
(a) Vol.116,No.2, K137-K140

61.20. Classical, semiclassical, and quantum theories of liquid structure (for electron states, see 71; for liquid helium, see 67)

J. MAĆKOWIAK
A Hamiltonian Model of Liquefaction and Crystallization in Three Dimensions
61.20 and 64.60
(b) Vol.154,No.1, 115-126

61.25. Studies of specific liquid structures

H. KRESSE, H. STETTIN, F. GOUDA, and G. ANDERSSON
Dielectric Relaxation of a Mixture of a Nematic Liquid Crystal and a Non-Liquid Crystalline Compound
61.25 and 77.20; S12
(a) Vol.111,No.2, K265-K268

R.V. GOPALA RAO and A. SATPATHY
Structural Studies of Charged Hard-Sphere Mixtures through Blums Generalized Gamma Parameter. Application to ...
61.25 and 61.50; S9.11
(b) Vol.153,No.2, 421-434

R.V. GOPALA RAO and A. SATPATHY
Application of Statistical Mechanics Using Asymmetric Electrolytes. Model Evaluation of Structural Properties of Ionic Gold-Cesium Melts
61.25; S1.3; S2
(b) Vol.154,No.2, 427-438

R.V. GOPALA RAO and U. BANDYOPADHYAY
Evaluation of Partial Structures and Radial Distribution Functions of CaF_2 and $BaCl_2$
61.25; S9.12
(a) Vol.115,No.2, 357-368

61.30. Liquid crystals

S.A. RÓŻAŃSKI, W. KUCZYŃSKI, and J. MAŁECKI
Biaxiality of the Ferroelectric Liquid Crystal DOBAMBC in Electric Field
77.80 and 78.20; 61.30; S12
(a) Vol.112,No.2, 617-624

YUAN-CHAO CHU, T. TSANG, and L. YIN
Structures of Cyano-Biphenyl Liquid Crystals
61.30; S12
(a) Vol.114,No.1, K1-K6

D. MONSELESAN and H.-R. TREBIN
Temperature Dependence of the Elastic Constants for Biaxial Nematic Liquid Crystals
61.30 and 62.10
(b) Vol.155,No.2, 349-358

G. HAUCK, E. KUKLEVA, and H.D. KOSWIG
Electrically Induced Deformations in Cholesteric Systems with Pretilt
61.30; 64.70; S12
(a) Vol.116,No.2, 545-554

61.40. Amorphous and polymeric materials

M. KŁOSEK, J.W. MOROŃ, and T. POLOCZEK
Activation Energy of Linear Heating Crystallization in $Ni_{70.3}Fe_{4.7}Si_{15}B_{10}$ Glassy Alloy Determined Using the Internal ...
61.40 and 64.70; 62.40; S1.1; S1.2
(a) Vol.111,No.1, 127-136

I. MUEHLING and P. BLAUDECK
Multiple Scattering Approach to the Density of States and Electron Energy Loss Spectra of Amorphous Carbon
61.40 and 71.20; S5
(b) Vol.151,No.2, 545-554

GONG YU, GUANGHUA CHEN, and FANGQING ZHANG
Theoretical Study of Raman Spectrum in a-$Si_{1-x}N_x$:H Films
61.40 and 68.30; 78.30; S5.11
(b) Vol.152,No.1, 73-78

J. VAN DER STEL, H.B. VAN VELDHUIZEN, G.W. KOEBRUGGE, J. SIETSMA, and A. VAN DEN BEUKEL
Change of Electrical Resistivity and Youngs Modulus during Crystallization of Amorphous $Fe_{40}Ni_{40}B_{20}$
61.40 and 64.70; 62.20; 72.15; S1.1; S1.2
(a) Vol.112,No.1, 35-40

KH.YA. MULYUKOV, R.Z. VALIEV, G.F. KORZNIKOVA, and V.V. STOLYAROV
The Amorphous $Fe_{83}Nd_{13}B_4$ Alloy Crystallization Kinetics and High Coercivity State Formation
75.50; 61.40; S1.1
(a) Vol.112,No.1, 137-144

YU.G. CHUKALKIN, V.R. SHTIRZ, and B.N. GOSHCHITSKII
The Structure and Magnetism of Amorphous $Y_3Fe_5O_{12}$
61.40 and 75.50; S11.2
(a) Vol.112,No.1, 161-174

K. HANDRICH and J. RESCH
On the Theory of Roton-Like Excitations in Amorphous Solids
61.40; 63.50
(b) Vol.152,No.2, 377-394

G.B. SMITH, D.R. MCKENZIE, and P.J. MARTIN
An XPS Study of Chemical Order in Hydrogenated Amorphous Silicon-Carbon Alloy Films
61.40 and 68.55; 78.70; S5; S5.11
(b) Vol.152,No.2, 475-480

O. UEMURA, T. AKAI, Y. KAMEDA. and T. SATOW
The Molten Structure of $CuTlSe_2$ and $CuTlTe_2$
61.40; 61.12; S8.16
(a) Vol.112,No.2, 467-472

WEN-CAI WANG and YU CHEN
Study on the Short-Range Order Structures of Fe-Ge Amorphous Thin Films by EXAFS
61.40 and 68.55; S1.1; S5.12
(a) Vol.112,No.2, 541-550

M. KISZA and E. OLESZKIEWICZ
Modified PMMA-Polymers for Masks in Ion Beam Lithography
61.40 and 61.80; S12
(a) Vol.112,No.2, 761-764

W. RUDOLPH, D. GRAMBOLE, C. HEISER, F. HERRMANN, and C. NEELMEIJER
Compositional Changes of PMMA Layers during ^{15}N Bombardment
61.40 and 61.80; S12
(a) Vol.112,No.2, 765-768

R.R. NIGMATULLIN
The Generalized Fractals and Statistical Properties of the Pore Space of the Sedimentary Rocks
61.40
(b) Vol.153,No.1, 49-58

C. LANDRON, J.P. COUTURES, R. ERRE, and P. LEHUEDE
Etude par ESCA de l'oxygene dans les verres silicatés alcalins
61.40; 64.75; S10.15
(a) Vol.113,No.1, 35-42

O.A. GOLIKOVA
Amorphous Boron and Amorphous-Like Borides
61.40 and 72.20; S5
(a) Vol.113,No.1, K1-K4

M. BOURROUS and H. KRONMUELLER
Structural Transition from the As-Quenched to the Relaxed State in Metallic Glasses
61.40 and 64.60; 75.60; S1.2
(a) Vol.113,No.2, 383-392

P. ROTH and E. HEGENBARTH
Connection between the Density of Localized States, Their Volume, and the Glass Transition Temperature of Amorphous Solids
61.40
(b) Vol.154,No.2, K93-K98

K.N. MADHUSOODANAN, K. NANDAKUMAR, J. PHILIP, S.S.K. TITUS, S. ASOKAN, and E.S.R. GOPAL
Photoacoustic Investigation of Glass Transition in As_xTe_{1-x} Glasses
64.70; 61.40; 65.70; 72.80; S8
(a) Vol.114,No.2, 525-530

R.K. SHUKLA, S. SWARUP, A. KUMAR, and A.N. NIGAM
Structural Studies in Glassy Semiconducting Se-Te Alloys
61.40 and 78.70; S4
(a) Vol.115,No.1, K105-K108

J. SERNA, J.CH. ABBÉ, and G. DUPLÂTRE
Size of Voids in Polyethylene
61.40 and 61.70; 78.70; S12
(a) Vol.115,No.2, 389-398

S. ENZO, G. COCCO, I. SOLETTA, C. ANTONIONE, and L. BATTEZZATI
Structuring Effects in FeSiB Metallic Glasses. An EXAFS Approach
61.40 and 64.60; S1.1
(a) Vol.115,No.2, 459-466

YUANSHENG WANG, MING MAO, and YIZHEN HE
Influence of a Third Metallic Component on the Crystallization of Metallic Glasses Fe-Ni-Si-B
61.40 and 64.70; S1.1; S1.2
(a) Vol.115,No.2, 467-474

V.N. NOVIKOV
Precision Dilatometric Studies of Metallic Glasses
61.40 and 64.70; S1.1; S1.2
(a) Vol.115,No.2, 475-482

D. MEIMARIS, F. VIRAS, and P. BEKRIS
A Study of the Structural Changes in Cd-Ge-Sb-As Glass during Transition to the Crystalline State Induced by Thermal Treatment
64.70 and 72.80; 61.40; S7.16
(a) Vol.115,No.2, 483-490

A. SCHAPER, D. ZENKE, E. SCHULZ, R. HIRTE, and M. TAEGE
Structure-Property Relationships of High-Performance Polyethylene Fibres
61.40; 61.16; 62.20; S12
(a) Vol.116,No.1, 179-196

DECHENG TIAN, KAIYANG LI, and DIHUA DING
Magnetic Aftereffect and Mossbauer Studies of Structural Relaxation in $(Fe_{1-x}Ce_x)_{84}B_{16}$ Amorphous Alloys
61.40; 75.60; 76.80; S1.1
(a) Vol.116,No.1, 203-212

V.V. LOPATIN and A.V. KABYSHEV
On the Connection between the Building of Aluminium Nitride Ceramics and Its Dielectric Properties
61.40; 72.80; 77.20; S1.62
(a) Vol.116,No.1, 221-230

HONG ZHANG, XIAO-GANG WANG, LE LUO, and SHU-HUI HU
A Study on the Positron Annihilation of Glassy $Bi_1Sr_1Ca_1Cu_2O_x$ Quenched from the Molten Liquid
61.40 and 78.70; 74.70; S10.15
(a) Vol.116,No.1, K7-K12

V.I. DIMITROV, YU.I. KLECHIN, and A.V. APOSTOLOV
Calculation of Elastic Moduli for Amorphous $Fe_{80}B_{20}$
62.20; 61.40; S1.1
(a) Vol.116,No.2, 627-632

61.50. Crystalline state (including molecular motion in solids)
(for magnetic structure and spin systems, see 75.25)

E. KOENIGSBERGER and H. SCHRUNNER
On Lattice Parameters and Enthalpies of Mixing of Alkali Halide Solid Solutions
61.50 and 64.75; 65.50; S9.11
(b) Vol.151,No.1, 101-110

J. KOLLÁR and A. SÜTŐ
Quasiperiodic Lattice Model in Two Dimensions
61.50
(b) Vol.151,No.2, K101-K104

C. ASCHERON, A. SCHINDLER, R. FLAGMEYER, and G. OTTO
Characterization of Irradiation-Induced Lattice Disorder by Dilatometric Measurements
61.50 and 61.80; S5.11; S5.12; S7.11; S7.12
(a) Vol.112,No.1, 391-394

G.M. CHECHIN, T.I. IVANOVA, and V.P. SAKHNENKO
Complete Order Parameter Condensate of Low-Symmetry Phases upon Structural Phase Transitions
61.50 and 64.60
(b) Vol.152,No.2, 431-446

M. KOHYAMA, R. YAMAMOTO, Y. EBATA, and M. KINOSHITA
Atomic Forces in the Self-Consistent Tight-Binding Model
61.50 and 71.45
(b) Vol.152,No.2, 533-542

M. VATSKICHEVA, L. VATSKICHEV, J. MUCHA, and M. KOLENDA
On Some Structural Properties of $(Ni_{80}Fe_{20})_{100-x}Ge_x$ Evaported Thin Films
61.50 and 68.55; S1.61
(a) Vol.112,No.2, K57-K60

M. D. TURNER and P. MOHAZZABI
Continuum Approximation of the General Inverse-Power Lattice Sums in Cubic Crystals
61.50
(b) Vol.153,No.1, 59-64

R.V. GOPALA RAO and A. SATPATHY
Structural Studies of Charged Hard-Sphere Mixtures through Blums Generalized Gamma Parameter. Application to ...
61.25 and 61.50; S9.11
(b) Vol.153,No.2, 421-434

W. KORNETA and Z. PYTEL
The Bond Percolation near the Surface
61.50
(b) Vol.154,No.1, K1-K4

M. TORRES, G. PASTOR, I. JIMÉNEZ, and J. FAYOS
Continuous Evolutions which Connect Quasilattices with Lattices
61.50
(b) Vol.154,No.2, 439-452

A.I. GUSEV and A.A. REMPEL
Local Static and Dynamic Atomic Displacements in Disordered Niobium Carbide
61.50; 61.12; S1.61
(b) Vol.154,No.2, 453-460

A. MEYER and W.H. YOUNG
Interatomic Forces and Pseudopotential Parameterization
61.50; S2
(b) Vol.154,No.2, 469-474

A.N. PILYANKEVICH, A.V. KURDYUMOV, and N.F. OSTROVSKAYA
On the Crystallographic Irreversibility of Some Phase Transformation in Boron Nitride
61.50 and 64.70; S1.62
(a) Vol.116,No.1, K1-K6

E.N. IZAKOVICH, N.A. ZHELTOVA, YU.M. KOROLEV, ZH.D. SOKOLOVSKAYA, V.N. SPECTOR, and M.L. KHIDEKEL
Study of Isostructural Compounds Y_2BaCuO_5 and $(Y_{1.4}Cu^{3+}_{0.6})BaCuO_5$
61.50; 74.70; S10.15
(a) Vol.116,No.1, K13-K18

OM PARKASH, CH.DURGA PRASAD, and DEVENDRA KUMAR
Preparation and Characterisation of the System $Sr_{1-x}La_xTi_{1-x}Co_xC_3$
61.50 and 72.20; S10.15
(a) Vol.116,No.1, K81-K84

T. SHIOZAWA and T. KOBAYASHI
High Resolution Electron Microscopy of the Orientational Growth of Silver Precipitates to the Silver Gold Sulfide Specks Formed on ...
61.50 and 61.60; S9
(a) Vol.116,No.2, 513-520

H. BOERNER, M. WURLITZER, H.-CH. SEMMELHACK, B. LIPPOLD, R.-M. BOETTCHER, and M. KROETZSCH
A Method for the Preparation of Single-Phase $YBa_2Cu_3O_x$ Ceramics
61.50; 74.70; S10.15
(a) Vol.116,No.2, 537-544

61.55. Specific structures of elements and alloys

S. POPOVIĆ, H. LOEFFLER, B. GRŽETA, G. WENDROCK, and P. CZURRATIS
Lattice Constant of Al-Rich Matrix of Two Al-(8 and 24)at% Zn Alloys in Equilibrium with (Metastable) Guinier Preston Zones
61.55; 62.20; 64.70; S4
(a) Vol.111,No.2, 417-430

S. ERKOC
A New Empirical Many-Body Potential Energy Function: Application to Microclusters
63.10; 61.55; S1.2; S1.3; S2; S5
(b) Vol.152,No.2, 447-454

S. ERKOÇ and S. KATIRCIOGLU
Molecular-Dynamics Simulation of Aluminum Microclusters
61.55; S4
(b) Vol.152,No.2, K37-K42

M. STRÓŻAK, M. SUBOTOWICZ, and P. MIKOLAJCZAK
Formation of Unusual Alloys by Means of Ion Beam Mixing of the Pd/V Superlattice
61.55 and 61.80; 68.48; S1.2
(a) Vol.112,No.2, 739-742

G. GŁADYSZEWSKI, P. MIKOŁAJCZAK, and M. SUBOTOWICZ
Ion Beam Mixing in Bi-Sb Superlattices
61.55 and 61.80; 68.48; S4
(a) Vol.112,No.2, 753-756

B.X. LIU, X. ZHOU, and H.-D. LI
Thermodynamics and Growth Kinetical Consideration of Metal-Nitride Formation by Nitrogen Implantation (Review Article)
61.80 and 64.60; 61.55; S1.62
(a) Vol.113,No.1, 11-22

H. LOEFFLER, C. ESCHRICH, and O. SIMMICH
Recalculation of Several Metastable Phase Lines of the Al-Zn System
61.55 and 64.75; S1.2
(a) Vol.113,No.2, 269-276

E. BURZO, N. PLUGARU, V. POP, and I. CREANGĂ
Magnetic Properties of $YCo_{4-x}Fe_xB$ Compounds
61.55 and 75.50; S1.1
(a) Vol.113,No.2, K253-K256

S. MOTIEJUNAS, D. BALTRŪNAS, N. GEČIAUSKAITĖ, and K. MAKARIŪNAS
Correlation between the Moessbauer and Semiconductor Parameters of Tin Telluride
76.80; 61.55; S1.63
(b) Vol.154,No.1, 341-346

A. KOWALCZYK, P. STEFAŃSKI, A. WRZECIONO, and A. SZLAFEREK
Structure and Magnetic Properties of $R_2Fe_{14-x}Ni_xB$ Compounds (R = Nd and Gd)
75.30; 61.55; S1.1; S1.2
(a) Vol.114,No.1, 355-360

K. SUMIYAMA, Y. HIROSE, and Y. NAKAMURA
Magnetic and Electrical Properties of Nonequilibrium Ni-Al Alloys Produced by Vapor Quenching
72.15 and 75.50; 61.55; S1.2
(a) Vol.114,No.2, 693-704

S. ERKOÇ
A New Empirical Many-Body Potential Energy Function. Application to Microclusters: Elements in B.C.C., F.C.C., and H.C.P. Structures
63.10; 61.55; S1.2; S2; S5
(b) Vol.155,No.2, 461-466

V.G. TSIRELSON, V.A. STRELTSOV, R.P. OZEROV, and K. YVON
Electron Density Distribution in 3d-Metal Sesquioxides
71.20; 61.55; S10
(a) Vol.115,No.2, 515-522

L. TERZIEV, J. LECOMTE-BECKERS, P. RASHEV, and T. GREDAY
Study of Transition Phases during Isothermal Ageing at 250 °C of Zn-27%Al Alloy
61.55; 61.14; 64.70; S1.2
(a) Vol.116,No.2, 521-528

V. PIERRON-BOHNES, C. LEROUX, J.P. AMBROISE, A. MENELLE, and P. BASTIE
Measurement of the Mean Static Displacement and of the Debye Waller Factor between 5 and 1464 K in Fe-19.5at%Al
61.55; 61.12; 76.80; S1.1
(a) Vol.116,No.2, 529-536

61.60. Specific structures of inorganic compounds

J.J. BARA and B.F. BOGACZ
Investigation of ^{57}Fe-Doped theta-$Al_xV_2O_5$ Bronzes
61.60; 76.80; S10
(a) Vol.111,No.1, 53-58

T. KANOMATA and T. GOTO
Note on the Crystal Structure of FeCoGe and FeNiGe
61.60; S1.61
(a) Vol.111,No.1, K1-K6

H. NEGISHI, S. OHARA, and M. INOUE
Effect of Pressure on Lattice Parameters of Intercalation Compounds M_xTiS_2 (M=Fe and Ni)
61.60; S8.16
(b) Vol.151,No.2, 441-444

R. TOVAR, M. QUINTERO, P. GRIMA, and J.C. WOOLLEY
Phase Diagram, Lattice Parameter, and Optical Energy Gap Values for the $Zn_{2x}(AgIn)_{1-x}Te_2$ Alloys
61.60; 64.75; S8.16
(a) Vol.111,No.2, 405-410

C. RAZZETTI, T. BESAGNI, S. BINI, and P.P. LOTTICI
Defective Pseudoternary Alloys. The $CdGa_2Se_4$-$CdIn_2Se_4$ System
61.60; 64.60; S8.15
(a) Vol.111,No.2, 411-416

G. ANTONIOLI, P.P. LOTTICI, and C. RAZZETTI
The Structure of the Defect Chalcopyrite $ZnGa_2Se_4$ Studied by EXAFS
61.60 and 78.70; S8.16
(b) Vol.152,No.1, 39-50

J. PAVLOVSKÝ and T. ZEMČÍK
Comparison of Surface and Bulk Crystallization of the Amorphous $Fe_{70}Co_{10}B_{20}$ Alloy
61.60; 76.80; S1.1
(a) Vol.112,No.1, K1-K4

V. RIEDE, H. NEUMANN, N. SHARIF, F.S. HASOON, and H. SOBOTTA
Structural and Infrared Optical Properties of $(CuInSe_2)_{1-x}(2InAs)_x$ Solid Solutions
61.60 and 78.30; S7.15; S8.15
(a) Vol.112,No.2, K139-K144

V.K. DUGAEV and P.P. PETROV
Spinodal Decomposition in Semiconductor Alloys
64.60; 61.60; S8.15
(b) Vol.153,No.1, 115-122

Z.C. KANG, D. MORIN, L. PIERRE, O. MONNEREAU, and C. BOULESTEIX
AEM Study of the High Temperature Superconductor $Bi_2Sr_2CaCu_2O_8$
61.60 and 74.70; 61.16; S10
(a) Vol.113,No.1, 23-34

M. GUYMONT, A. TOMAS, M.-P. PARDO, and M. GUITTARD
Electron Microscopy Study of gamma-Ga_2S_3
61.16 and 61.60; S8
(a) Vol.113,No.1, K5-K8

D. BAHADUR, RAM BILAS, and B.S.R. MURTY
Electronic Properties of the Defect Perovskite System $La_{1-x}Li_xMO_3$
61.60 and 72.20; 75.50; S11
(b) Vol.154,No.1, 297-304

G. BABONAS, R.R. GAŁĄZKA, G. PUKINSKAS, and A. ŠILEIKA
Crystal Structure and Magnetooptical Properties of $Cd_{1-x}Mn_xTe$ with High Mn Concentrations
78.20; 61.60; S8.15
(b) Vol.154,No.1, 389-396

Z.C. KANG, C. BOULESTEIX, G. NIHOUL, O. MONNEREAU, F. REMY, and A. CASALOT
High Resolution Image of a New Mixed Valence Compound: $Al_2Ti_7O_{15}$
61.60; 61.16; S10
(a) Vol.114,No.1, 73-82

W. PAULUS, G.A. WIEGERS, A. MEETSMA, S. VAN SMAALEN, and J.L. DE BOER
The Structure of $Cu_{0.38}NbSe_2$. An Orthorhombic Superstructure of a Hexagonal Basic Structure Occurring in Three Orientation Variants
61.60; S1.63
(a) Vol.114,No.1, 91-98

M.M. KORSUKOVA, V.N. GURIN, YU.B. KUZMA, N.F. CHABAN, S.I. CHYKHRII, V.V. MOSHCHALKOV, N.B. BRANDT, A.A. GIPPIUS, and KHO KHYU NYAN
Crystal Structure, Electrical, and Magnetic Properties of the New Ternary Compounds $LnAlB_{14}$
61.60 and 72.80; 75.50; S1.4
(a) Vol.114,No.1, 265-272

E. WOLSKA and U. SCHWERTMANN
Selective X-Ray Line Broadening in the Goethite-Derived Hematite Phase
61.10 and 64.70; 61.60; S10.1
(a) Vol.114,No.1, K11-K16

G. ANTONIOLI, P.P. LOTTICI, and C. RAZZETTI
A Study of the Second and Third Neighbour Structure in Chalcopyrite $CuGaSe_2$ by EXAFS
61.60; 78.70; S8.16
(b) Vol.154,No.2, 461-468

Z.C. KANG, F. REMY, O. MONNEREAU, A. CASALOT, and C. BOULESTEIX
Plane Defects in a New Mixed Valence Compound $Al_2Ti_7O_{15}$
61.60; 61.16; 61.70; S10
(a) Vol.114,No.2, 441-450

JINGKUI LIANG, XIAOLONG CHEN, YULING ZHANG, SISHEN XIE, JINQI HUANG, and YONGMIN NI
The Crystal Structure and Superconductivity of the Superconducting Phase $TlSr_2(Ca, La)Cu_2O_7$
61.60 and 74.70; S10.15
(a) Vol.114,No.2, 651-658

V.M. TALANOV
Anion Ordering in Spinels
61.60; S11.2
(a) Vol.115,No.1, K1-K4

A. CONAN, A. BONNET, and M. MORSLI
Electrical and Structural Properties of the Tungsten Bronze $Bi_4W_{20}O_{62}$
72.20; 61.60; 71.20; S10
(b) Vol.155,No.2, 609-614

A.I. GUSEV
Short-Range Order in Nonstoichiometric Transition Metal Carbides Nitrides and Oxides (Review Article)
61.60 and 75.30; 76.60; S1.61; S1.62; S10
(b) Vol.156,No.1, 11-40

S.K. PLACHKOVA, O. GEORGIEVA, K. PETROV, and D. KOVACHEVA
X-Ray Study of Defects in $GeTe-Ag_2Te-Bi_2Te_3$ System
61.60 and 61.70; S8.16
(a) Vol.116,No.1, 213-220

F.G. REQUEJO, A.G. BIBILONI, C.P. MASSOLO, J. DESIMONI, and M. RENTERIA
TDPAC Study of Stoichiometric and Reduced MoO_3. Localization of Indium Trapped Oxygen Single Vacancies
61.60; 61.70; 76.80; S10.1
(a) Vol.116,No.2, 503-512

T. SHIOZAWA and T. KOBAYASHI
High Resolution Electron Microscopy of the Orientational Growth of Silver Precipitates to the Silver Gold Sulfide Specks Formed on ...
61.50 and 61.60; S9
(a) Vol.116,No.2, 513-520

61.65. Specific structures of organic compounds

J. MERTSCHING
Double Periodic Peierls Phases
61.65 and 71.25; S12.1
(b) Vol.152,No.2, 551-562

61.70. Defects in crystals

(see also 61.80. Radiation damage,
62. Mechanical and acoustical properties,
71.55. Impurity and defect levels,
76.30. EPR of color centres and other defects,
78.50. Impurity and defect absorption in solids)

V. HOLÝ and J. KUBĚNA
On the Integrated Intensity of X-Ray Diffraction in Crystals with Randomly Distributed Defects
61.10 and 61.70; S5.11
(b) Vol.151,No.1, 23-28

A. STOLOVICH and G. LIIDJA
V_K Centres in Alkali Halides: Direct Spin-Lattice Relaxation and Electron-Phonon Interaction
61.70; S9.11
(b) Vol.151,No.1, 29-38

R. CYWIŃSKI, E. MUGEŃSKI, W. NOWY-WIECHUŁA, and J. WIECHUŁA
Suzuki Phase Precipitates in Doubly Doped KCl:Pb^{2+},Mn^{2+} Crystals
61.70; 76.30; 78.50; S9.11
(b) Vol.151,No.1, 47-52

E. GERLACH and S. KISTINGER
Plasma Excitation in Semiconductors Containing Extended Scattering Centers
72.20; 61.70; 71.45
(b) Vol.151,No.1, 241-248

I.A. OVIDKO
Polyhedra, Disclination Ensembles, and Structure of Grain Boundaries
61.70
(b) Vol.151,No.1, K7-K10

WEN-CHEN ZHENG
A Study of Distortion of the Immediate Environment of the Chromium Ion in Alums
61.70; 62.20; 76.30; S11.1
(b) Vol.151,No.1, K11-K16

WEN-CHEN ZHENG
A Further Study of Tetragonal Distortion of the Cr^{3+}-V_c Centre in MgO from Optical and EPR Spectra
61.70 and 71.70; 76.30; S10.1
(b) Vol.151,No.1, K17-K22

H.S. KIM and S.S. SHEININ
An Electron Microscope Method for Determing the Nature of Stacking Faults in F.C.C. Materials which Does not Rely on ...
61.16 and 61.70; S1.2
(a) Vol.111,No.1, 47-52

F. HAIDER
Calculation of TEM Contrast of Small Coherent Precipitates
61.16; 61.70; S1.2; S1.3
(a) Vol.111,No.1, 71-80

A.F. VLADIMIROV and E.N. MOOS
Development of Eshelby Concept in Solution of Boundary Problems for Elastic Solids with Defects
61.70; 68.25
(a) Vol.111,No.1, 99-108

A. KRELL and W. POMPE
Grain Boundary Microdamage Due to Visco-Elastic Relaxation of Residual Stresses in Alumina Ceramics
61.70 and 62.20; S10.15
(a) Vol.111,No.1, 109-118

D.M. VANDERWALKER
Fracture in Titanium Diboride
61.70 and 62.50; S1.2
(a) Vol.111,No.1, 119-126

Y. MERA, K. MAEDA, and S. HYODO
Microscopic Origin of Electric Conduction Anisotropy in Plastically Deformed CdS Single Crystals
72.20; 61.70; S8.11
(a) Vol.111,No.1, 223-232

A.A. KARANOVICH and A.V. DVURECHENSKII
The EPR Study of Inhomogeneous Deformations in Neutron-Irradiated Silicon
61.70; 76.30; S5.11
(a) Vol.111,No.1, 257-264

F.M. KAMARLY, N.G. GUSEINOV, and A.YU. YANGIROV
Magnetic Susceptibility and Electrophysical Properties of SiTe Single Crystals
61.70; 72.80; 75.50; S8
(a) Vol.111,No.1, K17-K20

J. GONG and S.H. YANG
Annealing Temperature Dependence of the 1/f Noise in Si-Implanted GaAs
61.70 and 72.70; S7.12
(a) Vol.111,No.1, K53-K56

R. SUBRAHMANYAM and Y.V.G.S. MURTI
Laser Induced F-F^- Conversions in KCl Crystals
61.70; 71.55; 78.50; S9.11
(b) Vol.151,No.2, 445-452

F. DE MATTEIS, A. SCACCO, F. SOMMA, G. BALDACCHINI, E. GIOVENALE, and U.M. GRASSANO
Complex Centres in Additively Coloured RbCl:Li^+
61.70; 78.50; 78.55; S9.11
(b) Vol.151,No.2, 453-462

P. STEGGER and G. LEHMANN
Dynamic Effects in a New Substitutional Center of Trivalent Iron in Quartz
61.70; 76.30; S10.1
(b) Vol.151,No.2, 463-468

J.D. COLLINS, G.A. GLEDHILL, R. MURRAY, P.S. NANDHRA, and R.C. NEWMAN
The Selective Trapping of Arsenic Interstitial Atoms by Impurities in Gallium Arsenide
61.70; 78.50; S7.12
(b) Vol.151,No.2, 469-478

U. KRAUSE, J.P. KUSKA, and R. WEDELL
Monovacancy Formation Energies in Cubic Crystals
61.70
(b) Vol.151,No.2, 479-494

A.I. GUSEV
Structural Stability Boundaries for Nonstoichiometric Compounds
61.70; S1.61
(a) Vol.111,No.2, 443-450

V.M. AROLE, M.G. TAKWALE, V.G. BHIDE, A.D. SHALIGRAM, and S.K. DAVID
Hydrogen-Induced Structural Changes in Polycrystalline Silicon as Revealed by Positron Lifetime Spectroscopy
61.70; 72.20; S5.11
(a) Vol.111,No.2, 457-462

R.M. DE LA CRUZ and R. PAREJA
Porous Silicon Investigated by Positron Annihilation
61.70; 78.70; S5.11
(a) Vol.111,No.2, 463-468

G.CH. CHRISTAKUDIS, S.K. PLACHKOVA, and L.E. SHELIMOVA
Influence of Defects on the $(GeTe)_{1-x}[(Ag_2Te)_{1-y}(Sb_2Te_3)_y]_x$ Phonon Thermal Resistivity
61.70 and 66.70; S8.16
(a) Vol.111,No.2, 469-476

V.V. TOKAREV, V.E. BORISENKO, T.M. PYATKOVA, and D.I. ZAROVSKII
Argon Ion Implantation Influence on Phase Formation and Structure of Nickel Silicides
61.70 and 68.55; S1.61
(a) Vol.111,No.2, 499-506

V.V. PETROV and V.V. CHORNYI
Recharging and Negative Annealing of Divacancies in Neutron-Irradiated Silicon
61.70 and 78.50; S5.11
(a) Vol.111,No.2, K141-K144

V.I. ALEKSEENKO, V.I. BARBASHOV, L.A. ZILBERMAN, V.M. MOSTOVOI, and M.YA. SKOROKHOD
Effects of Dislocation Loop Interaction in Indium Antimonide. Dynamics and Geometric Aspects
61.70; S7.13
(a) Vol.111,No.2, K145-K148

V.P. MARKEVICH and L.I. MURIN
Thermal Donor Formation in Pre-Heat-Treated n-Si:O Crystals
61.70; S5.11
(a) Vol.111,No.2, K149-K154

M.J. MARCINKOWSKI
Correspondence between Vortices and Dislocations (Review Article)
61.70
(b) Vol.152,No.1, 9-32

YIFAN CHENG and RENHUI WANG
Dynamical Theory of Electron Diffraction for Quasicrystals
61.14; 61.70; S1.2; S4
(b) Vol.152,No.1, 33-38

V. TOPA, A. IOAN, B. ILIESCU, and G. MITROAICA
The Conversion of Electronic Centres into U and OH^- Centres
61.70; S9.11
(b) Vol.152,No.1, 51-60

L. BOSI and M. NIMIS
On the Decay Properties of the M′-Center in LiF Crystals
61.70; 78.50; S9.11
(b) Vol.152,No.1, 67-72

B.C. DE COOMAN and C.B. CARTER
The Formation of Faulted Dipoles in Plastically Deformed GaAs
61.70; 61.16; 62.20; S7.12
(a) Vol.112,No.1, 41-54

D.M. VANDERWALKER
Hydride Formation on Dislocations in Titanium
61.70; 61.16; S1.2
(a) Vol.112,No.1, 73-78

C. LEVADE, J.J. COUDERC, and A. KARA
Plastic Deformation of Natural Sphalerite Single Crystals between 300 and 620 K
62.20; 61.70; S8.11
(a) Vol.112,No.1, 89-98

A. TUROS
Ion Implantation Studies of Collision Cascades in Solids
61.70; S1.1; S1.2; S1.3; S1.4
(a) Vol.112,No.1, 253-262

J. HLÁVKA and J. GRŇO
Annealing Stages of Implanted Silicon
61.70; 72.40; S5.11
(a) Vol.112,No.1, 327-330

J. JAGIELSKI, G. GAWLIK, A. PODGORSKI, A. TUROS, A. DYGO, and N. MADI
Thermal Effects on Depth Distribution of Nitrogen Atoms Implanted in Ferrous Alloys
61.80; 61.70; S1.1
(a) Vol.112,No.1, 343-346

J. KRYNICKI and H. RZEWUSKI
Ionization Assisted Transformations of Defects in Carbon Implanted Silicon
61.70; S5.11
(a) Vol.112,No.1, 405-410

D. HOEHL and H. BARTSCH
Results of In-Situ Annealing of Ion-Implanted Silicon in a High Voltage Electron Microsope
61.70 and 61.80; 61.14; 68.48; S5.11
(a) Vol.112,No.1, 419-424

V.S. LYSENKO, A.N. NAZAROV, G.A. NAUMOVETS, V.B. POPOV, and A.S. TKACHENKO
Manifestation of Hydrogen in Al-SiO_2-Si Structures Subjected to a RF Plasma Annealing
61.70; 71.55; 73.40; S5.11
(a) Vol.112,No.1, K9-K12

S. DZHUMANOV and P.K. KHABIBULLAEV
The Coexistence of Unstable,Metastable, and Separated Frenkel Pair Defects in Solids. III. Theory of the Non-Impact Mechanism for ...
61.70; 71.35; 71.38; S9.11
(b) Vol.152,No.2, 395-414

D. GOURIER, T. GBEHI, R. VISOCEKAS, J. THERY, and D. VIVIEN
Color Centres in Barium Hexaaluminate (Phase I). Study by EPR, Optical Absorption, and Thermally Stimulated Luminescence
61.70; 76.30; 78.50; 78.60; S11
(b) Vol.152,No.2, 415-430

B.C. DE COOMAN and C.B. CARTER
High-Resolution Microscopy of Dissociated Screw Dislocations in GaAs
61.70; 61.16; S7.12
(a) Vol.112,No.2, 473-492

P. CORDIER, B. BOULOGNE, N. DOUKHAN, and J.C. DOUKHAN
Lattice Defects in Lithium Tantalate
61.70; 77.60; S11.1
(a) Vol.112,No.2, 493-510

E.L. MATHÉ and J.C. DESOYER
Implantation of Aluminum in Silicon: Simultaneous Showing of Self-Si Defects and of Al Precipitates
61.70; S5.11
(a) Vol.112,No.2, 511-518

G. WAGNER, V. GOTTSCHALCH, H. RHAN, and P. PAUFLER
Pseudomorphic Growth and Nucleation of Misfit Dislocations in the Epitaxial System (001) InP/$In_{1-x}Ga_xAs$. I. Pseudomorphic Growth ...
61.70 and 68.55; S7.11; S7.12; S7.15
(a) Vol.112,No.2, 519-532

V.V. PETROV and V.S. PROSOLOVICH
Thermal Defects in Silicon Doped with Rare-Earth Elements
61.70 and 71.55; 72.20; 78.30; S5.11
(a) Vol.112,No.2, 561-568

R.K. BAKANAS, F.G. BASS, and V.V. KONOTOP
Interaction of Depletion and Accumulation Waves with Localized Defect
72.20; 61.70
(a) Vol.112,No.2, 579-588

W. SKORUPA, R. GROETZSCHEL, and H. BARTSCH
Recent Advances of Ion Beam Synthesis for Silicon-on-Insulator Structures
61.70 and 61.80; S5.11
(a) Vol.112,No.2, 661-666

M. SETVAK
The Present State of Ion Implantation Equipment
61.70 and 61.80
(a) Vol.112,No.2, 689-694

P.W. ŻUKOWSKI, S.B. KANTOROW, D. MĄCZKA, and W.F. STELMAKH
Procecces of Radiation Defect Interaction and Amorphisation of Silicon at Large Implantation Doses
61.70; 79.20; S5.11
(a) Vol.112,No.2, 695-698

R. MUEHLE and T. HARTUNG
Ion Beam Exposure Characteristics of Nitrocellulose
61.70 and 61.80; S12
(a) Vol.112,No.2, 769-772

R.A. ILKHAMOV, LI JEN HO, S.S. KHUSMURODOV, D. MACHAJDIK, G.M. OSETINSKIJ, M. PAJEK, R. SANDRIK, and D.M. SHIROKOV
Hydrogen Depth Profiling by Elastic Recoil Detection
61.70 and 61.80; S5.11
(a) Vol.112,No.2, 781-784

R. KUDUK, M. KULIK, D. MĄCZKA, K. PYSZNIAK, and J. ŻUK
Depth Profile Measurements of Ions Implanted in Semiconductor Materials by Particle Induced Photon Emission
61.70; 78.60; 79.20
(a) Vol.112,No.2, 799-804

V.I. GAVRILENKO, N.I. KLYUI, V.G. LITOVCHENKO, and B.N. ROMANYUK
Investigation of the Ion Implanted Silicon by Electroreflectance Spectroscopy
61.70; 78.20; S5.11
(a) Vol.112,No.2, 805-810

I.D. KOZMIK, I.I. GRIGORCHAK, Z.D. KOVALYUK, and I.Z. MARCHUK
Kinetic and Optical Properties of H_xInSe and H_xGaSe Proton Intercalates and Their Thermodynamic Model
61.70; 72.80; 78.50; S8
(a) Vol.112,No.2, K61-K64

K.D. GLINCHUK, N.M. LITOVCHENKO, and V.YU. PTITSIN
Influence of Germanium on the Formation of Oxygen-Induced Recombination Centres in Heat-Treated Silicon
61.70; 71.55; S5.11
(a) Vol.112,No.2, K65-K70

M.J. PONNAMBALAM
Stark Effect and the F-Center Excited States
61.70 and 71.55; 78.20
(b) Vol.153,No.1, 79-88

S.N. ROMANOV, A.I. MELKER, and D.B. MIZANDRONTSEV
Athermal Clusters of Vacancies in B.C.C. and H.C.P. Metals
61.70; 61.80; S1; S1.1
(b) Vol.153,No.1, K5-K10

S.G. PSAKHIE, S.V. ALEKSEEV, and V.E. PANIN
Influence of Shear on the Binding Energy of Complexes of Impurities and Vacancy in Aluminium
61.70; 62.20; S4
(b) Vol.153,No.1, K11-K16

S. SUZUKI
Hydrogen Cold-Work Peak in High Purity Molybdenum
61.70 and 62.40; S1
(a) Vol.113,No.1, 51-56

G. WAGNER, V. GOTTSCHALCH, H. RHAN, and P. PAUFLER
Pseudomorphic Growth and Nucleation of Misfit Dislocations in the Epitaxial System (001) $InP/In_{1-x}Ga_xAs$. II. Critical Thickness and ...
61.70 and 68.55; S7.11; S7.15
(a) Vol.113,No.1, 71-82

S. NOWAK, T. KACZOR, and J. ŻMIJA
Photochromic Phenomenon Kinetics
61.70 and 78.50; 71.55; S10.15
(a) Vol.113,No.1, 199-206

T. FUKUMORI and K. FUTAGAMI
Nonuniformity on GaAs Wafers Revealed by an X-Ray Double-Crystal Method
61.10 and 61.70; S7.12
(a) Vol.113,No.1, K19-K22

N.I. BORGARDT, S.K. MAKSIMOV, and V.V. OVCHAROV
Defocused Electron Microscopy Images of Small Defects with Black-and-White Contrast
61.16 and 61.70
(a) Vol.113,No.1, K23-K26

L. KOVÁCS, R. SOMMERFELDT, YE MING, and E. KRAETZIG
Light-Induced Charge Transport in $LiNbO_3$:Mg,Cr Crystals
61.70 and 72.40; 78.50; S11.1
(a) Vol.113,No.1, K75-K78

J. HORN and H.-J. ZACHARIAS
The Lines of Constant Vacancy Content in the Phase Diagram of the Al-Fe Ferrite System
61.70 and 75.50; S11.2
(a) Vol.113,No.1, K103-K106

N.M. FICZA and A.L. TOTH
The Effect of Lead Impurity on Amorphous Si Thin Film Transistors
61.70 and 85; 68.55; S5.11; S10.1
(a) Vol.113,No.1, K147-K148

R.R. GALIMOV and S.B. GORYACHEV
The Sink Strength of Dislocation Dipole and Low-Angle Tilt Boundary
61.70
(b) Vol.153,No.2, 443-454

H.T. ZHANG, E. MANN, and A. SEEGER
Approximate Solution of the Peierls Equation for Screw Dislocations with Threefold Symmetry
61.70; 62.20; 62.40
(b) Vol.153,No.2, 465-478

M.E. POLYAKOV
Fundamental Solution of the Problem on the Relief Shape of a Kink on Dislocation in an External Force Field Equal to the Barrier Reaction
61.70
(b) Vol.153,No.2, 479-488

Y. MADHAVAN, K. RAMACHANDRAN, and T.M. HARIDASAN
On the Mean Square Displacements of Hg and Te in HgTe
63.20; 61.70; S8.13
(b) Vol.153,No.2, 579-582

G.L. GUTSEV, G.S. MYAKENKAYA, V.V. FROLOV, and V.B. GLAZMAN
Nature of Hydrogen Bonding in Si-A Center
71.55; 61.70; S5.11
(b) Vol.153,No.2, 659-666

N.I. BORGARDT, V.N. KUKIN, S.K. MAKSIMOV, and D.I. PISKUNOV
Defocused Images of Stacking Faults
61.16 and 61.70
(a) Vol.113,No.2, 251-268

A.V. ZASTAVNOY and V.M. KOROL
Radiation Defects in n-Silicon Doped with Lithium and Sodium
61.70; 71.55; S5.11
(a) Vol.113,No.2, 277-284

L.A. KAZAKEVICH, P.F. LUGAKOV, and I.M. FILIPPOV
The Role of Dislocations during Charge-Carrier Recombination at Radiation Defects in p-Si
61.70 and 72.20; 61.80; S5.11
(a) Vol.113,No.2, 307-314

R. KOEGLER, E. WIESER, J. ALBRECHT, and P. KNOTHE
Investigation of As-O Clustering in Si
61.70; 72.20; S5.11
(a) Vol.113,No.2, 321-330

D. BAITHER and D. PANKNIN
Residual Defects in Implanted GaAs after Rapid Thermal Annealing with Incoherent Light
61.70; S7.12
(a) Vol.113,No.2, 331-336

M.YU. GUTKIN, M. MILITZER, A.E. ROMANOV, and V.I. VLADIMIROV
Equilibrium Position of Misfit Dislocations
61.70 and 68.48
(a) Vol.113,No.2, 337-344

Y. LI and Q.P. KONG
On the Relationship between Creep Rate and Stacking Fault Energy
62.20; 61.70
(a) Vol.113,No.2, 345-352

SHAM-TSONG SHIUE and SANBOH LEE
Elastic Interaction between Screw Dislocations and Blunt Crack
61.70 and 62.20
(a) Vol.113,No.2, 365-374

C. Q. TANG and X. CHEN
Positron Lifetimes in Deformed Two-Phase Brass
61.70 and 78.70; S1.2
(a) Vol.113,No.2, K161-K164

G.A. KACHURIN, I.E. TYSHCHENKO, V.P. POPOV, and S.A. TIJS
Interaction of Nitrogen Implanted into Silicon with Defects in the Buried Layer
61.70; S5.11
(a) Vol.113,No.2, K165-K170

L. BISCHOFF and R. KOEGLER
Rapid Thermal Annealing of Neutron Transmutation Doped Silicon
61.70; 72.80; S5.11
(a) Vol.113,No.2, K185-K188

S.L. PINJARE, N. BALASUBRAMANYAM, and VIKRAM KUMAR
Photoluminescence at 0.944 eV from Heat Treated n-Type Silicon
78.55; 61.70; 71.55; S5.11
(a) Vol.113,No.2, K261-K264

M.O. MANARESH, D.W. FISCHER, and W.C. MITCHEL
The EL2 Defect in GaAs: Some Recent Developments (Review Article)
61.70 and 71.55; 72.40; 78.30; 78.50; 78.55; S7.12
(b) Vol.154,No.1, 11-42

R.R. GALIMOV and S.B. GORYACHEV
The Sink Strength of Dislocation Multipole and Dislocation Wall
61.70
(b) Vol.154,No.1, 43-54

XIAOYU HU and H. MARGOLIN
Dislocation Arrangements in Stress Gradients
61.70; 62.20; S4
(b) Vol.154,No.1, 61-70

A.S. KROCHUK, O.R. ONUFRIV, and Z.P. CHORNIY
Characteristic Properties of the Radiation Colouring Mechanism in MeFX Compounds (Me = Sr,Ba, X = Cl,Br)
61.70; 78.50; S9.16
(b) Vol.154,No.1, K9-K12

FANG-QING ZHANG and GUANG-HUA CHEN
Study of the Hydrogen Content and Profiles in a-Ge:H Films by Nuclear Resonant Reactor Analysis
61.70 and 68.60; S5.12
(b) Vol.154,No.1, K39-K42

K.Z. BOTROS and A.S. TITI
On Assessing Deviations from Symmetrical Laue Diffraction Conditions Using TEM Images of Stacking Faults
61.16 and 61.70; S1.2
(a) Vol.114,No.1, 83-90

M.V. GALUSTASHVILI, D.G. DRIYAEV, I.A. POLITOV, and Z.K. SARALIDZE
Changes in the Electrical Properties of LiF Crystals in the Process of Post-Radiation Annealing
61.70; S9.11
(a) Vol.114,No.1, 99-104

N.D. THEODORE, B.C. DE COOMAN, and C.B. CARTER
Extrinsic Dissociation of a Dislocation in GaAs in the Presence of Point Defects
61.70; S7.12
(a) Vol.114,No.1, 105-112

I.I. KHODOS, M.SH. SHIKHSAIDOV, I.I. SNIGHIREVA, A.P. USHAKOVA, and V.I. NIKOLAICHIK
The Dissociation of Dislocations in Pure and Doped GaAs Crystals
61.70; S7.12
(a) Vol.114,No.1, 113-126

K. SWIATEK, M. GODLEWSKI, D. HOMMEL, and H. HARTMANN
On the Incorporation of Rare Earth Ions into II-VI Compounds: ZnS:Eu
61.70; 76.30; S8.11
(a) Vol.114,No.1, 127-134

V.A. MEZRIN
Phase Transitions in High-Temperature Fluorite-Structure Superionic Conductors
64.60; 61.70; S9.12; S10.1
(a) Vol.114,No.1, 145-156

N.T. BAGRAEV, I.S. POLOVTSEV, and K. SCHMALZ
An ONP Investigation of Deep Level Defects Produced by Heat Treatment at 600 and 800 °C in Cz-Si
61.70 and 71.55; 76.60; S5.11
(a) Vol.114,No.1, 233-246

A.D. STEVENS, M.C.R. SYMONS, and R.S. EACHUS
ESR Study of Silver Clusters in AgCl and AgBr. An Attempt to Detect Photographic Latent Images
61.70; 76.30; S9
(b) Vol.154,No.2, 483-496

V.A. BORODIN and A.I. RYAZANOV
Kinetics of an Ensemble of Jogs on the Edge Dislocation for Inefficient Capture of Point Defects by Jogs
61.70
(b) Vol.154,No.2, 497-506

J. TATARKIEWICZ and W. WITTHUHN
Hydrogen Vibrations in Oxygen-Doped Silicon
61.70 and 63.20; S5.11
(b) Vol.154,No.2, K99-K102

Z.C. KANG, F. REMY, O. MONNEREAU, A. CASALOT, and C. BOULESTEIX
Plane Defects in a New Mixed Valence Compound $Al_2Ti_7O_{15}$
61.60; 61.16; 61.70; S10
(a) Vol.114,No.2, 441-450

T.-M. WANG, B.-Y. WANG, B.-X. FENG, C.-L. LIU, B.-H. JIANG, and Z.-Y. XU
The Recovery Behaviour of Quenched-in Vacancies in CuZnAl Alloy
61.70; 78.70; S1.2
(a) Vol.114,No.2, 451-456

V.A. GANSHIN, YU.N KORKISHKO, T.V. MOROZOVA, and V.V. SARAIKIN
The Study of Proton Exchange in Lithium Tantalate Crystals
61.70; 78.20; S11.1
(a) Vol.114,No.2, 457-466

L. TESAŘ and R. LENHARD
Evaluation of Silicon Wafers with Different Concentrations of Interstitial Oxygen
61.70 and 71.55; 73.40; S5.11
(a) Vol.114,No.2, 475-480

V.N. BRUDNYI, A.D. POGREBNYAK, YU.P. SUROV, and A.S. RUDNEV
Positron Trapping at Point Defects in Electron-Irradiated GaAs
61.70; 61.80; 78.70; S7.12
(a) Vol.114,No.2, 481-490

D. TRIANTIS, C. ANASTASIADIS, and S. ORFANAKOS
Evaluation of Activation Energy from the Ionic Thermocurrent Spectrum
61.70; 77.30
(a) Vol.114,No.2, 491-496

S. LAKSHMI RAGHAVAN, N. RAMESH, and K. SRINIVASAN
Dispersion of the Quadratic Electro-Optical (Kerr) Constants in $KCl:Cd^{2+}$ and $KCl:Sr^{2+}$
78.20; 61.70; S9.11
(a) Vol.114,No.2, 749-756

R. RICHTER, T. BREMER, P. HERTEL, and E. KRAETZIG
Refractive Index and Concentration Profiles of Proton-Exchanged $LiNbO_3$ Waveguides
78.20; 61.70; S10.15
(a) Vol.114,No.2, 765-774

H. BUDINOV, V. STAVROV, and R. BURKOVA
Flash Lamp Annealing of Phosphorus-Implanted Silicon
61.70; 79.20; S5.11
(a) Vol.114,No.2, K131-K134

DECHENG TIAN, ZHU-XIN YU, CAI-GUO ZHAN, and SHAO-LING WANG
Positron Annihilation Study of Ageing of an Al-Mg-Si-RE Alloy
61.70 and 78.70; 64.70; S4
(a) Vol.114,No.2, K135-K138

XIAOYU HU and H. MARGOLIN
The Stress Fields on Secondary Slip Systems Due to a Group of Dislocations in a Stress Gradient
61.70; 62.20; S4
(b) Vol.155,No.1, 53-64

A.G. IVANCHIN
Phase Model of the Point-Defect Core
61.70
(b) Vol.155,No.1, K1-K2

R.S. DE BIASI
Comments on Concentration Peculiarities of a Set of Impurity Centres in Fluorite Crystals Activated by Gadolinium
61.70; 76.30; 78.60; S9.15
(b) Vol.155,No.1, K3-K4

F. MAES, F. CALLENS, P. MATTHYS, and E. BOESMAN
Electron Paramagnetic Resonance of an O_3^- Centre in KBr
76.30; 61.70; S9.11
(b) Vol.155,No.1, K55-K58

H.S. KIM and S.S. SHEININ
The Many Beam Dynamical Theory of Contrast in Electron Microscope Images of Microtwins for the Non-Symmetrical Laue ...
61.14; 61.70
(a) Vol.115,No.1, 51-60

P. LOŠŤÁK, Z. STARY, J. HORÁK, and J. PANCÍŘ
Substitutional Defects in Sb_2Te_3 Crystals
61.70; 72.20; S8
(a) Vol.115,No.1, 87-96

J.D. STEPHENSON
Simulated Quenching of Silicon WSR Monochromators Using WSR Section Topography
61.70; 65.70; S5.11; S5.12
(a) Vol.115,No.1, 97-106

F. FERMI, A. NAGORNYI, and C. PARACCHINI
Defect Annealing in Pure CdF_2 Crystals
61.70 and 78.60; S9.12
(a) Vol.115,No.1, 325-334

R. DUBOURG, M. SCHVOERER, R. BERGER, and B. DUMERCQ
Mécanismes de thermoluminescence dans des fluorines CaF_2 naturelles et de synthèse
78.60; 61.70; S9.12
(a) Vol.115,No.1, 335-346

V.V. LITVINOV, G.V. PALCHIK, and V.I. URENEV
On the Radiation-Enhanched Mechanism of the Thermal Donor Formation Process in Germanium
61.70 and 71.55; S5.12
(a) Vol.115,No.1, K9-K12

E.M. GOLOLOBOV, N.A. PRYTKOVA, ZH.M. TOMILO, D.M. TURTSEVICH, M.S. TSELUEVSKII, and N.M. SHIMANSKAYA
Influence of Aluminium and Indium Addition on the Properties of Superconducting $Bi_2Sr_2CaCu_2O_y$ Metal Oxide
61.70 and 74.70; S10.15
(a) Vol.115,No.1, K79-K82

V. HOLÝ and J. KUBĚNA
Characterization of Microdefects in Silicon by Means of X-Ray Reflection Curves
61.10 and 61.70; S5.11
(b) Vol.155,No.2, 339-348

A.A. MIKHALENKO, V.M. KHULUGUROV, E.I. DMITRIEVA, and V.N. SALOMATOV
F_2^+ Centers in LiF-Mg Crystals
61.70 and 71.55; S9.11
(b) Vol.155,No.2, 359-364

A.V. EGRANOV, A.I. NEPOMNYACHIKH, V.V. OTROSHOK, and B.P. CHERNYAGO
H_2-Centre in LiF-H^-
61.70; 76.30; 78.40; S9.11
(b) Vol.155,No.2, 365-372

V.G. KAPINOS, YU.N. OSETSKII, and P.A. PLATONOV
Simulation of Perfect Vacancy Loops in B.C.C. Metals
61.70
(b) Vol.155,No.2, 373-384

D.L. BEKE, G. NEUMANN, and I.A. SZABÓ
Relations between Characteristics of Point Defects and Bulk Properties of Solids. A Criticism of the Varotsos-Alexopoulos Model.
61.70
(b) Vol.155,No.2, 385-396

V.S. KISHAN KUMAR, S.B.S. SASTRY, and B.S.V.S.R. ACHARYULU
Optical Studies on Europium Doped K_2SO_4
61.70 and 78.50; 78.55; S11.1
(b) Vol.155,No.2, 679-684

J. SERNA, J.CH. ABBÉ, and G. DUPLÂTRE
Size of Voids in Polyethylene
61.40 and 61.70; 78.70; S12
(a) Vol.115,No.2, 389-398

A.F. BURENKOV, V.S. VARYCHENKO, A.M. ZAITSEV, F.F. KOMAROV,
G.G. KONOPLYANIK, A.A. MELNIKOV, V.F. STELMAKH, and V.V. TKACHEV
Depth Distribution of Defects and Impurities in Diamond and Cubic Boron Nitride after High-Energy Ion Implantation
61.80; 61.70; 78.60; S5; S7.1
(a) Vol.115,No.2, 427-436

L. LEJČEK, A. FOUSKOVÁ, and M. PAVEL
Note on Light Scattering by Dislocations in TGS
61.70 and 78.20; S12
(a) Vol.115,No.2, 445-450

R.Z. VALIEV, R.SH. MUSALIMOV, and N.K. TSENEV
The Non-Equilibrium State of Grain Boundaries and the Grain Boundary Precipitations in Aluminium Alloys
61.70; S4
(a) Vol.115,No.2, 451-458

A. KLIMAKOW and M. SCHENK
Grain-Boundary Diffusion of Mercury in p-Type (Hg,Cd)Te
61.70 and 66.30; S8.15
(a) Vol.115,No.2, K135-K138

V.I. BARBASHOV and T.V. RODZINA
Polarity of the Screw-Dislocation Motion in InSb Crystals
61.70; S7.13
(a) Vol.115,No.2, K143-K146

K. NATARAJAN, K. RAMKUMAR, and M. SATYAM
Reverse Characteristics of Phosphorous Passivated Polysilicon p-n Junction Diodes
73.40 and 85; 61.70; S5.11
(a) Vol.115,No.2, K265-K268

M. SOUIRI, A. GOLTZENÉ, and C. SCHWAB
Extension of the Crystal Chemistry Approach for the Location of Muonium in Crystals
61.70; 76.90
(b) Vol.156,No.1, 71-76

L. BOSI and M. NIMIS
On the Criterions for F_A(II) Center Formation in Alkali Halide Crystals
61.70; S9.11
(b) Vol.156,No.1, K5-K10

S. FUJITA
On the Feasibility of Nuclear Fusion in F.C.C. Metals
61.70 and 66.30
(b) Vol.156,No.1, K17-K22

YU.M. ANNENKOV, V.F. STOLARENKO, and T.S. FRANGULIAN
The Effect of Temperature on the Formation and Accumulation of X_3^- Centers in KBr Crystals under Powerful Pulsed Electron Irradiation
61.70; S9.11
(b) Vol.156,No.1, K23-K26

K.-E. PEIPONEN, P. SILFSTEN, P. KETOLAINEN, and T. TSUBOI
A Remark on the Thermo-Optical Properties of the F_A(II) Centre in KCl:Li Crystals
61.70 and 78.50; S9.11
(b) Vol.156,No.1, K81-K84

R. KERN and M. KROHN
Interaction of Metal Clusters with Surface Steps
68.20; 61.70; 62.20; S9.11
(a) Vol.116,No.1, 23-38

N.Y. JIN, F. PHILLIPP, and A. SEEGER
Investigation of Defect Ordering in Heavily Irradiated Metals by High-Voltage Electron Microscopy
61.70; 61.16; S1.2; S1.3
(a) Vol.116,No.1, 91-112

K. SCHEERSCHMIDT, R. HILLEBRAND, and J. HEYDENREICH
Computer Simulation of Diffraction Contrast Images and Lattice Fringe Patterns of Small Spherical Inclusions
61.16; 61.70
(a) Vol.116,No.1, 123-144

F. APPEL
On the Intersection of Oblique Screw Dislocations in NaCl Single Crystals
61.70; 62.20; S9.11
(a) Vol.116,No.1, 153-164

E. PIPPEL and J. WOLTERSDORF
Electron Microscopy of TiC Containing Alumina Ceramics
61.70; 61.16; S1.61; S10
(a) Vol.116,No.1, 165-178

S.K. PLACHKOVA, O. GEORGIEVA, K. PETROV, and D. KOVACHEVA
X-Ray Study of Defects in GeTe-Ag_2Te-Bi_2Te_3 System
61.60 and 61.70; S8.16
(a) Vol.116,No.1, 213-220

JUN CHEN and DUAN FENG
Crystallographic Shear Planes and Stacking Disorders in Perovskites $KNbO_3$ and $NaNbO_3$
61.70; 77.80; S10
(a) Vol.116,No.1, 231-236

I.S. SMIRNOVA
Twins and Dislocations in $RBa_2Cu_3O_7$ Crystals
61.70; 74.70; S10.15
(a) Vol.116,No.1, 237-254

V.V. TOKAREV, V.E. BORISENKO, A.I. DEMCHENKO, and T.M. PYATKOVA
Growth of Epitaxial Nickel Disilicide during Rapid Thermal Processing of Argon-Implanted Nickel Films on Silicon
68.55; 61.70; S1.61; S5.11
(a) Vol.116,No.1, 331-336

B. POHORYLES
Helium Pressure Effects on Photoconductivity and Position of Dislocation Deep Electron States in Plastically Deformed Ge
61.70 and 71.55; 72.40; S5.12
(a) Vol.116,No.1, 349-356

V.K. GOYAL, D. KUMAR, and K. SEN
Effect of Grain Boundary States Density in the Polycrystalline MIS Solar Cells
73.20 and 85; 61.70; 72.40
(a) Vol.116,No.1, 425-432

I.V. ANTONOVA, A.V. VASILEV, V.I. PANOV, and S.S. SHAIMEEV
Formation of Clusters in Gold Doped Silicon
61.70; S5.11
(a) Vol.116,No.1, K33-K36

K. SCHMALZ, K. TITTELBACH, V.V. EMTSEV, and YU.N. DALUDA
Effects of Oxygen Clustering at 600 °C on the Annealing of A-Centers in Cz Silicon
61.70; S5.11
(a) Vol.116,No.1, K37-K42

V.V. LOPATIN, A.V. KABYSHEV, and L.S. BUSHNEV
Modification of Polycrystalline BN, AlN, and Si_3N_4 Surfaces by Ion Beams
61.70 and 68.20; S6; S7
(a) Vol.116,No.1, K69-K72

W. HOERIG, V.F. ORLENKO, H. NEUMANN, and C. ASCHERON
Near-Edge Optical Absorption Spectra of Proton-Implanted GaAs
61.70 and 78.50; S7.12
(a) Vol.116,No.1, K119-K124

A.V. GRACHEV, A.I. NESTEROV, and S.G. OVCHINNIKOV
The Gauge Theory of Point Defects
61.70
(b) Vol.156,No.2, 403-410

GH. STOICESCU, S.V. NISTOR, and C.D. MATEESCU
Aggregation of Bismuth in NaCl Crystals
61.70; 66.30; 78.50; S9.11
(b) Vol.156,No.2, 411-418

R.A. VARDANYAN and S.G. VESELKO
Dislocation Recharging in the Plastically Deformed Crystals of the A_2B_6-Type
61.70 and 71.55; 72.90; S8.1
(b) Vol.156,No.2, 527-536

D. VIGNAUD and J.L. FARVACQUE
Free-Carrier Optical Absorption Induced by Dislocation Scattering Mechanisms in Semiconductors: Application to GaAs
78.20; 61.70; S7.12
(b) Vol.156,No.2, 717-724

M. KUMEDA, M. KIWAKI, and T. SHIMIZU
Erratum to Temperature Dependence of the E′ Center Creation in Silica Glasses (phys. stat. sol. (b) 147, K1 (1988))
61.70 and 76.30; S10.1
(b) Vol.156,No.2, K163-K164

F.G. REQUEJO, A.G. BIBILONI, C.P. MASSOLO, J. DESIMONI, and M. RENTERIA
TDPAC Study of Stoichiometric and Reduced MoO_3. Localization of Indium Trapped Oxygen Single Vacancies
61.60; 61.70; 76.80; S10.1
(a) Vol.116,No.2, 503-512

D.K. AN, K. MADL, A. BARNA, G. BATTISTIG, and J. GYULAI
The Simultaneous Diffusion of Gold and Boron into Silicon. Push Effect of Gold to Boron
66.30; 61.70; S5.11
(a) Vol.116,No.2, 561-570

H.U. JAEGER, T. FEUDEL, and S. ULBRICHT
Modeling of Defect-Phosphorus Pair Diffusion in Phosphorus-Implanted Silicon
61.70; 66.30; S5.11
(a) Vol.116,No.2, 571-582

S.A. LITVINENKO and V.I. SOKOLOV
Formation of Pores in Thermal SiO_2 Due to Internal Mechanical Stress Relaxation
61.70 and 64.75; S10.1
(a) Vol.116,No.2, 615-620

KEQIN ZHENG, SHENGYAN LIU, and DECHENG TIAN
Investigation of the Behavior of Internal Friction of Al-Mg-Si-RE Alloys
62.40; 61.70; S1.4; S3
(a) Vol.116,No.2, 621-626

V.I. KHRUPA, V.V. NIKOLAEV, and M.YA. SKOROKHOD
On the Transition between Dynamical and Kinematical X-Ray Diffraction in Thin Crystals with Randomly Distributed Dislocations
61.10; 61.70; S5.11
(a) Vol.116,No.2, K141-K146

T. KANADANI and A. SAKAKIBARA
Effect of an Addition of Ge on the Formation of GP Zones in Al-4mass%Zn Alloy
61.70; S4
(a) Vol.116,No.2, K153-K156

61.80. Radiation damage and other irradiation effects
(for techniques of structure determination, see 61.10 to 61.16; for electron and ion impact phenomena, see 79.20)

J. HEROLD, H. BARTSCH, W. WESCH, and G. GOETZ
Investigation of Low-Temperature Epitaxial Regrowth of Ion-Implanted Amorphous GaAs
68.48; 61.16; 61.80; S7.12
(a) Vol.111,No.1, 59-70

S.V. KOVESHNIKOV, E.B. YAKIMOV, N.A. YARYKIN, and V.A. YUNKIN
Plasma Stimulated Impurity Redistribution in Silicon.
61.80 and 71.55; 73.40; S5.11
(a) Vol.111,No.1, 81-88

R. PASIANOT, G. GALAMBOS, E.J. SAVINO, L. AMARAL, M. BEHAR, D. FARKAS, and S. RAO
Argon Irradiation of Sn Thin Layers Deposited on Fe Substrates
61.80 and 68.20; S1.1; S4
(a) Vol.111,No.1, 173-180

S. WAPLAK and T. KRAJEWSKI
EPR Studies of $LiNaSO_4$ Chromium Cr^{6+} Doped Monocrystals after X-Ray Irradiation
76.30; 61.80; 66.70; S11
(a) Vol.111,No.1, 271-278

V.P. GOLTSEV, V.V. KHODASEVICH, G. GOETZ, I.I. PRIKHODKO, V.M. DRAKO, and V.V. UGLOV
Study of Ion Bombardment of Double-Layer Thin Zr/Ni Films
61.80 and 68.48; S1.2
(a) Vol.111,No.1, K13-K16

N. AZZIZ and P.C. MURLEY
Effective Charge of Moving Ions in Matter
61.80; S5.11
(b) Vol.151,No.2, K105-K108

R.A. YAKSHIBAEV, V.N. ZABOLOTSKII, and R.F. ALMUKHAMATOV
Structural Features and Ionic Transport in the Ag_xNbSe_2 Two-Dimensional Superionic Conductor
61.80 and 66.90; S8.15
(a) Vol.111,No.2, 431-442

L. ZAKHARENKO, V. KOZLOVSKII, and YA. PILKEVICH
Role of Irradiation Temperature in the Conductivity Compensation of n-GaAs by Radiation Defects
61.80 and 72.80; S7.12
(a) Vol.111,No.2, K215-K218

V.V. KAPLIN, V.I. GRIDNEV, V.G. KHLABUTIN, E.I. ROZUM, and S.A. VOROBIEV
Volume Capture of 1.5 to 5.7 MeV Electrons in Quantum Channeling States
61.80; S5.11
(b) Vol.152,No.1, 61-66

P.S.S. PRASAD and S. RADHAKRISHNA
Chromium Centres in 3 $CdSO_4 \cdot 8\ H_2O$ Single Crystals
61.80 and 71.55; 78.50; S11.1
(b) Vol.152,No.1, 171-180

E.A. LITTLE, D.J. BUTTLE, and C.B. SCRUBY
Radation Damage Studies in Model Ferritic Alloys Using Micromagnetic Techniques
61.80; 75.50; S1.1
(a) Vol.112,No.1, 55-72

V. GEIST and D. STEPHAN
Investigation of Lattice Strain in Proton-Irradiated Semiconductor Materials
61.80; 61.10; S5.11; S5.12; S7.1
(a) Vol.112,No.1, 79-88

J. GYULAI
Some Recently Emerged Problems of Ion Implantation: A Review
61.80; 68.55; S5.11
(a) Vol.112,No.1, 237-244

W. ROSIŃSKI and J. KACZMAREK
Present Position and Further Developments of Ion Implantation in Metals
61.80 and 68.20; S1.1
(a) Vol.112,No.1, 245-252

M. POSSELT
Linear Cascade Theory of Inhomogeneous Targets
61.80
(a) Vol.112,No.1, 263-272

F.F. KOMAROV and A.P. NOVIKOV
Self-Annealing in Ion-Implanted Silicon
61.80; S5.11
(a) Vol.112,No.1, 273-288

E. WENDLER, W. WESCH, and G. GOETZ
Defects in Weakly Damaged Ion Implanted GaAs and Other III-V Semiconductors
61.80; 78.50; S7.11; S7.12
(a) Vol.112,No.1, 289-300

K. PAPROCKI, I. BRYŁOWSKA, W. ANDRAE, and J. KRYNICKI
Recoil Implantation from Sb Thin Films under Ion Bombardment
61.80 and 79.20; S5.11
(a) Vol.112,No.1, 301-304

A.V. SIMASHKEVICH, L.V. GORCHAK, E.V. GILAN, and K.D. SUSHKEVICH
Study of p-n Junctions Fabricated by Phosphorus-Ion Implantation into CdTe
61.80 and 73.40; S8.13
(a) Vol.112,No.1, 305-310

G. BARDOS
An Approximation for Solving the Transport Equation of the Angular Flux of Ions in Matter
61.80; S1.2; S5.11
(a) Vol.112,No.1, 311-314

W. SZYSZKO and J. SIELANKO
Monte Carlo Calculation of Depth Distribution of Implanted Ions and Sputtering Yield. Comparison for Some of the ...
61.80 and 79.20; S5; S5.11
(a) Vol.112,No.1, 315-318

J. SIELANKO
Influence of the Primary Ion Beam on the Depth Distribution of Implanted Atoms Measured by SIMS
61.80; 79.20; S5.11
(a) Vol.112,No.1, 319-322

F.F. KOMAROV, A.P. NOVIKOV, E.V. KOTOV, and E.A. PODLIPKO
The Behaviour of Arsenic in Crystals of Silicon during Self-Annealing Ion Implantation
61.80; S5.11
(a) Vol.112,No.1, 323-326

V.F. KOROTOV, V.I. KHITKO, V.A. YURCHENKO, and V.S. PIVOVAR
On the Specific Features of Gallium Arsenide Doping with Donor by Ion Implantation
61.80 and 73.40; S7.12
(a) Vol.112,No.1, 331-334

M. BUCKI, A. GEMPERLE, S.A. KONSTANTINOV, M.M. KANTOR, F. ČERNÝ, K. KULT, G.S. BURCHANOV, and E.V. OTTENBERG
Modification of Surface Layers of Molybdenum by B^+ and N^+ Ion Implantation
61.80 and 68.20; S1
(a) Vol.112,No.1, 335-338

T. OTMIANOWSKI and B. KAMIENSKA-STEFANIUK
Some Aspects of the Surface Wear Process of Steel Implanted with Nitrogen
61.80 and 68.20; S1.1
(a) Vol.112,No.1, 339-342

J. JAGIELSKI, G. GAWLIK, A. PODGORSKI, A. TUROS, A. DYGO, and N. MADI
Thermal Effects on Depth Distribution of Nitrogen Atoms Implanted in Ferrous Alloys
61.80; 61.70; S1.1
(a) Vol.112,No.1, 343-346

T. BARBASZEWSKI, M. DABROWSKI, M. DRWIEGA, J. GAWLIK, and E. LIPINSKA
Implantation Profiles of Nitrogen and Titanium in a Low Energy Bombarded High Speed Tool Steel
61.80 and 68.20; S1.1
(a) Vol.112,No.1, 347-352

P. TARKOWSKI, I. BRYŁOWSKA, and M. SUBOTOWICZ
Influence of Nitrogen Implantation on Corrosion Process in Steel
61.80; 68.20; S1.1
(a) Vol.112,No.1, 353-356

J. JĘDRKOWIAK, J. MARTAN, J. MASALSKI, and D.B. BOGOMOLOV
Influence of Ar, Al, and N Ion Implantation on the Surface Properties of the OOH18M2Nb Steel
61.80; 68.20; S1.1
(a) Vol.112,No.1, 357-360

V.L. KUTUZOV, M.YU. OVSYANNIKOV, I.G. ROMANOV, A.D. POGREBNYAK, and G.E. REMNEV
Mechanical and Frictional Properties of Tool Steels Exposed to HPIB Irradiation
62.20; 61.80; S1.1
(a) Vol.112,No.1, 361-364

J. ŁUNARSKI and M. ZIELECKI
Relationship between Microhardness and Fatigue Strength after Glass Micro-Bead Peening and Ion Implantation Processes
62.20; 61.80; S1.1
(a) Vol.112,No.1, 365-374

V.F. STELMAKH, YU.R. SUPRUN-BELEVICH, and A.R. CHELYADINSKII
Effect of Radiation Defects and Elastic Incompatibility Stresses on the Electrical Activation and Diffusion of Boron ...
66.30; 61.80; 62.20; S5.11
(a) Vol.112,No.1, 381-384

D. JAWORSKA, E. TARNOWSKA, A.P. KOBZEV, and R.A. ILKHAMOV
Depth Distribution of the Gettered Au Atoms Measured by Means of RBS and Radioactive Methods
61.80; 66.30; S5.11
(a) Vol.112,No.1, 385-390

C. ASCHERON, A. SCHINDLER, R. FLAGMEYER, and G. OTTO
Characterization of Irradiation-Induced Lattice Disorder by Dilatometric Measurements
61.50 and 61.80; S5.11; S5.12; S7.11; S7.12
(a) Vol.112,No.1, 391-394

J.TH. ZETTLER, J. BOLLMANN, K.H. HECKNER, and H.A. KLOSE
Radiation-Induced Defects in Silicon Due to Low-Energetic Ion-Assisted Deposition of Indium Tin Oxide
61.80 and 68.55; 73.40; S10.15
(a) Vol.112,No.1, 395-404

A.V. KONDRACHUK
Model of Subsurface Redistribution of Impurity Atoms and Radiation Defects upon Light Ion Irradiation at Elevated Temperature
61.80; 68.20; S5.11
(a) Vol.112,No.1, 411-414

K. MOJEJKO-KOTLIŃSKA, J. KOTLIŃSKI, and M. SUBOTOWICZ
Deep Traps in Doubly Implanted p-Type Silicon. DLTS and C-U Measurements
61.80; 71.55; S5.11
(a) Vol.112,No.1, 415-418

D. HOEHL and H. BARTSCH
Results of In-Situ Annealing of Ion-Implanted Silicon in a High Voltage Electron Microsope
61.70 and 61.80; 61.14; 68.48; S5.11
(a) Vol.112,No.1, 419-424

W. SOSZKA
Krypton Saturation of the Metal Surface during Kr^+ Ion Bombardment Determined Due to Autoionization Electron Analysis
61.80; 68.20; 79.20; S1
(a) Vol.112,No.1, 425-430

W. SKORUPA, R. GROETZSCHEL, and H. BARTSCH
Recent Advances of Ion Beam Synthesis for Silicon-on-Insulator Structures
61.70 and 61.80; S5.11
(a) Vol.112,No.2, 661-666

W. SCHARFF, K. HAMMER, B. EIBISCH, O. STENZEL, S. ROTH, and T. FRAUENHEIM
Deposition of Hyrdogenated Amorphous Carbon Films by a ECR Microwave Plasma
61.80 and 68.55; 73.60; 78.65; S5
(a) Vol.112,No.2, 667-678

J. AULEYTNER
Structural Changes in Solids Induced by Heavy Ions at Energies of MeV/n. Results of X-Ray and Electron Microscopic Studies
61.80; 61.10; 61.16; S5.11; S11.2
(a) Vol.112,No.2, 679-688

M. SETVAK
The Present State of Ion Implantation Equipment
61.70 and 61.80
(a) Vol.112,No.2, 689-694

P.I. GAIDUK, F.F. KOMAROV, and V.S. SOLOVEV
Structure Changes in Ion Implanted (111)-Si Layers during Pulsed Nanosecond Annealing and Following Thermal Treatment
61.80 and 64.75; S5.11
(a) Vol.112,No.2, 703-706

A.V. SUVOROV, K.K. BOURDELLE, N.G. CHECHENIN, and V.N. MAKAROV
Studies of Amorphous Layer Formation in SiC under Ga^+ Bombardment
61.80; 64.70; S1.61
(a) Vol.112,No.2, 707-714

H. BARTSCH, D. HOEHL, and P. WERNER
TEM Cross-Sectional Investigations of the Amorphous/Crystalline Phase Transition of Ion-Implanted Silicon
61.80 and 68.55; 68.48; S5.11
(a) Vol.112,No.2, 715-720

J. DANILOVICH, K. GAERTNER, G. GOETZ, A. NOVIKOV, and B. WEBER
Ion Implanted Buried Si_3N_4 Layers below Epitaxial $NiSi_2$ Layers
68.55; 61.80; S1.61
(a) Vol.112,No.2, 733-738

M. STRÓŻAK, M. SUBOTOWICZ, and P. MIKOŁAJCZAK
Formation of Unusual Alloys by Means of Ion Beam Mixing of the Pd/V Superlattice
61.55 and 61.80; 68.48; S1.2
(a) Vol.112,No.2, 739-742

I. BRYŁOWSKA, K. PAPROCKI, and R. ILKHAMOV
The RBS Study of Ion Induced Mixing in Sb-Si System
61.80 and 68.48; S1.61
(a) Vol.112,No.2, 743-746

G. GŁADYSZEWSKI, P. MIKOŁAJCZAK, and M. SUBOTOWICZ
Ion Beam Mixing in Bi-Sb Superlattices
61.55 and 61.80; 68.48; S4
(a) Vol.112,No.2, 753-756

J. PIEKOSZEWSKI, Z. WERNER, C. POCHRYBNIAK, J. LANGNER, M. GRYZINSKI, and A. HORODEŃSKI
Pulse Ion Beam Doping and Modification of Solids
61.80 and 68.20
(a) Vol.112,No.2, 757-760

M. KISZA and E. OLESZKIEWICZ
Modified PMMA-Polymers for Masks in Ion Beam Lithography
61.40 and 61.80; S12
(a) Vol.112,No.2, 761-764

W. RUDOLPH, D. GRAMBOLE, C. HEISER, F. HERRMANN, and C. NEELMEIJER
Compositional Changes of PMMA Layers during ^{15}N Bombardment
61.40 and 61.80; S12
(a) Vol.112,No.2, 765-768

R. MUEHLE and T. HARTUNG
Ion Beam Exposure Characteristics of Nitrocellulose
61.70 and 61.80; S12
(a) Vol.112,No.2, 769-772

J. CZERBNIAK, M. MONETA, and B. PAWLOWSKI
The Equilibrium Charge of Planar Channeled 50 to 100 keV Protons
61.80; S5.11
(a) Vol.112,No.2, 773-776

A. KOZANECKI
Backscattering and Optical Studies of Ytterbium Implanted into $A^{III}B^{V}$ Compounds
61.80; 78.55; S7.11; S7.15
(a) Vol.112,No.2, 777-780

R.A. ILKHAMOV, LI JEN HO, S.S. KHUSMURODOV, D. MACHAJDIK, G.M. OSETINSKIJ, M. PAJEK, R. SANDRIK, and D.M. SHIROKOV
Hydrogen Depth Profiling by Elastic Recoil Detection
61.70 and 61.80; S5.11
(a) Vol.112,No.2, 781-784

M. KULIK, and K. ŻUKOWSKA
The Multilayer Model of Implanted GaAs. Ellipsometric Investigation
61.80; 78.20; S7.12
(a) Vol.112,No.2, 785-788

M. KULIK and J. ŻUK
Ellipsometric Investigation of the Kinetics of Native Oxide Growth on Aluminium and Argon Implanted GaAs
61.80 and 68.55; 78.20; S7.12
(a) Vol.112,No.2, 789-792

M. KULIK, D. JAWORSKA, and J. TRZCIANKA
Determination of the Damage Profiles in Implanted GaAs from the Change of (n,k) Optical Parameters
61.80; 78.20; S7.12
(a) Vol.112,No.2, 793-798

W. DZIURDA, L. GABLA, and M. TULETA
Bombardment-Induced Photon Emission from Modified Surface of Glass
61.80 and 78.55; S10
(a) Vol.112,No.2, 811-814

J. MITTENBACHER and R. MUEHLE
A Scanning Ion Microbeam System with a Liquid Metal Ion Source
61.80 and 79.70; 61.16; S1.3; S4
(a) Vol.112,No.2, 825-828

R. RIESENBERG, A. UNKROTH, and D. DIETRICH
Stress Controlled Adhesion and Magnetic Properties of Ion Implanted Co Films
62.20; 61.80; 75.60; S12
(a) Vol.112,No.2, K85-K90

J. TATARKIEWICZ
Optical Effects of High Energy Implantations in Semiconductors
61.80 and 78.20; 78.30; 78.50; 78.55; S5.11; S7.12
(b) Vol.153,No.1, 11-48

A.R. URMANOV, V.G. GRIGORIEV, and A.A. PUZANOV
A Kinetic Theory for the Motion of Accelerated Particles in Solids. II. Dechanneling in Complex Crystals
61.80 and 79.20; S10
(b) Vol.153,No.1, 65-78

N.A. DOUMANIS and CH.G. TRICALINOS
Derivation of the Fokker-Planck Kinetic Equation for Planar Channeling of Relativistic Positively Charged Particles
61.80; S4
(b) Vol.153,No.1, K1-K4

S.N. ROMANOV, A.I. MELKER, and D.B. MIZANDRONTSEV
Athermal Clusters of Vacancies in B.C.C. and H.C.P. Metals
61.70; 61.80; S1; S1.1
(b) Vol.153,No.1, K5-K10

B.X. LIU, X. ZHOU, and H.-D. LI
Thermodynamics and Growth Kinetical Consideration of Metal-Nitride Formation by Nitrogen Implantation (Review Article)
61.80 and 64.60; 61.55; S1.62
(a) Vol.113,No.1, 11-22

S.K. NIKOGOSYAN and V.A. SAAKYAN
The Effect of Electron Radiation on Thermal and Electrophysical Properties of n-InP
61.80; 66.70; 72.20; S7.11
(a) Vol.113,No.1, 43-50

A.A. HOSSEINI and I.P. JONES
The Radiation Induced Segregation of Nickel in B.C.C. Iron Alloys
64.75; 61.80; S1.1
(a) Vol.113,No.1, 57-62

Y. SHI, F.M. WU, D.X. SHEN, M.K. DENG, K.J. CHENG, and C.H. WANG
Influence of Previous Defects on the Formation of Irradiated Defects in NTD Si
61.80; S5.11
(a) Vol.113,No.1, K15-K18

S. SATPATHY and A.P. PATHAK
Planar Channeling of Electrons and Positrons in Crystals. Local-Density Calculation of Planar Averaged Potential and ..
61.80 and 78.70; S5; S5.11; S5.12
(b) Vol.153,No.2, 455-464

S.D. KOUIMTZI
Localized States in Crystalline Semiconductors
61.80 and 71.55; S5.11; S5.12; S7.1
(b) Vol.153,No.2, 675-682

N.YU. LEBEDEV and K. LENKEIT
Azimuthal Angular Dependence of the Backscattering Yield in the Axial-to-Planar Channeling Transition Case
61.80; S1.3
(b) Vol.153,No.2, K97-K102

YU.V. LILENKO, K.V. SHASTOV, A.S. PETROV, A.V. VOITSEKHOVSKII, V.S. KULIKAUSKAS, V.N. KUZNETSOV, and A.P. MAMONTOV
Radation-Induced Defects in Implanted $Hg_{1-x}Cd_xTe$ Crystals
61.80; 72.20; 79.20; S8.15
(a) Vol.113,No.2, 285-294

S. KOPTA, R. HAJDUK, J. LEKKI, B. RAJCHEL, and A.Z. HRYNKIEWICZ
Experimental Determination of the Energy Loss of Protons Channeled along the (111) Axis through a Silicon Single Crystal
61.80; S5.11
(a) Vol.113,No.2, 295-306

L.A. KAZAKEVICH, P.F. LUGAKOV, and I.M. FILIPPOV
The Role of Dislocations during Charge-Carrier Recombination at Radiation Defects in p-Si
61.70 and 72.20; 61.80; S5.11
(a) Vol.113,No.2, 307-314

V.V. BOLOTOV, V.M. EMEKSUZYAN, V.N. SPIRIDONOV, K. SCHMALZ, and M. TRAPP
Radiation-Induced Redistribution of Gold in the SiO_2-Si Structures
61.80 and 73.20; 68.48; S5.11
(a) Vol.113,No.2, 315-320

K.D. GLINCHUK, N.S. ZAYATS, and A.V. PROKHOROVICH
Effect of Proton Bombardment on GaAs Luminescence
71.55 and 78.55; 61.80; S7.12
(a) Vol.113,No.2, 497-502

M. OLSZOWY
Electret State in the Polyethylene Terephthalate Films Implanted by Oxygen and Bismuth Ions
77.30; 61.80; S12
(a) Vol.113,No.2, 609-614

G. KLUGE, A. THOMAS, R. KLABES, and P. SUEPTITZ
Applicability of Silver Implanted Amorphous $GeSe_2$ as Inorganic Negative Resist
61.80; 66.30; 79.20; S8
(a) Vol.113,No.2, K171-K176

B.B. BOIKO, F.P. KORSHUNOV, G.V. GATALSKII, A.I. AKIMOV, V.I. GATALSKAYA, and E.K. STRIBUK
Effect of Irradiation by Fast Electrons and ^{60}Co-gamma Rays on Superconducting and Electrical Properties of Thallium ...
61.80 and 74.70; S10.15
(a) Vol.113,No.2, K235-K240

A.L. SARCE and E.J. SAVINO
Thermodynamics of Deformation under Irradiation
62.20; 61.80; 65.50
(b) Vol.154,No.1, 105-114

M. MONETA and J. CZERBNIAK
Charge State of Planar Channeled Ions
61.80
(b) Vol.154,No.1, K13-K18

E. SOBESLAVSKY and W. SKORUPA
Modeling of Nitrogen High Dose Implantation into Silicon in the Energy Range of 150 to 330 keV
61.80; S5.11
(a) Vol.114,No.1, 135-144

J. BUDINAVIČIUS, L. PRANEVIČIUS, and S. TAMULEVIČIUS
Peculiarities of the Ag-Si Interface Formed under Simultaneous Deposition and Ion Beam Irradiation
61.80 and 68.48; S5.11
(a) Vol.114,No.1, K25-K28

P. BROGOWSKI, H. MUCHA, and J. PIOTROWSKI
Modification of Mercury Cadmium Telluride, Mercury Manganese Telluride, and Mercury Zinc Telluride by Ion Etching
61.80 and 72.80; 85; S8.15
(a) Vol.114,No.1, K37-K40

G. E. ZARDAS, Z. HATZOPOULOS, and P.H. YANNAKOPOULOS
Effect of Low Dose Electron Irradiation on AlGaAs/GaAs at Low Temperatures
61.80 and 72.40; S7.12; S7.15
(a) Vol.114,No.1, K41-K44

A.K. ABIEV, E.Z. ALIEV, B.A. NATIG, and R.M. SARDARLY
Electron Irradiation Effect on Dielectric Properties of $TlInS_2$ Single Crystals
77.80; 61.80; S8.16
(a) Vol.114,No.1, K119-K122

V.L. LITVINOV, N.A. UKHIN, and YU.I. CHURAKOV
Potential of Fast Neutron Damage Regions in Heavily Doped Semiconductors
61.80
(b) Vol.154,No.2, 507-512

V.G. MALESH, YU.M. AZHNYUK, D.B. GOYER, and A.V. GOMONNAI
Disorder-Activated First-Order Raman Scattering by Acoustic Phonons in Electron-Irradiated GaP Crystals
61.80 and 78.30; S7.11
(b) Vol.154,No.2, K197-K200

M. TOULEMONDE, J. DURAL, G. NOUET, P. MARY, J.F. HAMET, M.F. BEAUFORT, J.C. DESOYER, C. BLANCHARD, and J. AULEYTNER
High Energy Heavy Ion Irradiation of Silicon
61.80; S5.11
(a) Vol.114,No.2, 467-474

V.N. BRUDNYI, A.D. POGREBNYAK, YU.P. SUROV, and A.S. RUDNEV
Positron Trapping at Point Defects in Electron-Irradiated GaAs
61.70; 61.80; 78.70; S7.12
(a) Vol.114,No.2, 481-490

M. MIHALİK, J. KOVÁČ, A. ZENTKO, and A. LOVAS
Some Magnetic Properties of Neutron Irradiated $Fe_{85-x}Cr_xB_{15}$ Metallic Glasses
61.80 and 75.30; 75.40; 75.60; S1.1; S1.2
(a) Vol.114,No.2, 679-684

V.N. BRUDNYI, V.V. PESHEV, and S.V. SMORODINOV
New Metastable W-Center in Electron-Irradiated n-Type InP
61.80; S7.11
(a) Vol.114,No.2, K139-K142

B.M.S. LEWERENTZ and B.G. SVENSSON
Capacitance-Voltage Characteristics of Au-Si Schottky Diodes: Influence of Electron Irradiation
61.80 and 73.30; S5.11
(a) Vol.114,No.2, K147-K152

A.G. BUDREVICH, A.S. LYUTOVICH, and B.L. OKSENGENDLER
Radiation Sieve Effect during Epitaxy from Low Energy Ion-Molecular Flow
68.55; 61.80; S5.11
(a) Vol.114,No.2, K163-K166

V. BICHEVIN, G. ALSEITOV, M. GAPANOV, and H. KAAMBRE
Excimer-Laser-Induced Exoemission and Thermoluminescence of Some Alkali Halides
78.60 and 79.75; 61.80; S9.11
(a) Vol.114,No.2, K241-K244

T.D. DZHAFAROV
Radiation-Stimulated Diffusion in Semiconductors (Review Article)
61.80 and 66.30; S5.11
(b) Vol.155,No.1, 11-52

V.S. LYSENKO, A.N. NAZAROV, I.M. ZARITSKII, G. SERFOZO, G. BATTISTIG, J. GYULAI, and L. DÓZSA
RF Plasma Modification of Heavy Destroyed Ion Implanted Subsurface Silicon Layers
61.80; 73.40; S5.11
(a) Vol.115,No.1, 75-80

KH.R. KAZDAEV, M.T. AKCHULAKOV, E.M. BAYADILOV, V.I. ENGĖLKO, A.V. LAZARENKO, and E.S. CHEBUKOV
Influence of Annealing Regimes on Phase Transitions in Nitrogen and Carbon Ion Implanted Molybdenum
61.80 and 68.20; S1.61; S1.62
(a) Vol.115,No.1, 179-184

I.G. MARCHENKO
The Effect of Radiation Defects on the Differential Resistance of p-n Junctions during Avalanche Breakdown
61.80 and 73.40; S5.11
(a) Vol.115,No.1, K39-K42

M. HRABCAK, M. MIHALIK, J. KOVAC, and A. LOVAS
Saturation Magnetostriction and Curie Temperature of Neutron Irradiated $Fe_{75}Cr_{10}B_{15}$ and $Fe_{77}Cr_{8}B_{15}$ Amorphous Alloys
75.80; 61.80; S1.1
(a) Vol.115,No.1, K83-K86

K. LENKEIT, A.K. GUTAKOVSKII, YU.O. KANTER, R. FLAGMEYER, S.M. PINTUS, S.V. RUBANOV, A.A. FEDOROV, and V.P. POPOV
Study of InGaAs-GaAs Strained-Layer Superlattices by TEM and RBS Techniques
68.48; 61.80; S7.12; S7.15
(a) Vol.115,No.2, 413-426

A.F. BURENKOV, V.S. VARYCHENKO, A.M. ZAITSEV, F.F. KOMAROV, G.G. KONOPLYANIK, A.A. MELNIKOV, V.F. STELMAKH, and V.V. TKACHEV
Depth Distribution of Defects and Impurities in Diamond and Cubic Boron Nitride after High-Energy Ion Implantation
61.80; 61.70; 78.60; S5; S7.1
(a) Vol.115,No.2, 427-436

R. FABBRI, M. SERVIDORI, A. ZANI, and F. CEMBALI
Two-Stage Recovery of Lattice Damage Induced in Silicon by Ion Implantation below Amorphization Threshold. Analysis by...
61.80; 61.10; S5.11
(a) Vol.115,No.2, 437-444

E.M. IBRAGIMOVA and V.N. STEPANOV
Surface Defect States at Silica-Polymer Interfaces in Fiber-Optic Waveguides
61.80 and 78.60; S10.1; S12
(a) Vol.115,No.2, K255-K260

T. KANEKO
Partial and Total Electronic Stoppings of Solids and Atoms for Energetic Ions
61.80; 71.10
(b) Vol.156,No.1, 49-64

A.A. ALYBAKOV, V.A. GUBANOVA, M.ZH. ZHANIBEKOV, and N. TOICHEV
EPR Spectra of X-Irradiated $KBr:CrO_4^{2-}$, Ba^{2+} Single Crystals
76.30; 61.80; S9.11
(b) Vol.156,No.1, K77-K80

N.A. MIRONOVA, V.N. SKVORTSOVA, A.S. SMIRNOV, D.V. RIEKSTINYA, G.A. GRINVALDS, and L.A. LITVINOV
Anomalous Behaviour of $MgO \cdot 2.5 Al_2O_3$ under Radiation
61.80 and 78.40; 78.55; S11.2
(a) Vol.116,No.1, K125-K128

P.C. EUTHYMIOU, P.C. BANBURY, G.J. PAPAIOANNOU, and G.E. ZARDAS
Dependence of I-V Characteristics of GaAs MESFET on Temperature and Eletron Irradiation
73.40 and 85; 61.80; S7.12
(a) Vol.116,No.1, K133-K136

YU.F. BLINOV and P.V. SERBA
Theory of the Recoil Implantation
61.80
(a) Vol.116,No.2, 555-560

M. OKUMURA, H. MIZUBAYASHI, S. OKUDA, and Y. TAGISHI
Internal Friction Study on Anomalous Self Interstitial Atom-Dislocation Interaction in Nb after Low Temperature Irradiation
61.80 and 62.40; S1
(a) Vol.116,No.2, 583-594

A. THOMAS, A. ZEHE, B. SCHRECKENBACH, J.-W. ERBEN, and R. GROETZSCHEL
Ion Beam Lithography of Polycrystalline CaF_2 Films Deposited on Silicon
61.80 and 68.55; 77.55; S9.12
(a) Vol.116,No.2, 735-744

Y. UMAKOSHI, W. TAKAHARA, T. YAMANE, and K.A. TANAKA
Effect of Laser Irradiation on the Superconductive Properties of $(Y_{0.95}Sm_{0.05})Ba_2Cu_3O_x$
74.70; 61.80; S10.15
(a) Vol.116,No.2, 787-792

V. RIEDE, H. SOBOTTA, H. NEUMANN, C. ASCHERON, C. NEELMEIJER, and A. SCHINDLER
Depth Dependence of Localized Mode Absorptions in Proton-Implanted InP
61.80 and 78.30; 63.20; S7.11
(a) Vol.116,No.2, K147-K152

H. DUŃSKI and H. SUGIER
Temperature-Programmed Desorption of Oxygen from gamma-Irradiated Polycrystalline Magnesium Oxide
61.80 and 68.45; S10.1
(a) Vol.116,No.2, K165-K168

62. Mechanical and Acoustical Properties of Condensed Matter
(see also 61.70. Defects in crystals, 68.30. Surfaces and interfaces)

62.10. Mechanical properties of liquids
(for viscosity of liquids, see also 66.20)

D. MONSELESAN and H.-R. TREBIN
Temperature Dependence of the Elastic Constants for Biaxial Nematic Liquid Crystals
61.30 and 62.10
(b) Vol.155,No.2, 349-358

62.20. Mechanical properties of solids
(related to microscopic structure)

F. POVOLO and E.B. HERMIDA
Analysis of Stress Relaxation and Creep Curves for a Log-Normal Distribution of Relaxation Times
62.20; S4; S11
(b) Vol.151,No.1, 71-84

J. POKLUDA and P. ŠANDERA
On the Ideal Strength of Crystals
62.20
(b) Vol.151,No.1, 85-94

WEN-CHEN ZHENG
A Study of Distortion of the Immediate Environment of the Chromium Ion in Alums
61.70; 62.20; 76.30; S11.1
(b) Vol.151,No.1, K11-K16

A. KRELL and W. POMPE
Grain Boundary Microdamage Due to Visco-Elastic Relaxation of Residual Stresses in Alumina Ceramics
61.70 and 62.20; S10.15
(a) Vol.111,No.1, 109-118

T.H. YOUSSEF, F.A. SAADALAH, and I.K. BISHAY
Plastic Instability during Combined Torsion-Tension Deformation in Al-0.72% Si Alloy
62.20; S4
(a) Vol.111,No.1, 137-144

T. OHGAKU and N. TAKEUCHI
Relation between Plastic Deformation and the Blaha Effect for Alkali Halide Crystals
62.20; 62.40; S9.11
(a) Vol.111,No.1, 165-172

B. GAMBIN and E. KROENER
Higher-Order Terms in the Homogenized Stress-Strain Relation of Periodic Elastic Media
62.20
(b) Vol.151,No.2, 513-520

S. POPOVIĆ, H. LOEFFLER, B. GRŽETA, G. WENDROCK, and P. CZURRATIS
Lattice Constant of Al-Rich Matrix of Two Al-(8 and 24)at% Zn Alloys in Equilibrium with (Metastable) Guinier Preston Zones
61.55; 62.20; 64.70; S4
(a) Vol.111,No.2, 417-430

L.E. SOSHNIKOV and A.U. SHELEG
Elastic Properties and Phase Transitions in beta-CdP_2
62.20 and 64.60; S7
(a) Vol.111,No.2, 485-490

K.T. AVETIAN, S.H. ANCHARAKIAN, and M.M. ARAKELIAN
Stress Distribution at Biaxial Stressed Conditions
62.20; S5.11
(a) Vol.111,No.2, K161-K164

J. VAN DER STEL, H.B. VAN VELDHUIZEN, G.W. KOEBRUGGE, J. SIETSMA, and A. VAN DEN BEUKEL
Change of Electrical Resistivity and Youngs Modulus during Crystallization of Amorphous $Fe_{40}Ni_{40}B_{20}$
61.40 and 64.70; 62.20; 72.15; S1.1; S1.2
(a) Vol.112,No.1, 35-40

B.C. DE COOMAN and C.B. CARTER
The Formation of Faulted Dipoles in Plastically Deformed GaAs
61.70; 61.16; 62.20; S7.12
(a) Vol.112,No.1, 41-54

C. LEVADE, J.J. COUDERC, and A. KARA
Plastic Deformation of Natural Sphalerite Single Crystals between 300 and 620 K
62.20; 61.70; S8.11
(a) Vol.112,No.1, 89-98

V.L. KUTUZOV, M.YU. OVSYANNIKOV, I.G. ROMANOV, A.D. POGREBNYAK, and G.E. REMNEV
Mechanical and Frictional Properties of Tool Steels Exposed to HPIB Irradiation
62.20; 61.80; S1.1
(a) Vol.112,No.1, 361-364

J. ŁUNARSKI and M. ZIELECKI
Relationship between Microhardness and Fatigue Strength after Glass Micro-Bead Peening and Ion Implantation Processes
62.20; 61.80; S1.1
(a) Vol.112,No.1, 365-374

V.F. STELMAKH, YU.R. SUPRUN-BELEVICH, and A.R. CHELYADINSKII
Effect of Radiation Defects and Elastic Incompatibility Stresses on the Electrical Activation and Diffusion of Boron ...
66.30; 61.80; 62.20; S5.11
(a) Vol.112,No.1, 381-384

R. RAMJI RAO and A. PADMAJA
Effective Second Order Elastic Constants and the Second Pressure Derivatives of a Hexagonal Crystal
62.20; S1.2; S3
(b) Vol.152,No.2, 455-462

M. SOWA, W. SZYSZKO, J. SIELANKO, L. GŁUSIEC, and B. KAMIEŃSKA
Microhardness of Boron, Titanium, and Nitrogen Implanted Steel
62.20; 66.30; S1.1
(a) Vol.112,No.2, 839-844

R. RIESENBERG, A. UNKROTH, and D. DIETRICH
Stress Controlled Adhesion and Magnetic Properties of Ion Implanted Co Films
62.20; 61.80; 75.60; S12
(a) Vol.112,No.2, K85-K90

M.B. ZIKRY and K.H. GEORGY
Creep Rate Sensitivities of Copper in an .alpha Phase Copper-Zinc Solid Solution at 473 K
62.20; S1.2
(a) Vol.112,No.2, K91-K96

R. SINGH and J. SHANKER
Three-Body Force and Shell Model Contributions to Third-Order-Elastic Constants for Fluorite Structure Crystals
62.20; S9.12
(b) Vol.153,No.1, 123-130

S.G. PSAKHIE, S.V. ALEKSEEV, and V.E. PANIN
Influence of Shear on the Binding Energy of Complexes of Impurities and Vacancy in Aluminium
61.70; 62.20; S4
(b) Vol.153,No.1, K11-K16

H.-MATSUO KAGAYA, H. KOTOKU, and T. SOMA
Thermal Pressure and Pressure-Volume Relations of III-V Covalent Crystals
62.20; 63.20; S7.1
(b) Vol.153,No.1, K17-K20

XIAO HUA CHEN, YE NING WANG, and HUI MING SHEN
Elastic Behaviour Related to the Superconductivity in $YBa_2Cu_3O_x$
62.20 and 74.70; S10.1
(a) Vol.113,No.1, K85-K88

H.T. ZHANG, E. MANN, and A. SEEGER
Approximate Solution of the Peierls Equation for Screw Dislocations with Threefold Symmetry
61.70; 62.20; 62.40
(b) Vol.153,No.2, 465-478

Y. LI and Q.P. KONG
On the Relationship between Creep Rate and Stacking Fault Energy
62.20; 61.70
(a) Vol.113,No.2, 345-352

SHAM-TSONG SHIUE and SANBOH LEE
Elastic Interaction between Screw Dislocations and Blunt Crack
61.70 and 62.20
(a) Vol.113,No.2, 365-374

V.V. VOROBEV, M.YA. KRUPOTKIN, and V.A. FINKEL
Anomalous Behaviour of Elastic Moduli of Terbium and Dysprosium in Magneto-Ordered State
62.20 and 64.60; 75.30; S1.4
(a) Vol.113,No.2, 375-382

M.Z. BUTT and M.A. KHAN
Short-Range Order in Copper-Manganese Alloy Single Crystals
62.20; S1.2
(a) Vol.113,No.2, K189-K194

T. KANADANI and A. SAKAKIBARA
GP Zone Formation by Nucleation and Growth in Al-4mass%Zn Alloy
62.20 and 64.75; S4
(a) Vol.113,No.2, K199-K202

XIAOYU HU and H. MARGOLIN
Dislocation Arrangements in Stress Gradients
61.70; 62.20; S4
(b) Vol.154,No.1, 61-70

D.Y. LI, X.F. WU, and T. KO
The Effect of Stress on the Lattice Instability of a TiNi Alloy
64.70; 61.10; 62.20; 72.15; S1.2
(b) Vol.154,No.1, 85-96

A.L. SARCE and E.J. SAVINO
Thermodynamics of Deformation under Irradiation
62.20; 61.80; 65.50
(b) Vol.154,No.1, 105-114

K.P. GHATAK
On the Carrier Contribution to Elastic Constants of $A_3^{II}B_2^V$ Semiconductors
62.20; S7
(b) Vol.154,No.1, K29-K34

B. MITRA and K.P. GHATAK
Electronic Effects in Elastic Constants of II-VI Semiconductors
62.20; S8.11
(b) Vol.154,No.1, K35-K38

A. MORAWIEC
Calculation of Polycrystal Elastic Constants from Single-Crystal Data
62.20; S1; S10
(b) Vol.154,No.2, 535-542

E. MALISZEWSKI, J. SOSNOWSKI, and S. BEDNARSKI
Lattice Dynamics of the Force Constant Disordered Crystal $Pd_{0.90}Ru_{0.10}$
63.20; 62.20; S1.3
(b) Vol.154,No.2, 543-552

H.-G. FRITSCHE
Particle Size Effects of Ion Crystallites
68.40; 62.20; S9.11
(b) Vol.154,No.2, 603-608

B.I. CRAIG and P.V. SMITH
Determination of the Bulk Properties of Cubic SiC by the Periodic LUC-MINDO Method
71.10; 62.20; S1.61
(b) Vol.154,No.2, K127-K130

R. LACROIX
Theoretical Considerations on the Splitting of the Ruby R_1 Line under Stress
71.70; 62.20; S10.1
(b) Vol.154,No.2, K143-K146

M.S. ANAND, B.M. PANDE, and R.P. AGARWALA
Recovery of Plastically Deformed A-203 Steel
62.20; S1.1
(a) Vol.114,No.2, K159-K162

XIAOYU HU and H. MARGOLIN
The Stress Fields on Secondary Slip Systems Due to a Group of Dislocations in a Stress Gradient
61.70; 62.20; S4
(b) Vol.155,No.1, 53-64

T. SOMA, Y. HORIKAGO, and H.-MATSUO KAGAYA
Temperature Effect on the Elastic Coefficients of Noble Metals
62.20; 63.20; S1.3
(b) Vol.155,No.1, K5-K8

T. KUZUBA and M. ISHII
Elastic Constant c_{44} of Uniaxial Layered Crystals
62.20; 63.20; 78.30; S1.63
(b) Vol.155,No.1, K13-K16

C.N. MISHRA, R.P. GOYAL, and B.R.K. GUPTA
Photoelastic Behaviour of Alkaline Earth Fluoride Crystals
62.20 and 78.20; S9.12
(b) Vol.155,No.1, K81-K84

D.O. NORTHWOOD and I.O. SMITH
Steady-State Creep and Strain Transients for Stress Change Tests in an Al-0.5wt%Zn Solid Solution Alloy
62.20; S4
(a) Vol.115,No.1, 125-134

S.N. BALYAKIN, V.N. EROFEEV, D.N. MOGILYANSKII, and V.SH. SHEKHTMAN
Investigation of the Structure and Properties of KCl-NaCl Crystals at Elevated Temperatures
64.70; 62.20; S9.16
(a) Vol.115,No.1, 199-206

N. GANEV, I. KRAUS, and J. TRPČEVSKÁ
X-Ray Measurement of Residual Stresses Induced by Laser Surface Treatment
62.20; 68.20; S1.1
(a) Vol.115,No.1, K13-K16

B. MRÓZ and R. LAIHO
Elastic Properties of $K_4LiH_3(SO_4)_4$ Crystals
77.60; 62.20; S11.1
(a) Vol.115,No.2, 575-580

T. KANADANI
Effect of Soft Surface Layer on the Fatigue Strength of Age-Hardened Al-12mass%Zn Alloy
62.20; S4
(a) Vol.115,No.2, K147-K148

A.E. ROMANOV and M. HECKER
Elastic Fields of Straight Linear Somigliana Dislocations near a Free Surface
62.20
(a) Vol.115,No.2, K159-K162

K. ARADHANA and R.P.S. RATHORE
Two- and Three-Body Forces in Lattice Dynamics and Elastic Behaviour of B.C.C. Calcium and Strontium
63.20; 62.20; S3
(b) Vol.156,No.1, 77-82

J. SHANKER, R. SINGH, and M. KUMAR
Analysis of Combination Rules for Elastic Constants of Mixed Ionic Crystals
62.20; S9.15
(b) Vol.156,No.1, 131-136

S.D. CHATURVEDI, S.B. SHARMA, P. PALIWAL, and M. KUMAR
Analysis of Crystal Binding and Structural Phase Transition in Alkaline-Earth and Alkali Chalcogenides
62.20 and 64.60; S8
(b) Vol.156,No.1, 171-180

B. SRINIVASA RAO and S.P. SANYAL
High Pressure Elastic Constants of NaCl
62.20; S9.11
(b) Vol.156,No.1, K27-K32

R. KERN and M. KROHN
Interaction of Metal Clusters with Surface Steps
68.20; 61.70; 62.20; S9.11
(a) Vol.116,No.1, 23-38

F. APPEL
On the Intersection of Oblique Screw Dislocations in NaCl Single Crystals
61.70; 62.20; S9.11
(a) Vol.116,No.1, 153-164

A. SCHAPER, D. ZENKE, E. SCHULZ, R. HIRTE, and M. TAEGE
Structure-Property Relationships of High-Performance Polyethylene Fibres
61.40; 61.16; 62.20; S12
(a) Vol.116,No.1, 179-196

O.G. VLOKH, A.V. KITYK, O.M. MOKRY, and V.G. GRYBYK
Pressure-Temperature Phase Diagram of $\{N(CH_3)_4\}_2CuCl_4$ Crystals. Birefringent and Elastic Properties
64.60; 62.20; 78.20; S12
(a) Vol.116,No.1, 287-294

G.A. HASSAN and H.H. HASSAN
Temperature Dependence of the Stress-Strain Characteristics of Technically Pure Silver Subjected to Prior Alternating Torsion
62.20; S1.3
(a) Vol.116,No.1, K43-K46

T. KUZUBA and M. ISHII
Erratum to Elastic Constant c_{44} of Uniaxial Layered Crystals
(phys. stat. sol. (b) 155, K13 (1989))
62.20; 63.20; 78.30; S1.63
(b) Vol.156,No.2, K167-K168

G.A. ANDREEV and M. HARTMANOVA
Flotation Method of Precise Density Measurements (Review Article)
62.20
(a) Vol.116,No.2, 457-468

E.YA. BAIGUZIN, A.I. MELKER, and A.I. MIKHAILIN
Atomic Mechanisms of Fracture Nucleation and Fracture Development in Two-Dimensional Crystals in Thermodynamic Equilibrium (II)
62.20
(a) Vol.116,No.2, 595-602

V.I. DIMITROV, YU.I. KLECHIN, and A.V. APOSTOLOV
Calculation of Elastic Moduli for Amorphous $Fe_{80}B_{20}$
62.20; 61.40; S1.1
(a) Vol.116,No.2, 627-632

A.B. LEBEDEV and S.B. KUSTOV
Effect of Temperature on Amplitude-Dependent Internal Friction and Yield Stress in Crystals
62.20 and 62.40
(a) Vol.116,No.2, 645-656

H. RUPPERSBERG, I. DETEMPLE, and J. KRIER
Evaluation of Strongly Non-Linear Surface-Stress Fields $\sigma_{xx}(z)$ and $\sigma_{yy}(z)$ from Diffraction Experiments
62.20 and 68.25; S1.2
(a) Vol.116,No.2, 681-688

C. LEXCELLENT and G. SOMÉ
Comportement en fluage à haute température de la phase γ du Cu-16,6%atSn. Tentative de corrélation avec la diffusion en volume
62.20; 66.30; S1.2
(a) Vol.116,No.2, 689-696

ZHONGMING LIU, XIAOHUA CHEN, HUIMIN SHEN, YENING WANG, HUIFANG YOUNG, and P.C.W. FUNG
Ferroelectric Domain Wall Motion and Related Internal Friction in TGS Crystal
62.20 and 77.80; S12
(a) Vol.116,No.2, K199-K204

62.30. Mechanical and elastic waves

A.G. FOKIN and T.D. SHERMERGOR
The Theory of Elastic Wave Propagation in Inhomogeneous Media
62.30
(b) Vol.153,No.1, 89-100

E.V. ZAROCHENTSEV, S.M. OREL, and YU.S. STEPANCHUK
Elastic Properties of Inhomogeneously Deformed Crystals
62.30
(a) Vol.116,No.1, 295-304

62.40. Anelasticity, internal friction, and mechanical resonances

For thermomechanical effects, see 65.70
For magnetomechanical effects, see 75.80
For piezoelectric effects, see 77.60
For elasto-optical effects, see 78.20

M. KŁOSEK, J.W. MOROŃ, and T. POLOCZEK
Activation Energy of Linear Heating Crystallization in $Ni_{70.3}Fe_{4.7}Si_{15}B_{10}$ Glassy Alloy Determined Using the Internal ...
61.40 and 64.70; 62.40; S1.1; S1.2
(a) Vol.111,No.1, 127-136

T. OHGAKU and N. TAKEUCHI
Relation between Plastic Deformation and the Blaha Effect for Alkali Halide Crystals
62.20; 62.40; S9.11
(a) Vol.111,No.1, 165-172

LANPING YUE
Effect of Heat Treatment on the Internal Friction Peaks of a-PdNiSi near T_g and T_x
62.40; S1
(a) Vol.111,No.2, K155-K160

S. SUZUKI
Hydrogen Cold-Work Peak in High Purity Molybdenum
61.70 and 62.40; S1
(a) Vol.113,No.1, 51-56

H.T. ZHANG, E. MANN, and A. SEEGER
Approximate Solution of the Peierls Equation for Screw Dislocations with Threefold Symmetry
61.70; 62.20; 62.40
(b) Vol.153,No.2, 465-478

A.W. ZHU and T.S. KÊ
Characteristics of the Internal Friction Peak Associated with Bamboo Grain Boundaries
62.40; S4
(a) Vol.113,No.2, 393-402

T.S. KÊ and A.W. ZHU
Experimental Evidence on the Existence of the Bamboo Boundary Internal Friction Peak
62.40; S4
(a) Vol.113,No.2, K195-K198

G.A. BIELIG
The Peaking Effect in Nominally Pure KCl Single Crystals
62.40; S9.11
(a) Vol.114,No.2, K153-K158

T.S. KÊ and B.L. CHENG
Mechanical Model of the Bamboo Boundary Internal Friction Peak
62.40; S4
(a) Vol.115,No.1, 119-124

FANCHUN ZHENG, JIANYI JIANG, YUPIN SUN, HUAQING YIN, and JIAJU DU
Effects of Cooling Rate on the Properties and the Oxygen Content of the Superconductor $Bi_2Sr_2CaCu_2O_y$
74.70; 62.40; S10.15
(a) Vol.115,No.1, K75-K78

A.B. KUKLOV
On Non-Debye Linear Mechanical Response of Condensed Media
62.40
(b) Vol.155,No.2, 407-418

I.V. ZOLOTUKHIN, V.I. BELYAVSKII, V.A. KHONIK, and T.N. RYABTSEVA
Internal Friction in Cold-Rolled Metallic Glasses $Cu_{50}Ti_{50}$ and $Ni_{78}Si_8B_{14}$
62.40; S1.2
(a) Vol.116,No.1, 255-266

M. OKUMURA, H. MIZUBAYASHI, S. OKUDA, and Y. TAGISHI
Internal Friction Study on Anomalous Self Interstitial Atom-Dislocation Interaction in Nb after Low Temperature Irradiation
61.80 and 62.40; S1
(a) Vol.116,No.2, 583-594

KEQIN ZHENG, SHENGYAN LIU, and DECHENG TIAN
Investigation of the Behavior of Internal Friction of Al-Mg-Si-RE Alloys
62.40; 61.70; S1.4; S3
(a) Vol.116,No.2, 621-626

H. MIZUBAYASHI, S. OKUDA, and Y. IKEDA
Inelastic Properties of Amorphous CuTi
62.40; S1.2
(a) Vol.116,No.2, 633-644

A.B. LEBEDEV and S.B. KUSTOV
Effect of Temperature on Amplitude-Dependent Internal Friction and Yield Stress in Crystals
62.20 and 62.40
(a) Vol.116,No.2, 645-656

62.50. High-pressure and shock-wave effects in solids

D.M. VANDERWALKER
Fracture in Titanium Diboride
61.70 and 62.50; S1.2
(a) Vol.111,No.1, 119-126

QIGUANG LI, SHAN JIANG, WEI LU, XIAOMING FANG, S.C. CHEN, and W. GIRIRAT
Optical Absorption Experiments of Cycling through the Structural Transition under Hydrostatic Pressure in $Cd_{1-x}Mn_xSe$
62.50 and 78.40; S8.15
(b) Vol.152,No.1, K33-K36

P. KONSIN and T. OERD
Influence of Hydrostatic Pressure on the Ferroelectric Phase Transition Temperature in (Pb,Sn)Te
62.50 and 63.75; 63.70; 77.80; S8
(b) Vol.156,No.1, K37-K40

62.60. Acoustical properties of liquids

For lattice dynamics, phonons, see 63
For second sound in quantum fluids, see 67.40

J. BODZENTA and Z. KLESZCZEWSKI
Theoretical Desciption of Light Diffraction on Acoustic Pulses in Braggs Region
62.65 and 78.70; 62.60
(b) Vol.153,No.2, 573-578

62.65. Acoustical properties of solids

For magnetoacoustic effects in metals, see 75.80
For acoustoelectric effects, see 72.50
For acoust-optical effects, see 78.20

R. MICHALEC, B. CHALUPA, and J. VÁVRA
Neutron Acoustic Resonance in Perfect InSb Single Crystals
61.12 and 62.65; S7.13
(a) Vol.112,No.1, 29-34

J. KLEJST and J. BRACINIK
Photoacoustic Study of Semi-Insulating GaAs:Cr
62.65 and 66.70; S7.12
(a) Vol.113,No.1, K37-K40

J. BODZENTA and Z. KLESZCZEWSKI
Theoretical Desciption of Light Diffraction on Acoustic Pulses in Braggs Region
62.65 and 78.70; 62.60
(b) Vol.153,No.2, 573-578

S. MUKHERJEE, S. SENGUPTA, and A.N. BASU
On the Nature of Distribution Function for Two-Level Systems in Amorphous Solids
62.65 and 65.70; 63.20
(b) Vol.154,No.2, K107-K112

A.P. ZDEBSKII, G. GARAYAGDEV, K. ORAZOV, and G.N. POLISSKII
Ultrasound-Stimulated Increase of Photosensitivity and Luminescence Intensity of II-VI Compound Solid Solutions
62.65 and 72.40; S8.15
(a) Vol.116,No.1, K91-K96

62.80. Ultrasonic relaxation
(see also 74.30. Ultrasonic attenuation in superconductors)

MOHEB H. YOUSSEF, M. ROUSHDY, and N.K. GOBRAN
Origin of Periodic Fluctuation with Temperature of Ultrasonic Attenuation in Zinc, Aluminium, and Iron
62.80; S1.1; S1.2; S4
(b) Vol.155,No.2, 453-460

63. Lattice Dynamics and Crystal Statistics

(see also 65. Thermal properties,
66.70. Thermal conduction,
68.30. Dynamics of surface and interface vibrations,
78.30. Infrared and Raman spectra)

63.10. General theory

I.P. DZYUB and YU.E. ZEROV
Dynamic Structure Factor of a Slowly Moving Soliton in a Thermal Bath
63.10 and 63.50
(b) Vol.151,No.1, 61-70

S. ERKOÇ
A New Empirical Many-Body Potential Energy Function: Application to Microclusters
63.10; 61.55; S1.2; S1.3; S2; S5
(b) Vol.152,No.2, 447-454

J.O.A. IDIODI
A Nonlocal Continuum Approach to the Dynamics of Crystals. I. S.C. and F.C.C. Lattices
63.10; 63.20
(b) Vol.153,No.2, 489-500

J.O.A. IDIODI and YINGGANG LEE
A Nonlocal Continuum Approach to the Dynamics of Crystals. II. B.C.C. Lattice
63.10; 63.20
(b) Vol.153,No.2, 501-516

J. MONECKE
On the Correct Use of Perturbation Theory for the Calculation of One-Particle Green Functions
71.10; 63.10
(b) Vol.153,No.2, 623-632

J. MONECKE
Collective Excitations in Microcrystal Aggregates and Composites
68.30 and 78.30; 63.10
(b) Vol.155,No.2, 437-444

S. ERKOÇ
A New Empirical Many-Body Potential Energy Function. Application to Microclusters: Elements in B.C.C., F.C.C., and H.C.P. Structures
63.10; 61.55; S1.2; S2; S5
(b) Vol.155,No.2, 461-466

V.B. KOKSHENEV and N.N. ZHOLONKO
Effects of Correlated Rotation of Linear Molecules in the Low-Temperature Dynamics and Thermodynamics of Weak Molecular Solution
63.10 and 65.40; 65.70; S5.2
(b) Vol.156,No.1, 119-130

63.20. Phonons and vibrations in crystal lattices

S. ICHINOSE
Two-Component Solitary Excitations in Molecular-Bonded Chains
63.20 and 71.45
(b) Vol.151,No.1, 95-100

T. TSUBOI
Vibronic Structure in the Absorption Spectra of $CaF_2:Sm^{2+}$
63.20 and 78.40; S9.12
(b) Vol.151,No.1, K91-K96

W.P. AMBROSE and A.J. SIEVERS
Determination of the Orientation of NO_2^- in KI by Persistent IR Spectral Hole Burning
63.20 and 78.50; S9.11
(b) Vol.151,No.1, K97-K100

M. KLENNER, C. FALTER, and W. LUDWIG
A Double Bond Charge Model for Si
63.20; S5.11
(b) Vol.151,No.2, 503-512

E.L. ALBUQUERQUE and P.W. MAURIZ
Debye-Waller Factor for Moessbauer Interface Atoms: Elastic Approximation
63.20 and 76.80
(b) Vol.151,No.2, K109-K114

N. KRISTOFFEL and M. KLOPOV
Influence of Soft Modes on the Local Dynamics of Defects ($SrTiO_3$)
63.20; S11.1
(b) Vol.151,No.2, K119-K122

T.K. KASUMOV and F.I. MAMEDOV
Photocurrent Oscillations Due to Optical Phonons in $AgGaSe_2$ Single Crystals
72.40; 63.20; S8.16
(a) Vol.111,No.2, K205-K208

V.D. LAKHNO and G.N. CHUEV
Local Phonons in Crystals with Electron Centres
63.20; 71.38; S9.11
(b) Vol.152,No.1, 79-88

G.A. KOUROUKLIS and E. ANASTASSAKIS
Pressure-Induced Phase Transitions and Anharmonicity Study of Alkaline-Earth Fluorides
64.60; 63.20; 78.30; S9.12
(b) Vol.152,No.1, 89-100

HUA XIA, X.K. ZHANG, A. HU, W. ZHANG, S.N. ZHU, and C. TIEN
Brillouin Scattering from Surface Phonons in Fibonacci Metallic Superlattices
63.20; 78.35; S1.2
(a) Vol.112,No.1, 99-104

M. ISHII, K. SHIBATA, and H. WADA
Raman Study of $M_xMo_6S_{8-y}$ (M = Yb, Fe, Cr)
78.30; 63.20; S1.63
(b) Vol.152,No.2, K63-K68

A.K. PRADHAN and R.N.P. CHOUDHARY
Raman and Infrared Effect of Sm^{3+} in $YAsO_4$ Crystals
78.30; 63.20; S11.1
(b) Vol.152,No.2, K69-K72

P. ČAPKOVÁ, M. MERISALO, P. BOUKAL, M. BLOMBERG, and W. LENGAUER
Thermal Vibrations in Vanadium Nitride Studied by X-Ray Diffraction
63.20; S1.62
(a) Vol.112,No.2, K81-K84

J.M. CHEN, L.M. XIE, G.R. BAI, and Y.G. ZHAO
Temperature Effects of the Several Intense IR Absorption Peaks in NTD Si:H
63.20 and 78.30; S5.11
(b) Vol.153,No.1, 107-114

H.-MATSUO KAGAYA, H. KOTOKU, and T. SOMA
Thermal Pressure and Pressure-Volume Relations of III-V Covalent Crystals
62.20; 63.20; S7.1
(b) Vol.153,No.1, K17-K20

N.M. GASANLY, B.A. NATIG, A.E. BAKHYSHOV, and K.E. SHIRINOV
Long-Wave Optical Phonons in $InSbSe_3$ Layer Single Crystals
63.20 and 78.30; S8.16
(b) Vol.153,No.1, K89-K92

S.E. SARKISOV, K.K. PUKHOV, A.A. KAMINSKII, A.G. PETROSYAN, and T.I. BUTAEVA
Manifestation of Electron-Phonon Interaction in Insulating Crystals Doped with Pr^{3+} Ions
63.20 and 78.55; 71.55; S10.15
(a) Vol.113,No.1, 193-198

J.O.A. IDIODI
A Nonlocal Continuum Approach to the Dynamics of Crystals. I. S.C. and F.C.C. Lattices
63.10; 63.20
(b) Vol.153,No.2, 489-500

J.O.A. IDIODI and YINGGANG LEE
A Nonlocal Continuum Approach to the Dynamics of Crystals.
II. B.C.C. Lattice
63.10; 63.20
(b) Vol.153,No.2, 501-516

Y. MADHAVAN, K. RAMACHANDRAN, and T.M. HARIDASAN
On the Mean Square Displacements of Hg and Te in HgTe
63.20; 61.70; S8.13
(b) Vol.153,No.2, 579-582

J. GALIBERT, P. PERRIER, and S. ASKENAZY
Résonance magnétophonon avec retournement de spin dans l'antimoniure d'indium
63.20 and 71.70; 72.20; S7.13
(b) Vol.153,No.2, 653-658

S.K. MISHRA and T.N. SINGH
Crystal Dynamics of F.C.C. Inert Gas Solids
63.20; S5.2
(b) Vol.154,No.1, 127-134

C.P. Chang, K.F. PAI, C.S. FANG, and W.S. TSE
Dispersion Curves of Phonon-Plasmon Coupled Polaritons in n-Type $Al_xGa_{1-x}As$
63.20; 71.36; 78.30; S7.15
(b) Vol.154,No.1, 135-142

J. SCHLEIFER, J. KALUS, U. SCHMELZER, and G. ECKOLD
Phonon Dispersion in an alpha-Perylen d_{12}-Crystal at 10 K
63.20; S12
(b) Vol.154,No.1, 153-166

N. SINGH, N.S. BANGER, and S.P. SINGH
Lattice Dynamics of B.C.C. Sr: A Dynamical Treatment
63.20; S2
(b) Vol.154,No.1, K19-K22

E. MALISZEWSKI, J. SOSNOWSKI, and S. BEDNARSKI
Lattice Dynamics of the Force Constant Disordered Crystal $Pd_{0.90}Ru_{0.10}$
63.20; 62.20; S1.3
(b) Vol.154,No.2, 543-552

R.K. RAM, S.S. KUSHWAHA, and A. SHUKLA
Phonon Assignments in II-VI and III-V Semiconductor Compounds Having Zincblende-Type Structure
63.20; 78.30; S7.1; S8.1
(b) Vol.154,No.2, 553-564

F. BECHSTEDT and H. GERECKE
Long-Wavelength Optical Phonons in GaAs-AlAs Superlattices
63.20 and 68.30; 73.40; S7.12
(b) Vol.154,No.2, 565-582

J. TATARKIEWICZ and W. WITTHUHN
Hydrogen Vibrations in Oxygen-Doped Silicon
61.70 and 63.20; S5.11
(b) Vol.154,No.2, K99-K102

K. SADAIYANDI and K. RAMACHANDRAN
Variation of the Mean Square Amplitude with Concentration in $Hg_{1-x}Cd_xTe$
63.20; S8.15
(b) Vol.154,No.2, K113-K116

P.P. LOTTICI
Breathing-Like A-Symmetry Vibrations in $ZnGa_2(S_xSe_{1-x})_4$ and $CdGa_2(S_xSe_{1-x})_4$
63.20; 78.30; S8.16
(b) Vol.154,No.2, K117-K120

J. TATARKIEWICZ, M. CARDONA, A. BREITSCHWERDT, C.-T. LIN, R. GROETZSCHEL, and W. WITTHUHN
In Search of Hydrogen Local Vibrational Modes in CuCl, CuBr, CuI, ZnS, and AlSb
63.20; 78.30; S7.13; S8.11; S9
(b) Vol.155,No.1, 77-88

J.F.J. JORDAN, J. KALUS, U. SCHMELZER, and G. ECKOLD
Anharmonic Phonon Frequency Shifts and Widths in Solid Deuterated Anthracene-d10
63.20; S10
(b) Vol.155,No.1, 89-106

T.D. IBRAGIMOV, G.G. KURBANOVA, and V.S. GORELIK
Raman Scattering in Gallium Selenide and Sulfide Films
63.20 and 68.30; 78.30; S8
(b) Vol.155,No.1, 113-116

A.M. VAINRUB, G.S. ZAVT, and A. GRAJA
Absence of the Gap in the Optical Excitation Spectrum of TCNQ Salts with Methyl Derivatives of Pyridinium
71.45; 63.20; S12.1
(b) Vol.155,No.1, 147-154

T. SOMA, Y. HORIKAGO, and H.-MATSUO KAGAYA
Temperature Effect on the Elastic Coefficients of Noble Metals
62.20; 63.20; S1.3
(b) Vol.155,No.1, K5-K8

S. FLACH, N.M. PLAKIDA, and V.L. AKSENOV
A Theoretical Study of Soft Mode Degeneracy in La_2CuO_4
63.20 and 63.75; S10
(b) Vol.155,No.1, K9-K12

T. KUZUBA and M. ISHII
Elastic Constant c_{44} of Uniaxial Layered Crystals
62.20; 63.20; 78.30; S1.63
(b) Vol.155,No.1, K13-K16

O.S. BINBREK, J.F. HIGGS, and A. ANDERSON
Lattice Dynamics of Nitrogen and Carbon Dioxide
63.20; S5.2
(b) Vol.155,No.2, 427-436

A.V. BELOUSOV, V.A. KOVARSKII, V.N. CHEBOTAR, and T.V. PISHKOVA
Quantum Calculation of Multiphonon Capture Cross-Sections in the Model of Tangent Adiabatic Potentials
71.55; 63.20
(b) Vol.155,No.2, 535-540

K. ARADHANA and R.P.S. RATHORE
Two- and Three-Body Forces in Lattice Dynamics and Elastic Behaviour of B.C.C. Calcium and Strontium
63.20; 62.20; S3
(b) Vol.156,No.1, 77-82

B.K. RATH and P.N. RAM
Gap Modes and Local Density of States of Impurities in RbCl and KBr
63.20 and 71.55; 63.50; S9.11
(b) Vol.156,No.1, 137-144

YI-JIAN JIANG, LI-JI LIAO, GANG CHEN, and PENG-XIANG ZHANG
Generalized Formulae of Frequency Dispersion and Raman Scattering Cross Section of Phonon-Polaritons
63.20 and 78.30; 71.36
(b) Vol.156,No.1, 145-150

F. BECHSTEDT and H. GERECKE
Relationship of Microscopic and Macroscopic Theories for Long-Wavelength Optical Phonons in GaAs-AlAs Superlattices
63.20; 68.30; S7.12; S7.15
(b) Vol.156,No.1, 151-170

N. SINGH
Lattice Dynamics of gamma-Iron at 1428 K
63.20; S1.1
(b) Vol.156,No.1, K33-K36

T. KUZUBA and M. ISHII
Erratum to Elastic Constant c_{44} of Uniaxial Layered Crystals
(phys. stat. sol. (b) 155, K13 (1989))
62.20; 63.20; 78.30; S1.63
(b) Vol.156,No.2, K167-K168

V. RIEDE, H. SOBOTTA, H. NEUMANN, C. ASCHERON, C. NEELMEIJER, and A. SCHINDLER
Depth Dependence of Localized Mode Absorptions in Proton-Implanted InP
61.80 and 78.30; 63.20; S7.11
(a) Vol.116,No.2, K147-K152

63.50. Vibrational states in disordered systems

I.P. DZYUB and YU.E. ZEROV
Dynamic Structure Factor of a Slowly Moving Soliton in a Thermal Bath
63.10 and 63.50
(b) Vol.151,No.1, 61-70

H. BOETTGER, I. BRADL, and G. KASNER
Vibrational Spectra of Fibonacci Chains with Defects
63.50
(b) Vol.151,No.2, 521-530

K. HANDRICH and J. RESCH
On the Theory of Roton-Like Excitations in Amorphous Solids
61.40; 63.50
(b) Vol.152,No.2, 377-394

M. RANI, A. PRATAP, and N.S. SAXENA
Dynamics of Liquid Potassium
63.50; S2
(b) Vol.154,No.1, K23-K28

B.K. RATH and P.N. RAM
Gap Modes and Local Density of States of Impurities in RbCl and KBr
63.20 and 71.55; 63.50; S9.11
(b) Vol.156,No.1, 137-144

H. BOETTGER and M. THEUERKAUF
Phonon Localization in Disordered Systems. Application of a Probabilistically-Based Self-Consistent Theory
63.50
(b) Vol.156,No.2, 431-440

63.70. Statistical mechanics of lattice vibrations
(see also 65. Thermal properties of condensed matter, 66.70. Thermal conduction)

M.S. WARTAK
On the Use of Coherent States for the Anharmonic Phonon Systems
63.70
(b) Vol.152,No.2, K43-K46

P. KONSIN and T. OERD
Influence of Hydrostatic Pressure on the Ferroelectric Phase Transition Temperature in (Pb,Sn)Te
62.50 and 63.75; 63.70; 77.80; S8
(b) Vol.156,No.1, K37-K40

A. GHAZI, J. BASHIR, M.M. BEG, N.M. BUTT, and Q.H. KHAN
Temperature Factors of RbCl by the Powder Elastic Neutron Diffraction Method
63.70; S9.11
(a) Vol.116,No.1, K47-K50

63.75. Statistical mechanics of displacive phase-transitions
For order-disorder and statistsical mechanics of model systems, see 64.60
For crystallographic aspects of polymorphic and order-disorder transformations, see 61.50

G. BUSIELLO, L. DE CESARE, K. LUKIERSKA-WALASEK, and K. WALASEK
Exotic Glassy Phase in the Site-Diluted Transverse Ising Model at Zero Temperature?
63.75 and 64.60; 75.50
(b) Vol.153,No.2, 561-572

S. FLACH, N.M. PLAKIDA, and V.L. AKSENOV
A Theoretical Study of Soft Mode Degeneracy in La_2CuO_4
63.20 and 63.75; S10
(b) Vol.155,No.1, K9-K12

P. KONSIN and T. OERD
Influence of Hydrostatic Pressure on the Ferroelectric Phase Transition Temperature in (Pb,Sn)Te
62.50 and 63.75; 63.70; 77.80; S8
(b) Vol.156,No.1, K37-K40

63.90. Other topics in lattice dynamics and crystal statistics

T.M. HARIDASAN and P. SEKAR
Thermodynamic Parameters of Anti-Frenkel Formation in Alkaline Earth Fluorides
65.50; 63.90; S9.12
(b) Vol.153,No.1, 131-138

I.YA. FUGOL, A.M. RATNER, YU.I. RYBALKO, and YU.L. STESHENKO
On the Mechanism of Vibrational Relaxation of Two-Centre Excitations in Solid Neon
63.90; 71.45; 78.55; S5.2
(b) Vol.153,No.2, 539-546

K.N. SHRIVASTAVA
Physical Properties of Fractals. Moessbauer Recoilless Fraction, Raman Scattering, and Spin-Lattice Relaxation
63.90; 76.80; 78.30
(b) Vol.153,No.2, 547-560

A.K. PACHAURI, A.K. SHARMA, and H.C. SAXENA
Volume Dependence of Short Range Force Constant and Dielectric Properties of Alkali Halides
63.90; 77.20; S9.11
(b) Vol.155,No.1, 73-76

64. Equations of State, Phase Equilibria, and Phase Transitions

64.10. General theory of equations of state and phase equilibria

W. VOGELSBERGER and E. MUELLER
On the Location of the Gibbs Surface of Tension in Thermodynamic Considerations of Interfaces
68.40; 64.10
(b) Vol.155,No.1, K17-K22

S. ICHINOSE
Two-Component Solitary Excitations in alpha-Helical Protein Molecules
64.10 and 71.45; S12
(b) Vol.155,No.2, 467-474

M. MARKOWSKI, J. MURAKOWSKI, and Z. ONYSZKIEWICZ
Thermodynamic Properties of Binary Alloy and Lattice Gas in Gaussian Fluctuation Approximation
64.10; 64.60
(b) Vol.156,No.1, 105-110

F.A. KHWAJA, M.S.K. RAZMI, and S.H. NAQVI
A Comparative Study of the Phenomenological Potentials for Ionic Crystals of Alkali Halides (II)
64.10; 64.60; S9.11
(b) Vol.156,No.1, 111-118

G.V. BEZRUKOV, A.N. MEN, and V.M. TALANOV
Theory of Isostructural Phase Transitions Described by Two Order Parameters
64.10
(a) Vol.116,No.2, 603-614

64.30. Equations of state of specific substances
(see also 65.70. Thermal expansion)

L.N. ALEKSANDROV, S.L. BABENKOVA, and V.S. ZINOVEV
Thermoelastic Stresses in Multilayered Structures of Silicon-on-Insulator Type (Computational Experiment)
65.70 and 68.25; 64.30; S5.11; S10.1
(a) Vol.116,No.1, K57-K60

64.60. General studies of phase transitions
(for magnetic, superconducting, and quantum fluid critical phenomena, see 75, 74, and 67, respectively)

M. VALLADE and M. PETIT
Stress-Induced Defects in the Incommensurate Phase of Quartz
64.60; S10.1
(a) Vol.111,No.1, 155-164

C. RAZZETTI, T. BESAGNI, S. BINI, and P.P. LOTTICI
Defective Pseudoternary Alloys. The $CdGa_2Se_4$-$CdIn_2Se_4$ System
61.60; 64.60; S8.15
(a) Vol.111,No.2, 411-416

L.E. SOSHNIKOV and A.U. SHELEG
Elastic Properties and Phase Transitions in beta-CdP_2
62.20 and 64.60; S7
(a) Vol.111,No.2, 485-490

G.A. KOUROUKLIS and E. ANASTASSAKIS
Pressure-Induced Phase Transitions and Anharmonicity Study of Alkaline-Earth Fluorides
64.60; 63.20; 78.30; S9.12
(b) Vol.152,No.1, 89-100

K.R. ALLAKHVERDIEV, S.S. BABAEV, M.M. TAGIEV, and M.M. SHIRINOV
Low-Temperature IR and Raman Scattering Spectra of $TlInS_2$ Layered Crystal
64.60 and 78.30; S8.16
(b) Vol.152,No.1, 317-328

G.M. CHECHIN, T.I. IVANOVA, and V.P. SAKHNENKO
Complete Order Parameter Condensate of Low-Symmetry Phases upon Structural Phase Transitions
61.50 and 64.60
(b) Vol.152,No.2, 431-446

C. MARTI, C. DE BEAUVAIS, T. CEVA, B. CROSET, and M. GOLDMANN
Incommensurate-Commensurate Transition Driven by Concentration in Rare Gas Mixtures Adsorbed on Graphite
64.60; 68.45; S5
(b) Vol.152,No.2, 463-474

M. VERWERFT, D. VAN DYCK, V.A.M. BRABERS, J. VAN LANDUYT, and S. AMELINCKX
Electron Microscopic Study of the Phase Transformations in $AgNbO_3$
61.16 and 64.60; S11.1
(a) Vol.112,No.2, 451-466

K. GAERTNER, G. GOETZ, C. KASCHNER, I. KASKO, A. WITZMANN, and A. ZENTGRAF
Laser Induced Silicide Formation
68.55 and 79.20; 64.60; S1.61
(a) Vol.112,No.2, 747-752

P.K. BAJPAI
Ferroelectric Phase Transition in $(NH_4)_2BeF_4$. Possibility of Molecular Distortion Type of Transition
64.60; 77.80; S11
(b) Vol.153,No.1, 101-106

V.K. DUGAEV and P.P. PETROV
Spinodal Decomposition in Semiconductor Alloys
64.60; 61.60; S8.15
(b) Vol.153,No.1, 115-122

QIANG WANG, XIAO-WEN ZHANG, and BING-LIN GU
Order-Disorder Model of Phase Transition in A_xB_{1-x} Binary Systems of F.C.C. Structure
64.60; 64.75; S1.1; S1.2
(b) Vol.153,No.1, 139-144

B.X. LIU, X. ZHOU, and H.-D. LI
Thermodynamics and Growth Kinetical Consideration of Metal-Nitride Formation by Nitrogen Implantation (Review Article)
61.80 and 64.60; 61.55; S1.62
(a) Vol.113,No.1, 11-22

O. SIMMICH, M. MACHMER, and H. WILL
Approximate Solutions for the Kinetics of Diffusion-Limited Dissolution and Growth of Planar Precipitates
64.60; 66.30
(a) Vol.113,No.1, K27-K30

YU.I. DURNEV, B.S. KULBUZHEV, A.U. MALSAGOV, L.M. RABKIN, V.I. TORGASHEV, and YU.I. YUZYK
Vibrational Spectra and Phase Transitions in Layered Semiconducting Ferroelectrics with $TlGaSe_2$ Structure. I. Temperatur Dependence ...
64.60 and 78.30; S8.16
(b) Vol.153,No.2, 517-528

YU.G. GONCHAROV, G.V. KOZLOV, B.S. KULBUZHEV, V.B. SHIROKOV, V.I. TORGASHEV, A.A. VOLKOV, and YU.I. YUZYK
Vibration Spectra and Phase Transitions in Layered Semiconducting Ferroelectrics with $TlGaSe_2$ Structure. II. Thermodynamic ...
64.60 and 78.30; S8.16
(b) Vol.153,No.2, 529-538

G. BUSIELLO, L. DE CESARE, K. LUKIERSKA-WALASEK, and K. WALASEK
Exotic Glassy Phase in the Site-Diluted Transverse Ising Model at Zero Temperature?
63.75 and 64.60; 75.50
(b) Vol.153,No.2, 561-572

I.R. YUKHNOVSKII and N.A. KORYNEVSKII
The Investigation of the Ferroelectric Phase Transition in Cluster Systems of Order-Disorder Type. I.Partition Function Functional
64.60; 77.80
(b) Vol.153,No.2, 583-594

V.V. VOROBEV, M.YA. KRUPOTKIN, and V.A. FINKEL
Anomalous Behaviour of Elastic Moduli of Terbium and Dysprosium in Magneto-Ordered State
62.20 and 64.60; 75.30; S1.4
(a) Vol.113,No.2, 375-382

M. BOURROUS and H. KRONMUELLER
Structural Transition from the As-Quenched to the Relaxed State in Metallic Glasses
61.40 and 64.60; 75.60; S1.2
(a) Vol.113,No.2, 383-392

R.A. RUPP, A.E. KRUMINS, and K. KERPELIN
Dielectric, Electrooptic, and Holographic Properties Close to the Diffuse Phase Transition of PLZT 10/65/35
78.20; 64.60; 77.80; S10
(a) Vol.113,No.2, 637-652

J. MAĆKOWIAK
A Hamiltonian Model of Liquefaction and Crystallization in Three Dimensions
61.20 and 64.60
(b) Vol.154,No.1, 115-126

A. LÓPEZ-ECHARRI, I. RUIZ-LARREA, and M.J. TELLO
Thermodynamics of the Phase Transition Sequence in the Incommensurate Compound $[N(CH_3)_4]_2CuBr_4$
64.60; 65.40; S9.16
(b) Vol.154,No.1, 143-152

ZHENGJU YANG and F. LUTY
Theoretical Investigation of Ferroelastic Phase Transition of Pure and Mixed Alkali Cyanide Systems by Elastic Dipole Model (I)
64.60; S2
(b) Vol.154,No.1, 167-180

ZHENGJU YANG and F. LUTY
Theoretical Investigation of Ferroelastic Phase Transition of Pure and Mixed Alkali Cyanide Systems by Elastic Dipole Model (II)
64.60; S9
(b) Vol.154,No.1, 181-194

V.A. MEZRIN
Phase Transitions in High-Temperature Fluorite-Structure Superionic Conductors
64.60; 61.70; S9.12; S10.1
(a) Vol.114,No.1, 145-156

I.R. YUKHNOVSKII and N.A. KORYNEVSKII
The Investigation of the Ferroelectric Phase Transition in Cluster Systems of Order-Disorder Type. II. Two-Particle Cluster System
64.60; 77.80
(b) Vol.154,No.2, 519-534

C. TUIJN and H. BAKKER
The Influence of the Vibrational Entropy on the Specific Heat of the Order-Disorder Transition in Alloys
65.50; 64.60
(b) Vol.155,No.1, 107-112

L.A. MANAKOVA and S.P. IONOV
Correlation Mechanism to Change Lattice Parameters and Phase Transitions in Compounds with Intermediate Valency
64.60; 71.25; S8.15
(b) Vol.155,No.2, 397-406

R.I. ZAINULLINA, K.B. VLASOV, M.A. MILYAEV, E.V. USTELEMOVA, and V.N. SYROMYATNIKOV
Oscillatory Temperature Dependence of Crystal Structure and Magnetic Parameters in $FeGe_2$
64.60; 75.30; S1.61
(b) Vol.155,No.2, 419-426

K.P. THAKUR and S. AHMAD
Phase Transformation in F.C.C. Thorium and Its Theoretical Stability on the Path of Transformation
64.60; S1.5
(b) Vol.155,No.2, 445-452

S. ENZO, G. COCCO, I. SOLETTA, C. ANTONIONE, and L. BATTEZZATI
Structuring Effects in FeSiB Metallic Glasses. An EXAFS Approach
61.40 and 64.60; S1.1
(a) Vol.115,No.2, 459-466

V.E. ZUBKUS, S. LAPINSKAS, and E.E. TORNAU
Calculation of Phase Diagrams in the Lattice-Gas Model by a Modified Cluster Field Method. Application to High-Temperature Superconductors
64.60 and 74.20
(b) Vol.156,No.1, 93-104

M. MARKOWSKI, J. MURAKOWSKI, and Z. ONYSZKIEWICZ
Thermodynamic Properties of Binary Alloy and Lattice Gas in Gaussian Fluctuation Approximation
64.10; 64.60
(b) Vol.156,No.1, 105-110

F.A. KHWAJA, M.S.K. RAZMI, and S.H. NAQVI
A Comparative Study of the Phenomenological Potentials for Ionic Crystals of Alkali Halides (II)
64.10; 64.60; S9.11
(b) Vol.156,No.1, 111-118

S.D. CHATURVEDI, S.B. SHARMA, P. PALIWAL, and M. KUMAR
Analysis of Crystal Binding and Structural Phase Transition in Alkaline-Earth and Alkali Chalcogenides
62.20 and 64.60; S8
(b) Vol.156,No.1, 171-180

D. BRACH, P. GLAREMIN, and H.-J. WEBER
Optical Investigations of the Phase Transition in Pure and Cr^{3+}-Doped Potassium Alum
64.60; 78.20; 78.55; S11.1
(b) Vol.156,No.1, 195-204

O.G. VLOKH, A.V. KITYK, O.M. MOKRY, and V.G. GRYBYK
Pressure-Temperature Phase Diagram of $\{N(CH_3)_4\}_2CuCl_4$ Crystals. Birefringent and Elastic Properties
64.60; 62.20; 78.20; S12
(a) Vol.116,No.1, 287-294

A.O. MEKHRABOV, Z.M. BABAEV, and Z.A. MATYSINA
The Effect of a Third Component Impurity C on the Heat Capacity of A_3B-Type Ordered Alloys
64.60 and 65.40; S1.1; S1.2
(b) Vol.156,No.2, 419-430

M. KRUPSKI
Rigid Sphere Model of the Phase Transitions in Complex Antifluorite and Elpasolite Compounds
64.60; S9.16
(a) Vol.116,No.2, 657-668

64.70. Phase equilibria, phase transitions, and critical points of specific substances

T. OVSENIK, R. BLINC, M. ČOPIČ, G. LAHAJNAR, I. ZUPANČIČ, P. GUENTER, H. AREND, and CH. BOSSHARD
Self-Diffusion in an Undercooled Organic Melt near the Liquid-Glass Transition
66.30; 64.70; S12
(b) Vol.151,No.1, K1-K6

V. DVOŘÁK
Landau Theory of the Structural Phase Transition in $YBa_2Cu_3O_{7-x}$
64.70; 74.70; S10.15
(b) Vol.151,No.1, K23-K28

M. KŁOSEK, J.W. MOROŃ, and T. POLOCZEK
Activation Energy of Linear Heating Crystallization in $Ni_{70.3}Fe_{4.7}Si_{15}B_{10}$ Glassy Alloy Determined Using the Internal ...
61.40 and 64.70; 62.40; S1.1; S1.2
(a) Vol.111,No.1, 127-136

T. KIYOSAWA, S. TAKAHASHI, and N. KOGUCHI
Solid Solubility Range and Lattice Parameter of Semiconducting $Pb_{1-x}Cd_xS_{1-y}Se_y$ for Mid-Infra-Red Lasers
64.70 and 64.75; S8.15
(a) Vol.111,No.1, K21-K26

R. JAKUBAS, P.E. TOMASZEWSKI, and L. SOBCZYK
Phase Transition in $(CH_3NH_3)_3Bi_2Cl_9$
64.70; 77.20; S12
(a) Vol.111,No.1, K27-K30

S. POPOVIĆ, H. LOEFFLER, B. GRŽETA, G. WENDROCK, and P. CZURRATIS
Lattice Constant of Al-Rich Matrix of Two Al-(8 and 24)at% Zn Alloys in Equilibrium with (Metastable) Guinier Preston Zones
61.55; 62.20; 64.70; S4
(a) Vol.111,No.2, 417-430

A.A. SUKHOVSKII, I.P. ALEKSANDROVA, and YU. N. MOSKVICH
NMR Confirmation of the Existence of Superstructure in the Ferroelectric Phase of NH_4HSeO_4
64.70; 76.60; S11.1
(a) Vol.111,No.2, 477-484

N. VOUROUTZIS and C. MANOLIKAS
Phase Transformations in Cuprous Telluride
64.70; S1.63
(a) Vol.111,No.2, 491-498

G.V. BICH, V.M. KAMYSHOV, A.N. MEN, and YU.A. PETROV
Structural Transformation during Spinel Phase Oxidation in the Ge-Fe-O System
64.70; S11.2
(a) Vol.111,No.2, K171-K174

J. VAN DER STEL, H.B. VAN VELDHUIZEN, G.W. KOEBRUGGE, J. SIETSMA, and A. VAN DEN BEUKEL
Change of Electrical Resistivity and Youngs Modulus during Crystallization of Amorphous $Fe_{40}Ni_{40}B_{20}$
61.40 and 64.70; 62.20; 72.15; S1.1; S1.2
(a) Vol.112,No.1, 35-40

P.E. TOMASZEWSKI
On the Effect of Doping on the Mechanism of Phase Transition of Some Inorganic Compounds (Comment)
64.70 and 77.80; 77.70; S11.1
(a) Vol.112,No.1, K13-K14

S.P. ZHVAVYI and O.L. SADOVSKAYA
Nanosecond Laser Annealing of Implanted Silicon.
Simulation of Dynamics
64.70; 66.70; S5.11
(a) Vol.112,No.1, K19-K22

YA.V. FATTACHOV, I.B. KHAIBULLIN, R.M. BAYAZITOV, and E.M. MISYURYOV
Anisotropic Local Melting of Semiconductors by Incoherent Light Pulses
79.20; 64.70; S5.11; S5.12
(a) Vol.112,No.2, 699-702

A.V. SUVOROV, K.K. BOURDELLE, N.G. CHECHENIN, and V.N. MAKAROV
Studies of Amorphous Layer Formation in SiC under Ga^+
Bombardment
61.80; 64.70; S1.61
(a) Vol.112,No.2, 707-714

R. NAVIL KUMAR and C.P.G. VALLABHAN
Effect of Deuteration on the Electrical Conductivity, Dielectric
Constant, and Phase Transition in $(NH_4)_3H(SO_4)_2$
72.80 and 77.20; 64.70; S11.1
(a) Vol.112,No.2, K113-K118

C.N.W. DARLINGTON
On the Changes in Structure of PLZT(8.7/65/35) between 80 and 750K
64.70; 61.10; 77.80
(a) Vol.113,No.1, 63-70

R. POPRAWSKI and S.A. TARASKIN
Calorimetric Studies of the Phase Transition in Nd_4DSeO_4 Crystals
64.70; 65.40; S11
(a) Vol.113,No.1, K31-K36

M.YA. GAMARNIK
On a Possibility of Sphalerite-Wurtzite Phase Transition Caused by
Change in the Crystal Shape
64.70; S8.11
(a) Vol.113,No.1, K45-K48

G.A. GOVOR, K. BAERNER, and J.-W. SCHUENEMANN
Phase Transitions in the $MnAs_{1-x}Sb_x$ Alloy System
64.70; 75.30; S1.2
(a) Vol.113,No.2, 403-420

D.Y. LI, X.F. WU, and T. KO
The Effect of Stress on the Lattice Instability of a TiNi Alloy
64.70; 61.10; 62.20; 72.15; S1.2
(b) Vol.154,No.1, 85-96

C. GIOVANELLA
The Superconductor Glasses (SuG). A Fractal Description
74.50; 64.70; S10.15
(b) Vol.154,No.1, 273-280

L. FIRLEJ, A. GRAJA, J. WOLAK, R.N. LYUBOVSKAYA, and L.M. GOLDENBERG
ESR Studies of Single Crystals of Bis(ethylenedithio)tetrahiafulvalene Based Conductor with Octaiodinemercurate Anion $(BEDT\text{-}TTF)_4Hg_3I_8$
64.70 and 76.30; 72.20; S12.1
(b) Vol.154,No.1, 333-340

M. TAYA, T. ASAJI, and D. NAKAMURA
A Low-Temperature Phase Transition in Ferroelectric $(CH_3NH_3)HgCl_3$ Definitely Located by Pyroelectric and Dielectric Measurements
64.70 and 77.80; S12
(a) Vol.114,No.1, 157-160

E. WOLSKA and U. SCHWERTMANN
Selective X-Ray Line Broadening in the Goethite-Derived Hematite Phase
61.10 and 64.70; 61.60; S10.1
(a) Vol.114,No.1, K11-K16

A.G. ARESHKIN, L.G. SUSLINA, and D.L. FEDOROV
On Concentration Dependences of the Band Parameters in $A^{II}B^{VI}$ Solid Solutions with Structural Phase Transitions
64.70 and 71.25; 71.35; 78.40; S8.15
(b) Vol.154,No.2, 609-622

M. POŁOMSKA and F. SMUTNÝ
$Rb_4LiH_3(SeO_4)_4$: A New Tetragonal Crystal with Low-Temperature Ferroelastic Phase
64.70; 77.20; S11
(b) Vol.154,No.2, K103-K106

J. WĄSICKI, W. NAWROCIK, Z. PAJĄK, I. NATKANIEC, and A.V. BELUSHKIN
Neutron Scattering Study of Some Pyridinium Salts
64.70; 61.12; S9
(a) Vol.114,No.2, 497-504

P. PISKUNOWICZ, T. BRECZEWSKI, and T. WOLEJKO
Thermal Expansion of $Rb_4LiH_3(SO_4)_4$ and $K_4LiH_3(SO_4)_4$ Crystals in the Temperature Range from 80 to 300 K
64.70 and 65.70; S11.1
(a) Vol.114,No.2, 505-510

K.N. MADHUSOODANAN, K. NANDAKUMAR, J. PHILIP, S.S.K. TITUS, S. ASOKAN, and E.S.R. GOPAL
Photoacoustic Investigation of Glass Transition in As_xTe_{1-x} Glasses
64.70; 61.40; 65.70; 72.80; S8
(a) Vol.114,No.2, 525-530

I.P. ALEKSANDROVA, PH. COLOMBAN, F. DÉNOYER, N. LE CALVÉ, A. NOVAK, B. PASQUIER, and A. ROZYCKI
Phase Transitions in Ammonium Hydrogen Selenates NH_4HSeO_4 and ND_4DSeO_4
64.70; 78.30; S11.1
(a) Vol.114,No.2, 531-544

I.N. ZAKHARCHENKO, E.S. NIKITIN, V.M. MUKHORTOV, YU.I. GOLOVKO, M.G. RADCHENKO, and V.P. DUDKEVICH
Peculiarities of Ferroelectric Phase Transition in $(Ba, Sr)TiO_3$ Heteroepitaxial Films Obtained by HF Cathode Sputtering
64.70 and 68.55; 77.80; S10.15
(a) Vol.114,No.2, 559-566

DECHENG TIAN, ZHU-XIN YU, CAI-GUO ZHAN, and SHAO-LING WANG
Positron Annihilation Study of Ageing of an Al-Mg-Si-RE Alloy
61.70 and 78.70; 64.70; S4
(a) Vol.114,No.2, K135-K138

H. ZANGAR, J.L. MIANE, C. COURSEILLE, N.B. CHANH, M. COUZI, and Y. MLIK
Structural Phase Transition in the Perovskite-Type Layer Compound $(C_3H_7NH_3)_2PbCl_4$
64.70; 61.14; 77.20; S11
(a) Vol.115,No.1, 107-118

O. BUDKE and P. ŠVEC
Application of Thermal Conductivity Measurement by the Relaxation Method to Crystallization Kinetics of Glassy As_2Se_3 +1mol%In
64.70; 66.70; S8
(a) Vol.115,No.1, 143-148

M. HIDAKA, Z.Y. ZHOU, and B.M. WANKLYN
Structural Phase Transitions of Layer Compounds $KFeF_4$, $KTiF_4$, and KVF_4
64.70; S9.16
(a) Vol.115,No.1, 149-156

M. QUINTERO, R. TOVAR, M. DHESI, and J.C. WOOLLEY
$(CuIn)_x(AgIn)_yMn_{2z}Te_2$ Alloys: T(z) Phase Diagram and Optical Energy Gap Values
64.70; S8.15
(a) Vol.115,No.1, 157-164

C.A. KLEINT, U. STOEPEL, and A. ROST
X-Ray Diffraction and Conductivity Investigations of Lanthanum-Doped Barium Titanate Ceramics
64.70 and 77.80; 64.80; S10.15
(a) Vol.115,No.1, 165-172

S.N. BALYAKIN, V.N. EROFEEV, D.N. MOGILYANSKII, and V.SH. SHEKHTMAN
Investigation of the Structure and Properties of KCl-NaCl Crystals at Elevated Temperatures
64.70; 62.20; S9.16
(a) Vol.115,No.1, 199-206

R. JAKUBAS and H. PYKACZ
Dilatometric Study of the Ferroelectric Phase Transition in $(CH_3NH_3)_5Bi_2Br_{11}$
64.70 and 65.70; S12
(a) Vol.115,No.1, K17-K18

O.I. BARKALOV, I.T. BELASH, and A.F. GUROV
Thermally-Activated Amorphization of the Quenched High-Pressure Phase in the Cd-Sb System
64.70; S4
(a) Vol.115,No.1, K19-K22

J.L. ZHANG and B.R. LI
Anomalous Electrical Conduction of $Cd_8Sb_2O_{10}$ Ceramics
64.70 and 72.80; S10
(a) Vol.115,No.1, K67-K70

N. VOUROUTZIS and C. MANOLIKAS
Phase Transformations and Discommensurations in Hexagonal Cuprous Telluride
61.14 and 64.70; S1.63
(a) Vol.115,No.2, 399-412

YUANSHENG WANG, MING MAO, and YIZHEN HE
Influence of a Third Metallic Component on the Crystallization of Metallic Glasses Fe-Ni-Si-B
61.40 and 64.70; S1.1; S1.2
(a) Vol.115,No.2, 467-474

V.N. NOVIKOV
Precision Dilatometric Studies of Metallic Glasses
61.40 and 64.70; S1.1; S1.2
(a) Vol.115,No.2, 475-482

D. MEIMARIS, F. VIRAS, and P. BEKRIS
A Study of the Structural Changes in Cd-Ge-Sb-As Glass during Transition to the Crystalline State Induced by Thermal Treatment
64.70 and 72.80; 61.40; S7.16
(a) Vol.115,No.2, 483-490

S. AHMED, A.M. SHAMAH, A. IBRAHIM, and F. HANNA
X-Ray Studies of the High Temperature Phase Transition of Ammonium Sulphate Crystals
64.70; 77.80; S11.1
(a) Vol.115,No.2, K149-K154

V.A. RUBTSOV, E.M. SMOLARENKO, V.M. TRUKHAN, V.N. YAKIMOVICH, and L.K. ORLIK
Phase Diagram of the CdP_2-$CdAs_2$ System
64.70; S7
(a) Vol.115,No.2, K155-K158

WENHUI SU, CHANGQING JIN, LANYING HE, YIFENG WANG, HONGJIAN LIU, and XIANG LONG
Study on Oxygen Hole and Superconductivity in $Y_1Ba_2Cu_3O_{7-\delta}$ Material
64.70 and 74.70; 61.10; S10.15
(a) Vol.116,No.1, 267-274

P.S.R. PRASAD and H.D. BIST
Vibrational Spectroscopic Study of Order-Disorder Transition in Ethylammonium Tetrachloro Cadmate
64.70 and 78.30; S12
(a) Vol.116,No.1, 275-286

A.N. PILYANKEVICH, A.V. KURDYUMOV, and N.F. OSTROVSKAYA
On the Crystallographic Irreversibility of Some Phase Transformation in Boron Nitride
61.50 and 64.70; S1.62
(a) Vol.116,No.1, K1-K6

L.J. HUANG, Q.M. CHEN, B.X. LIU, Y.D. FAN, and H.D. LI
Icosahedral Incommensurate Fe-Mo Phase Formed by Solid State Transformation
64.70; S1.1
(a) Vol.116,No.1, K51-K56

V.A. KOVARSKII and M.A. CHECHELNITSKII
A Microscopical Model of Optical Spectra of Local Crystal Centres Caused by Order-Disorder Phase Transition
64.70 and 78.50
(b) Vol.156,No.2, K157-K162

L. TERZIEV, J. LECOMTE-BECKERS, P. RASHEV, and T. GREDAY
Study of Transition Phases during Isothermal Ageing at 250 °C of Zn-27%Al Alloy
61.55; 61.14; 64.70; S1.2
(a) Vol.116,No.2, 521-528

G. HAUCK, E. KUKLEVA, and H.D. KOSWIG
Electrically Induced Deformations in Cholesteric Systems with Pretilt
61.30; 64.70; S12
(a) Vol.116,No.2, 545-554

N. MLIKI, M. BEN SALEM, and B. YANGUI
Parois de domaines ferroelastiques dans B-Ln_2O_3: Description en terme de densité de dislocations
64.70; S10
(a) Vol.116,No.2, 669-680

W. ZAPART, M.B. ZAPART, and A.I. OTKO
EPR Detection of the Ferroelastic Phase Transition in $KSc(WO_4)_2$
76.30; 64.70; S11.1
(a) Vol.116,No.2, K191-K194

64.75. Solubility, segregation, and mixing

E. KOENIGSBERGER and H. SCHRUNNER
On Lattice Parameters and Enthalpies of Mixing of Alkali Halide Solid Solutions
61.50 and 64.75; 65.50; S9.11
(b) Vol.151,No.1, 101-110

T. KIYOSAWA, S. TAKAHASHI, and N. KOGUCHI
Solid Solubility Range and Lattice Parameter of Semiconducting $Pb_{1-x}Cd_xS_{1-y}Se_y$ for Mid-Infra-Red Lasers
64.70 and 64.75; S8.15
(a) Vol.111,No.1, K21-K26

R. TOVAR, M. QUINTERO, P. GRIMA, and J.C. WOOLLEY
Phase Diagram, Lattice Parameter, and Optical Energy Gap Values for the $Zn_{2x}(AgIn)_{1-x}Te_2$ Alloys
61.60; 64.75; S8.16
(a) Vol.111,No.2, 405-410

A. SAKAKIBARA, M. YAMADA, T. KANADANI, and M. OHTA
Effect of the Thickness of the Specimen on the Formation of GP Zones in Al-4wt%Zn Alloy
64.75; S4
(a) Vol.112,No.1, K15-K18

U. BECKER, H. ZIMMERMANN, P. RUDOLPH, and R. BOYN
Optical Study of Impurity Distribution in Vertical-Bridgman Grown CdTe Crystals
71.55; 64.75; 78.50; 78.55; S8.13
(a) Vol.112,No.2, 569-578

P.I. GAIDUK, F.F. KOMAROV, and V.S. SOLOVEV
Structure Changes in Ion Implanted (111)-Si Layers during Pulsed Nanosecond Annealing and Following Thermal Treatment
61.80 and 64.75; S5.11
(a) Vol.112,No.2, 703-706

QIANG WANG, XIAO-WEN ZHANG, and BING-LIN GU
Order-Disorder Model of Phase Transition in A_xB_{1-x} Binary Systems of F.C.C. Structure
64.60; 64.75; S1.1; S1.2
(b) Vol.153,No.1, 139-144

C. LANDRON, J.P. COUTURES, R. ERRE, and P. LEHUEDE
Etude par ESCA de l'oxygene dans les verres silicatés alcalins
61.40; 64.75; S10.15
(a) Vol.113,No.1, 35-42

A.A. HOSSEINI and I.P. JONES
The Radiation Induced Segregation of Nickel in B.C.C. Iron Alloys
64.75; 61.80; S1.1
(a) Vol.113,No.1, 57-62

L.F. ZAKHARENKOV, V.F. KOBELEV, A.YA. NIKOLAICH, and V.A. OPARIN
Analysis of Precipitate Phases in Fe-Doped Indium Phosphide Crystals
64.75; S7.11
(a) Vol.113,No.1, K41-K44

H. LOEFFLER, C. ESCHRICH, and O. SIMMICH
Recalculation of Several Metastable Phase Lines of the Al-Zn System
61.55 and 64.75; S1.2
(a) Vol.113,No.2, 269-276

V.I. KUZNETSOV, E.N. YURCHENKO, M.T. PROTASOVA, E.A. TARABAN, O.P. KRIVORUCHKO, and R.A. BUYANOV
Moessbauer Investigation of the Influence of Al(III) Ions on the Properties of Mixed Fe(III)-Al(III) Hydroxides and Oxides
64.75; 76.80; S10
(a) Vol.113,No.2, 359-364

T. KANADANI and A. SAKAKIBARA
GP Zone Formation by Nucleation and Growth in Al-4mass%Zn Alloy
62.20 and 64.75; S4
(a) Vol.113,No.2, K199-K202

T. KANADANI and A. SAKAKIBARA
Short-Range Clustering in Dilute Al-Cu Alloy
64.75 and 72.15; S4
(a) Vol.114,No.1, K17-K20

S. ABIS and G. RIONTINO
A Scanning Electrical Resistivity (SER) Study of Phase Separation in an Al-22at%Zn Alloy
64.75; 72.15; S1.2; S4
(a) Vol.114,No.2, 511-518

R.K. SHARMA, S.K. MEHTA, V.K. SINGH, G. PRASAD, and B.B. SHARMA
Stoichiometric Annealing and Electrical Properties of $Hg_{0.8}Cd_{0.2}Te$ Grown by Solid State Recrystallization
64.75; 72.20; S8.15
(a) Vol.115,No.1, 135-142

S.N. MAITI, S. SENGUPTA, and D. ROY
Theoretical Interpretation of EXAFS Results for the Relaxation in Mixed Crystals
64.75; 61.10; S9.11
(b) Vol.156,No.1, 83-92

V.E. ADAMIAN, A.A. ARTSRUNI, A. BENAISSA, A.N. KOCHARIAN, M.A. MELIKIAN, and A.G. TONEIAN
Magnetic Properties of Model Multicomponent Compounds of CsCl Structure
64.75 and 75.30; S1.4
(b) Vol.156,No.2, 633-640

S.A. LITVINENKO and V.I. SOKOLOV
Formation of Pores in Thermal SiO_2 Due to Internal Mechanical Stress Relaxation
61.70 and 64.75; S10.1
(a) Vol.116,No.2, 615-620

64.80. Other phase properties of systems

D. VLADIKOVA, L. ILKOV, and S. KARBANOV
Influence of the Microstructure on Some Microwave Properties of Substituted Nickel Ferrites
64.80; 72.20; 75.50; S11.2
(a) Vol.111,No.1, 145-154

V.G. BESHENKOV, CH.V. KOPETSKII, and YU.A. SHIYANOV
Phase Composition of Thin Films as Revealed by Auger Depth Profiling
64.80 and 68.55; 79.20; S1.61
(a) Vol.114,No.1, 191-198

J. MONECKE
Microstructure Dependence of Material Properties of Composites
77.20; 64.80
(b) Vol.154,No.2, 805-814

C.A. KLEINT, U. STOEPEL, and A. ROST
X-Ray Diffraction and Conductivity Investigations of Lanthanum-Doped Barium Titanate Ceramics
64.70 and 77.80; 64.80; S10.15
(a) Vol.115,No.1, 165-172

65. Thermal Properties of Condensed Matter
*(see also 63. Lattice dynamics;
for thermodynamic properties of quantum fluids, see 67.40;
for thermal properties of solid helium, see 67.80)*

65.20. Heat capacities of liquids

R.N. JOARDER and S.M. MUJIBUR RAHMAN
Temperature Dependence of Entropies and Specific Heats of Liquid Na and K
65.50; 65.20; S2
(b) Vol.156,No.1, 181-194

65.40. Heat capacities of solids
*(for specific heat of superconductors, see 74.30;
for specific heat of magnetic systems, see 75.40)*

S.M. PATTALWAR, R.N. DIXIT, S.Y. SHETE, and B.K. BASU
Low-Temperature Heat Capacity Measurement in a Few Alloys near $PdPb_2$ Composition
65.40; S4
(b) Vol.151,No.2, K115-K118

K. BARTKOWSKI, G. POMPE, and E. HEGENBARTH
Specific Heat of Single-Crystalline Cd_3As_2, Cd_3P_2, and Zn_3P_2 at Low Temperatures
65.40; S7
(a) Vol.111,No.2, K165-K170

K. KOKKO, E. OJALA, and K. MANSIKKA
Fermi Level Density of States in Ag-Au Alloys. Nuclear Magnetic Spin-Lattice Relaxation Rate and Low-Temperature Specific Heat
71.20; 65.40; 76.60; S1.3
(b) Vol.153,No.1, 235-242

R. POPRAWSKI and S.A. TARASKIN
Calorimetric Studies of the Phase Transition in Nd_4DSeO_4 Crystals
64.70; 65.40; S11
(a) Vol.113,No.1, K31-K36

A.A. REMPEL and A.I. GUSEV
Heat Capacity of Niobium Carbide in Different Structural States
65.40; S1.61
(a) Vol.113,No.2, 353-358

A.A. ABDURRAGIMOV, M.A. ALYANOV, N.G. GUSEINOV, and A.G. GUSEINOV
Heat Capacity of $CoCr_2S_4$
65.40; S8.16
(a) Vol.113,No.2, K207-K210

A. LOPEZ-ECHARRI, I. RUIZ-LARREA, and M.J. TELLO
Thermodynamics of the Phase Transition Sequence in the Incommensurate Compound $[N(CH_3)_4]_2CuBr_4$
64.60; 65.40; S9.16
(b) Vol.154,No.1, 143-152

D.A. GUSEINOV, A.M. ABDULLAEV, A.B. ABDULLAEV, and S.M. YAVADOVA
Thermodynamic Properties and Heat Capacity of $CoFe_2Se_4$
65.40 and 65.50; S8.16
(a) Vol.115,No.1, K5-K8

V.B. KOKSHENEV and N.N. ZHOLONKO
Effects of Correlated Rotation of Linear Molecules in the Low-Temperature Dynamics and Thermodynamics of Weak Molecular Solution
63.10 and 65.40; 65.70; S5.2
(b) Vol.156,No.1, 119-130

R. KUENTZLER, E. MURRAY, Y. DOSSMANN, and D. AST
Electronic Properties of AlMn and AlMnSi Quasicrystals
75.40; 65.40; 71.20; S1.2
(a) Vol.116,No.1, 399-404

A.O. MEKHRABOV, Z.M. BABAEV, and Z.A. MATYSINA
The Effect of a Third Component Impurity C on the Heat Capacity of A_3B-Type Ordered Alloys
64.60 and 65.40; S1.1; S1.2
(b) Vol.156,No.2, 419-430

M. ROESLER and W. STOLZMANN
Electronic Specific Heat of Coulombic Systems
65.40 and 71.45
(b) Vol.156,No.2, 515-526

65.50. Thermodynamic properties and entropy

C. MALINOWSKA-ADAMSKA
The Effect of Pressure on the Thermodynamic Properties of Crystals near the Lattice Instability Point
65.50; S1.2; S1.3; S2; S4
(b) Vol.151,No.1, 53-60

E. KOENIGSBERGER and H. SCHRUNNER
On Lattice Parameters and Enthalpies of Mixing of Alkali Halide Solid Solutions
61.50 and 64.75; 65.50; S9.11
(b) Vol.151,No.1, 101-110

T.M. HARIDASAN and P. SEKAR
Thermodynamic Parameters of Anti-Frenkel Formation in Alkaline Earth Fluorides
65.50; 63.90; S9.12
(b) Vol.153,No.1, 131-138

W.D. KRAEFT, W. STOLZMANN, I. FROMBOLD-TREU, and T. ROTHER
Thermodynamics of Two Component Gaseous and Solid State Plasma at Any Degeneracy
65.50
(b) Vol.153,No.2, K109-K114

T.A. MARSAGISHVILI and M.N. MACHAVARIANI
Free Energies of the Polyatomic Polarizable Dipoleactive Particles in Polar Media
65.50
(b) Vol.154,No.1, 97-104

A.L. SARCE and E.J. SAVINO
Thermodynamics of Deformation under Irradiation
62.20; 61.80; 65.50
(b) Vol.154,No.1, 105-114

C. TUIJN and H. BAKKER
The Influence of the Vibrational Entropy on the Specific Heat of the Order-Disorder Transition in Alloys
65.50; 64.60
(b) Vol.155,No.1, 107-112

D.A. GUSEINOV, A.M. ABDULLAEV, A.B. ABDULLAEV, and S.M. YAVADOVA
Thermodynamic Properties and Heat Capacity of $CoFe_2Se_4$
65.40 and 65.50; S8.16
(a) Vol.115,No.1, K5-K8

R.N. JOARDER and S.M. MUJIBUR RAHMAN
Temperature Dependence of Entropies and Specific Heats of Liquid Na and K
65.50; 65.20; S2
(b) Vol.156,No.1, 181-194

R.B. MCLELLAN
The Thermodynamics of Pd-(Ce,Y,Er)-H Ternary Solutions
65.50; S1.2; S1.4
(a) Vol.116,No.2, K157-K160

65.70. Thermal expansion and thermomechanical effects
(see also 64.30. Equations of state)
For thermal conduction in nonmetallic liquids, see 66.60; in nonmetallic solids, see 66.70
For electronic thermal conduction, see 72.10 and 72.20
For thermal conductivity in superconductors, see 74.30
For pyroelectric and electrocaloric effects, see 77.70

WEN-CHEN ZHENG
Local Thermal Expansion around the Impurity in Chromium-doped alpha-$LiIO_3$ Crystals
65.70; 71.70; 76.30; S11.1
(b) Vol.153,No.2, K103-K108

U. PIETSCH and J. WOLF
The Thermal Expansion Coefficient of Lattice Matched $Ga_{1-x}Al_xAs$ Layers Grown on a [001] GaAs Substrate
65.70; S7.15
(a) Vol.113,No.2, K203-K206

A.A. POVZNER, S.V. KORTOV, and P.V. GELD
Magneto-Volume Effects in the Weak Itinerant Ferromagnets on the Basis of Solid Solutions of $Fe_{1-y}Mn_ySi$ and $Fe_{1-x}Co_xSi$
65.70 and 75.40; S1.61
(a) Vol.114,No.1, 315-326

S. MUKHERJEE, S. SENGUPTA, and A.N. BASU
On the Nature of Distribution Function for Two-Level Systems in Amorphous Solids
62.65 and 65.70; 63.20
(b) Vol.154,No.2, K107-K112

P. PISKUNOWICZ, T. BRECZEWSKI, and T. WOLEJKO
Thermal Expansion of $Rb_4LiH_3(SO_4)_4$ and $K_4LiH_3(SO_4)_4$ Crystals in the Temperature Range from 80 to 300 K
64.70 and 65.70; S11.1
(a) Vol.114,No.2, 505-510

G. AMATO, G. BENEDETTO, R. SPAGNOLO, and M. TURNATURI
Photoacoustic Measurements of Doped Silicon Wafers
65.70 and 66.70; 72.80; S5.11
(a) Vol.114,No.2, 519-524

K.N. MADHUSOODANAN, K. NANDAKUMAR, J. PHILIP, S.S.K. TITUS, S. ASOKAN, and E.S.R. GOPAL
Photoacoustic Investigation of Glass Transition in As_xTe_{1-x} Glasses
64.70; 61.40; 65.70; 72.80; S8
(a) Vol.114,No.2, 525-530

J.D. STEPHENSON
Simulated Quenching of Silicon WSR Monochromators Using WSR Section Topography
61.70; 65.70; S5.11; S5.12
(a) Vol.115,No.1, 97-106

R. JAKUBAS and H. PYKACZ
Dilatometric Study of the Ferroelectric Phase Transition in $(CH_3NH_3)_5Bi_2Br_{11}$
64.70 and 65.70; S12
(a) Vol.115,No.1, K17-K18

N.P. KOLMAKOVA, R.Z. LEVITIN, and V.N. ORLOV
Anomalies of Crystal Lattice Parameters in Paramagnetic Rare-Earth Garnets at Low Temperatures
65.70 and 75.80; S11.2
(a) Vol.115,No.1, K87-K90

V.B. KOKSHENEV and N.N. ZHOLONKO
Effects of Correlated Rotation of Linear Molecules in the Low-Temperature Dynamics and Thermodynamics of Weak Molecular Solution
63.10 and 65.40; 65.70; S5.2
(b) Vol.156,No.1, 119-130

L.N. ALEKSANDROV, S.L. BABENKOVA, and V.S. ZINOVEV
Thermoelastic Stresses in Multilayered Structures of Silicon-on-Insulator Type (Computational Experiment)
65.70 and 68.25; 64.30; S5.11; S10.1
(a) Vol.116,No.1, K57-K60

66. Transport Properties of Condensed Matter (Nonelectronic)

66.30. Diffusion in solids

J. KVASNICA
Possible Anomalies in Multicomponent Diffusion
66.30
(b) Vol.151,No.1, 39-46

T. OVSENIK, R. BLINC, M. ČOPIČ, G. LAHAJNAR, I. ZUPANČIČ, P. GUENTER, H. AREND, and CH. BOSSHARD
Self-Diffusion in an Undercooled Organic Melt near the Liquid-Glass Transition
66.30; 64.70; S12
(b) Vol.151,No.1, K1-K6

K. YAMAKAWA, B. HOHLER, and H. KRONMUELLER
Diffusion of Hydrogen in Nickel Based Alloys
66.30; 72.15; S1.2
(a) Vol.111,No.1, 89-98

H. KAHNT and T. POSSNER
Diffusion from a Gaussian Half-Profile at an Absorbing Boundary in the Presence of an Electric Field
66.30
(a) Vol.111,No.1, K7-K12

T.O. BUDKO, E.V. GUSHCHINSKAYA, YU.S. EMELYANENKO, and S.A. MALYSHEV
Diffusion of Zinc into Unpassivated Surface of Indium Phosphide
66.30; S7.11
(a) Vol.111,No.2, 451-456

V.F. STELMAKH, YU.R. SUPRUN-BELEVICH, and A.R. CHELYADINSKII
Effect of Radiation Defects and Elastic Incompatibility Stresses on the Electrical Activation and Diffusion of Boron ...
66.30; 61.80; 62.20; S5.11
(a) Vol.112,No.1, 381-384

D. JAWORSKA, E. TARNOWSKA, A.P. KOBZEV, and R.A. ILKHAMOV
Depth Distribution of the Gettered Au Atoms Measured by Means of RBS and Radioactive Methods
61.80; 66.30; S5.11
(a) Vol.112,No.1, 385-390

R. SELLMANN and J. MIMKES
Diffusion of Thallium in Silicon
66.30; S5.11
(a) Vol.112,No.1, K5-K8

J. PIPREK
Numerical Analysis of TVS Ionic Current Applied to the Na^+-Motion in Implanted MOS Oxide Layers
66.30 and 73.40; S10.1
(a) Vol.112,No.2, 551-560

M. SOWA, W. SZYSZKO, L. GŁUSIEC, J. TRZCIANKA, and J. ZINKIEWICZ
Diffusion of Indium Implanted into Steel
66.30; S1.1
(a) Vol.112,No.2, 835-838

M. SOWA, W. SZYSZKO, J. SIELANKO, L. GŁUSIEC, and B. KAMIEŃSKA
Microhardness of Boron, Titanium, and Nitrogen Implanted Steel
62.20; 66.30; S1.1
(a) Vol.112,No.2, 839-844

V.N. SHAMBALEV, L.A. BASHKIROV, V.V. PANKOV, A.G. DUTOV, and A.N. IGUMENTSEV
Self-Diffusion of Cations in the Solid Solution System. Ti-Substituted Mn-Zn Ferrites
66.30; S11.2
(a) Vol.112,No.2, K71-K74

H. GDANITZ and K. SCHMALZ
On the Shape of Au_s Depth Profile in Heat Treated Si Wafers
66.30; S5.11
(a) Vol.112,No.2, K75-K80

R.A. YAKSHIBAEV, M.KH. BALAPANOV, N.N. MUKHAMADEVA, and G.R. AKMANOVA
Partial Conductivity of Cations of Different Kinds in the Alloys of Cu_2X-Ag_2X (X=Se,Te) Mixed Conductors
66.30; S8
(a) Vol.112,No.2, K97-K100

K. SCHMALZ and K. TITTELBACH-HELMRICH
On the Hydrogen Diffusion in Silicon at about 100 °C
66.30; S5.11
(a) Vol.113,No.1, K9-K14

O. SIMMICH, M. MACHMER, and H. WILL
Approximate Solutions for the Kinetics of Diffusion-Limited Dissolution and Growth of Planar Precipitates
64.60; 66.30
(a) Vol.113,No.1, K27-K30

G. KLUGE, A. THOMAS, R. KLABES, and P. SUEPTITZ
Applicability of Silver Implanted Amorphous $GeSe_2$ as Inorganic Negative Resist
61.80; 66.30; 79.20; S8
(a) Vol.113,No.2, K171-K176

R. ČERNÝ and D. RAFAJA
Solution of Diffusion Equation in a Finite Single Phase Sample with Variable Coefficient
66.30
(a) Vol.113,No.2, K181-K184

Y. MADHAVAN, K. RAMACHANDRAN, and T.M. HARIDASAN
Diffusion Studies in CdTe and HgTe
66.30; S8.13
(b) Vol.154,No.1, 55-60

H. HAGENSCHULTE and TH. HEUMANN
On the Determination of Enhancement Factors of Solvent Diffusion
66.30; S1.3
(b) Vol.154,No.1, 71-84

P. ENDERS and U. SCHADE
Addendum to: Atomic Diffusion in Tetrahedral Semiconductors. Improved Approximate Diffusion Profiles
66.30; S7.1
(b) Vol.154,No.1, K5-K8

E.M. SLYUSARENKO and V.S. SOLDATOV
Description of Interdiffusion Processes in Multicomponent Systems by the First Fick's Law
66.30
(b) Vol.154,No.2, 475-482

I.P. KOGUTYUK, P.N. KEIVAN, and I.M. RARENKO
A Study of Interstitial Impurity Diffusion by the Linear Response Theory
66.30; S5.12
(b) Vol.154,No.2, 513-518

K. SOMOGYI, D.N. KORBUTYAK, L.N. LASHKEVICH, and I. POZSGAI
Cu Diffusion into GaAs from Water Based Solution
66.30 and 73.40; 78.65; S7.12
(a) Vol.114,No.2, 635-642

T. TANABE, R. IKEDA, and D. NAKAMURA
The Highest-Temperature Crystalline Phase of Tetramethylammonium Thiocyanate with the Cations Performing Three-Dimensional ...
66.30 and 76.60; S12
(a) Vol.114,No.2, K143-K146

T.D. DZHAFAROV
Radiation-Stimulated Diffusion in Semiconductors (Review Article)
61.80 and 66.30; S5.11
(b) Vol.155,No.1, 11-52

D. LUPU
Hydrogen Diffusion and Radiationless Electron Transfer in Metal Hydrides
66.30; 71.38; 78.20; S1.2; S1.4; S3
(b) Vol.155,No.1, 65-72

P. WESOŁOWSKI, W. JAKUBOWSKI, and J.L. NOWIŃSKI
Electrical Properties of Superionic Silver-Borate Glasses Doped with AgI
66.30; S10
(a) Vol.115,No.1, 81-86

K. SADAIYANDI and K. RAMACHANDRAN
Diffusion Studies in ZnTe for an Active Role of Vacancies
66.30; S8.13
(b) Vol.155,No.2, K85-K88

A. KLIMAKOW and M. SCHENK
Grain-Boundary Diffusion of Mercury in p-Type (Hg,Cd)Te
61.70 and 66.30; S8.15
(a) Vol.115,No.2, K135-K138

A.V. KOLOBOV, V.M. LYUBIN, and J. TROELTZSCH
On the Mechanism of Anomalous Surface Diffusion of Silver in an Amorphous Chalcogenide/Silver Structure
66.30; S8
(a) Vol.115,No.2, K139-K142

B. FRITZSCH, R. FRITZSCH, and A. ZEHE
Simulation of Vacancy Migration in B.C.C. Metals
66.30
(b) Vol.156,No.1, 65-70

HUI-FANG DENG
Generalized Fick Law, Its Microscopic Derivation, and Dispersive Diffusion
66.30
(b) Vol.156,No.1, K11-K16

S. FUJITA
On the Feasibility of Nuclear Fusion in F.C.C. Metals
61.70 and 66.30
(b) Vol.156,No.1, K17-K22

H. HAEFKE and H. STENZEL
Cross-Section Decoration. A Method for Determination of Diffusion Coefficients
66.30 and 68.48; S9.11
(a) Vol.116,No.1, 39-46

K. EFTAXIAS and V. HADJICONTIS
Analysis of Self-Diffusion Data in V and Nb
66.30; S1.2
(b) Vol.156,No.2, 393-402

GH. STOICESCU, S.V. NISTOR, and C.D. MATEESCU
Aggregation of Bismuth in NaCl Crystals
61.70; 66.30; 78.50; S9.11
(b) Vol.156,No.2, 411-418

YU.N. MOSKVICH, B.I. CHERKASOV, A.M. POLYAKOV, A.A. SUKHOVSKII, and R.L. DAVIDOVICH
NMR Study of Internal Motions in a New Family of Ionic Conductors M_2SbF_5
76.60; 66.30; 72.20; S9.16
(b) Vol.156,No.2, 615-632

D.K. AN, K. MADL, A. BARNA, G. BATTISTIG, and J. GYULAI
The Simultaneous Diffusion of Gold and Boron into Silicon. Push Effect of Gold to Boron
66.30; 61.70; S5.11
(a) Vol.116,No.2, 561-570

H.U. JAEGER, T. FEUDEL, and S. ULBRICHT
Modeling of Defect-Phosphorus Pair Diffusion in Phosphorus-Implanted Silicon
61.70; 66.30; S5.11
(a) Vol.116,No.2, 571-582

C. LEXCELLENT and G. SOMÉ
Comportement en fluage à haute température de la phase γ du Cu-16,6%atSn. Tentative de corrélation avec la diffusion en volume
62.20; 66.30; S1.2
(a) Vol.116,No.2, 689-696

L. PRANEVIČIUS, V. SARGŪNAS, and G. ZUBAUSKAS
Component Autodiffusion into Protective Coatings in Ion-Implanted InP
66.30 and 68.48; S7.11
(a) Vol.116,No.2, K161-K164

66.70. Nonelectronic thermal conduction and heat-pulse propagation in nonmetallic solids

(for thermal conduction in solid metals, see 72.15; for statistical mechanics of lattice vibrations, see 63.70)

S. WAPLAK and T. KRAJEWSKI
EPR Studies of $LiNaSO_4$ Chromium Cr^{6+} Doped Monocrystals after X-Ray Irradiation
76.30; 61.80; 66.70; S11
(a) Vol.111,No.1, 271-278

A.H. AWAD and J.S. SAIF
The Correction Term and Core Dislocations in the Frame of the Generalized Callaway Integral
66.70
(b) Vol.151,No.2, 495-502

G.CH. CHRISTAKUDIS, S.K. PLACHKOVA, and L.E. SHELIMOVA
Influence of Defects on the $(GeTe)_{1-x}[(Ag_2Te)_{1-y}(Sb_2Te_3)_y]_x$ Phonon Thermal Resistivity
61.70 and 66.70; S8.16
(a) Vol.111,No.2, 469-476

S.P. ZHVAVYI and O.L. SADOVSKAYA
Nanosecond Laser Annealing of Implanted Silicon. Simulation of Dynamics
64.70; 66.70; S5.11
(a) Vol.112,No.1, K19-K22

S.K. NIKOGOSYAN and V.A. SAAKYAN
The Effect of Electron Radiation on Thermal and Electrophysical Properties of n-InP
61.80; 66.70; 72.20; S7.11
(a) Vol.113,No.1, 43-50

J. KLEJST and J. BRACINÍK
Photoacoustic Study of Semi-Insulating GaAs:Cr
62.65 and 66.70; S7.12
(a) Vol.113,No.1, K37-K40

M.M. KHALIFA, M.N. AVASTHI, and S.M. EL-MASHRI
Heat of Transport of an Organo-Metallic Superionic Material
66.70; S12
(a) Vol.113,No.2, K177-K180

T. KOPTÉ and E. HEGENBARTH
Behaviour of the Thermal Conductivity of High Temperature Superconductors around the Critical Temperature
66.70 and 74.70; S10.15
(a) Vol.114,No.1, K67-K70

G. AMATO, G. BENEDETTO, R. SPAGNOLO, and M. TURNATURI
Photoacoustic Measurements of Doped Silicon Wafers
65.70 and 66.70; 72.80; S5.11
(a) Vol.114,No.2, 519-524

O. BUDKE and P. ŠVEC
Application of Thermal Conductivity Measurement by the Relaxation Method to Crystallization Kinetics of Glassy As_2Se_3+1mol%In
64.70; 66.70; S8
(a) Vol.115,No.1, 143-148

G.CH. CHRISTAKUDIS
Phonon Thermal Resistivity in $(GeTe)_{1-x}[(Ag_2Te)_{1-y}(Sb_2Te_3)_y]_x$ Solid Solutions
66.70; S8.16
(b) Vol.155,No.2, K89-K92

T. KOPTE and E. HEGENBARTH
Thermal Conductivity of $BaCuO_{2+x}$ at Low Temperatures and High Magnetic Fields
66.70; S10.15
(b) Vol.156,No.2, K97-K100

66.90. Other topics in nonelectronic transport properties

R.A. YAKSHIBAEV, V.N. ZABOLOTSKII, and R.F. ALMUKHAMATOV
Structural Features and Ionic Transport in the Ag_xNbSe_2 Two-Dimensional Superionic Conductor
61.80 and 66.90; S8.15
(a) Vol.111,No.2, 431-442

E. MIHÓKOVÁ
On Negative Probabilities from the Agranovich and Galanin Model of Luminescence and Energy Transfer
78.55; 66.90
(b) Vol.154,No.2, K183-K186

68. Surfaces and Interfaces; Thin Films and Whiskers
(for impact phenomena, see 79;
for crystal growth, see 61.50)

68.20. Solid surface structures

A.M. ROSHCHUPKIN, I.L. BATARONOV, O.A. TROITSKII, and M.M. MOISEENKO
Electric Current Effect on Metal Surface Layers
68.20 and 72.15
(b) Vol.151,No.1, 121-126

R. PASIANOT, G. GALAMBOS, E.J. SAVINO, L. AMARAL, M. BEHAR, D. FARKAS, and S. RAO
Argon Irradiation of Sn Thin Layers Deposited on Fe Substrates
61.80 and 68.20; S1.1; S4
(a) Vol.111,No.1, 173-180

C. ENGLER
Potential Surfaces for Chemisorption on Semiconductor Surfaces from Pair Potential Considerations
68.20; 71.10; 73.20; S5.11
(b) Vol.151,No.2, 555-564

A. SCHUBERT and B. MICHEL
Simultaneous Line Broadening and Texture Analysis
61.10 and 68.20; S1.1
(a) Vol.111,No.2, K137-K140

W. ROSIŃSKI and J. KACZMAREK
Present Position and Further Developments of Ion Implantation in Metals
61.80 and 68.20; S1.1
(a) Vol.112,No.1, 245-252

M. BUCKI, A. GEMPERLE, S.A. KONSTANTINOV, M.M. KANTOR, F. ČERNÝ, K. KULT, G.S. BURCHANOV, and E.V. OTTENBERG
Modification of Surface Layers of Molybdenum by B^+ and N^+ Ion Implantation
61.80 and 68.20; S1
(a) Vol.112,No.1, 335-338

T. OTMIANOWSKI and B. KAMIEŃSKA-STEFANIUK
Some Aspects of the Surface Wear Process of Steel Implanted with Nitrogen
61.80 and 68.20; S1.1
(a) Vol.112,No.1, 339-342

T. BARBASZEWSKI, M. DABROWSKI, M. DRWIEGA, J. GAWLIK, and E. LIPINSKA
Implantation Profiles of Nitrogen and Titanium in a Low Energy Bombarded High Speed Tool Steel
61.80 and 68.20; S1.1
(a) Vol.112,No.1, 347-352

P. TARKOWSKI, I. BRYŁOWSKA, and M. SUBOTOWICZ
Influence of Nitrogen Implantation on Corrosion Process in Steel
61.80; 68.20; S1.1
(a) Vol.112,No.1, 353-356

J. JĘDRKOWIAK, J. MARTAN, J. MASALSKI, and D.B. BOGOMOLOV
Influence of Ar, Al, and N Ion Implantation on the Surface Properties of the 00H18M2Nb Steel
61.80; 68.20; S1.1
(a) Vol.112,No.1, 357-360

A.V. KONDRACHUK
Model of Subsurface Redistribution of Impurity Atoms and Radiation Defects upon Light Ion Irradiation at Elevated Temperature
61.80; 68.20; S5.11
(a) Vol.112,No.1, 411-414

W. SOSZKA
Krypton Saturation of the Metal Surface during Kr^+ Ion Bombardment Determined Due to Autoionization Electron Analysis
61.80; 68.20; 79.20; S1
(a) Vol.112,No.1, 425-430

J. PIEKOSZEWSKI, Z. WERNER, C. POCHRYBNIAK, J. LANGNER, M. GRYZIŃSKI, and A. HORODEŃSKI
Pulse Ion Beam Doping and Modification of Solids
61.80 and 68.20
(a) Vol.112,No.2, 757-760

J.S. NKOMA
Radiation Induced Luminescence from Dipoles Immersed in a Solid with a Rough Surface
68.20 and 78.55; S7.11
(b) Vol.153,No.1, 383-390

D.V. NOVIKOV, E.A. KONDRASHKINA, and S.A. STEPANOV
Study of X-Ray Surface Backscattering
61.10; 68.20; S5.12
(a) Vol.114,No.1, K7-K10

R. HEYNER and G. MARX
Depth Profiles and Surface Characterization by Glow Discharge Optical Emission Spectroscopy
68.20; 79.20; S1.2
(a) Vol.114,No.2, 609-620

KH.R. KAZDAEV, M.T. AKCHULAKOV, E.M. BAYADILOV, V.I. ENGELKO, A.V. LAZARENKO, and E.S. CHEBUKOV
Influence of Annealing Regimes on Phase Transitions in Nitrogen and Carbon Ion Implanted Molybdenum
61.80 and 68.20; S1.61; S1.62
(a) Vol.115,No.1, 179-184

S. BAUNACK and A. ZEHE
A Study of UV/Ozone Cleaning Procedure for Silicon Surfaces
68.20; 68.55; 79.20; S5.11
(a) Vol.115,No.1, 223-228

N. GANEV, I. KRAUS, and J. TRPČEVSKÁ
X-Ray Measurement of Residual Stresses Induced by Laser Surface Treatment
62.20; 68.20; S1.1
(a) Vol.115,No.1, K13-K16

E. TARANKO, G.M. GAVRILENKO, R. TARANKO, and V.K. FEDYANIN
On Hydrogen Chemisorption on Simple Metals
68.20 and 68.45; 73.20
(b) Vol.155,No.2, 481-490

R.M. IMAMOV, A.A. LOMOV, and D.V. NOVIKOV
Grazing Incidence Diffraction X-Ray Topography
61.10; 68.20; S5.11
(a) Vol.115,No.2, K133-K134

V.I. LYGIN, A.D. SERAZETDINOV, E.A. RYABENKO, and O.I. CHERTIKHINA
Quantum-Chemical Calculations of Models for Surface Defects in Dehydroxylated Silica
68.20; S10.1
(b) Vol.156,No.1, 205-210

R. KERN and M. KROHN
Interaction of Metal Clusters with Surface Steps
68.20; 61.70; 62.20; S9.11
(a) Vol.116,No.1, 23-38

F. HEINRICHT
Reflection Electron Microscopy of Ultra Fast Surface Processes
61.16; 68.20
(a) Vol.116;No.1, 145-152

V.V. LOPATIN, A.V. KABYSHEV, and L.S. BUSHNEV
Modification of Polycrystalline BN, AlN, and Si_3N_4 Surfaces by Ion Beams
61.70 and 68.20; S6; S7
(a) Vol.116,No.1, K69-K72

F. BECHSTEDT and D. REICHARDT
Calculation of Surface-Induced Core-Level Shifts for Covalent Semiconductors C, Si, Ge, and alpha-Sn. I.(111)2x1 Surfaces
68.20 and 73.20; S5; S5.11; S5.12
(b) Vol.156,No.2, 471-486

C. HAMANN, R. LAIHO, and A. MRWA
Surface Analysis on Lead Phthalocyanine with STM
68.20 and 68.55; S12
(a) Vol.116,No.2, 729-734

68.25. Mechanical and acoustical properties of solid surface and interfaces

A.F. VLADIMIROV and E.N. MOOS
Development of Eshelby Concept in Solution of Boundary Problems for Elastic Solids with Defects
61.70; 68.25
(a) Vol.111,No.1, 99-108

WEI LIU, ZHONGGUANG WANG, and YUEBO XIA
Crystallographic Characteristics of Slip Band Continuity across Grain Boundaries in Multicrystalline Aluminum during Cyclic Deformation
68.25; S4
(a) Vol.114,No.1, 177-184

M. TOMÁŠEK, Š. PICK, G.V. GADIYAK, and YU.N. MOROKOV
Simple Analysis of Atomic Forces Acting on Silicon Surfaces
68.25 and 68.40; 73.20; S5.11
(b) Vol.156,No.1, 243-252

L.N. ALEKSANDROV, S.L. BABENKOVA, and V.S. ZINOVEV
Thermoelastic Stresses in Multilayered Structures of Silicon-on-Insulator Type (Computational Experiment)
65.70 and 68.25; 64.30; S5.11; S10.1
(a) Vol.116,No.1, K57-K60

P. KOŠTIAL, I. JAMNICKÝ, and J. ĎURČEK
Contribution to the Experimental Methods of Deep Level Acoustic Transient Spectroscopy
68.25 and 71.55; 72.50; S5.11; S10
(a) Vol.116,No.1, K77-K80

H. RUPPERSBERG, I. DETEMPLE, and J. KRIER
Evaluation of Strongly Non-Linear Surface-Stress Fields $\sigma_{xx}(z)$ and $\sigma_{yy}(z)$ from Diffraction Experiments
62.20 and 68.25; S1.2
(a) Vol.116,No.2, 681-688

68.30. Dynamics of solid surfaces and interface vibrations

GONG YU, GUANGHUA CHEN, and FANGQING ZHANG
Theoretical Study of Raman Spectrum in a-$Si_{1-x}N_x$:H Films
61.40 and 68.30; 78.30; S5.11
(b) Vol.152,No.1, 73-78

F. BECHSTEDT and H. GERECKE
Long-Wavelength Optical Phonons in GaAs-AlAs Superlattices
63.20 and 68.30; 73.40; S7.12
(b) Vol.154,No.2, 565-582

T.D. IBRAGIMOV, G.G. KURBANOVA, and V.S. GORELIK
Raman Scattering in Gallium Selenide and Sulfide Films
63.20 and 68.30; 78.30; S8
(b) Vol.155,No.1, 113-116

V.G. PODOPRIGORA, I.A. REMIZOV, V.F. SHABANOV, and L.V. SHMELEVA
Dynamic and Electrooptic Parameters of Monolayers in Ethane Adsorbed on Graphite
68.30; S12
(b) Vol.155,No.1, 117-126

K.S. SRIVASTAVA and A. TANDON
Dispersion Relation for Coupled Surface Waves in Thin Polar Semiconducting Foils
68.30; 71.45; S7.12
(b) Vol.155,No.1, 127-136

XING-KUI ZHANG, HUA XIA, and AN HU
Raman Scattering in Nb-Cu Metallic Multilayers
68.30; 78.30; S1.2; S1.3
(b) Vol.155,No.1, 137-146

J. MONECKE
Collective Excitations in Microcrystal Aggregates and Composites
68.30 and 78.30; 63.10
(b) Vol.155,No.2, 437-444

F. BECHSTEDT and H. GERECKE
Relationship of Microscopic and Macroscopic Theories for Long-Wavelength Optical Phonons in GaAs-AlAs Superlattices
63.20; 68.30; S7.12; S7.15
(b) Vol.156,No.1, 151-170

68.40. Surface energy of solids; thermodynamic properties

M. BABIKER and M.G. COTTAM
A Quantized Screening Theory for the Surface Energy of Solids in the Hydrodynamic Jellium Model
68.40; S1.2; S3; S4
(b) Vol.153,No.1, 145-152

H.-G. FRITSCHE
Particle Size Effects of Ion Crystallites
68.40; 62.20; S9.11
(b) Vol.154,No.2, 603-608

W. VOGELSBERGER and E. MUELLER
On the Location of the Gibbs Surface of Tension in Thermodynamic Considerations of Interfaces
68.40; 64.10
(b) Vol.155,No.1, K17-K22

M. TOMÁŠEK, Š. PICK, G.V. GADIYAK, and YU.N. MOROKOV
Simple Analysis of Atomic Forces Acting on Silicon Surfaces
68.25 and 68.40; 73.20; S5.11
(b) Vol.156,No.1, 243-252

T. HALICIOĞLU, H.O. PAMUK, S. ERKOÇ, and W.A. TILLER
The Effect of a Potential Function Range on Surface Properties
68.40
(b) Vol.156,No.2, K113-K118

68.45. Solid-fluid interface processes

I.D. MAKUTA, S.K. POZNYAK, A.I. KULAK, and E.A. STRELTSOV
Photocurrent Generation and Optical Transitions on Degenerate Cadmium Oxide Photoanodes
68.45 and 72.40; S10.1
(a) Vol.111,No.1, 193-200

A. SZÁSZ and D.J. FABIAN
Subsurface Hydrogen Exciton-Like States in Solid Catalysts
68.45 and 73.20; 71.35; S1
(b) Vol.152,No.1, 117-122

C. MARTI, C. DE BEAUVAIS, T. CEVA, B. CROSET, and M. GOLDMANN
Incommensurate-Commensurate Transition Driven by Concentration in Rare Gas Mixtures Adsorbed on Graphite
64.60; 68.45; S5
(b) Vol.152,No.2, 463-474

H. DUŃSKI and H. SUGIER
The Study of Temperature Programmed Desorption of Oxygen from Magnesium Oxide
68.45; S10.1
(a) Vol.113,No.1, K49-K54

E. TARANKO, G.M. GAVRILENKO, R. TARANKO, and V.K. FEDYANIN
On Hydrogen Chemisorption on Simple Metals
68.20 and 68.45; 73.20
(b) Vol.155,No.2, 481-490

E. KESSLER
On the Measurement of the Heat of Condensation of Ag, Bi, Sb, and Pb
68.45; S1.3; S4
(a) Vol.116,No.1, K61-K64

N.C. BACALIS and A.B. KUNZ
Ab Inition Calculations on the Adsorption of H, O, and CO atop Cu^{2+} on MgO(001)
68.45; S10.1
(b) Vol.156,No.2, 545-560

A.E. KIV, V.V. KOVALCHUK, and V.A. YANCHUK
H Adsorption by Disordered Si Clusters
68.45; S5.11
(b) Vol.156,No.2, K101-K104

S. ERKOÇ, T. HALICIOǦLU, and W.A. TILLER
Adsorption of Gold Adatom on the GaAs(110) Surface: A Molecular-Dynamics Simulation
68.45; S7.12
(b) Vol.156,No.2, K105-K108

H. DUŃSKI and H. SUGIER
Temperature-Programmed Desorption of Oxygen from gamma-Irradiated Polycrystalline Magnesium Oxide
61.80 and 68.45; S10.1
(a) Vol.116,No.2, K165-K168

68.48. Solid-solid interfaces (including bicrystals)
(for grain boundaries, see 61.70)

J. HEROLD, H. BARTSCH, W. WESCH, and G. GOETZ
Investigation of Low-Temperature Epitaxial Regrowth of Ion-Implanted Amorphous GaAs
68.48; 61.16; 61.80; S7.12
(a) Vol.111,No.1, 59-70

V.P. GOLTSEV, V.V. KHODASEVICH, G. GOETZ, I.I. PRIKHODKO, V.M. DRAKO, and V.V. UGLOV
Study of Ion Bombardment of Double-Layer Thin Zr/Ni Films
61.80 and 68.48; S1.2
(a) Vol.111,No.1, K13-K16

L.B. GUOBA, A.A. VITKAUSKAS, J.V. KAMENECKAS, V.R. SARGUNAS, and A.P. SALALAS
The Influence of Ion-Irradiation on Interdiffusion and Annealing Behavior of Ohmic Contacts to GaAs
68.48; 79.20; S7.12
(a) Vol.111,No.2, 507-514

M.A.Z. VASCONCELLOS, J.A.T. BORGES DA COSTA, W.H. SCHREINER, and I.J.R. BAUMVOL
A Comparison of Thermal Annealing and Ion Beam Mixing in the Fe/Al Bilayer
68.48 and 68.55; S1.1; S4
(a) Vol.112,No.1, 105-114

D. HOEHL and H. BARTSCH
Results of In-Situ Annealing of Ion-Implanted Silicon in a High Voltage Electron Microsope
61.70 and 61.80; 61.14; 68.48; S5.11
(a) Vol.112.No.1, 419-424

E. HENSEL, H. SOMMER, P. KNOTHE, and E. RICHTER
Silicon Nitride Layers on Tool Steel Produced by Ion Beam Mixing and Ion Beam Assisted Deposition
68.48 and 68.55; S1.1; S7
(a) Vol.112,No.2, 533-540

H. BARTSCH, D. HOEHL, and P. WERNER
TEM Cross-Sectional Investigations of the Amorphous/Crystalline Phase Transition of Ion-Implanted Silicon
61.80 and 68.55; 68.48; S5.11
(a) Vol.112,No.2, 715-720

M. STRÓŻAK, M. SUBOTOWICZ, and P. MIKOŁAJCZAK
Formation of Unusual Alloys by Means of Ion Beam Mixing of the Pd/V Superlattice
61.55 and 61.80; 68.48; S1.2
(a) Vol.112,No.2, 739-742

I. BRYŁOWSKA, K. PAPROCKI, and R. ILKHAMOV
The RBS Study of Ion Induced Mixing in Sb-Si System
61.80 and 68.48; S1.61
(a) Vol.112,No.2, 743-746

G. GŁADYSZEWSKI, P. MIKOŁAJCZAK, and M. SUBOTOWICZ
Ion Beam Mixing in Bi-Sb Superlattices
61.55 and 61.80; 68.48; S4
(a) Vol.112,No.2, 753-756

E. FARRÉS, J. SUÑÉ, I. PLACENCIA, N. BARNIOL, and X. AYMERICH
$Si-SiO_2$ Interfacial Atomic Scale Roughness Caused by Inhomogeneous Thermal Oxidation
68.55 and 73.40; 68.48; S10.1
(a) Vol.113,No.1, 83-96

G. RAATZ and J. WOLTERSDORF
Structure of Metal Deposits on Ceramic Materials Studied in the Ni/MgO System
68.55; 68.48; S1.2; S10.1
(a) Vol.113,No.1, 131-142

V.V. BOLOTOV, V.M. EMEKSUZYAN, V.N. SPIRIDONOV, K. SCHMALZ, and M. TRAPP
Radiation-Induced Redistribution of Gold in the SiO_2-Si Structures
61.80 and 73.20; 68.48; S5.11
(a) Vol.113,No.2, 315-320

M.YU. GUTKIN, M. MILITZER, A.E. ROMANOV, and V.I. VLADIMIROV
Equilibrium Position of Misfit Dislocations
61.70 and 68.48
(a) Vol.113,No.2, 337-344

XIN-FAN HUANG, XI-MAO BAO, and JIAN CHEN
Hydrogen Passivation and CO_2 Laser Annealing of Interface between Ar^+ Laser Recrystallized Poly-Si and SiO_2 Substrate
68.48 and 73.40; S5.11
(a) Vol.113,No.2, 453-458

J. BUDINAVIČIUS, L. PRANEVIČIUS, and S. TAMULEVIČIUS
Peculiarities of the Ag-Si Interface Formed under Simultaneous Deposition and Ion Beam Irradiation
61.80 and 68.48; S5.11
(a) Vol.114,No.1, K25-K28

V. HEERA
A Critical Discussion of the Vacancy Diffusion Model of Ion Beam Induced Epitaxial Crystallization
68.48 and 68.55; S5.11
(a) Vol.114,No.2, 599-608

W. UNGER and W. POSSART
An XPS Investigation of the Interface between Thin Polymethylmethacrylate (PMMA) Films and Natively Oxidized Al
68.48 and 79.60; S4; S12
(a) Vol.114,No.2, K175-K178

K. LENKEIT, A.K. GUTAKOVSKII, YU.O. KANTER, R. FLAGMEYER, S.M. PINTUS, S.V. RUBANOV, A.A. FEDOROV, and V.P. POPOV
Study of InGaAs-GaAs Strained-Layer Superlattices by TEM and RBS Techniques
68.48; 61.80; S7.12; S7.15
(a) Vol.115,No.2, 413-426

H. HAEFKE and H. STENZEL
Cross-Section Decoration. A Method for Determination of Diffusion Coefficients
66.30 and 68.48; S9.11
(a) Vol.116,No.1, 39-46

K. MEINEL, O. LICHTENBERGER, and M. KLAUA
Growth, Stability, and Desorption of Ag Films on Ni(111)
68.55; 68.48; S1.3
(a) Vol.116,No.1, 47-54

P.B. BARNA, M. EL HITI, L. MALICSKÓ, and F.M. REICHA
On the Orientation Relationships of Decorating Gold Particles on Alkali Halides
68.48 and 68.55; S1.3; S9.11
(a) Vol.116,No.1, 55-66

D. HESSE and R. MATTHEIS
The Growth Behaviour of Epitaxial $NiSi_2$ Islands of A- and B-Types during the Reaction of Nickel Vapour with the Si(111) Surface
68.48 and 68.55; S1.61
(a) Vol.116,No.1, 67-80

P. WERNER, R. MATTHEIS, D. HESSE, R. HILLEBRAND, and J. HEYDENREICH
HREM Investigations of the $NiSi_2$/Si{111} Interfaces Bounding A-Type $NiSi_2$ Islands on Si(111)
68.48 and 68.55; S1.61
(a) Vol.116,No.1, 81-90

L. PRANEVIČIUS, V. SARGŪNAS, and G. ZUBAUSKAS
Component Autodiffusion into Protective Coatings in Ion-Implanted InP
66.30 and 68.48; S7.11
(a) Vol.116,No.2, K161-K164

68.55. Thin film growth, structure, and epitaxy

C.D. LOKHANDE and S.H. PAWAR
Electrodeposition of Thin Film Semiconductors(Review Article)
68.55; 73.40; 85; S8
(a) Vol.111,No.1, 17-40

M.S. MURALI SASTRY
Large Scale Coalescence in Terms of a New Agglomeration Rate
68.55; 73.60; S1.3; S4
(a) Vol.111,No.1, K37-K42

V.V. TOKAREV, V.E. BORISENKO, T.M. PYATKOVA, and D.I. ZAROVSKII
Argon Ion Implantation Influence on Phase Formation and Structure of Nickel Silicides
61.70 and 68.55; S1.61
(a) Vol.111,No.2, 499-506

U. PAL, S. SAHA, B.K. SAMANTARAY, H.D. BANERJEE, and A.K. CHOUDHURY
X-Ray Line Broadening and Electron Microscopic Studies on ZnTe Thin Films
68.55; S8.12
(a) Vol.111,No.2, 515-522

S.A. AL-KUHAIMI
Optical and Compositional Properties of $CuInSe_2$ Thin Films at High Temperature
68.55 and 78.65; S8.16
(a) Vol.111,No.2, 523-528

S.N. SAHU and A. BOURDILON
First Non-Aqueous Electrocodeposition in InP, In, and P Films
68.55; 78.65; S4; S5; S7.11
(a) Vol.111,No.2, K179-K184

A. SINGH
Two-Step Annealing for Cobalt Disilicide Formation with Lowest Sheet Resistance
68.55 and 73.60; S1.61
(a) Vol.111,No.2, K191-K194

M.A.Z. VASCONCELLOS, J.A.T. BORGES DA COSTA, W.H. SCHREINER, and I.J.R. BAUMVOL
A Comparison of Thermal Annealing and Ion Beam Mixing in the Fe/Al Bilayer
68.48 and 68.55; S1.1; S4
(a) Vol.112,No.1, 105-114

W. THIELEMANN, B. RHEINLAENDER, P. NEVERMANN, and G. OELGART
Determination of the Internal Quantum Yield and of the Injection Effiency of Red-Light Emitting (Al,Ga)As-DHS Diodes
73.40 and 78.60; 68.55; 78.55; S7.15
(a) Vol.112,No.1, 229-236

J. GYULAI
Some Recently Emerged Problems of Ion Implantation: A Review
61.80; 68.55; S5.11
(a) Vol.112,No.1, 237-244

J.TH. ZETTLER, J. BOLLMANN, K.H. HECKNER, and H.A. KLOSE
Radiation-Induced Defects in Silicon Due to Low-Energetic Ion-Assisted Deposition of Indium Tin Oxide
61.80 and 68.55; 73.40; S10.15
(a) Vol.112,No.1, 395-404

J. BASZYNSKI
Superconducting $YBa_2Cu_3O_{7-\delta}$ Textured Films on MgO Substrates
68.55 and 74.70; S10.15
(a) Vol.112,No.1, K39-K42

G.B. SMITH, D.R. MCKENZIE, and P.J. MARTIN
An XPS Study of Chemical Order in Hydrogenated Amorphous Silicon-Carbon Alloy Films
61.40 and 68.55; 78.70; S5; S5.11
(b) Vol.152,No.2, 475-480

G. WAGNER, V. GOTTSCHALCH, H. RHAN, and P. PAUFLER
Pseudomorphic Growth and Nucleation of Misfit Dislocations in the Epitaxial System (001) InP/$In_{1-x}Ga_xAs$. I. Pseudomorphic Growth ...
61.70 and 68.55; S7.11; S7.12; S7.15
(a) Vol.112,No.2, 519-532

E. HENSEL, H. SOMMER, P. KNOTHE, and E. RICHTER
Silicon Nitride Layers on Tool Steel Produced by Ion Beam Mixing and Ion Beam Assisted Deposition
68.48 and 68.55; S1.1; S7
(a) Vol.112,No.2, 533-540

WEN-CAI WANG and YU CHEN
Study on the Short-Range Order Structures of Fe-Ge Amorphous Thin Films by EXAFS
61.40 and 68.55; S1.1; S5.12
(a) Vol.112,No.2, 541-550

W. SCHARFF, K. HAMMER, B. EIBISCH, O. STENZEL, S. ROTH, and T. FRAUENHEIM
Deposition of Hyrdogenated Amorphous Carbon Films by a ECR Microwave Plasma
61.80 and 68.55; 73.60; 78.65; S5
(a) Vol.112,No.2, 667-678

H. BARTSCH, D. HOEHL, and P. WERNER
TEM Cross-Sectional Investigations of the Amorphous/Crystalline Phase Transition of Ion-Implanted Silicon
61.80 and 68.55; 68.48; S5.11
(a) Vol.112,No.2, 715-720

R. BANISCH, B. ŢILLACK, H.H. RICHTER, B. HUNGER, P. BARNA, V. SCHILLER, A. ÁDAM, and J. GYULAI
On the Characteristics of CMOS-Transistors in Thick SOI Films
73.40; 68.55; S5.11
(a) Vol.112,No.2, 721-726

D. WOLFF, H.D. GEILER, and G. GOETZ
Nucleation Controlled Ignition of Explosive Crystallization in SOI Structures
68.55
(a) Vol.112,No.2, 727-732

J. DANILOVICH, K. GAERTNER, G. GOETZ, A. NOVIKOV, and B. WEBER
Ion Implanted Buried Si_3N_4 Layers below Epitaxial $NiSi_2$ Layers
68.55; 61.80; S1.61
(a) Vol.112,No.2, 733-738

K. GAERTNER, G. GOETZ, C. KASCHNER, I. KASKO, A. WITZMANN, and A. ZENTGRAF
Laser Induced Silicide Formation
68.55 and 79.20; 64.60; S1.61
(a) Vol.112,No.2, 747-752

M. KULIK and J. ŻUK
Ellipsometric Investigation of the Kinetics of Native Oxide Growth on Aluminium and Argon Implanted GaAs
61.80 and 68.55; 78.20; S7.12
(a) Vol.112,No.2, 789-792

M. VATSKICHEVA, L. VATSKICHEV, J. MUCHA, and M. KOLENDA
On Some Structural Properties of $(Ni_{80}Fe_{20})_{100-x}Ge_x$ Evaported Thin Films
61.50 and 68.55; S1.61
(a) Vol.112,No.2, K57-K60

G. WAGNER, V. GOTTSCHALCH, H. RHAN, and P. PAUFLER
Pseudomorphic Growth and Nucleation of Misfit Dislocations in the Epitaxial System (001) $InP/In_{1-x}Ga_xAs$. II. Critical Thickness and ...
61.70 and 68.55; S7.11; S7.15
(a) Vol.113,No.1, 71-82

E. FARRÉS, J. SUÑÉ, I. PLACENCIA, N. BARNIOL, and X. AYMERICH
$Si-SiO_2$ Interfacial Atomic Scale Roughness Caused by Inhomogeneous Thermal Oxidation
68.55 and 73.40; 68.48; S10.1
(a) Vol.113,No.1, 83-96

M.E. ABU-ZEID, Y.A. YOUSSEF, A.E. RAKHSHANI, and J. VARGHESE
Investigation of Cu_2O Thin Film Using Thermal Interferometry
68.55; 78.65; S10
(a) Vol.113,No.1, 123-130

G. RAATZ and J. WOLTERSDORF
Structure of Metal Deposits on Ceramic Materials Studied in the Ni/MgO System
68.55; 68.48; S1.2; S10.1
(a) Vol.113,No.1, 131-142

K. SCHMIDT, L. BAUCH, G. KLUGE, and P. SUEPTITZ
Preparation and Properties of Spin Coated Amorphous $GeSe_2$ Layers
68.55; 78.65; S8
(a) Vol.113,No.1, K55-K60

N.M. FICZA and A.L. TOTH
The Effect of Lead Impurity on Amorphous Si Thin Film Transistors
61.70 and 85; 68.55; S5.11; S10.1
(a) Vol.113,No.1, K147-K148

T. DRUESEDAU, M. ECKLER, R. BINDEMANN, and H. FIEDLER
Enhancement of Photoconductivity in Magnetron-Sputtered a-Si:H Doped with Nitrogen
68.55 and 73.60; 72.40; 78.65; S5.11
(b) Vol.153,No.2, K119-K124

A.V. LATYSHEV, A.L. ASEEV, A.B. KRASILNIKOV, and S.I. STENIN
Initial Stages of Silicon Homoepitaxy Studied by in situ Reflection Electron Microscopy
68.55; 61.16; S5.11
(a) Vol.113,No.2, 421-430

N.N. FALEEV, L.I. FLAKS, and S.G. KONNIKOV
Double Crystal X-Ray Diffraction Characterization of Multilayer Heteroepitaxial Structures with Submicron Layers
68.55; 61.10; S7.15
(a) Vol.113,No.2, 431-438

K. PONNURAJU, P.R. VAYA, and J. SOBHANADRI
The Effect of Substrate Temperature on the Optical Properties of $Pb_{1-x}Sn_xTe$ Films Formed by Molecular Beam Epitaxy
78.65; 68.55; S8.15
(a) Vol.113,No.2, 477-482

M. ANWAR, C.A. HOGARTH, and K.A.K. LOTT
ESR Studies of MoO_3/In_2O_3 Amorphous Thin Films Deposited by Co-Evaporation
68.55 and 76.30; S10.15
(a) Vol.113,No.2, 483-488

R. FLAGMEYER, U. PIETSCH, H. RHAN, M. HOERICKE, and B. JENICHEN
Strain Relaxation in $In_xGa_{1-x}As/GaAs$ Heterostructures
68.55; S7.15
(a) Vol.113,No.2, K211-K216

M. GASGNIER
Rare Earth Compounds (Oxides, Sulfides ...)as Thin Films and Thin Crystals. Physico-Chemical Properties and Applications(Review Article)
68.55 and 73.40; 75.70; 78.65; S1.4; S1.63; S10
(a) Vol.114,No.1, 11-72

O.I. VASIN, A.E. KLIMOV, I.G. NEIZVESTNYI, and V.N. SHUMSKII
The Three-Layer Structure of $Pb_{0.8}Sn_{0.2}Te$ Epitaxial Films Prepared by MBE Technique on BaF_2
73.60; 68.55; S8.15
(a) Vol.114,No.1, 161-166

E. FARRÉS, J. SUÑÉ, I. PLACENCIA, N. BARNIOL, and X. AYMERICH
The Thermal Growth of Very Thin SiO_2 Films. A Diffusion-Controlled Process
68.55; S10.1
(a) Vol.114,No.1, 167-176

V.G. BESHENKOV, CH.V. KOPETSKII, and YU.A. SHIYANOV
Phase Composition of Thin Films as Revealed by Auger Depth Profiling
64.80 and 68.55; 79.20; S1.61
(a) Vol.114,No.1, 191-198

O.S. RAJORA and A.E. CURZON
The Use of Electron-Image Contrast to Measure the Composition and Thickness of Cu-Mo Thin Films
68.60; 68.55; S1.2
(a) Vol.114,No.1, 199-206

J. NOTHOLT and J. LIESE
The Influence of the Preparation Conditions on the Structure of Evaporated As_2Se_3 Layers
68.55; S8
(a) Vol.114,No.1, 207-214

M. ANWAR and C.A. HOGARTH
A Study of Optical Absorption in MoO_3/SiO Amorphous Thin Films
78.65; 68.55; S10.15
(a) Vol.114,No.1, 215-224

M. ANWAR, C.A. HOGARTH, and K.A.K. LOTT
Electron Spin Resonance Study of Thermally Evaporated Amorphous Thin Films of MoO_3/SiO
68.55 and 76.30; S10.15
(a) Vol.114,No.1, 225-232

A. HEILMANN, C. HAMANN, G. KAMPFRATH, and DO NGOC UAN
TEM Investigations on Plasma Polymer Silver Composite Films
68.55; S1.3; S12
(a) Vol.114,No.2, 551-558

I.N. ZAKHARCHENKO, E.S. NIKITIN, V.M. MUKHORTOV, YU.I. GOLOVKO, M.G. RADCHENKO, and V.P. DUDKEVICH
Peculiarities of Ferroelectric Phase Transition in (Ba,Sr)TiO_3 Heteroepitaxial Films Obtained by HF Cathode Sputtering
64.70 and 68.55; 77.80; S10.15
(a) Vol.114,No.2, 559-566

Š. CHROMIK, H. TOLLET, C. CREMERS, D. MACHAJDÍK, V. ŠMATKO, Y. BRUYNSERAEDE, and Š. BEŇAČKA
The Composition of High T_c Superconducting Y-Ba-Cu-O Thin Films Prepared by Vacuum Coevaporation
68.55 and 74.70; S10.15
(a) Vol.114,No.2, 567-572

E. HAJTO, P.J.S. EWEN, P.G. HILL, and A.E. OWEN
Spectroscopic Investigation of Spin-Deposited As-S Amorphous Films
78.65; 68.55; S8
(a) Vol.114,No.2, 587-598

V. HEERA
A Critical Discussion of the Vacancy Diffusion Model of Ion Beam Induced Epitaxial Crystallization
68.48 and 68.55; S5.11
(a) Vol.114,No.2, 599-608

J.W. TOMM, J. GRIESCHE, H. SCHMIDT, D. GENZOW, and J. SIEMON
Strain in Lead Salt Layers. A Luminescence Study
68.55; 78.55; S8
(a) Vol.114,No.2, 621-628

A.G. BUDREVICH, A.S. LYUTOVICH, and B.L. OKSENGENDLER
Radiation Sieve Effect during Epitaxy from Low Energy Ion-Molecular Flow
68.55; 61.80; S5.11
(a) Vol.114,No.2, K163-K166

A. BROZI
A New Computer Model of the MOCVD Process
68.55; S7.12
(a) Vol.114,No.2, K167-K170

M.S. MURALI SASTRY and MANJUNATHA PATTABI
Variation of the Tunnelling Barrier in Island Cu Films
73.60; 68.55; S1.3
(a) Vol.114,No.2, K179-K182

H. WEINERT, K. JARAŠIŪNAS, and B. SCHUMANN
Carrier Diffusion and Recombination in Type II GaAs/$Ca_xSr_{1-x}F_2$ Quantum Well Heterostructures Studied by the Transient Gating Technique
73.40; 68.55; S7.12; S9.15
(a) Vol.114,No.2, K183-K188

H.-J. DIETZE and S. BECKER
Laser-Induced Plasma Deposition of Superconducting Bi-Sr-Ca-Cu-O Thin Films on Silicon
68.55 and 74.70; S10.15
(a) Vol.114,No.2, K193-K196

B.I. BAYRACHNYI, L.V. LIASHOK, V.P. GOMOZOV, and L.I. SKATKOV
Changes in Structure and Properties of Nb_2O_5 Anodic Films Caused by Generating Anion Defects on Their Surface
68.55; 73.60; S10
(a) Vol.115,No.1, 207-212

S. BAUNACK and A. ZEHE
A Study of UV/Ozone Cleaning Procedure for Silicon Surfaces
68.20; 68.55; 79.20; S5.11
(a) Vol.115,No.1, 223-228

Z. ŠIMŠA and P. GOERNERT
Preparation and Faraday Rotation Spectra of YIG:Pb,Pt Garnet Films
68.55 and 78.20; S11.2
(a) Vol.115,No.1, 229-236

B.B. ISMAIL and R.D. GOULD
Structural and Electronic Properties of Evaporated Thin Films of Cadmium Telluride
68.55 and 73.60; S8.13
(a) Vol.115,No.1, 237-246

L.N. ALEKSANDROV, V.YU. BALANDIN, A.V. DVURECHENSKII, and O.A. KULYASOVA
A Two-Dimensional Model of SOI Structure Crystallization by Pulse Nanosecond Heating
68.55 and 79.20; S5.11; S10.1
(a) Vol.115,No.1, K23-K26

G.I. ZHOVNIR, S.V. KLETSKII, and N.V. SOCHINSKII
Computer Simulation of the Inverse Stefan Problem in Conformity with the Liquid Phase Epitaxy of $Hg_{1-x}Mn_xTe$
68.55; S8.15
(a) Vol.115,No.1, K31-K34

B.C. CHAKRAVARTY, S.K. SHARMA, S.N. SINGH, B.K. DAS, RAVI KUMAR, and B.R. CHAKRABORTY
Effect of the Growth of Thin SiO Interlayers on the Performance of Si Solar Cell
68.55 and 85; 73.40; S5.11
(a) Vol.115,No.1, K125-K128

A.A. ZAHAB, M. ABD-LEFDIL, and M. CADENE
Effect of Annealing on Structure and Electrical Conductivity of Evaporated Indium Telluride
68.55 and 73.60; S8
(a) Vol.115,No.2, 491-496

H. LANGE, H. REIS, and F. FENSKE
AES Study of the Interaction of Ni and Al Overlayers with W-Ti and W-Re Diffusion Barriers
68.55 and 79.20; S1.2
(a) Vol.115,No.2, 497-504

N.N. DINH, V.T. BICH, L.Q. MINH, and N.T. TOAN
Preparation of Lu-Diphthalocyanine Thin Films by Electron Beam Deposition and Electrochromic Effect
68.55 and 78.65; S12
(a) Vol.115,No.2, K163-K166

G. GŁADYSZEWSKI, A. SMAL, Z. MITURA, and M. SUBOTOWICZ
Formation of Superconducting Bi-Sr-Ca-Cu-O Thin Films by Flash Evaporation
68.55 and 74.70; S10.15
(a) Vol.115,No.2, K207-K210

K. MEINEL, O. LICHTENBERGER, and M. KLAUA
Growth, Stability, and Desorption of Ag Films on Ni(111)
68.55; 68.48; S1.3
(a) Vol.116,No.1, 47-54

P.B. BARNA, M. EL HITI, L. MALICSKÓ, and F.M. REICHA
On the Orientation Relationships of Decorating Gold Particles on Alkali Halides
68.48 and 68.55; S1.3; S9.11
(a) Vol.116,No.1, 55-66

D. HESSE and R. MATTHEIS
The Growth Behaviour of Epitaxial $NiSi_2$ Islands of A- and B-Types during the Reaction of Nickel Vapour with the Si(111) Surface
68.48 and 68.55; S1.61
(a) Vol.116,No.1, 67-80

P. WERNER, R. MATTHEIS, D. HESSE, R. HILLEBRAND, and J. HEYDENREICH
HREM Investigations of the $NiSi_2$/Si{111} Interfaces Bounding A-Type $NiSi_2$ Islands on Si(111)
68.48 and 68.55; S1.61
(a) Vol.116,No.1, 81-90

R.B. BECK and B. MAJKUSIAK
The Initial Growth Rate of Thermal Silicon Oxide
68.55; S10
(a) Vol.116,No.1, 313-330

V.V. TOKAREV, V.E. BORISENKO, A.I. DEMCHENKO, and T.M. PYATKOVA
Growth of Epitaxial Nickel Disilicide during Rapid Thermal Processing of Argon-Implanted Nickel Films on Silicon
68.55; 61.70; S1.61; S5.11
(a) Vol.116,No.1, 331-336

R. MAMY and X. ZAOUI
Growth of Au on MoS_2 Studied by Photoemission
68.55 and 79.60; S8
(b) Vol.156,No.2, K109-K112

M. PESSA, H. ASONEN, J. VARRIO, and A. SALOKATVE
A Comparative Study of Molecular Beam Epitaxy Methods
(Review Article)
68.55; S7.11; S7.12
(a) Vol.116,No.2, 443-456

D.A. PAVLOV, A.F. KHOKHLOV, R.V. KUDRYAVTSEVA, and A.V. ERSHOV
Structure and Properties of Amorphous Silicon Doped with Isovalent Impurities
68.55 and 73.60; S5.11
(a) Vol.116,No.2, 697-702

C. HAMANN, R. LAIHO, and A. MRWA
Surface Analysis on Lead Phthalocyanine with STM
68.20 and 68.55; S12
(a) Vol.116,No.2, 729-734

A. THOMAS, A. ZEHE, B. SCHRECKENBACH, J.-W. ERBEN, and R. GROETZSCHEL
Ion Beam Lithography of Polycrystalline CaF_2 Films Deposited on Silicon
61.80 and 68.55; 77.55; S9.12
(a) Vol.116,No.2, 735-744

L.I. BEREZHINSKII, S.I. VLASKINA, V.E. RODIONOV, and H.A. SHAMURATOV
Initial Stages of the Epitaxial Silicon Carbide Film Growth on Silicon Substrates
68.55; S6
(a) Vol.116,No.2, K169-K172

68.60. Physical properties of thin films: nonelectronic

FANG-QING ZHANG and GUANG-HUA CHEN
Study of the Hydrogen Content and Profiles in a-Ge:H Films by Nuclear Resonant Reactor Analysis
61.70 and 68.60; S5.12
(b) Vol.154,No.1, K39-K42

O.S. RAJORA and A.E. CURZON
The Use of Electron-Image Contrast to Measure the Composition and Thickness of Cu-Mo Thin Films
68.60; 68.55; S1.2
(a) Vol.114,No.1, 199-206

GUANGHUA CHEN, FANGQING ZHANG, WEI JIA, and WEI CHEN
Study on Defects in Reactively Sputtered a-GeN_x:H Films
68.60 and 76.30; S6
(a) Vol.114,No.2, K171-K174

68.90. Other topics in the structure and nonelectronic properties of surfaces and thin films

M.YA. GAMARNIK and YU.YU. SIDORIN
Change of the Unit Cell Parameters in Highly Dispersed Gold, Silver, and Copper Powders
61.10 and 68.90; S1.3
(b) Vol.156,No.1, K1-K4

70. Condensed Matter: Electronic Structure, Electrical, Magnetic, and Optical Properties

71. Electronic States

(see also 63. Lattice dynamics,
73. Electronic structure and electrical properties of surfaces, interfaces, and thin films)

71.10. General theories and computational techniques

K. KOKKO
Determination of Electron Momentum Density from Electron Number Density. A New Version of Semiclassical Transformation
71.10; S2
(b) Vol.151,No.1, 139-144

C. ENGLER
Potential Surfaces for Chemisorption on Semiconductor Surfaces from Pair Potential Considerations
68.20; 71.10; 73.20; S5.11
(b) Vol.151,No.2, 555-564

N.J. CASTELLANI and W.R.L. LAMBRECHT
Core Hole Relaxation Energy in Semiconductors. Using a Pseudopotential Description of the Core Hole Perturbation
71.10; S5.11; S7.1; S8.1
(b) Vol.151,No.2, 565-570

R.M. KRISHNA, J.L. RAO, PREM CHAND, and S.V.J. LAKSHMAN
Electron Paramagnetic Resonance Study of Copper(II) Ions in Single Crystals of Thallium Magnesium Sullphate Hexahydrate
71.10 and 76.30; S11
(b) Vol.151,No.2, 615-622

E. HEINER
General Properties of Highly Excited Direct Semiconductors in the Athermal Stage
72.10; 71.10; 78.20
(b) Vol.151,No.2, 645-654

P. STREITENBERGER, P. ZIESCHE, and N. NAFARI
The Force Constant of a Bijellic Interface
71.10 and 73.40
(b) Vol.152,No.1, 111-116

J. MONECKE
Band Tails in the Density of States of Mixed Crystals
71.10; 71.20; 71.55
(b) Vol.152,No.1, 123-134

M. TROTT and CH. SCHNITTLER
Construction and Application of the Greens Function for Stepwise Constant Potentials
71.10; 71.20; 71.25; 71.50
(b) Vol.152,No.1, 153-162

P. FANTUCCI and G. PACCHIONI
Theoretical INDO Investigation of the Electronic Structure of Aluminum Clusters
71.10; 71.25; S4
(b) Vol.153,No.1, 193-206

YUAN LI and P.J. LIN-CHUNG
Semiempirical Construction of LCAO Parameters for II-VI Compound Semiconductors
71.10; 71.25; S8.1
(b) Vol.153,No.1, 215-224

M. TROTT and CH. SCHNITTLER
Extension of the Local Density Approximation to a New Class of Potentials
71.10; 71.50
(b) Vol.153,No.1, K39-K44

J. MONECKE
On the Correct Use of Perturbation Theory for the Calculation of One-Particle Green Functions
71.10; 63.10
(b) Vol.153,No.2, 623-632

R. PICKENHAIN, B. FLIETNER, and K. UNGER
The Quadratic Response of a Fermi Gas for Arbitrary Degree of Degeneracy
71.10
(b) Vol.154,No.1, 219-228

F.N. YOUSIF, R.C. GREENHOW, and J.A.D. MATTHEW
Pseudopotentials in the Presence of Core Holes
71.10
(b) Vol.154,No.1, K47-K50

B.I. CRAIG and P.V. SMITH
Determination of the Bulk Properties of Cubic SiC by the Periodic LUC-MINDO Method
71.10; 62.20; S1.61
(b) Vol.154,No.2, K127-K130

H. AOURAG, A. BELAIDI, T. KOBAYASI, R.N. WEST, and B. KHELIFA
Positron Annihilation in Si and Ge
71.10 and 78.70; S5.11; S5.12
(b) Vol.155,No.1, 191-200

J. MONECKE
Can Bound States be Obtained by Perturbation Theory?
71.10; 71.55
(b) Vol.155,No.2, K103-K104

T. KANEKO
Partial and Total Electronic Stoppings of Solids and Atoms for Energetic Ions
61.80; 71.10
(b) Vol.156,No.1, 49-64

V.V. KARYAGIN and I.I. LYAPILIN
Thermo-Galvanomagnetic Phenomena in Spin Quantum Wells
71.10 and 73.40; 72.20; S8.15
(b) Vol.156,No.1, 235-242

71.20. Electronic density of states determinations (including energy states of liquid semiconsuctors)
(see also 65.40. Electronic heat capacity)

M. LEVALOIS and G. ALLAIS
Etude par diffraction de R-X de la densité de charge de valence dans les deux semi-conducteurs tétraédriques $ZnSiAs_2$ et $ZnGeAs_2$
71.20; S7.16
(a) Vol.111,No.1, 211-222

I. MUEHLING and P. BLAUDECK
Multiple Scattering Approach to the Density of States and Electron Energy Loss Spectra of Amorphous Carbon
61.40 and 71.20; S5
(b) Vol.151,No.2, 545-554

R.B. PRABHU
Spectral Function for the Mix-Valence System
71.20
(b) Vol.151,No.2, K141-K144

J. MONECKE
Band Tails in the Density of States of Mixed Crystals
71.10; 71.20; 71.55
(b) Vol.152,No.1, 123-134

M. TROTT and CH. SCHNITTLER
Construction and Application of the Greens Function for Stepwise Constant Potentials
71.10; 71.20; 71.25; 71.50
(b) Vol.152,No.1, 153-162

A. HUPFER, D. HIRSCH, and S. SCHULZE
Photoemission on $A_3^{II}B_2^V$ Semiconductor Material
71.20 and 79.60; S7
(b) Vol.152,No.2, 505-518

F. LÓPEZ-AGUILAR, J. COSTA-QUINTANA, and S. BALLE
Hubbard Systems with Two Channels for the Localization
71.20 and 71.25; 71.50; S1.4
(b) Vol.152,No.2, 543-550

H. SOMMER, D.W. BULLETT, H. NEUMANN, and G. KUEHN
Valence Band Density of States in $LiInSe_2$
71.20; 79.60; S8.16
(b) Vol.152,No.2, K47-K50

B.S. ACHARYA, S.K. SINGH, and B.B. NAYAK
X-Ray Emission Spectroscopy of the Electronic States in $Pb_{1-x}Hg_xS$ Thin Films
71.20 and 73.60; 78.70; S8.15
(b) Vol.153,No.1, 153-160

A. JEZIERSKI
Density of States in Transition Metal-Metalloid Amorphous Alloys
71.20; S1.2
(b) Vol.153,No.1, 185-192

K. KOKKO, E. OJALA, and K. MANSIKKA
Fermi Level Density of States in Ag-Au Alloys. Nuclear Magnetic Spin-Lattice Relaxation Rate and Low-Temperature Specific Heat
71.20; 65.40; 76.60; S1.3
(b) Vol.153,No.1, 235-242

YANG-FANG CHEN
Comment on the Determination of the Midgap Density of States in Amorphous Semiconductors Using the Space-Charge-Limited-Current ...
71.20 and 72.20; S5.11
(b) Vol.153,No.2, 695-702

V.E. EGORUSHKIN and V.B. LAPIN
Electronic Structure of Disordered TiNi and Pseudobinary $Ti(Ni_{0.8}Cu_{0.2})$
71.20 and 71.25; S1.2
(b) Vol.154,No.1, K43-K46

S.R. HOFFMAN and L.M. SCARFONE
A Two-Band Model of Substitutionally Disordered Ternary Metal Alloys in the Coherent Potential Approximation
71.20; 72.15; S1.2; S1.3
(b) Vol.155,No.1, K33-K38

O.A. GOLIKOVA and A.M. GREKHOV
Structural Network and Some Photoelectrical and Optical Parameters of Amorphous Hydrogenated Silicon
71.20; 73.60; 78.65; S5.11
(b) Vol.155,No.2, 475-480

G.P. NIZHNIKOVA and O.V. FARBEROVICH
Theoretical Investigation of XPS, UPS, and BIS in Rare-Earth Compounds - Thulium Monochalcogenides
71.20; 71.50; 79.60; S1.63
(b) Vol.155,No.2, 491-500

EN-GE WANG, LI-YUAN ZHANG, and HUAI-YU WANG
Electronic Properties of Vacancy-Induced Structures in AlAs/GaAs Superlattices
71.55; 71.20; 73.20; S7.12
(b) Vol.155,No.2, 587-596

A. CONAN, A. BONNET, and M. MORSLI
Electrical and Structural Properties of the Tungsten Bronze $Bi_4W_{20}O_{62}$
72.20; 61.60; 71.20; S10
(b) Vol.155,No.2, 609-614

N.S. SULLIVAN and Y. LIN
Low-Energy Excitations in the Orientational Glass State of Solid Hydrogen
71.20; 76.60; S5.2
(b) Vol.155,No.2, K109-K112

V.G. TSIRELSON, V.A. STRELTSOV, R.P. OZEROV, and K. YVON
Electron Density Distribution in 3d-Metal Sesquioxides
71.20; 61.55; S10
(a) Vol.115,No.2, 515-522

R. ENDERLEIN, T. HOLZ, and J.L. GONDAR
The Quantum Well in an Electric Field. A Density of States Approach
71.20
(b) Vol.156,No.1, 259-274

I.V. BAZHIN, A.B. KOLPACHEV, and I.YA. NIKIFOROV
Electronic Energy Structure of Substoichiometric Molybdenum Carbides in Different Crystallographic Modifications
71.20; S1.61
(b) Vol.156,No.1, 309-318

R. KUENTZLER, E. MURRAY, Y. DOSSMANN, and D. AST
Electronic Properties of AlMn and AlMnSi Quasicrystals
75.40; 65.40; 71.20; S1.2
(a) Vol.116,No.1, 399-404

B. RADJENOVIC and D. TJAPKIN
The Influence of Hybridization of the Impurity and the Conduction Bands on the Density of States of Doped Semiconductors
71.20; 71.30; S5.11
(b) Vol.156,No.2, 487-496

H. AOURAG, A. BELAIDI, T. KOBAYASI, R.N. WEST, and B. KHELIFA
Electron and Positron Distribution for the Plane (1$\bar{1}$0) in Si and GaAs
71.20 and 71.60; S5.11; S7.12
(b) Vol.156,No.2, 497-506

K. KOKKO
Effects of Inhomogeneity in the Calculation of the Electron Momentum Density from the Electron Number Density
71.20
(b) Vol.156,No.2, K119-K122

71.25. Nonlocalized single-particle electronic states

J. SABÍN DEL VALLE, J. LÓPEZ GONDAR, and M. DE DIOS LEYVA
Landau Levels and Cyclotron Masses in GaAs-AlGaAs Quantum Wells
71.25; S7.12; S7.15
(b) Vol.151,No.1, 127-132

J.-P. ZOELLNER and G. PAASCH
Conditions for the Occurence of the Electric Quantum Limit in Inversion Layers
71.25 and 73.40; S5.11; S5.12; S7.1
(b) Vol.151,No.1, 145-158

S. BISWAS and R. BHATTACHARYA
Two Valence Subbands in Single Crystals of Bismuth Telluride Doped with Lead and Its Electrical Properties
72.20; 71.25; S8
(b) Vol.151,No.1, 193-202

R.A. EVARESTOV, V.P. SMIRNOV, and S.A. EGOROV
Band Corepresentations of Magnetic Space Groups
75.10; 71.25; 71.50; S11
(b) Vol.151,No.1, 275-282

SH.M. EFENDIEV, A.M. MAMEDOV, and N.G. DARVISHOV
VUV Reflection Spectra of $PbWO_4$ Single Crystals
78.40; 71.25; S11.1
(b) Vol.151,No.1, K87-K90

E. SCHULZ, G. DRAEGER, W. CZOLBE, and O. BRUEMMER
X-Ray Spectroscopic Investigations of Titanium Oxides and Basic Materials for Ferroelectric Ceramics
78.70; 71.25; S10
(a) Vol.111,No.1, 361-366

V.P. ZHUKOV, N.I. MEDVEDEVA, and V.A. GUBANOV
The Electronic Band Structure and Non-Empirical Calculations of Cohesive Properties of Refractory Compounds (Review Article)
71.25; S1.61; S1.62
(b) Vol.151,No.2, 407-440

I.I. GEGUSIN, V.N. DATSYUK, A.D. YUMATOFF, and R.V. VEDRINSKII
Atomic and Solid State Effects in X-Ray Absorption Spectra of Ionic Crystals
71.25 and 78.70; S9.11
(b) Vol.151,No.2, 581-590

M. TROTT, S. TROTT, and CH. SCHNITTLER
Normalization, Orthogonality, and Completeness for the Finite Step Potential
71.25 and 73.90
(b) Vol.151,No.2, K123-K128

K.P. GHATAK and A. GHOSAL
On the Effective Electron Mass in n-Channel Inversion Layers on Ternary Semiconductors in the Presence of a Parallel ...
71.25 and 73.40; S8.15
(b) Vol.151,No.2, K135-K140

F.M. MOHAMMAD, B.K. SHARMA, H. SINGH, and B.L. AHUJA
Study of Electron Distribution in Rhodium by Compton Scattering Technique
71.25 and 78.70; S1
(b) Vol.152,No.1, 145-152

M. TROTT and CH. SCHNITTLER
Construction and Application of the Greens Function for Stepwise Constant Potentials
71.10; 71.20; 71.25; 71.50
(b) Vol.152,No.1, 153-162

V. DETCHEVA, M. METODIEVA, T. KHRISTOV, and B.D. KANDILAROV
One-Dimensional Dirac Equation Approach to the Effective Mass Concept in Diatomic Crystals
71.25
(b) Vol.152,No.1, 163-170

P.A. APANASEVICH, V.D. ASIMOVA, V.A. ZAPOROZHCHENKO, R.G. ZAPOROZHCHENKO, and I.V. PILIPOVICH
Energetic and Temporal Characteristics of Picosecond Pulse Self-Diffraction in Silicon
78.70; 71.25; S5.11
(b) Vol.152,No.1, 347-356

HONG CHEN and XIANG WU
Effective Mass Approximation for Tunneling States with Dissipation
71.25; 73.40
(b) Vol.152,No.2, 499-504

F. LÓPEZ-AGUILAR, J. COSTA-QUINTANA, and S. BALLE
Hubbard Systems with Two Channels for the Localization
71.20 and 71.25; 71.50; S1.4
(b) Vol.152,No.2, 543-550

J. MERTSCHING
Double Periodic Peierls Phases
61.65 and 71.25; S12.1
(b) Vol.152,No.2, 551-562

M. DEY(BOSE) and S. GHOSH
Amplificaion of Acoustic Waves in Magnetised Piezoelectric Semiconductors. Effect of Nonlinearity in Electron Effective Mass ...
72.50; 71.25; S7.13
(b) Vol.152,No.2, 625-632

R.A. EVARESTOV and V.P. SMIRNOV
Tensor Fields in Crystals and Band Representations of Space Groups
75.10; 71.25; 71.50; S11
(b) Vol.152,No.2, 633-638

N.A. GUSHCHINA, V.K. NIKULIN, and YU.N. TZAREV
On the Band Structure of Ultrathin Layer Semiconductor Superlattices
73.40; 71.25; S7.15
(b) Vol.153,No.1, 161-166

H.N. NAZARENO and J.C. GALLARDO
Bloch Oscillations of an Electron in a Crystal under the Action of an Electric Field
71.25; 78.20
(b) Vol.153,No.1, 179-184

P. FANTUCCI and G. PACCHIONI
Theoretical INDO Investigation of the Electronic Structure of Aluminum Clusters
71.10; 71.25; S4
(b) Vol.153,No.1, 193-206

YUAN LI and P.J. LIN-CHUNG
Semiempirical Construction of LCAO Parameters for II-VI Compound Semiconductors
71.10; 71.25; S8.1
(b) Vol.153,No.1, 215-224

L. WANG and R.F. O'CONNELL
Ideal Two-Dimensional Electron Gas in a Magnetic Field and at Non-Zero Temperatures. An Alternative Approach
71.25 and 75.10
(b) Vol.153,No.1, 343-360

M. ZAŁUŻNY and M. PIŁAT
On the Energy Spectrum of Size-Quantized Films in an Oriented Magnetic Field
71.25 and 73.60
(b) Vol.153,No.1, K21-K24

W. MAGNUS, C. SALA, and K. DE MEYER
An Application of Feynman's Variational Principle to the Calculation of Subband Energies and Wave Functions
71.25
(b) Vol.153,No.1, K31-K34

N.R. LOKHANDE, V.B. SAPRE, and C. MANDE
Energy Gap and Chemical Shift of X-Ray K Absorption Discontinuity of Arsenic in Its Compounds
71.25 and 78.70; S8; S10.1
(b) Vol.153,No.1, K35-K38

H. RUECKER, R. ENDERLEIN, and F. BECHSTEDT
Strain Effects on the Band Structure of $Si/Si_{1-x}Ge_x(001)$ Superlattices
71.25 and 73.60; S5.11; S5.12
(b) Vol.153,No.2, 595-610

J. SCHLOESSER and A. STAHL
A Symmetry-Adapted RPA: Dynamics of the Electronic Density Matrix in a Semiconductor
71.25 and 78.20; S8.11
(b) Vol.153,No.2, 773-786

M. ZAŁUZNY and W. OKULSKI
On Electron States of Quantum Wells with Anisotropic and Tilted Valleys in Magnetic Field
71.25 and 73.60; S8
(b) Vol.153,No.2, K115-K118

G. ALLAIS, A. HAIRIE, M. LEVALOIS, R. MADELON, and E. PAUMIER
La densité de charge de valence dans $ZnSiAs_2$ et $ZnGeAs_2$. Comparison des resultats de pseudo-potentieal avec l experience
71.25; S7.16
(a) Vol.113,No.2, 489-496

C. PRASAD and M. SAHAY
Electronic Structure and Properties of Boron Phosphide and Boron Arsenide
71.25; S7.1
(b) Vol.154,No.1, 201-208

DOAN NHAT QUANG
On the Theory of Auger Recombination in Disordered Semiconductors. A Green's Function Approach
71.25 and 72.20; S7.13
(b) Vol.154,No.1, 237-248

V.V. SOBOLEV, A.I. KOZLOV, YU.I. POLYGALOV, V.E. TUPITSYN, and A.S. POPLAVNOI
Reflectivity Spectra and Band Structure of the Zinc and Cadmium Diphosphides
71.25 and 78.40; 71.35; S7
(b) Vol.154,No.1, 377-388

V.E. EGORUSHKIN and V.B. LAPIN
Electronic Structure of Disordered TiNi and Pseudobinary $Ti(Ni_{0.8}Cu_{0.2})$
71.20 and 71.25; S1.2
(b) Vol.154,No.1, K43-K46

M. SINGH and S.Y. TANG
Magneto-Optics of Holes in the Presence of Internal Strains and Temperature in Single Quantum Well Structures
71.25 and 72.10; 78.20
(b) Vol.154,No.2, 589-602

A.G. ARESHKIN, L.G. SUSLINA, and D.L. FEDOROV
On Concentration Dependences of the Band Parameters in $A^{II}B^{VI}$ Solid Solutions with Structural Phase Transitions
64.70 and 71.25; 71.35; 78.40; S8.15
(b) Vol.154,No.2, 609-622

L.M. SANDRATSKII and P.G. GULETSKII
Comparative Analysis of Tendencies in the Temperature-Induced Changes in Electronic Structure of Iron and Nickel
71.25; 75.50; S1.1; S1.2
(b) Vol.154,No.2, 623-636

V.S. STEPANYUK, A. SZÁSZ, B.L. GRIGORENKO, O.V. FARBEROVICH, and A.A. KATSNELSON
Electronic Structure and Optical Properties of MgO: Band Structure Calculation and Cluster Model
71.25; 78.20; S10.1
(b) Vol.155,No.1, 179-184

V.S. STEPANYUK, A. SZÁSZ, O.V. FARBEROVICH, A.A. GRIGORENKO, A.V. KOZLOV, and V.V. MIKHAILIN
An Electronic Band Structure Calculation and the Optical Properties of Alkaline-Earth Sulphides
71.25; 78.20; S8
(b) Vol.155,No.1, 215-220

E. FLACH, H.-E. GUMLICH, CH. JUNG, and M. KRAUSE
Reflectance Spectroscopy on $Zn_{1-x}Mn_xS$ in the Energy Range from 11 to 20 eV by Use of Synchrotron Radiation
78.40; 71.25; 71.35; S8.15
(b) Vol.155,No.1, 317-328

B. MITRA, A. GHOSHAL, and K.P. GHATAK
On the Electric Susceptibility Mass in Bi under Arbitrary Magnetic Quantization
71.25; S4
(b) Vol.155,No.1, K23-K28

L.A. MANAKOVA and S.P. IONOV
Correlation Mechanism to Change Lattice Parameters and Phase Transitions in Compounds with Intermediate Valency
64.60; 71.25; S8.15
(b) Vol.155,No.2, 397-406

V.K. ADAMCHUK, A.E. KUCHMA, and S.L. MOLODTSOV
Effects of Low-Energy Electron Transport in Insulators
71.25; 79.60; S9.11
(b) Vol.155,No.2, 525-534

A. DARGYS
Hole Velocity in an Electric Field for Valence Bands Degenerate at k=0
72.10; 71.25
(b) Vol.155,No.2, 615-626

W. OKULSKI and M. ZAŁUŻNY
Intersubband Raman Scattering in Size-Quantized Narrow-Gap Semiconductor Films
71.25 and 78.30
(b) Vol.155,No.2, 709-716

V.I. GAVRILENKO, J. HUMLIČEK, N.I. KLYUI, and V.G. LITOVCHENKO
Electron Energy Structure and Optical Properties of Microcrystalline Silicon
78.20; 71.25; S5.11
(b) Vol.155,No.2, 723-732

L.M. SANDRATSKII
On a Peculiarity of the Optical Characteristics of Rare-Earth Metals
71.25; 78.20; S1.4
(b) Vol.155,No.2, K105-K108

V.E. BORISENKO, S.YU. NIKITIN, A.B. FILONOV, and A.L. PUSHKARCHUK
Electronic Property Simulation of the $YBa_2Cu_3O_7$ Films
71.25 and 74.70; S10.15
(b) Vol.155,No.2, K127-K130

S. NOOR MOHAMMAD and M. ABDUS SOBHAN
Doping Dependent Bandgap Narrowing in Silicon and Germanium
71.25; 71.55; S5.11; S5.12
(b) Vol.156,No.1, 287-302

E. DĘBOWSKA and J. RUDZINSKA-GIRULSKA
Densities of Interstitial Electrons in Pd(Ni)-Noble Metal Alloys
71.25; 78.70; S1.2; S1.3
(b) Vol.156,No.1, 303-308

K. KUMAZAKI, L. VIÑA, C. UMBACH, and M. CARDONA
Interband Critical Point Parameters Determined by Ellipsometry in $Cd_xHg_{1-x}Se$
71.25 and 78.20; 77.20; S8.15
(b) Vol.156,No.1, 371-376

Ľ. HRIVNÁK
Compositional Dependences of Band-Gap Energy and Electron and Light Hole Effective Masses in $In_xGa_{1-x}As$
71.25; S7.12; S7.15
(a) Vol.116,No.1, K73-K76

SH. EFENDIEV, V. GAVRYUSHIN, G. RACIUKAITIS, G. PUZONAS, A. KAZLAUSKAS, N. DARVISHOV, and S. SHAKHDGAN
Two-Photon Spectroscopy of $PbMoO_4$ Single Crystals. Indirect Interband Transitions
78.40; 71.25; S11.1
(b) Vol.156,No.2, 697-704

71.30. Metal-insulator transitions

V.A. KASIYAN, D.D. NEDEOGLO, A.V. SIMASHKEVICH, and I.N. TIMCHENKO
Conductivity and the Metal-Dielectric Transition in Compensated n-ZnSe
72.20; 71.30; 71.55; S8.12
(b) Vol.154,No.1, 287-296

A.B. FAZAKAS and R. PITIS
Non-Zero Momentum Pairing and Peierls Order in One Dimensional Systems
71.30; S12.1
(b) Vol.154,No.2, 637-642

V.A. KASIYAN, D.D. NEDEOGLO, A.V. SIMASHKEVICH, and I.N. TIMCHENKO
Critical Behaviour of n-ZnSe Parameters in the Vicinity of the Metal-Insulator Transition
72.20; 71.30; S8.12
(b) Vol.154,No.2, 691-702

G.A. MATVEEV, I.M. TSIDILKOVSKII, and A.T. LONCHAKOV
Piezomagnetoresistance Anisotropy of n-Ge in Strong Magnetic Fields near the Metal-Insulator Transition
71.30 and 72.20; S5.12
(b) Vol.154,No.2, 703-712

B.M. ANGHELOV
Energy Levels and Intensity Parameters of $LiTbP_4O_{12}$
71.30 and 78.20; S10
(b) Vol.155,No.2, K165-K168

R. KIEMEL, W. WISCHERT, and S. KEMMLER-SACK
Dependence of T_c on the Hole Concentration in the High-T_c Superconductors $Bi_{1.5}Pb_{0.5}Sr_2Ca_{1-x}Gd_xCu_2O_{8+z}$
74.70; 71.30; 72.20; S10.15
(b) Vol.156,No.1, 339-344

B. RADJENOVIC and D. TJAPKIN
The Influence of Hybridization of the Impurity and the Conduction Bands on the Density of States of Doped Semiconductors
71.20; 71.30; S5.11
(b) Vol.156,No.2, 487-496

71.35. Excitons and related phenomena (including electron-hole drops)

R. MARASAS and H.N. SPECTOR
Binding Energy of Two-Dimensional Excitons in Strong Magnetic Fields
71.35
(b) Vol.151,No.1, 133-138

C.P. HILTON, J.J. DAVIES, J.E. NICHOLLS, and O. GOEDE
Optically-Detected Magnetic Resonance of Excitons Bound at Te_2 Isoelectronic Centres in $CdS_{1-x}Te_x$
71.35 and 71.55; 76.90; 78.55; S8.11; S8.15
(b) Vol.151,No.1, 175-184

G. MIZUTANI and N. NAGASAWA
Exciton-Grating Induced Light Mixing at the Giant Two-Photon Resonances of Excitonic Molecules in CuBr
71.35 and 78.20; S9
(b) Vol.151,No.2, 591-598

V.A. BIBIK and N.A. DAVYDOVA
Transformation of Excitonic Spectra of PbI_2 Layer Crystals under Order-Disorder-Order Structural Transitions
71.35 and 78.20; S9
(b) Vol.151,No.2, 733-738

A. MANAR, M.J.M. GOMES, B. HOENERLAGE, and R. LEVY
Time Evolution of the Refractive Index of CuCl under Picosecond Pulsed Excitation
71.35 and 78.40; S9
(b) Vol.151,No.2, 747-758

C. GHEZZI, D. MARTIN, A. PARISINI, J.L. STAEHLI, and L. TARRICONE
Investigation of (AlGa)As/GaAs Quantum Wells by Optical and Phototransport Measurements
73.40; 71.35; 72.40; 78.55; S7.15
(b) Vol.152,No.1, 101-110

A. SZÁSZ and D.J. FABIAN
Subsurface Hydrogen Exciton-Like States in Solid Catalysts
68.45 and 73.20; 71.35; S1
(b) Vol.152,No.1, 117-122

C. TRALLERO-GINER and R. RIERA
Multiple LO-Phonon Raman Scattering and Hot Excitons
71.35 and 78.30; S9
(b) Vol.152,No.1, 357-368

S. DZHUMANOV and P.K. KHABIBULLAEV
The Coexistence of Unstable,Metastáble, and Separated Frenkel Pair Defects in Solids. III. Theory of the Non-Impact Mechanism for ...
61.70; 71.35; 71.38; S9.11
(b) Vol.152,No.2, 395-414

G. MIZUTANI and N. NAGASAWA
Exciton-Grating Induced Metamorphoses of the Degenerate Four-Wave Mixing Spectra in the Giant Two-Photon Resonances of the Excitonic ..
71.35 and 78.20; S9
(b) Vol.152,No.2, 489-498

G. IADONISI, F. BASSANI, and G. STRINATI
Polaronic Effects on Exciton States with Different Angular Momenta
71.35 and 71.38
(b) Vol.153,No.2, 611-622

N.I. GRIGORCHUK and L.G. GRECHKO
Lineshape of Light Absorption by Thermalized Frenkel Excitons under Band-to-Band Transitions
71.35 and 78.20
(b) Vol.153,No.2, 633-640

C. GOURDON and P. LAVALLARD
Exiton Transfer Between Localized States in $CdS_{1-x}Se_x$ Alloys
71.35 and 78.55; S8.15
(b) Vol.153,No.2, 641-652

P.E. MOZOL, N.A. SKUBENKO, P.A. SKUBENKO, YU.P. GNATENKO, E.A. SALKOV, and Z.D. KOVALYUK
Effect of Laser Irradiation on Low-Temperature Photoconductivity and Photoluminescence Spectra of Gallium Selenide
71.35 and 71.55; 72.40; 78.55; S8
(b) Vol.153,No.2, 667-674

S.S. CAMARGO,JR. and J.P. VON DER WEID
Triplet Molecular Excitations in Potassium Halide Cyanide Crystals. I. The CN^- Impurity Centre
71.35 and 78.60; S9.11
(b) Vol.153,No.2, 757-764

S.S. CAMARGO,JR. and J.P. VON DER WEID
Triplet Molecular Excitations in Potassium Halide Cyanide Crystals. II. The $K(CN)_xCl_{1-x}$ Mixed System
71.35 and 78.60; S9.15
(b) Vol.153,No.2, 765-772

B. ČABRIĆ and LJ. RISTOVSKI
Analysis of Elementary Excitations in a Crystal with Point Defects in the Inverse Space
71.35
(b) Vol.153,No.2, K125-K130

H. WEINERT, A. ŽUKAUSKAS, G. TAMULAITIS, and V. STEPANKEVIČIUS
Some Hints on the Influence of the Barrier on Excitonic Optical Properties of GaAs-GaAlAs Superlattices Obtained by ...
78.55; 71.35; 78.30; S7.12; S7.15
(b) Vol.153,No.2, K193-K198

M.I. SHMIGLYUK, P.I. BARDETSKII, and S.D. TIRON
Nonlinear Optical Nutation on Interexciton Transition in Cu_2O. An Analogue of Equidistant Three-Level System
71.35; 78.20; S10.1
(b) Vol.154,No.1, 209-218

V.V. SOBOLEV, A.I. KOZLOV, YU.I. POLYGALOV, V.E. TUPITSYN, and A.S. POPLAVNOI
Reflectivity Spectra and Band Structure of the Zinc and Cadmium Diphosphides
71.25 and 78.40; 71.35; S7
(b) Vol.154,No.1, 377-388

A.G. ARESHKIN, L.G. SUSLINA, and D.L. FEDOROV
On Concentration Dependences of the Band Parameters in $A^{II}B^{VI}$ Solid Solutions with Structural Phase Transitions
64.70 and 71.25; 71.35; 78.40; S8.15
(b) Vol.154,No.2, 609-622

A.A. O'CONNEL-BRONIN and A.B. TRESHCHALOV
Dynamics of Exciton Reflection Spectra on the Change of the Incidence Angle in NaI Crystals
78.40; 71.35; S9.11
(b) Vol.154,No.2, 763-768

A. BECKMANN, K. DETTMER, and F.R. KESSLER
Thermoabsorption Studies of the Indirect Exciton Bandgap of Si and GaP
78.30 and 78.40; 71.35; S5.11; S7.11
(b) Vol.154,No.2, K191-K196

E. FLACH, H.-E. GUMLICH, CH. JUNG, and M. KRAUSE
Reflectance Spectroscopy on $Zn_{1-x}Mn_xS$ in the Energy Range from 11 to 20 eV by Use of Synchrotron Radiation
78.40; 71.25; 71.35; S8.15
(b) Vol.155,No.1, 317-328

F.R. KESSLER and J. NIEDER
Impact Ionization of Indirect Excitons in Crossed Electric and Magnetic Fields in Germanium
71.35; 72.20; 72.40; S5.12
(b) Vol.156,No.1, K45-K50

V.P. GRIBKOVSKII, L.G. ZIMIN, S.V. GAPONENKO, I.E. MALINOVSKII, P.I. KUZNETSOV, and G.G. YAKUSHCHEVA
Nonlinear Laser Spectroscopy of Exciton Absorption in MOCVD Grown ZnSe Monocrystalline Films
71.35 and 78.65; S8.12
(b) Vol.156,No.2, 449-454

U. STUTENBAEUMER, H.-E. GUMLICH, and H. ZUBER
The Energy Transfer between Free Excitons, Bound Excitons, and the 3d-Shell of Mn^{2+} in $Zn_{1-x}Mn_xSe$ Mixed Crystals and s-d Type Exchange
71.35 and 71.70; 78.55; S8.15
(b) Vol.156,No.2, 561-568

71.36. Polaritons (including photon-phonon and photon-magnon interactions)

N. RAJ and D.R. TILLEY
Green Functions for Polariton Modes in Magnetic Superlattices
71.36; 71.45; 75.10
(b) Vol.152,No.1, 135-144

C.P. Chang, K.F. PAI, C.S. FANG, and W.S. TSE
Dispersion Curves of Phonon-Plasmon Coupled Polaritons in n-Type $Al_xGa_{1-x}As$
63.20; 71.36; 78.30; S7.15
(b) Vol.154,No.1, 135-142

G.F. GLINSKII and Z.G. KOINOV
Polariton Theory of Light Propagation in Crystals.
I. Connection between the Microscopic Theory of Exciton-Phonon ...
71.36
(b) Vol.155,No.2, 501-512

Z.G. KOINOV and G.F. GLINSKII
Polariton Theory of Light Propagation in Crystals.
II. Crystal Optics Approximation
71.36
(b) Vol.155,No.2, 513-524

YI-JIAN JIANG, LI-JI LIAO, GANG CHEN, and PENG-XIANG ZHANG
Generalized Formulae of Frequency Dispersion and Raman Scattering Cross Section of Phonon-Polaritons
63.20 and 78.30; 71.36
(b) Vol.156,No.1, 145-150

V.H. ARAKELIAN and L.A. BAGDASSARIAN
Theory of Surface Polariton Excitation by Charged Particles in a Layer with Account of Retardation
71.36 and 78.70
(b) Vol.156,No.1, 275-286

71.38. Polarons and electron-phonon interactions
(see also 63.20. Phonon-electron interactions in lattices)

P. NAGY and P. MARKOŠ
On the Nature of the Self-Trapping of Different Types of Polarons
71.38
(b) Vol.151,No.2, 571-580

V.D. LAKHNO and G.N. CHUEV
Local Phonons in Crystals with Electron Centres
63.20; 71.38; S9.11
(b) Vol.152,No.1, 79-88

S. DZHUMANOV and P.K. KHABIBULLAEV
The Coexistence of Unstable,Metastable, and Separated Frenkel Pair Defects in Solids. III. Theory of the Non-Impact Mechanism for ...
61.70; 71.35; 71.38; S9.11
(b) Vol.152,No.2, 395-414

R.S. EACHUS, M.T. OLM, R. JANES, and M.C.R. SYMONS
Observation by ESR of Electrons Localised at Intrinsic Shallow Traps in Silver Halide Systems
71.50 and 76.30; 71.38; 76.40; S9
(b) Vol.152,No.2, 583-592

V.A. OSIPOV, G. BOCHNACKA, V.K. FEDYANIN, and J. MALEK
Polarons in Trans-Polyacetylene and Polyyne Chains
71.38; S12
(b) Vol.153,No.1, 207-214

V.V. ONOOCHIN and G.YU. YASHIN
On the Behaviour of the Large Polaron in Dispersive Media under an Applied Electric Field
71.38
(b) Vol.153,No.1, K25-K30

G. IADONISI, F. BASSANI, and G. STRINATI
Polaronic Effects on Exciton States with Different Angular Momenta
71.35 and 71.38
(b) Vol.153,No.2, 611-622

JIE LIU, DE-GANG ZHANG, and JIN YANG
Effects of Alternating Potential on the Dimerisation of Diatomic Peierls Systems
71.45; 71.38; S12.1
(b) Vol.153,No.2, K131-K134

W. LAI and P.N. BUTCHER
A Weak-Coupling Calculation of the Binding Energy and Effective Mass of a Polaron in a Multiple Quantum Well
71.38; 73.40
(b) Vol.154,No.1, 229-236

B. SUKUMAR and K. NAVANEETHAKRISHNAN
Shallow Donor Binding Energies in Semiconductors in a Magnetic Field
71.55; 71.38; S5.11; S7.11
(b) Vol.154,No.2, 643-648

CHUAN-YU CHEN, PEI-WAN JIN, and JIN-LIANG GAO
The Two-Dimensional Polaron in Polyatomic Crystals in a Magnetic Field
71.38
(b) Vol.154,No.2, K137-K142

D. LUPU
Hydrogen Diffusion and Radiationless Electron Transfer in Metal Hydrides
66.30; 71.38; 78.20; S1.2; S1.4; S3
(b) Vol.155,No.1, 65-72

O.V. SELYUGIN and M.A. SMONDYREV
Phase Transition and Padé Approximation for Froehlich Polarons
71.38
(b) Vol.155,No.1, 155-168

I.D. FERANCHUK, S.I. FISHER, and L.I. KOMAROV
Investigation of the Transition between Large- and Small-Radius Polarons in a One-Dimensional Crystal
71.38
(b) Vol.155,No.1, 169-178

A.P. POPOV and A.A. LAVRENTEV
The Rebuilding of the Electron Spectra in One-Dimensional Structures by Increasing the Electron-Phonon Binding
71.38; S12.1
(b) Vol.155,No.1, 185-190

R. MAŃKA and A. MOLAK
The Soliton Generalization of Bipolaronic Model Proposed for the High-Tc Superconductivity in Lanthanum and Yttrium Cupric Oxides
74.20; 71.38; S10
(b) Vol.155,No.1, 231-240

V.A. OSIPOV, G. BOCHNACKA, and V.K. FEDYANIN
Polarons in the Finite-Band Continuum Model of Diatomic Polymers
71.38; S12.1
(b) Vol.155,No.1, K29-K32

P. KLEINERT
Functional-Integral Treatment of the Disordered Tomonaga-Luttinger Model
71.38
(b) Vol.155,No.2, K99-K102

E.P. POKATILOV, S.I. BERIL, V.M. FOMIN, and G.YU. RIABUKHIN
Surface Polarons at the Contact of Two Polar Crystals
71.38 and 73.20
(b) Vol.156,No.1, 225-234

P. KLEINERT
Damping of Long-Wavelength Coupled Plasmon-Phonon Excitations
71.38
(b) Vol.156,No.1, K41-K44

P. ENDERS
Comment on: Effects of Alternating Potential on the Dimerisation of Diatomic Peierls Systems by Liu et al.(phys.stat.sol.(b)153,K131(1989)
71.45; 71.38; S12.1
(b) Vol.156,No.2, K123-K124

71.45. Collective effects

S. ICHINOSE
Two-Component Solitary Excitations in Molecular-Bonded Chains
63.20 and 71.45
(b) Vol.151,No.1, 95-100

E. GERLACH and S. KISTINGER
Plasma Excitation in Semiconductors Containing Extended Scattering Centers
72.20; 61.70; 71.45
(b) Vol.151,No.1, 241-248

G.M. FILIPPOV
Quasi-Bound States of Electrons in Solids in the Field of a Moving Ion
71.45; 71.50; 72.10
(b) Vol.151,No.1, K29-K34

N. RAJ and D.R. TILLEY
Green Functions for Polariton Modes in Magnetic Superlattices
71.36; 71.45; 75.10
(b) Vol.152,No.1, 135-144

A.E. BOCQUET, J.F. DOBSON, P.C. HEALY, S. MYHRA, and J.G. THOMPSON
X-Ray Photoelectron Spectra of High Tc Bismuth Compounds: General Trends for Layer Structures
71.45 and 79.60; 74.10; S11
(b) Vol.152,No.2, 519-532

M. KOHYAMA, R. YAMAMOTO, Y. EBATA, and M. KINOSHITA
Atomic Forces in the Self-Consistent Tight-Binding Model
61.50 and 71.45
(b) Vol.152,No.2, 533-542

M. HAASE and M. TAUT
Plasmon Dispersion in Trans-Polyacetylene
71.45; S12
(b) Vol.153,No.1, 225-234

I.YA. FUGOL, A.M. RATNER, YU.I. RYBALKO, and YU.L. STESHENKO
On the Mechanism of Vibrational Relaxation of Two-Centre Excitations in Solid Neon
63.90; 71.45; 78.55; S5.2
(b) Vol.153,No.2, 539-546

JIE LIU, DE-GANG ZHANG, and JIN YANG
Effects of Alternating Potential on the Dimerisation of Diatomic Peierls Systems
71.45; 71.38; S12.1
(b) Vol.153,No.2, K131-K134

K.S. SRIVASTAVA and A. TANDON
Dispersion Relation for Coupled Surface Waves in Thin Polar Semiconducting Foils
68.30; 71.45; S7.12
(b) Vol.155,No.1, 127-136

A.M. VAINRUB, G.S. ZAVT, and A. GRAJA
Absence of the Gap in the Optical Excitation Spectrum of TCNQ Salts with Methyl Derivatives of Pyridinium
71.45; 63.20; S12.1
(b) Vol.155,No.1, 147-154

YU.V. ALEKSANDROV, V.G. TSIRELSON, I.M. REZNIK, and R.P. OZEROV
The Crystal Electron Energy and Compton Profile Calculations from X-Ray Diffraction Data
71.45; 78.70
(b) Vol.155,No.1, 201-208

V. KUMAR and G.M.PRASAD
Electronic Properties of A_mB_n Complex Crystals
71.45; S9; S10
(b) Vol.155,No.1, 209-214

L.L. BALASHOVA, O.V. CHUMANOVA, and N.M. KABACHNIK
Z_2-Oscillations of the Proton Stopping Maximum in Gases and Solids
71.45 and 78.70; 79.20
(b) Vol.155,No.1, 289-294

A.A. KLYUKANOV, E.A. SENOKOSOV, and L.E. CHIBOTARU
The Shape of Multi-Plasmon Radiation Band in Direct Gap Semiconductors
71.45 and 78.55; S8.12
(b) Vol.155,No.1, 295-308

S. ICHINOSE
Two-Component Solitary Excitations in alpha-Helical Protein Molecules
64.10 and 71.45; S12
(b) Vol.155,No.2, 467-474

M. ZIMPEL and J. BARNAS
Plasmons in Randomly Layered Systems
71.45 and 71.70
(b) Vol.155,No.2, 581-586

V.N. ERMAKOV and S.P. KRUCHININ
One-Particle Excitation Spectrum in a Bisoliton Model of Superconductivity
71.45 and 74.20
(b) Vol.156,No.1, 333-338

H.K. SY
Plasma Oscillations at the Surface of a Type II Superlattice
71.45 and 73.20
(b) Vol.156,No.2, 455-460

A.I. MITSEK
Collective Excitations of Localized Electrons and Their Effect on Magnetic-Material Kinetics
71.45; 72.15; 75.80; S1.2; S1.4; S1.5
(b) Vol.156,No.2, 507-514

M. ROESLER and W. STOLZMANN
Electronic Specific Heat of Coulombic Systems
65.40 and 71.45
(b) Vol.156,No.2, 515-526

P. ENDERS
Comment on: Effects of Alternating Potential on the Dimerisation of Diatomic Peierls Systems by Liu et al.(phys.stat.sol.(b)153,K131(1989)
71.45; 71.38; S12.1
(b) Vol.156,No.2, K123-K124

71.50. Localized single-particle electron states (excluding impurities)

R.A. EVARESTOV, V.P. SMIRNOV, and S.A. EGOROV
Band Corepresentations of Magnetic Space Groups
75.10; 71.25; 71.50; S11
(b) Vol.151,No.1, 275-282

G.M. FILIPPOV
Quasi-Bound States of Electrons in Solids in the Field of a Moving Ion
71.45; 71.50; 72.10
(b) Vol.151,No.1, K29-K34

M. TROTT and CH. SCHNITTLER
Construction and Application of the Greens Function for Stepwise Constant Potentials
71.10; 71.20; 71.25; 71.50
(b) Vol.152,No.1, 153-162

P. MODRAK and R. WOJNECKI
Wannier Functions for d-Band of Ferromagnetic Nickel
71.50; S1.2
(b) Vol.152,No.1, 203-208

F. LÓPEZ-AGUILAR, J. COSTA-QUINTANA, and S. BALLE
Hubbard Systems with Two Channels for the Localization
71.20 and 71.25; 71.50; S1.4
(b) Vol.152,No.2, 543-550

R.S. EACHUS, M.T. OLM, R. JANES, and M.C.R. SYMONS
Observation by ESR of Electrons Localised at Intrinsic Shallow Traps in Silver Halide Systems
71.50 and 76.30; 71.38; 76.40; S9
(b) Vol.152,No.2, 583-592

R.A. EVARESTOV and V.P. SMIRNOV
Tensor Fields in Crystals and Band Representations of Space Groups
75.10; 71.25; 71.50; S11
(b) Vol.152,No.2, 633-638

M. TROTT and CH. SCHNITTLER
Extension of the Local Density Approximation to a New Class of Potentials
71.10; 71.50
(b) Vol.153,No.1, K39-K44

K. TOMPA, R. PIRLOT, L. GRENACS, and J. LEHMANN
Hyperfine Fields at Implanted ^{12}Boron in F.C.C. Metals
76.60; 71.50; S1; S1.3; S4
(b) Vol.153,No.2, 721-726

G.P. NIZHNIKOVA and O.V. FARBEROVICH
Theoretical Investigation of XPS, UPS, and BIS in Rare-Earth Compounds - Thulium Monochalcogenides
71.20; 71.50; 79.60; S1.63
(b) Vol.155,No.2, 491-500

71.55. Impurity and defect levels

A.A. BUGAI, I.N. KRAVTSOVA, V.M. MAKSIMENKO, B.D. SHANINA, and V.I. SHAKHOVTSOV
Properties of the Si-S1 Center in $Si_{1-x}Ge_x$ Solid Solutions
71.55; 76.30; S5.11; S5.12
(b) Vol.151,No.1, 159-166

C.P. HILTON, J.J. DAVIES, J.E. NICHOLLS, and O. GOEDE
Optically-Detected Magnetic Resonance of Excitons Bound at Te_2 Isoelectronic Centres in $CdS_{1-x}Te_x$
71.35 and 71.55; 76.90; 78.55; S8.11; S8.15
(b) Vol.151,No.1, 175-184

S.V. KOVESHNIKOV, E.B. YAKIMOV, N.A. YARYKIN, and V.A. YUNKIN
Plasma Stimulated Impurity Redistribution in Silicon
61.80 and 71.55; 73.40; S5.11
(a) Vol.111,No.1, 81-88

S.L. VINOGRADOV, O.K. MALINOVSKAYA, and V.A. TOLOKONNIKOV
Relaxation of Trapping Sites in ZnSe Films of Au-ZnSe-SiO_2-Si and Au-ZnSe-Ge Structures
71.55 and 73.30; S8.12
(a) Vol.111,No.1, 201-210

J. WIEHL, R. HIRRLE, W. WISCHERT, and S. KEMMLER-SACK
Cathodo- and Photoluminescence in the Mn^{2+} Activated Garnets $Cd_{3-x}Ca_xAl_2Ge_3O_{12}$
78.55 and 78.60; 71.55; S11.2
(a) Vol.111,No.1, 315-324

J. PASTRŇÁK, J. OSWALD, I. GREGORA, V. VORLÍČEK, and M. BABINSKÝ
Lateral Distribution of Si Related Defects and Carrier Density in GaAs Crystals
78.55; 71.55; 78.30; S7.12
(a) Vol.111,No.1, 345-360

MIN GONG and ZHI-PU YOU
The Deep Levels in Nickel-Doped Silicon
71.55; S5.11
(a) Vol.111,No.1, K49-K52

F. FIRSZT and H.L. OCZKOWSKI
Distribution of Traps in ZnSe:Cu Crystals Obtained by Fractional Glow Technique
71.55 and 78.60; S8.12
(a) Vol.111,No.1, K113-K118

R. SUBRAHMANYAM and Y.V.G.S. MURTI
Laser Induced F-F^- Conversions in KCl Crystals
61.70; 71.55; 78.50; S9.11
(b) Vol.151,No.2, 445-452

J.L. RAO, M.R. REDDY, and S.V.J. LAKSHMAN
Electronic Absorption Spectrum in Ni^{2+} Ions Doped in Zinc Thallium Sulphate Hexahydrate
71.55 and 78.50; S11
(b) Vol.151,No.2, 599-604

G. WEBER and L.E. OLIVEIRA
On the Binding Energies of D^- States in Doped Germanium. Effects of Valley-Orbit Corrections and Compressive Uniaxial Stress
71.55; S5.12
(b) Vol.151,No.2, 605-614

R. BALTRAMIEJŪNAS, V. GAVRYUSHIN, G. RAČIUKAITIS, V. KUBERTAVIČIUS, and A. KAZLAUSKAS
Two-Photon Absorption via Deep Local Levels as Virtual Intermediate States in Semiconductors
71.55 and 78.50; S8.12; S10.1
(b) Vol.151,No.2, 721-732

R. HIRRLE, J. WIEHL, W. WISCHERT, and S. KEMMLER-SACK
Cathodo- and Photoluminescence in the Tb^{3+} Activated Garnets of Type $Cd_{3-x}Ca_xAl_2Ge_3O_{12}$
78.55 and 78.60; 71.55; S11.2
(a) Vol.111,No.2, 629-638

K. SCHMALZ, F.-G. KIRSCHT, and J. REICHEL
On Quenching Induced Deep Levels in Czochralski p-Silicon
71.55; S5.11
(a) Vol.111,No.2, K195-K200

T. KUBOZOE and Y. WATANABE
Thermally Stimulated Exoelectron Emission from LiF Single Crystals Excited with Electrons and X-Rays and Subsequently with UV Light
71.55 and 79.75; S9.11
(a) Vol.111,No.2, K201-K204

J. MISIEWICZ, Z. GUMIENNY, R. CYWINSKI, and E. MUGENSKI
The Luminescence of Zn_3P_2 in the 0.84 to 1.1 μm Wavelength Range
78.55; 71.55; S7.1
(a) Vol.111,No.2, K249-K252

J. MONECKE
Band Tails in the Density of States of Mixed Crystals
71.10; 71.20; 71.55
(b) Vol.152,No.1, 123-134

P.S.S. PRASAD and S. RADHAKRISHNA
Chromium Centres in 3 $CdSO_4 \cdot 8\ H_2O$ Single Crystals
61.80 and 71.55; 78.50; S11.1
(b) Vol.152,No.1, 171-180

J. CALAS, S. ABDEL-MAKSOUD, and M. AVEROUS
Temperature Dependence of Carrier Trapping in Bulk Vitreous $GeSe_3$ from Transit Time Measurements
71.55 and 72.40; S8
(b) Vol.152,No.1, 209-216

M.G. OSTAPENKO and E.G. PETROV
The Donor-Acceptor Electron Transfer with Participation of Short Polymer Chains
72.80; 71.55; S12.1
(b) Vol.152,No.1, 239-248

R. CYWIŃSKI and M. MANFREDI
Aggregation Processes in the $NaCl:Eu^{2+}$ System Induced by the High-Temperature Annealing
78.20; 71.55; S9.11
(a) Vol.112,No.1, 215-220

L. JONIKAS, K. JARAŠIŪNAS, and J. VAITKUS
Study of Carrier Dynamics and Radiation Defects in Ion-Implanted Silicon by Transient Grating Technique
71.55 and 78.20; S5.11
(a) Vol.112,No.1, 375-380

K. MOJEJKO-KOTLIŃSKA, J. KOTLIŃSKI, and M. SUBOTOWICZ
Deep Traps in Doubly Implanted p-Type Silicon. DLTS and C-U Measurements
61.80; 71.55; S5.11
(a) Vol.112,No.1, 415-418

V.S. LYSENKO, A.N. NAZAROV, G.A. NAUMOVETS, V.B. POPOV, and A.S. TKACHENKO
Manifestation of Hydrogen in Al-SiO_2-Si Structures Subjected to a RF Plasma Annealing
61.70; 71.55; 73.40; S5.11
(a) Vol.112,No.1, K9-K12

S.R. BLIGHT and H. THOMAS
Observation of Bulk Traps in Ion-Implanted GaAs MESFETs
71.55; 85; S7.12
(a) Vol.112,No.1, K29-K34

G. LIIDJA, V. NAGIRNII, T. SOOVIK, and S. ZAZUBOVICH
New Off-Centre System: Jahn-Teller Mercury-Like Impurities in Alkali Halides
71.55 and 78.55; S9.11
(b) Vol.152,No.2, 563-576

V.YA. KOSTYUCHENKO, E.M. SKOK, and S.A. STUDENIKIN
Magneto-Resonant Absorption from the Local Center in Degenerate n-InSb
78.20; 71.55; S7.13
(b) Vol.152,No.2, K59-K62

V.V. PETROV and V.S. PROSOLOVICH
Thermal Defects in Silicon Doped with Rare-Earth Elements
61.70 and 71.55; 72.20; 78.30; S5.11
(a) Vol.112,No.2, 561-568

U. BECKER, H. ZIMMERMANN, P. RUDOLPH, and R. BOYN
Optical Study of Impurity Distribution in Vertical-Bridgman Grown CdTe Crystals
71.55; 64.75; 78.50; 78.55; S8.13
(a) Vol.112,No.2, 569-578

K.D. GLINCHUK, N.M. LITOVCHENKO, and V.YU. PTITSIN
Influence of Germanium on the Formation of Oxygen-Induced Recombination Centres in Heat-Treated Silicon
61.70; 71.55; S5.11
(a) Vol.112,No.2, K65-K70

M.J. PONNAMBALAM
Stark Effect and the F-Center Excited States
61.70 and 71.55; 78.20
(b) Vol.153,No.1, 79-88

SHENG-LI GUO and KE-QIN LU
Analysis of the Spectra of Mn^{2+} in Rhodonite
71.55 and 78.50; S11.1
(b) Vol.153,No.1, K45-K50

T. KAWAKUBO and M. OKADA
Optical and Electron Paramagnetic Resonance Studies on Neutron-Irradiated GaP Crystals
71.55; 76.30; 78.50; S7.11
(b) Vol.153,No.1, K93-K96

M.M. KAZANIN and V.V. KUTASOV
Fe Impurity in alpha- and beta-Boron and in Some Boron Compounds
71.55; 72.20; 76.70; S5
(a) Vol.113,No.1, 143-150

S.E. SARKISOV, K.K. PUKHOV, A.A. KAMINSKII, A.G. PETROSYAN, and T.I. BUTAEVA
Manifestation of Electron-Phonon Interaction in Insulating Crystals Doped with Pr^{3+} Ions
63.20 and 78.55; 71.55; S10.15
(a) Vol.113,No.1, 193-198

S. NOWAK, T. KACZOR, and J. ŻMIJA
Photochromic Phenomenon Kinetics
61.70 and 78.50; 71.55; S10.15
(a) Vol.113,No.1, 199-206

G.L. GUTSEV, G.S. MYAKENKAYA, V.V. FROLOV, and V.B. GLAZMAN
Nature of Hydrogen Bonding in Si-A Center
71.55; 61.70; S5.11
(b) Vol.153,No.2, 659-666

P.E. MOZOL, N.A. SKUBENKO, P.A. SKUBENKO, YU.P. GNATENKO, E.A. SALKOV, and Z.D. KOVALYUK
Effect of Laser Irradiation on Low-Temperature Photoconductivity and Photoluminescence Spectra of Gallium Selenide
71.35 and 71.55; 72.40; 78.55; S8
(b) Vol.153,No.2, 667-674

S.D. KOUIMTZI
Localized States in Crystalline Semiconductors
61.80 and 71.55; S5.11; S5.12; S7.1
(b) Vol.153,No.2, 675-682

S. EMURA and M. ISHIGURO
Effects of Hydrostatic Pressure on the Cu^{+} Center in Potassium Halides
71.55; 78.50; 78.55; S9.11
(b) Vol.153,No.2, 683-694

A.V. ZASTAVNOY and V.M. KOROL
Radiation Defects in n-Silicon Doped with Lithium and Sodium
61.70; 71.55; S5.11
(a) Vol.113,No.2, 277-284

K.D. GLINCHUK, N.S. ZAYATS, and A.V. PROKHOROVICH
Effect of Proton Bombardment on GaAs Luminescence
71.55 and 78.55; 61.80; S7.12
(a) Vol.113,No.2, 497-502

C.A. LONDOS
Defect States in Electron Bombarded n-Type Silicon
71.55; S5.11
(a) Vol.113,No.2, 503-510

L.L. KULYUK, N.F. PERELMAN, S.I. RADAUTSAN, A.V. SIMINEL, and E.E. STRUMBAN
About the Possibility of Double Centre Formation in InP:Fe:O
71.55 and 72.40; 76.30; 78.55; S7.11
(a) Vol.113,No.2, K227-K230

S.L. PINJARE, N. BALASUBRAMANYAM, and VIKRAM KUMAR
Photoluminescence at 0.944 eV from Heat Treated n-Type Silicon
78.55; 61.70; 71.55; S5.11
(a) Vol.113,No.2, K261-K264

M.O. MANARESH, D.W. FISCHER, and W.C. MITCHEL
The EL2 Defect in GaAs: Some Recent Developments (Review Article)
61.70 and 71.55; 72.40; 78.30; 78.50; 78.55; S7.12
(b) Vol.154,No.1, 11-42

V.A. KASIYAN, D.D. NEDEOGLO, A.V. SIMASHKEVICH, and I.N. TIMCHENKO
Conductivity and the Metal-Dielectric Transition in Compensated n-ZnSe
72.20; 71.30; 71.55; S8.12
(b) Vol.154,No.1, 287-296

N.T. BAGRAEV, I.S. POLOVTSEV, and K. SCHMALZ
An ONP Investigation of Deep Level Defects Produced by Heat Treatment at 600 and 800 °C in Cz-Si
61.70 and 71.55; 76.60; S5.11
(a) Vol.114,No.1, 233-246

G.A. ADEGBOYEGA, A. POGGI, E. SUSI, and M. REICHE
The Effects of High Temperature Anneal on the Electrical Activity of Iron in B-Doped Cz Silicon
71.55; 72.20; 78.30; S5.11
(a) Vol.114,No.1, 247-252

G. MICOCCI, A. RIZZO, P. SICILIANO, and A. TEPORE
Deep Level Transient Spectroscopy Measurements in InSe Single Crystals
71.55; S8
(a) Vol.114,No.1, 253-258

S.I. RADAUTSAN, I.M. TIGINYANU, V.N. FULGA, and YU.O. DERID
Quasi-Continuously Distributed Traps and Photoluminescence in $ZnGa_2Se_4$ Single Crystals
71.55; 78.55; S8.16
(a) Vol.114,No.1, 259-264

V.K. LUKYANYUK, M.V. TOVARNITSKII, and Z.D. KOVALYUK
Hall-Coefficient Sign Inversion, Electrical Conductivity, and Its Anisotropy in Highly Intercalated Na_xInSe Single Crystals
72.80; 71.55; S8.16
(a) Vol.114,No.1, 293-298

H. WIECZOREK and W. FUHS
Deep Trapping of Carriers in a-Si:H Solar Cells Studied by Transient Photocurrents
72.40 and 73.40; 71.55; S5.11
(a) Vol.114,No.1, 413-418

B. SUKUMAR and K. NAVANEETHAKRISHNAN
Shallow Donor Binding Energies in Semiconductors in a Magnetic Field
71.55; 71.38; S5.11; S7.11
(b) Vol.154,No.2, 643-648

YONG-LIANG WANG and H. TEICHLER
LCAO Recursion Approach for the Bound Electron States at the 90°-Partial Dislocation in Silicon
71.55; S5.11
(b) Vol.154,No.2, 649-660

E.R. LÓPEZ-TÉLLEZ, D. CÁRDENAS-GARCÍA, and C. RUÍZ-MEJÍA
F_A-Center Optical Absorption in CsF
71.55 and 78.50; S9.11
(b) Vol.154,No.2, K171-K174

L. TESAŘ and R. LENHARD
Evaluation of Silicon Wafers with Different Concentrations of Interstitial Oxygen
61.70 and 71.55; 73.40; S5.11
(a) Vol.114,No.2, 475-480

J. PROKEŠ and O. STEFAN,JR.
Determination of the Capture Cross Section from DLTS Measurements
71.55; S5.11
(a) Vol.114,No.2, 643-650

Y.N. MOHAPATRA and V. KUMAR
Temperature Dependence of Photocurrent in Undoped Semi-Insulating Gallium Arsenide
72.40; 71.55; S7.12
(a) Vol.114,No.2, 659-664

A.A. KAMINSKII, A. KHOLOV, P.V. KLEVTSOV, and S.KH. KHAFIZOV
Spectroscopy and Stimulated Emission of Nd^{3+}-Doped Tetragonal $NaBi(MoO_4)_2$ and $NaBi(WO_4)_2$ Disordered Crystals
78.45 and 78.55; 71.55; 78.50; S10
(a) Vol.114,No.2, 713-720

A.B. VASSILIKOU-DOVA, H. MATTHIES, K. RECKER, F. WALLRAFEN, and G. LEHMANN
EPR of Mn^{2+} in Ba_2ZnF_6 Single Crystals
76.30; 71.55; S9.16
(b) Vol.155,No.1, K59-K62

E.L. NAGAEV
Local Levels in Ferromagnetic Semiconductors near and above the Curie Point
71.55 and 75.30
(b) Vol.155,No.1, K71-K76

A. POEPPL and G. VOELKEL
ESR Investigation of the Oxygen Vacancy in Pure and Bi_2O_3-Doped ZnO Ceramics
71.55; 76.30; S10.1
(a) Vol.115,No.1, 247-256

G. OELGART, B. LIPPOLD, R. HEILMANN, H. NEUMANN, and B. JACOBS
Determination of the Germanium Acceptor Ionisation Energy of $Al_xGa_{1-x}As$ ($0<x<0.40$) by Hall Effect and Luminescence
71.55; 72.80; 78.55; S7.15
(a) Vol.115,No.1, 257-266

V.V. LITVINOV, G.V. PALCHIK, and V.I. URENEV
On the Radiation-Enhanched Mechanism of the Thermal Donor Formation Process in Germanium
61.70 and 71.55; S5.12
(a) Vol.115,No.1, K9-K12

S. SHIGETOMI, T. IKARI, H. NAKASHIMA, and H. OHKUBO
Deep Levels of Cd-Doped InSe
71.55; S8
(a) Vol.115,No.1, K43-K46

A.A. MIKHALENKO, V.M. KHULUGUROV, E.I. DMITRIEVA, and V.N. SALOMATOV
F^{2+} Centers in LiF-Mg Crystals
61.70 and 71.55; S9.11
(b) Vol.155,No.2, 359-364

A.V. BELOUSOV, V.A. KOVARSKII, V.N. CHEBOTAR, and T.V. PISHKOVA
Quantum Calculation of Multiphonon Capture Cross-Sections in the Model of Tangent Adiabatic Potentials
71.55; 63.20
(b) Vol.155,No.2, 535-540

A.F. KOVALENKO, E.N. SOVYAK, and M.F. GOLOVKO
Electronic States of a Hydrogen Atom near a Hard Wall
71.55 and 73.40
(b) Vol.155,No.2, 549-558

J. DEXPERT-GHYS, Y. CHARREIRE, P. ESTEBAN-PUGES, L. ALBERT, and H. DEXPERT
Optical and EXAFS Investigations of ZnS:Tm^{3+} Powders
71.55; 61.10; 78.55; S8.11
(b) Vol.155,No.2, 571-580

EN-GE WANG, LI-YUAN ZHANG, and HUAI-YU WANG
Electronic Properties of Vacancy-Induced Structures in AlAs/GaAs Superlattices
71.55; 71.20; 73.20; S7.12
(b) Vol.155,No.2, 587-596

J. KREISSL, W. ULRICI, and W. GEHLHOFF
Electron Paramagnetic Resonance Study of the Dynamical Jahn-Teller Effect on the $3d^1$ Impurities Ti^{3+} and V^{4+} in GaP
71.55; 76.30; S7.11
(b) Vol.155,No.2, 597-608

J. MONECKE
Can Bound States be Obtained by Perturbation Theory?
71.10; 71.55
(b) Vol.155,No.2, K103-K104

F.S. NASREDINOV, M.M. MEZDROGINA, P.P. SEREGIN, U.ZH. ABDUMANAPOV, and U.S. TURSUNOV
Nature of the Impurity States Arising from Transition Metals in Hydrogenated Amorphous Silicon
71.55; S5.11
(b) Vol.155,No.2, K113-K118

I.S. GORBAN, N.S. KORETS, TS.A. KRYSKOV, and M.V. CHUKICHEV
Cathodoluminescence in Doped CdP_2 and $CdSiP_2$ Crystals
78.60; 71.55; S7
(a) Vol.115,No.2, 555-560

E. GRILLI, M. GUZZI, and M. PASTORI
Time Resolved Photoluminescence of $CdIn_2S_4$ Single Crystals
78.55; 71.55; S8.16
(a) Vol.115,No.2, 599-606

S.I. RADAUTSAN, I.M. TIGINYANU, V.V. URSAKII, and N.B. PYSHNAYA
Investigation of Traps in InP:Zn Single Crystals Implanted by Phosphorus Ions
71.55; S7.11
(a) Vol.115,No.2, K191-K194

B.K. RATH and P.N. RAM
Gap Modes and Local Density of States of Impurities in RbCl and KBr
63.20 and 71.55; 63.50; S9.11
(b) Vol.156,No.1, 137-144

S. NOOR MOHAMMAD and M. ABDUS SOBHAN
Doping Dependent Bandgap Narrowing in Silicon and Germanium
71.25; 71.55; S5.11; S5.12
(b) Vol.156,No.1, 287-302

G.L. GUTSEV and G.S. MYAKENKAYA
The Acceptor States of Hydrogen in Silicon
71.55; S5.11
(b) Vol.156,No.1, 319-324

M. HOEHNE
EPR of Lithium-Induced Silicon-$5d^n$ Pairs in Silicon
71.55 and 76.30; S5.11
(b) Vol.156,No.1, 325-332

A. ZEHE and A. JUAREZ
Double-Peak Formation as a Doping-Superlattice Effect in Donor-Acceptor Pair-Luminescence Spectra
71.55 and 78.55; 73.40; S7.12
(b) Vol.156,No.1, K89-K92

A. OPANOWICZ
On the Kinetics Order of Thermoluminescence in Insulating Crystals
71.55 and 78.60
(a) Vol.116,No.1, 343-348

B. POHORYLES
Helium Pressure Effects on Photoconductivity and Position of Dislocation Deep Electron States in Plastically Deformed Ge
61.70 and 71.55; 72.40; S5.12
(a) Vol.116,No.1, 349-356

K. ŽĎÁNSKÝ and THAN BINH LE
Kinetics of DX Centers with Illumination. The Justification of the Band Structure Model
71.55; S7.15
(a) Vol.116,No.1, 357-366

P. KOŠTIAL, I. JAMNICKÝ, and J. ĎURČEK
Contribution to the Experimental Methods of Deep Level Acoustic Transient Spectroscopy
68.25 and 71.55; 72.50; S5.11; S10
(a) Vol.116,No.1, K77-K80

R.A. VARDANYAN and S.G. VESELKO
Dislocation Recharging in the Plastically Deformed Crystals of the A B -Type
61.70 and 71.55; 72.90; S8.1
(b) Vol.156,No.2, 527-536

P.V. CHERNOV, E.M. DIANOV, V.N. KARPECHEV, L.S. KORNIENKO, I.O. MOROZOVA, A.O. RYBALTOVSKII, V.O. SOKOLOV, and V.B. SULIMOV
Spectroscopic Manifestations of Self-Trapped Holes in Silica: Theory and Experiment
78.55; 71.55; S10.1
(b) Vol.156,No.2, 663-676

M.T. PRIMATAROWA
Kink Defections in Quantum Crystals
71.55
(b) Vol.156,No.2, K125-K128

XIANG-FENG JIN, BING MAO, SHU-QIN ZHOU, REN-YUAN QIAN, J.S. PARK, and P. EHRLICH
Carrier Transport Properties of Trans-Polyphenylacetylene
73.60; 71.55; S12
(a) Vol.116,No.2, 709-714

J. KANIEWSKI, M. KANIEWSKA, and M. JEŻEWSKI
Interpretation of DLTS Emission and Capture Data of Co Related Deep Level in GaP
71.55; S7.11
(a) Vol.116,No.2, 745-754

K. IRMSCHER and J. KREISSL
Electronic Characterization of the Manganese-Boron Pair in Silicon
71.55
(a) Vol.116,No.2, 755-768

V.A. GURTOV and P.A. RAIKERUS
Radiation Charge in Silicon Dioxide Considering Diffusion Currents
72.20; 71.55; 85; S5.11
(a) Vol.116,No.2, 769-776

71.60. Positron states
(see also 78.70. Positron annihilation)

S.Y. TANG and M. SINGH
Surface and Phonon Assisted Diffusion of Positrons in Semiconductors
71.60
(b) Vol.154,No.2, K131-K136

H. AOURAG, A. BELAIDI, T. KOBAYASI, R.N. WEST, and B. KHELIFA
Electron and Positron Distribution for the Plane $(1\bar{1}0)$ in Si and GaAs
71.20 and 71.60; S5.11; S7.12
(b) Vol.156,No.2, 497-506

71.70. Level splitting and interactions
(see also 75.10. ... magnetic phenomena,
75.30. Exchange and superexchange interactions,
73.20. Electronic surface states)

SANG-BO ZHAO, HUI-SU WANG, and JUN-KAI XIE
Coordination Symmetry and Energy Spectrum Analysis of Fe^{2+} in $ZnS:Fe^{2+}$ Crystals
71.70; S8.1
(b) Vol.151,No.1, 167-174

WEN-CHEN ZHENG
A Further Study of Tetragonal Distortion of the Cr^{3+}-V_c Centre in MgO from Optical and EPR Spectra
61.70 and 71.70; 76.30; S10.1
(b) Vol.151,No.1, K17-K22

WEN-CHEN ZHENG
A Study of the Spin-Hamiltonian Parameters of Isoelectronic Impurity Ions in the Same Host Crystal
71.70; 76.30; S9.12; S10; S10.1
(b) Vol.151,No.2, K145-K150

SANG-BO ZHAO and KANG-WEI ZHOU
Strong-Field Scheme for $C_{3v}(d^3)$ and the Spectra of Ruby and Cr_2O_3
71.70; 78.50; S10.1
(b) Vol.152,No.1, 181-190

L.K. AMINOV, A.V. VINOKUROV, I.N. KURKIN, YU.YU. KOSTETSKII, S.L. KORABLEVA, I.KH. SALIKHOV, S.I. SAKHAEVA, and M.A. TEPLOV
Distortion of a Crystal Field in the van Vleck Paramagnet $LiTmF_4$ under Activation by Isovalent Impurities
71.70; 76.30; 78.50; S9.16
(b) Vol.152,No.1, 191-202

KE-QIN LU, SHENG-LI GUO, and PING-FENG WU
An Interpretation of the Pressure Dependence of EPR Zero-Field Splitting of Mn^{2+} in $Na_2ZnCl_4 \cdot 3 H_2O$
71.70 and 76.30; S9.16
(b) Vol.152,No.2, 577-582

N. GUSKOS
Ground State of the Gd^{3+} Ion in MF_2 Single Crystals under the Influence of External Static Electric Field
71.70; 76.30; S9.12
(b) Vol.153,No.1, 243-248

MAO-LU DU and MIN-GUANG ZHAO
Gyromagnetic Factor and Spin-Orbit Coupling of Ligands in $CsVX_3$ (X=Cl,Br,I)
71.70; 76.30; S1.2
(b) Vol.153,No.1, 249-256

J.L. RAO, G.L. NARENDRA, B. SREEDHAR, and S.V.J. LAKSHMAN
Electron Spin Resonance and Optical Absorption Spectra of Mn^{2+} Ions in K_2SO_4-$ZnSO_4$ Glasses
71.70; 76.30; 78.50; S11
(b) Vol.153,No.1, 257-262

C. BENECKE, W. BUSSE, H.-E. GUMLICH, H. HOFFMANN, S. HOFFMANN, A. KROST, and H. WALDMANN
On the Excitation and Emission of Infrared Luminescence of Wide Band Gap Mn II-VI Semimagnetic Semiconductors
78.55; 71.70; S8.15
(b) Vol.153,No.1, 391-402

M. DIVIŠ, P. SVOBODA, Z. SMETANA, V. NEKVASIL, and A.V. ANDREEV
The Influence of Crystal-Field Splitting on the Paramagnetic Magnetisation of $TmCu_2$
71.70 and 75.60; S1.2
(b) Vol.153,No.1, K69-K72

J. GALIBERT, P. PERRIER, and S. ASKENAZY
Résonance magnétophonon avec retournement de spin dans l'antimoniure d'indium
63.20 and 71.70; 72.20; S7.13
(b) Vol.153,No.2, 653-658

WEN-CHEN ZHENG
Local Thermal Expansion around the Impurity in Chromium-doped alpha-$LiIO_3$ Crystals
65.70; 71.70; 76.30; S11.1
(b) Vol.153,No.2, K103-K108

SHENG-LI GUO, KE-QIN LIU, LIN-RUI PING, and GAO-CHAO LIU
Analysis of the EPR of Cr^{3+}-Vacancy Center MgO
71.70 and 76.30; S10.1
(b) Vol.153,No.2, K167-K170

YU.M. SEIDOV and A.M. SULEIMANOV
On the Theory of Raman Scattering by Surface Collective Excitations in Singlet-Triplet Ferromagnets
78.30; 71.70; 75.30; S1.62
(b) Vol.153,No.2, K203-K208

S.E. SARKISOV, A.A. KAMINSKII, and A.V. BUTASHIN
Optical Phonons and Lattice-Activator Interactions of Li^{3+} Ions in the Disordered Oxide Single Crystals with Ca-Gallogermanate ...
78.30; 71.70; S11.2
(a) Vol.113,No.2, 585-596

L.E. MISIAK, S.K. MISRA, and U. ORHUN
Study of Temperature Variation of EPR Linewidth of Gd^{3+}-Doped $LiYb_xY_{1-x}F_4$ Single Crystals. Estimation of Yb^{3+} Spin-Lattice
71.70 and 76.30; S9.16
(b) Vol.154,No.1, 249-258

S.K. MISRA and CHUNZHENG WANG
EPR and Optical-Absorption Studies of Cu^{2+}-Doped $Mg(CH_3COO)_2 \cdot 4\ H_2O$ Single Crystal. Core Polarization and Molecular-Orbital Coefficients
71.70; 76.30; 78.50; S12
(b) Vol.154,No.1, 259-272

WEN-CHEN ZHENG
Investigation of the Pressure and the Stress Dependence of Low-Temperature EPR Spectra for $ZnSiF_6 \cdot 6\ H_2O:V^{2+}$ Crystals
76.30; 71.70; S9.15
(b) Vol.154,No.1, K73-K78

R. LACROIX
Theoretical Considerations on the Splitting of the Ruby R_1 Line under Stress
71.70; 62.20; S10.1
(b) Vol.154,No.2, K143-K146

WEN-CHEN ZHENG
Theoretical Investigation of the Spin-Lattice Coupling Coefficients G_{11} and G_{44} for Cr^{3+} in Cubic $SrTiO_3$ Crystals
71.70 and 76.30; S11.1
(b) Vol.154,No.2, K167-K170

C.K. JAYASANKAR and F.S. RICHARDSON
Spin-Correlated Crystal-Field Analyses of the $4f^2(Pr^{3+})$ Energy Levels in $LiYF_4:Pr^{3+}$ and $LiBiF_4:Pr^{3+}$
71.70; S11.1
(b) Vol.155,No.1, 221-230

D. TALAPATRA and R.K. MUKHERJEE
Optical Studies of Electronic Transitions of Mn^{2+} Ion in $Na_2Mn_3Cl_8$ Single Crystal
71.70; 78.20; S9.16
(b) Vol.155,No.2, 541-548

C.K. JAYASANKAR, M.F. REID, and F.S. RICHARDSON
Comparative Crystal-Field Analyses of $4f^N$ Energy Levels in $LiYF_4:Ln^{3+}$ Systems
71.70; S9.16
(b) Vol.155,No.2, 559-570

M. ZIMPEL and J. BARNAS
Plasmons in Randomly Layered Systems
71.45 and 71.70
(b) Vol.155,No.2, 581-586

WEN-CHEN ZHENG
Theoretical Considerations on the Axial Crystal-Field Splitting and Its Dependence on the Bonding Angle Distortion in Tetragonal ...
71.70 and 76.30; S9.16
(b) Vol.155,No.2, K149-K152

M. TAIBI, J. ARIDE, E. ANTIC-FIDANCEV, M. LEMAITRE-BLAISE, and P. PORCHER
Crystal Field Parameters of Eu^{3+} and Nd^{3+} in Ln_2BaZnO_5 Phases (Ln=La,Nd,Y)
71.70; 78.50; 78.55; S10.15
(a) Vol.115,No.2, 523-532

H. DONKER, N. YAMASHITA, W.M.A. SMIT, and G. BLASSE
Luminescence Decay Times of the Sb^{3+}, Pb^{2+}, and Bi^{3+} Ions in Alkaline-Earth Sulfides
71.70; 78.55; S8.11
(b) Vol.156,No.2, 537-544

U. STUTENBAEUMER, H.-E. GUMLICH, and H. ZUBER
The Energy Transfer between Free Excitons, Bound Excitons, and the 3d-Shell of Mn^{2+} in $Zn_{1-x}Mn_xSe$ Mixed Crystals and s-d Type Exchange
71.35 and 71.70; 78.55; S8.15
(b) Vol.156,No.2, 561-568

72. Electronic Transport in Condensed Matter
(for surfaces, interfaces, and thin films, see 73)

72.10. Theory of electronic transport, scattering mechanism

S. KUMAR, P.K. AHLUWALIA, and K.C. SHARMA
Thermoelectric Power of Dilute Magnetic Alloys in the Presence of CEF and Exchange Scattering
72.10 and 72.15
(b) Vol.151,No.1, 225-230

J. HEINRICHS
Effect of a Finite Voltage on Statistical Properties of Resistance in One-Dimensional Disordered Systems
72.10
(b) Vol.151,No.1, 231-240

G.M. FILIPPOV
Quasi-Bound States of Electrons in Solids in the Field of a Moving Ion
71.45; 71.50; 72.10
(b) Vol.151,No.1, K29-K34

A.I. DMITRIEV and G.V. LASHKAREV
Nonthermal Spread of Quantum Levels in Solids Due to Defects
72.10; S8.15
(b) Vol.151,No.2, 633-644

E. HEINER
General Properties of Highly Excited Direct Semiconductors in the Athermal Stage
72.10; 71.10; 78.20
(b) Vol.151,No.2, 645-654

M. MOCKER, F. LEMKE, and P. SELBMANN
Direction Dependence of Auger Recombination in Low-Dimensional PbSnTe Structures
72.10; S8
(b) Vol.151,No.2, K151-K154

I.P. ZVYAGIN
Hopping Relaxation and Photoconductivity of Disordered Semiconductors
72.10; 72.20; 72.40
(b) Vol.152,No.1, 231-238

M. MACUCCI, B. PELLEGRINI, P. TERRENI, and L. REGGIANI
Non-Equilibrium Velocity Autocorrelation Function for Semiconductors in the Presence of Trapping Phenomena
72.10 and 72.20
(b) Vol.152,No.2, 601-616

B.V. KARPENKO, A.V. KUZNETSOV, and L.D. FALKOVSKAYA
On the Theory of Conduction of Antiferromagnetic Semiconductors
72.10; 72.20
(b) Vol.153,No.1, 271-280

E. HEINER
Screening of the Coulomb Potential in the Athermal Stage in Highly Excited Semiconductors
72.10; 72.20
(b) Vol.153,No.1, 295-306

G. GHIBAUDO
A Simple Derivation of the Kubo-Greenwood Formula
72.10
(b) Vol.153,No.2, K155-K158

M. SINGH and S.Y. TANG
Magneto-Optics of Holes in the Presence of Internal Strains and Temperature in Single Quantum Well Structures
71.25 and 72.10; 78.20
(b) Vol.154,No.2, 589-602

V.A. KUTASOV and L.N. LUKYANOVA
The Conduction Band Parameters and Scattering Mechanisms in Solid Solutions Based on Bi_2Te_3
72.10; S8
(b) Vol.154,No.2, 669-678

G.S. KOUSIK, C.M. VAN VLIET, G. BOSMAN, and HORNG-JYE LUO
Quantum 1/f Noise Associated with Intervalley Scattering in Nondegenerate Semiconductors. I. Analytical Calculations
72.10 and 72.70; S5.11
(b) Vol.154,No.2, 713-726

L.J. DUNNE
Electronic Degeneracies, Inelastic Processes, and Bohm-Aharonov Magnetoresistance Oscillations in Ordered Normal-Metal Rings
72.10; 72.15
(b) Vol.154,No.2, 727-732

R. LENK
First-Order Quantum Corrections to the Diffusivity and Current-Induced Density Fluctuations in the Three-Dimensional Bulk
72.10
(b) Vol.154,No.2, 733-742

A. SCHENK, M. STAHL, and H.-J. WUENSCHE
Calculation of Interband Tunneling in Inhomogeneous Fields
72.10 and 73.40; 77.50; S8.15
(b) Vol.154,No.2, 815-826

A. DARGYS
Hole Velocity in an Electric Field for Valence Bands Degenerate at k=0
72.10; 71.25
(b) Vol.155,No.2, 615-626

R. LENK
Theory of the Residual Resistivity Dipole. Diffusion of Non-Interacting Particles around a Fixed Point Scatterer
72.10
(b) Vol.155,No.2, 627-638

B.G. MULIMANI, C.H. RAMAKRISHNA, B.S. KRISHNAMURTHY, and C.K. SARKAR
Transient Response of Hot Electrons in Semiconductor Quantum Well Wires
72.10 and 72.40; S7.12
(b) Vol.156,No.1, 253-258

U. DILLNER and W. SCHNELLE
Electrical and Galvanomagnetic Properties of Undoped and Doped Polycrystalline Bismuth Films. II. Analysis in an Anisotropic ...
73.60; 72.10; S4
(a) Vol.116,No.1, 337-342

V.M. MIKHEEV
Impurity Scattering in Weakly Compensated Gapless Semiconductors at Low Temperatures
72.10; 72.20; S8.15
(b) Vol.156,No.2, 577-582

U. DILLNER and W. SCHNELLE
Electrical and Galvanomagnetic Properties of Undoped and Doped Polycrystalline Bismuth Films. III. Analysis in an Anisotropic ...
73.60; 72.10; S4
(a) Vol.116,No.2, 715-724

72.15. Electronic conduction in metals and alloys

A.M. ROSHCHUPKIN, I.L. BATARONOV, O.A. TROITSKII, and M.M. MOISEENKO
Electric Current Effect on Metal Surface Layers
68.20 and 72.15
(b) Vol.151,No.1, 121-126

S. KUMAR, P.K. AHLUWALIA, and K.C. SHARMA
Thermoelectric Power of Dilute Magnetic Alloys in the Presence of CEF and Exchange Scattering
72.10 and 72.15
(b) Vol.151,No.1, 225-230

TH. BORN
Calculation of the Electrical Resistivity of Liquid Sn-Te Alloys according to the Effective-Medium Theory
72.15; S4
(b) Vol.151,No.1, K41-K44

R.B. PRABHU and A.F. BARBOSA
Low Temperature Resistivity of Mix-Valent $Ce_{0.76}La_{0.14}Th_{0.1}$
72.15; S1.4
(b) Vol.151,No.1, K45-K48

K. YAMAKAWA, B. HOHLER, and H. KRONMUELLER
Diffusion of Hydrogen in Nickel Based Alloys
66.30; 72.15; S1.2
(a) Vol.111,No.1, 89-98

H. DINTNER and F. THRUM
The Homogeneous Distribution of Non-Metallic Bonds in As-Deposited NiCr-O Resistive Films
72.15 and 73.60; S1.2; S10.1
(a) Vol.111,No.2, 551-558

J. VAN DER STEL, H.B. VAN VELDHUIZEN, G.W. KOEBRUGGE, J. SIETSMA, and A. VAN DEN BEUKEL
Change of Electrical Resistivity and Youngs Modulus during Crystallization of Amorphous $Fe_{40}Ni_{40}B_{20}$
61.40 and 64.70; 62.20; 72.15; S1.1; S1.2
(a) Vol.112,No.1, 35-40

G. GOVINDARAJ
Electron Transport in Metallic Thin Films. A Comparative Study
72.15 and 73.60
(a) Vol.113,No.1, 113-122

S. ARAJS and T.A. JENKINS
Electrical Conduction in Some Directionally Solidified Eutectic Composites
72.15; S1.2
(a) Vol.113,No.1, K71-K74

V. BÂRSAN
The Conductivity of the Tomonaga-Luttinger Model
72.15
(b) Vol.153,No.2, K151-K154

D.Y. LI, X.F. WU, and T. KO
The Effect of Stress on the Lattice Instability of a TiNi Alloy
64.70; 61.10; 62.20; 72.15; S1.2
(b) Vol.154,No.1, 85-96

J.N. DAOU, J.P. BURGER, and P. VAJDA
Anomalous Behavior of the Low-Temperature Spin Disorder Resistivity in PrH_{2+x} Compounds
72.15; 75.50; S1.4
(b) Vol.154,No.1, 321-326

C.D. MUKHERJEE, R. RANGANATHAN, A.K. RAYCHAUDHURI, and N. CHATTERJEE
Thermoelectric Power of the Ni and Cd Substituted YBCO System
72.15 and 74.70; S10.15
(b) Vol.154,No.1, K55-K58

T. KANADANI and A. SAKAKIBARA
Short-Range Clustering in Dilute Al-Cu Alloy
64.75 and 72.15; S4
(a) Vol.114,No.1, K17-K20

L.J. DUNNE
Electronic Degeneracies, Inelastic Processes, and Bohm-Aharonov Magnetoresistance Oscillations in Ordered Normal-Metal Rings
72.10; 72.15
(b) Vol.154,No.2, 727-732

S. ABIS and G. RIONTINO
A Scanning Electrical Resistivity (SER) Study of Phase Separation in an Al-22at%Zn Alloy
64.75; 72.15; S1.2; S4
(a) Vol.114,No.2, 511-518

J. LEDZION and J. KIERUL
Thickness Dependence of Thin Film Conductivity
72.15 and 73.60; S4
(a) Vol.114,No.2, 545-550

K. SUMIYAMA, Y. HIROSE, and Y. NAKAMURA
Magnetic and Electrical Properties of Nonequilibrium Ni-Al Alloys Produced by Vapor Quenching
72.15 and 75.50; 61.55; S1.2
(a) Vol.114,No.2, 693-704

T. KANADANI and A. SAKAKIBARA
Effect of the Specimen Thickness on the Aging of an Al-10mass%Zn Alloy after Quenching from Low Temperature
72.15; S4
(a) Vol.114,No.2, K197-K200

S.R. HOFFMAN and L.M. SCARFONE
A Two-Band Model of Substitutionally Disordered Ternary Metal Alloys in the Coherent Potential Approximation
71.20; 72.15; S1.2; S1.3
(b) Vol.155,No.1, K33-K38

M.S.R. CHARI and R.G. SHARMA
The Electronic Lorenz Number and Magnon Thermal Conductivity of Cobalt and Iron
72.15; S1.1; S1.2
(a) Vol.116,No.1, 377-382

A.I. MITSEK
Collective Excitations of Localized Electrons and Their Effect on Magnetic-Material Kinetics
71.45; 72.15; 75.80; S1.2; S1.4; S1.5
(b) Vol.156,No.2, 507-514

O. AKINLADE and O. BELHACHEMI
The Electrical Resistivity of Some Lithium Based Liquid Alloys
72.15; S2
(b) Vol.156,No.2, 569-576

72.20. Conductivity phenomena in semiconductors and insulators
(for nonelectronic conduction, see 66.70)

S. BISWAS and R. BHATTACHARYA
Two Valence Subbands in Single Crystals of Bismuth Telluride Doped with Lead and Its Electrical Properties
72.20; 71.25; S8
(b) Vol.151,No.1, 193-202

E. GERLACH and S. KISTINGER
Plasma Excitation in Semiconductors Containing Extended Scattering Centers
72.20; 61.70; 71.45
(b) Vol.151,No.1, 241-248

D. VLADIKOVA, L. ILKOV, and S. KARBANOV
Influence of the Microstructure on Some Microwave Properties of Substituted Nickel Ferrites
64.80; 72.20; 75.50; S11.2
(a) Vol.111,No.1, 145-154

Y. MERA, K. MAEDA, and S. HYODO
Microscopic Origin of Electric Conduction Anisotropy in Plastically Deformed CdS Single Crystals
72.20; 61.70; S8.11
(a) Vol.111,No.1, 223-232

T. KOJIMA
Semiconducting and Thermoelectric Properties of Sintered Iron Silicide
72.60; 72.20; S1.61
(a) Vol.111,No.1, 233-242

P.P. BASU, A. BHATTACHARYYA, and D. CHATTOPADHYAY
AC Mobility of One-Dimensional Hot Electrons in Quantum-Well Wires
72.20 and 73.40
(b) Vol.151,No.2, K129-K134

B.M. ASKEROV, M.M. BABAEV, and M.I. DZHAFAROV
Thermopower in a Quantizing Longitudinal Magnetic Field
72.20
(b) Vol.151,No.2, K157-K160

V.M. AROLE, M.G. TAKWALE, V.G. BHIDE, A.D. SHALIGRAM, and S.K. DAVID
Hydrogen-Induced Structural Changes in Polycrystalline Silicon as Revealed by Positron Lifetime Spectroscopy
61.70; 72.20; S5.11
(a) Vol.111,No.2, 457-462

A. ROGALSKI and K. JOZWIKOWSKI
The Intrinsic Carrier Concentration in $Pb_{1-x}Sn_xTe$, $Pb_{1-x}Sn_xSe$, and $PbS_{1-x}Se_x$
72.20; S8.15
(a) Vol.111,No.2, 559-566

A.M. SANDAUER
Anomalous Hall Effect in PbSnMnTe
72.20; S8
(a) Vol.111,No.2, K219-K224

YU.G. OKLADNOY
Anomalous Magnetic Field Dependence of Electron Mobility in Anisotropic Ferromagnetic Semiconductors
72.20; 75.30
(b) Vol.152,No.1, 217-224

I.M. GOLBAN
Determination of Deformation Potential Constants of Arsenic
72.20; S5
(b) Vol.152,No.1, 225-230

I.P. ZVYAGIN
Hopping Relaxation and Photoconductivity of Disordered Semiconductors
72.10; 72.20; 72.40
(b) Vol.152,No.1, 231-238

M. PĘKAŁA, K. PĘKAŁA, and A. PEJĄCZKOWSKA
Normal State Thermoelectric Power of Bi-Sr-Ca-Cu-O Superconductors
72.20 and 74.70; S10.15
(b) Vol.152,No.1, K1-K4

R. HERRMANN, U. PREPPERNAU, CH. RELLIN, A. KRAPF, and H. DWELK
Normal State Thermoelectric Power of Bi-Sr-Ca-Cu-O High-Tc Superconductor
72.20 and 74.70; S10.15
(b) Vol.152,No.1, K5-K8

P.I. BARANSKII, I.S. BUDA, and V.V. SAVYAK
Mechanisms Responsible for Anisotropy of Thermal E.M.F. in Semiconductors (Review Article)
72.20
(a) Vol.112,No.1, 11-28

C. DERVOS, P.D. BOURKAS, and E.A. KAYAFAS
Potential Profile of Dielectrics during High-Voltage Pulse Application
72.20 and 77.50; S12
(a) Vol.112,No.1, 123-130

S. PHANJOUBAM, D. KOTHARI, and J.S. BAIJAL
Electrical Conductivity of Tetravalent Titanium Substituted Lithium Zinc Ferrite
72.20; S11.2
(a) Vol.112,No.1, 131-136

I.A. CHAIKOVSKII and A.I. GERMAN
AC Conductivity of Inhomogeneous Semiconductors in the Presence of Classically High Magnetic Field
72.20
(b) Vol.152,No.2, 593-600

M. MACUCCI, B. PELLEGRINI, P. TERRENI, and L. REGGIANI
Non-Equilibrium Velocity Autocorrelation Function for Semiconductors in the Presence of Trapping Phenomena
72.10 and 72.20
(b) Vol.152,No.2, 601-616

V.V. PETROV and V.S. PROSOLOVICH
Thermal Defects in Silicon Doped with Rare-Earth Elements
61.70 and 71.55; 72.20; 78.30; S5.11
(a) Vol.112,No.2, 561-568

R.K. BAKANAS, F.G. BASS, and V.V. KONOTOP
Interaction of Depletion and Accumulation Waves with Localized Defect
72.20; 61.70
(a) Vol.112,No.2, 579-588

I.I. LYAPILIN and V.V. KARYAGIN
Longitudinal Thermogalvanomagnetic Effects in Dilute Magnetic Semiconductors in a Quantizing Magnetic Field
72.20
(b) Vol.153,No.1, 263-270

B.V. KARPENKO, A.V. KUZNETSOV, and L.D. FALKOVSKAYA
On the Theory of Conduction of Antiferromagnetic Semiconductors
72.10; 72.20
(b) Vol.153,No.1, 271-280

E. HEINER
Screening of the Coulomb Potential in the Athermal Stage in Highly Excited Semiconductors
72.10; 72.20
(b) Vol.153,No.1, 295-306

E. PAVLIDOU and O. VALASSISDES
Multicarrier Investigation of $Pb_{0.95}Ge_{0.05}Te$ Crystals by the Magnetoconductivity Tensor Component $\sigma_{xy}(B)$ Analysis
72.20; S8
(b) Vol.153,No.1, K51-K56

S.K. NIKOGOSYAN and V.A. SAAKYAN
The Effect of Electron Radiation on Thermal and Electrophysical Properties of n-InP
61.80; 66.70; 72.20; S7.11
(a) Vol.113,No.1, 43-50

M.M. KAZANIN and V.V. KUTASOV
Fe Impurity in alpha- and beta-Boron and in Some Boron Compounds
71.55; 72.20; 76.70; S5
(a) Vol.113,No.1, 143-150

H. DWELK, D. FREUDE, R. HERRMANN, H. PFEIFFER, and N. PRUSS
Substitution of Yttrium for Boron in the Structure of $YBa_2Cu_3O_{7-\delta}$
72.20 and 74.70; 76.60; S10.15
(a) Vol.113,No.1, 159-162

O.A. GOLIKOVA
Amorphous Boron and Amorphous-Like Borides
61.40 and 72.20; S5
(a) Vol.113,No.1, K1-K4

J. GALIBERT, P. PERRIER, and S. ASKENAZY
Résonance magnétophonon avec retournement de spin dans l'antimoniure d'indium
63.20 and 71.70; 72.20; S7.13
(b) Vol.153,No.2, 653-658

YANG-FANG CHEN
Comment on the Determination of the Midgap Density of States in Amorphous Semiconductors Using the Space-Charge-Limited-Current ...
71.20 and 72.20; S5.11
(b) Vol.153,No.2, 695-702

K. BHATTACHARYYA
Alloy Scattering Limited Mobility of Two-Dimensional Electron Gas in SD-$Al_{0.3}Ga_{0.7}As/Ga_{0.87}In_{0.13}As/GaAs$ Structures
72.20 and 73.40; S7.12; S7.15
(b) Vol.153,No.2, K135-K140

E. MAJKOVÁ, J. ČERVENÁK, J. KREMPASKÝ, and P. DUHAJ
Temperature Dependence of the Seebeck Coefficient in InSb Prepared by Rapid Quenching
72.20; S7.13
(b) Vol.153,No.2, K147-K150

YU.V. LILENKO, K.V. SHASTOV, A.S. PETROV, A.V. VOITSEKHOVSKII, V.S. KULIKAUSKAS, V.N. KUZNETSOV, and A.P. MAMONTOV
Radiation-Induced Defects in Implanted $Hg_{1-x}Cd_xTe$ Crystals
61.80; 72.20; 79.20; S8.15
(a) Vol.113,No.2, 285-294

L.A. KAZAKEVICH, P.F. LUGAKOV, and I.M. FILIPPOV
The Role of Dislocations during Charge-Carrier Recombination at Radiation Defects in p-Si
61.70 and 72.20; 61.80; S5.11
(a) Vol.113,No.2, 307-314

R. KOEGLER, E. WIESER, J. ALBRECHT, and P. KNOTHE
Investigation of As-O Clustering in Si
61.70; 72.20; S5.11
(a) Vol.113,No.2, 321-330

A.V. DVURECHENSKII, V.A. DRAVIN, and A.I. YAKIMOV
Electrical Properties of Ion-Doped Amorphous Silicon
72.20; 72.60; S5.11
(a) Vol.113,No.2, 519-528

DOAN NHAT QUANG
On the Theory of Auger Recombination in Disordered Semiconductors. A Green's Function Approach
71.25 and 72.20; S7.13
(b) Vol.154,No.1, 237-248

D. CHATTOPADHYAY and A. KABASI
Warm Electron Transport in GaAs Quantum Wells
72.20 and 73.40; S7.12
(b) Vol.154,No.1, 281-286

V.A. KASIYAN, D.D. NEDEOGLO, A.V. SIMASHKEVICH, and I.N. TIMCHENKO
Conductivity and the Metal-Dielectric Transition in Compensated n-ZnSe
72.20; 71.30; 71.55; S8.12
(b) Vol.154,No.1, 287-296

D. BAHADUR, RAM BILAS, and B.S.R. MURTY
Electronic Properties of the Defect Perovskite System $La_{1-x}Li_xMO_3$
61.60 and 72.20; 75.50; S11
(b) Vol.154,No.1, 297-304

H. ISOTALO, J. LAAKSO, P. KUIVALAINEN, H. STUBB, J.-E. OESTERHOLM, and P. YLI-LAHTI
Charge Transport in $FeCl_3$-Doped Polyethylenevinylacetate-Polyoctylthiophene Blends
72.20; 72.60; 76.30; S12
(b) Vol.154,No.1, 305-314

L. FIRLEJ, A. GRAJA, J. WOLAK, R.N. LYUBOVSKAYA, and L.M. GOLDENBERG
ESR Studies of Single Crystals of Bis(ethylenedithio)tetrahiafulvalene Based Conductor with Octaiodinemercurate Anion $(BEDT\text{-}TTF)_4Hg_3I_8$
64.70 and 76.30; 72.20; S12.1
(b) Vol.154,No.1, 333-340

B.N. NARAHARI ACHAR and J.R. FERGUSON
Quantized Helicon Damping in a Semiconductor Superlattice in the Integral Quantum Hall Regime
72.20 and 73.60
(b) Vol.154,No.1, K59-K62

G.A. ADEGBOYEGA, A. POGGI, E. SUSI, and M. REICHE
The Effects of High Temperature Anneal on the Electrical Activity of Iron in B-Doped Cz Silicon
71.55; 72.20; 78.30; S5.11
(a) Vol.114,No.1, 247-252

M.N. KHAN
The Electronic and Spectroscopic Studies of Semiconducting Germanium-Tellurite Glasses Containing Iron Oxide
72.20 and 78.30; S10.15
(a) Vol.114,No.1, 285-292

V.A. KASIYAN, D.D. NEDEOGLO, A.V. SIMASHKEVICH, and I.N. TIMCHENKO
Critical Behaviour of n-ZnSe Parameters in the Vicinity of the Metal-Insulator Transition
72.20; 71.30; S8.12
(b) Vol.154,No.2, 691-702

G.A. MATVEEV, I.M. TSIDILKOVSKII, and A.T. LONCHAKOV
Piezomagnetoresistance Anisotropy of n-Ge in Strong Magnetic Fields near the Metal-Insulator Transition
71.30 and 72.20; S5.12
(b) Vol.154,No.2, 703-712

B. MITRA, A. GHOSHAL, and K.P. GHATAK
Influence of Quantizing Magnetic and Crossed Oscillatory Electric Fields on the Einstein Relation in Wide-Gap ...
72.20
(b) Vol.154,No.2, K147-K150

H. BOUDRIOT and H.A. SCHNEIDER
Influence of Dipole Scattering on the Free Carrier Absorption
78.30; 72.20; S7.1
(b) Vol.154,No.2, K175-K178

T.A. CHRISTAKUDI, G.CH. CHRISTAKUDIS, and L.D. BORISOVA
Thermoelectric Power of Bismuth Doped Lead Telluride
72.20; S8
(a) Vol.114,No.2, K201-K204

A. GALDIKAS, S. GREBINSKII, T. AMINOV, and L. OCHERTIANOVA
Interrelation of Piezo- and Magnetoresistance Anisotropy in Ferromagnetic Semiconductors p-$HgCr_2Se_4$ and p-$CdCr_2Se_4$
72.20 and 72.80; S8.16
(a) Vol.114,No.2, K211-K214

P. LOŠŤÁK, Z. STARÝ, J. HORÁK, and J. PANCÍŘ
Substitutional Defects in Sb_2Te_3 Crystals
61.70; 72.20; S8
(a) Vol.115,No.1, 87-96

R.K. SHARMA, S.K. MEHTA, V.K. SINGH, G. PRASAD, and B.B. SHARMA
Stoichiometric Annealing and Electrical Properties of $Hg_{0.8}Cd_{0.2}Te$ Grown by Solid State Recrystallization
64.75; 72.20; S8.15
(a) Vol.115,No.1, 135-142

G.V. BAZUEV, K.R. KRYLOV, A.I. PONOMAREV, V.I. TSIDILKOVSKII, I.M. TSIDILKOVSKII, and T.B. CHARIKOVA
Peculiarities of Galvanomagnetic and Thermoelectric Properties of $YBa_{2-x}La_xCu_3O_{7-\delta}$ Solid Solutions
74.70; 72.20; S10.15
(a) Vol.115,No.1, 267-272

A. CONAN, A. BONNET, and M. MORSLI
Electrical and Structural Properties of the Tungsten Bronze $Bi_4W_{20}O_{62}$
72.20; 61.60; 71.20; S10
(b) Vol.155,No.2, 609-614

M. PĘKAŁA, R. RATAJ, A. PAJĄCZKOWSKA, and B. GEGENHEIMER
Normal State Transport Study of the Ba-K-Bi-O Superconductor
72.80 and 74.70; 72.20; S10.15
(b) Vol.155,No.2, K123-K126

V.V. KARYAGIN and I.I. LYAPILIN
Thermo-Galvanomagnetic Phenomena in Spin Quantum Wells
71.10 and 73.40; 72.20; S8.15
(b) Vol.156,No.1, 235-242

R. KIEMEL, W. WISCHERT, and S. KEMMLER-SACK
Dependence of T_c on the Hole Concentration in the High-T_c Superconductors $Bi_{1.5}Pb_{0.5}Sr_2Ca_{1-x}Gd_xCu_2O_{8+z}$
74.70; 71.30; 72.20; S10.15
(b) Vol.156,No.1, 339-344

F.R. KESSLER and J. NIEDER
Impact Ionization of Indirect Excitons in Crossed Electric and Magnetic Fields in Germanium
71.35; 72.20; 72.40; S5.12
(b) Vol.156,No.1, K45-K50

G.I. URUSHADZE
Quantum Voltage Fluctuations in Normal Tunnel Junctions and Departure of Conductivity from the Ohmic Law
72.20 and 73.40
(b) Vol.156,No.1, K63-K66

SH.M. EFENDIEV, A.A. AGASIEV, V.E. BAGIEV, and YA.YU. GUSEINOV
Nonlinear Current-Voltage Characteristics in Me-Bi_2O_3-Me Structure
73.40; 72.20; S10
(a) Vol.116,No.1, 305-312

OM PARKASH, CH.DURGA PRASAD, and DEVENDRA KUMAR
Preparation and Characterisation of the System $Sr_{1-x}La_xTi_{1-x}Co_xC_3$
61.50 and 72.20; S10.15
(a) Vol.116,No.1, K81-K84

K. VAVŘINA and J. KODEŠ
Simulation of Transient Phenomena in n-Type Si
72.20; S5.11
(a) Vol.116,No.1, K85-K90

V.M. MIKHEEV
Impurity Scattering in Weakly Compensated Gapless Semiconductors at Low Temperatures
72.10; 72.20; S8.15
(b) Vol.156,No.2, 577-582

YU.N. MOSKVICH, B.I. CHERKASOV, A.M. POLYAKOV, A.A. SUKHOVSKII, and R.L. DAVIDOVICH
NMR Study of Internal Motions in a New Family of Ionic Conductors M_2SbF_5
76.60; 66.30; 72.20; S9.16
(b) Vol.156,No.2, 615-632

C.K. SARKAR and P. BANERJI
Estimation of the Alloy Scattering Potential in $Ga_{0.47}In_{0.53}As$ Using Cyclotron Resonance Measurements
76.30; 72.20; S7.15
(b) Vol.156,No.2, K145-K148

V.A. GURTOV and P.A. RAIKERUS
Radiation Charge in Silicon Dioxide Considering Diffusion Currents
72.20; 71.55; 85; S5.11
(a) Vol.116,No.2, 769-776

N. PRUSS, W. KRAAK, H.-U. MUELLER, A. JACOBI, H. DWELK, and R. HERRMANN
Influence of the Sintering Temperature on the Physical Properties of $YBa_2Cu_3O_{7-\delta}$ Ceramic Samples
72.20 and 74.20; S10.15
(a) Vol.116,No.2, 793-802

A.L. TÓTH
Monte-Carlo Simulation of Diffusion Length Measurements by a Wedge Absorber in the Scanning Electron Microscope (SEM)
61.16 and 72.20
(a) Vol.116,No.2, K137-K140

72.30. High-frequency effects; plasma effects

R. FUCHS and K. GHOSH
Anomalous Skin Effect in a Cylindrical Conductor
72.30
(a) Vol.111,No.1, K83-K88

72.40. Photoconduction and photovoltaic effects; photodielectric effects

I.D. MAKUTA, S.K. POZNYAK, A.I. KULAK, and E.A. STRELTSOV
Photocurrent Generation and Optical Transitions on Degenerate Cadmium Oxide Photoanodes
68.45 and 72.40; S10.1
(a) Vol.111,No.1, 193-200

RENYUAN QIAN, SONGQING SHI, TONGMING LIU, and DAOBEN ZHU
Carrier Transport Properties of Poly(1,6-Bis-9-Carbazolyl-2,4-Hexadinyne)
72.80; 72.40; 78.40; S12.1
(a) Vol.111,No.1, K67-K72

D.D. ASKEROV, T.D. DZHAFAROV, A.M. AGAEV, S.G. ABDINOVA, and M.S. SADYGOV
Photoelectric Properties of $CuInSe_2$-GaAs Heterojunctions
72.40 and 73.40; S8.16
(a) Vol.111,No.1, K73-K76

T.K. KASUMOV and F.I. MAMEDOV
Photocurrent Oscillations Due to Optical Phonons in $AgGaSe_2$ Single Crystals
72.40; 63.20; S8.16
(a) Vol.111,No.2, K205-K208

C. GHEZZI, D. MARTIN, A. PARISINI, J.L. STAEHLI, and L. TARRICONE
Investigation of (AlGa)As/GaAs Quantum Wells by Optical and Phototransport Measurements
73.40; 71.35; 72.40; 78.55; S7.15
(b) Vol.152,No.1, 101-110

J. CALAS, S. ABDEL-MAKSOUD, and M. AVEROUS
Temperature Dependence of Carrier Trapping in Bulk Vitreous $GeSe_3$ from Transit Time Measurements
71.55 and 72.40; S8
(b) Vol.152,No.1, 209-216

I.P. ZVYAGIN
Hopping Relaxation and Photoconductivity of Disordered Semiconductors
72.10; 72.20; 72.40
(b) Vol.152,No.1, 231-238

A.M. ELKORASHY
Photoconductivity in Germanium Selenide Single Crystals
72.40; S8
(b) Vol.152,No.1, 249-260

J. HLÁVKA and J. GRŇO
Annealing Stages of Implanted Silicon
61.70; 72.40; S5.11
(a) Vol.112,No.1, 327-330

P. ALTANZOG, D. BATSUUR, S. BOLD, B. BURMAA, SH. CHADRAABAL, and A.G. KASANSKII
Temperature Dependence of Dark Conductivity and Steady-State Photoconductivity of a-Si:H Samples
72.40 and 73.60; S5.11
(a) Vol.112,No.1, K23-K28

E.E. MATYAS, T.E. BORISENKO, N.N. KOREN, and V.A. IVANOV
Spectral Distribution of Photoresponse in PbTe-SnS Films
72.40; S8
(a) Vol.112,No.2, K111-K112

R. SIGNERSKI, J. KALINOWSKI, P. DI MARCO, and G. GIRO
Photoconductivity of Polycrystalline Thionaphtenindol
72.40 and 73.60; S12
(b) Vol.153,No.1, 281-294

T. BLUM and H. BAESSLER
Light-Induced Inter-Chain Charge Transfer in Crystalline Polydiacetylene (PTS)
72.40; S12
(b) Vol.153,No.1, K57-K62

L. SZARO
Steady-State Surface Photovoltage in Large-Gap Semiconductors under Sub-Bandgap Illumination
73.20 and 73.30; 72.40
(a) Vol.113,No.1, 97-106

V.YU. RUD, YU.V. RUD, M. SERGINOV, and M.A. TAIROV
Production of and Physical Processes in n-p-$CdSiAs_2$ Structures
72.40 and 73.40; 85; S7.16
(a) Vol.113,No.1, 207-214

L. KOVÁCS, R. SOMMERFELDT, YE MING, and E. KRAETZIG
Light-Induced Charge Transport in $LiNbO_3$:Mg,Cr Crystals
61.70 and 72.40; 78.50; S11.1
(a) Vol.113,No.1, K75-K78

U. BRAATZ, E. KRAUSE, and J. KUNSCH
Optoelectronic Switching in ZnSe and ZnS
72.40; S8.11; S8.12
(a) Vol.113,No.1, K79-K84

L. HOLTMANN
A Model for the Nonlinear Photoconductivity of $BaTiO_3$
72.40; S11.1
(a) Vol.113,No.1, K89-K94

P.E. MOZOL, N.A. SKUBENKO, P.A. SKUBENKO, YU.P. GNATENKO, E.A. SALKOV, and Z.D. KOVALYUK
Effect of Laser Irradiation on Low-Temperature Photoconductivity and Photoluminescence Spectra of Gallium Selenide
71.35 and 71.55; 72.40; 78.55; S8
(b) Vol.153,No.2, 667-674

T. DRUESEDAU, M. ECKLER, R. BINDEMANN, and H. FIEDLER
Enhancement of Photoconductivity in Magnetron-Sputtered a-Si:H Doped with Nitrogen
68.55 and 73.60; 72.40; 78.65; S5.11
(b) Vol.153,No.2, K119-K124

M.J. MAŁACHOWSKI, J. PIOTROWSKI, and A. ROGALSKI
Influence of Dislocaions on the Performance of 3 to 5 µm $Hg_{1-x}Cd_xTe$ Graded Gap Photoresistors
72.40 and 73.60; S8.15
(a) Vol.113,No.2, 467-476

V.D. USIKOV
Self-Stabilzed Avalanche Process in MIS Structures under Nonstationary Conditions. I. Generation of Carriers Due to Illumination
73.40; 72.40
(a) Vol.113,No.2, 667-676

WANLU WANG and KEJUN LIAO
Persistent Photoconductivity in a-Si:H/a-SiN_x:H Multilayer Films at Low Temperature
72.40 and 73.60; S5.11; S6
(a) Vol.113,No.2, K223-K226

L.L. KULYUK, N.F. PERELMAN, S.I. RADAUTSAN, A.V. SIMINEL, and E.E. STRUMBAN
About the Possibility of Double Centre Formation in InP:Fe:O
71.55 and 72.40; 76.30; 78.55; S7.1'
(a) Vol.113,No.2, K227-K230

T.K. KASUMOV and F.I. MAMEDOV
Combined Photoconductivity in $AgGaSe_2$ Single Crystals
72.40; S8.16
(a) Vol.113,No.2, K241-K244

M.O. MANARESH, D.W. FISCHER, and W.C. MITCHEL
The EL2 Defect in GaAs: Some Recent Developments (Review Article)
61.70 and 71.55; 72.40; 78.30; 78.50; 78.55; S7.12
(b) Vol.154,No.1, 11-42

H. WIECZOREK and W. FUHS
Deep Trapping of Carriers in a-Si:H Solar Cells Studied by Transient Photocurrents
72.40 and 73.40; 71.55; S5.11
(a) Vol.114,No.1, 413-418

G. E. ZARDAS, Z. HATZOPOULOS, and P.H. YANNAKOPOULOS
Effect of Low Dose Electron Irradiation on AlGaAs/GaAs at Low Temperatures
61.80 and 72.40; S7.12; S7.15
(a) Vol.114,No.1, K41-K44

Y.N. MOHAPATRA and V. KUMAR
Temperature Dependence of Photocurrent in Undoped Semi-Insulating Gallium Arsenide
72.40; 71.55; S7.12
(a) Vol.114,No.2, 659-664

G.J. PAPAIOANNOU, E. ANAGNOSTAKIS, P.C. EUTHYMIOU, C.D. KOURKOUTAS, and G. ZARDAS
On the Impurity Photoconductivity of Semi-Insulating GaP:Cr
72.40; 78.50; S7.11
(a) Vol.114,No.2, K215-K218

K.H. HERRMANN, U. JANK, and K.-P. MOELLMANN
Bipolar Generation within the Exponential Tail of the Interband Absorption Edge in HgCdTe
72.40 and 78.30; S8.15
(a) Vol.114,No.2, K249-K254

G.A. MEDVEDKIN, YU.V. RUD, and M.A. TAIROV
Photoelectric Anisotropy of II-IV-V_2 Ternary Semiconductors (Review Article)
72.40; 78.20; S7.16
(a) Vol.115,No.1, 11-50

H. BERGNER and T. DAMM
Binary States in Integrated Circuits Investigated by Laser Scanning Microscopy with Optical Beam Induced Current
72.40 and 85; S5.11
(a) Vol.115,No.2, 607-616

G. BABONAS and S. MARCINKEVIČIUS
Photovoltaic Effects in $CdGeP_2$
72.40; S7.16
(a) Vol.115,No.2, K211-K214

B.G. MULIMANI, C.H. RAMAKRISHNA, B.S. KRISHNAMURTHY, and C.K. SARKAR
Transient Response of Hot Electrons in Semiconductor Quantum Well Wires
72.10 and 72.40; S7.12
(b) Vol.156,No.1, 253-258

F.R. KESSLER and J. NIEDER
Impact Ionization of Indirect Excitons in Crossed Electric and Magnetic Fields in Germanium
71.35; 72.20; 72.40; S5.12
(b) Vol.156,No.1, K45-K50

B. POHORYLES
Helium Pressure Effects on Photoconductivity and Position of Dislocation Deep Electron States in Plastically Deformed Ge
61.70 and 71.55; 72.40; S5.12
(a) Vol.116,No.1, 349-356

V.K. GOYAL, D. KUMAR, and K. SEN
Effect of Grain Boundary States Density in the Polycrystalline MIS Solar Cells
73.20 and 85; 61.70; 72.40
(a) Vol.116,No.1, 425-432

A.P. ZDEBSKII, G. GARAYAGDEV, K. ORAZOV, and G.N. POLISSKII
Ultrasound-Stimulated Increase of Photosensitivity and Luminescence Intensity of II-VI Compound Solid Solutions
62.65 and 72.40; S8.15
(a) Vol.116,No.1, K91-K96

A.SH. ABDINOV, A.M. GUSEINOV, YU.G. NURULLAEV, and O.M. SADYKHOV
The Influence of Gadolinium Doping on the Switching Effect in Indium Selenide Single Crystals
72.40; S8
(a) Vol.116,No.2, K173-K178

U. BRAATZ, J. KUNSCH, E. KRAUSE, and E. JAEGER
Sweep Drive Circuit for Streak Cameras Based on ZnS and ZnSe Optoelectronic Switches
72.40 and 85; S8.11; S8.12
(a) Vol.116,No.2, K219-K222

72.50. Acoustoelectric effects

M. DEY(BOSE) and S. GHOSH
Amplificaion of Acoustic Waves in Magnetised Piezoelectric Semiconductors. Effect of Nonlinearity in Electron Effective Mass ...
72.50; 71.25; S7.13
(b) Vol.152,No.2, 625-632

VO HONG ANH and NGUYEN HONG SHON
Photostimulated Acoustoelectronic Effects in Semiconductors in the Long-Wavelength Region
72.50; 72.55
(b) Vol.154,No.2, 661-668

P. KOSTIAL, I. JAMNICKY, and J. DURCEK
Contribution to the Experimental Methods of Deep Level Acoustic Transient Spectroscopy
68.25 and 71.55; 72.50; S5.11; S10
(a) Vol.116,No.1, K77-K80

72.55. Magnetoacoustic effects

VO HONG ANH and NGUYEN HONG SHON
Photostimulated Acoustoelectronic Effects in Semiconductors in the Long-Wavelength Region
72.50; 72.55
(b) Vol.154,No.2, 661-668

72.60. Mixed conductivity and conductivity transitions

V.A. KULBACHINSKII
Galvanomagnetic Properties of Graphite Fibers
72.60 and 74.90; S5
(b) Vol.151,No.1, 185-192

T. KOJIMA
Semiconducting and Thermoelectric Properties of Sintered Iron Silicide
72.60; 72.20; S1.61
(a) Vol.111,No.1, 233-242

A.V. DVURECHENSKII, V.A. DRAVIN, and A.I. YAKIMOV
Electrical Properties of Ion-Doped Amorphous Silicon
72.20; 72.60; S5.11
(a) Vol.113,No.2, 519-528

H. ISOTALO, J. LAAKSO, P. KUIVALAINEN, H. STUBB, J.-E. OESTERHOLM, and P. YLI-LAHTI
Charge Transport in $FeCl_3$-Doped Polyethylenevinylacetate-Polyoctylthiophene Blends
72.20; 72.60; 76.30; S12
(b) Vol.154,No.1, 305-314

72.70. Noise processes and phenomena

J. ŠIKULA, P. VAŠINA, and I. HUETTEL
The Noise of the GaAlAs Super-LED Radiation
72.70 and 73.40; 78.60; S7.15
(a) Vol.111,No.1, 367-370

J. GONG and S.H. YANG
Annealing Temperature Dependence of the 1/f Noise in Si-Implanted GaAs
61.70 and 72.70; S7.12
(a) Vol.111,No.1, K53-K56

J. BELAN and V. MIKOLAJ
Voltage Noise of the Breakdown Region of Subsurface Zener Diodes
72.70; 85; S5.11
(a) Vol.112,No.2, K119-K120

G.S. KOUSIK, C.M. VAN VLIET, G. BOSMAN, and HORNG-JYE LUO
Quantum 1/f Noise Associated with Intervalley Scattering in Nondegenerate Semiconductors. I. Analytical Calculations
72.10 and 72.70; S5.11
(b) Vol.154,No.2, 713-726

72.80. Conductivity of specific semiconductors and insulators

F.M. KAMARLY, N.G. GUSEINOV, and A.YU. YANGIROV
Magnetic Susceptibility and Electrophysical Properties of SiTe Single Crystals
61.70; 72.80; 75.50; S8
(a) Vol.111,No.1, K17-K20

K. SIERAŃSKI and J. SZATKOWSKI
Electrical Properties of Zn_3P_2 at Room Temperature
72.80; S7.11
(a) Vol.111,No.1, K57-K60

B. MALLIK
Electrical Conduction in Ferrocene with Adsorbed Vapours
72.80; 73.25; S12
(a) Vol.111,No.1, K61-K66

RENYUAN QIAN, SONGQING SHI, TONGMING LIU, and DAOBEN ZHU
Carrier Transport Properties of
Poly(1,6-Bis-9-Carbazolyl-2,4-Hexadinyne)
72.80; 72.40; 78.40; S12.1
(a) Vol.111,No.1, K67-K72

V.I. GARASIM, D.M. ZAYACHUK, R.D. IVANCHUK, V.B. LOTOTSKII, P.M. STARIK, and V.A. SHENDEROVSKII
Peculiarities of Physical Properties of Lead and Tin Selenide Solid Solutions
72.80 and 75.50; S8
(a) Vol.111,No.1, K99-K104

K. TABATA and H. KIDO
Preparation and Magnetic and Electrical Properties of $La_{1-x}Ce_xCoO_3$ Compounds
72.80 and 75.50; S10.15
(a) Vol.111,No.1, K105-K108

D. ELEFANT, C. GLADUN, A. HEINRICH, J. SCHUMANN, and H. VINZELBERG
Comment on the Metal-Semiconductor Transition in Amorphous $Si_{1-x}Cr_x$ Films at Room Temperature
72.80; S1.61
(b) Vol.151,No.2, K155-K156

M.A. AHMED
Electrical Properties of Co-Zn Ferrites
72.80; S11.2
(a) Vol.111,No.2, 567-572

L. ZAKHARENKO, V. KOZLOVSKII, and YA. PILKEVICH
Role of Irradiation Temperature in the Conductivity Compensation of n-GaAs by Radiation Defects
61.80 and 72.80; S7.12
(a) Vol.111,No.2, K215-K218

M.G. OSTAPENKO and E.G. PETROV
The Donor-Acceptor Electron Transfer with Participation of Short Polymer Chains
72.80; 71.55; S12.1
(b) Vol.152,No.1, 239-248

W. SIEGEL, H. WITTE, and G. KUEHNEL
Inhomogeneity of Semi-Insulating GaAs Wafers with Nearly Uniform Resistivity Distribution
72.80; S7.12
(a) Vol.112,No.1, K35-K38

I.D. KOZMIK, I.I. GRIGORCHAK, Z.D. KOVALYUK, and I.Z. MARCHUK
Kinetic and Optical Properties of H_xInSe and H_xGaSe Proton Intercalates and Their Thermodynamic Model
61.70; 72.80; 78.50; S8
(a) Vol.112,No.2, K61-K64

R. NAVIL KUMAR and C.P.G. VALLABHAN
Effect of Deuteration on the Electrical Conductivity, Dielectric Constant, and Phase Transition in $(NH_4)_3H(SO_4)_2$
72.80 and 77.20; 64.70; S11.1
(a) Vol.112,No.2, K113-K118

L. BISCHOFF and R. KOEGLER
Rapid Thermal Annealing of Neutron Transmutation Doped Silicon
61.70; 72.80; S5.11
(a) Vol.113,No.2, K185-K188

K. GERMANOVA, V. DONCHEV, V. VALCHEV, and CH. HARDALOV
Characterization of Medium-Doped n-Type GaAs by Hall Measurements
72.80; S7.12
(a) Vol.113,No.2, K231-K234

M.M. KORSUKOVA, V.N. GURIN, YU.B. KUZMA, N.F. CHABAN, S.I. CHYKHRII, V.V. MOSHCHALKOV, N.B. BRANDT, A.A. GIPPIUS, and KHO KHYU NYAN
Crystal Structure, Electrical, and Magnetic Properties of the New Ternary Compounds $LnAlB_{14}$
61.60 and 72.80; 75.50; S1.4
(a) Vol.114,No.1, 265-272

S. BISWAS and R. BHATTACHARYA
Electrical Properties of Bismuth Telluride Doped with Bismuth Iodide
72.80; S8
(a) Vol.114,No.1, 277-284

V.K. LUKYANYUK, M.V. TOVARNITSKII, and Z.D. KOVALYUK
Hall-Coefficient Sign Inversion, Electrical Conductivity, and Its Anisotropy in Highly Intercalated Na_xInSe Single Crystals
72.80; 71.55; S8.16
(a) Vol.114,No.1, 293-298

T.P. BEALES, S. QUERESHI, and M.R. WILLIS
The Effect of Pressure on the Electrical Conductivity of $[(C_6H_5)_3PCH_3]^+(TCNQ)_2$
72.80; S12
(a) Vol.114,No.1, 299-304

P. BROGOWSKI, H. MUCHA, and J. PIOTROWSKI
Modification of Mercury Cadmium Telluride, Mercury Manganese Telluride, and Mercury Zinc Telluride by Ion Etching
61.80 and 72.80; 85; S8.15
(a) Vol.114,No.1, K37-K40

M.YA. BAKIROV, B.A. NAYAFOV, V.S. MAMEDOV, and R.S. MADATOV
Electrical and Optical Properties of Amorphous $Ge_{1-x}Si_x$ Semiconductors
72.80 and 78.40; S5.11; S5.12
(a) Vol.114,No.1, K45-K48

P.I. SAVITSKII, A.V. KONSTANTINOVICH, and Z.D. KOVALYUK
The Te Intercalation Effect on the Electrical Properties of Indium Monoselenide
72.80; S8
(b) Vol.154,No.2, K151-K156

G. AMATO, G. BENEDETTO, R. SPAGNOLO, and M. TURNATURI
Photoacoustic Measurements of Doped Silicon Wafers
65.70 and 66.70; 72.80; S5.11
(a) Vol.114,No.2, 519-524

K.N. MADHUSOODANAN, K. NANDAKUMAR, J. PHILIP, S.S.K. TITUS, S. ASOKAN, and E.S.R. GOPAL
Photoacoustic Investigation of Glass Transition in As_xTe_{1-x} Glasses
64.70; 61.40; 65.70; 72.80; S8
(a) Vol.114,No.2, 525-530

S.A. HUSSEIN and A.T. NAGAT
Characteristic Features of Electrical Properties of In_2Te_3 Single Crystals
72.80; S8
(a) Vol.114,No.2, K205-K210

A. GALDIKAS, S. GREBINSKII, T. AMINOV, and L. OCHERTIANOVA
Interrelation of Piezo- and Magnetoresistance Anisotropy in Ferromagnetic Semiconductors p-$HgCr_2Se_4$ and p-$CdCr_2Se_4$
72.20 and 72.80; S8.16
(a) Vol.114,No.2, K211-K214

G. OELGART, B. LIPPOLD, R. HEILMANN, H. NEUMANN, and B. JACOBS
Determination of the Germanium Acceptor Ionisation Energy of $Al_xGa_{1-x}As$ $(0<x<0.40)$ by Hall Effect and Luminescence
71.55; 72.80; 78.55; S7.15
(a) Vol.115,No.1, 257-266

A.V. VANNIKOV, A.YU. KRYUKOV, A.G. TYURIN, and T.S. ZHURAVLEVA
Influence of the Medium Polarity on Electron Transport in Polymer Systems
72.80; S12
(a) Vol.115,No.1, K47-K52

YU.A. VIDADI, V.E. BAGIEV, and YA.YU. GUSEINOV
Frequency-Compensating Effect in the Electric Conductivity of Bi_2O_3 Films
72.80; S10.1
(a) Vol.115,No.1, K53-K56

PARAMDEEP KHURANA and K.K. SRIVASTAVA
Pressure Dependence of Capacitance and Conductance for a-$Ge_{17}Te_{83}$ at Moderate Temperatures
72.80; S8
(a) Vol.115,No.1, K57-K62

M.K. JAYARAJ and C.P.G. VALLABHAN
AC Conductivity in Europium Oxide Thin Films
72.80; S10.1
(a) Vol.115,No.1, K63-K66

J.L. ZHANG and B.R. LI
Anomalous Electrical Conduction of $Cd_6Sb_2O_{10}$ Ceramics
64.70 and 72.80; S10
(a) Vol.115,No.1, K67-K70

I.A. AKSENOV, I.R. GULAKOV, V.I. LIPNITSKII, A.I. LUKOMSKII, and L.A. MAKOVETSKAYA
Cathodoluminescence and Electrical Properties of $CuAlS_2$
72.80 and 78.60; S8.16
(a) Vol.115,No.1, K113-K116

M. PĘKAŁA, R. RATAJ, A. PAJĄCZKOWSKA, and B. GEGENHEIMER
Normal State Transport Study of the Ba-K-Bi-O Superconductor
72.80 and 74.70; 72.20; S10.15
(b) Vol.155,No.2, K123-K126

D. MEIMARIS, F. VIRAS, and P. BEKRIS
A Study of the Structural Changes in Cd-Ge-Sb-As Glass during Transition to the Crystalline State Induced by Thermal Treatment
64.70 and 72.80; 61.40; S7.16
(a) Vol.115,No.2, 483-490

V.V. LOPATIN and A.V. KABYSHEV
On the Connection between the Building of Aluminium Nitride Ceramics and Its Dielectric Properties
61.40; 72.80; 77.20; S1.62
(a) Vol.116,No.1, 221-230

72.90. Other topics in electronic transport in condensed matter

R.A. VARDANYAN and S.G. VESELKO
Dislocation Recharging in the Plastically Deformed Crystals of the A_2B_6-Type
61.70 and 71.55; 72.90; S8.1
(b) Vol.156,No.2, 527-536

73. Electronic Structure and Electrical Properties of Surfaces, Interfaces, and Thin Films

73.20. Electronic surface states
(for emission and impact phenomena, see 79.)

C. ENGLER
Potential Surfaces for Chemisorption on Semiconductor Surfaces from Pair Potential Considerations
68.20; 71.10; 73.20; S5.11
(b) Vol.151,No.2, 555-564

A. SZÁSZ and D.J. FABIAN
Subsurface Hydrogen Exciton-Like States in Solid Catalysts
68.45 and 73.20; 71.35; S1
(b) Vol.152,No.1, 117-122

L. SZARO
Steady-State Surface Photovoltage in Large-Gap Semiconductors under Sub-Bandgap Illumination
73.20 and 73.30; 72.40
(a) Vol.113,No.1, 97-106

V.V. BOLOTOV, V.M. EMEKSUZYAN, V.N. SPIRIDONOV, K. SCHMALZ, and M. TRAPP
Radiation-Induced Redistribution of Gold in the SiO_2-Si Structures
61.80 and 73.20; 68.48; S5.11
(a) Vol.113,No.2, 315-320

T. HARA, A. SUGA, and R. ICHIKAWA
Properties of CVD WSi_x Films and WSi_x/GaAs Schottky Barriers
73.30 and 73.60; 73.20; S1.61; S7.12
(a) Vol.113,No.2, 459-466

H. YAMAMOTO, Y. KANIE, and K. TANIGUCHI
Transmission Coefficient and Resonance Condition in Rectangular N-Fold Barrier Structures
73.20 and 73.40
(b) Vol.154,No.1, 195-200

K. HEILIG, E. KOLBIG, and J. REINEKE
Electrical Properties of Silicon/Lead Boro-Alumino-Silicate Glass Interfaces: Insulator Charge and Distribution of Interface States
73.20 and 73.40; S5.11; S10
(a) Vol.114,No.2, 579-586

M. MEBARKI
GaSb(n)-$Ga_{0.83}Al_{0.17}Sb$(p) Heterojunction by Capacitance-Voltage Measurements
73.40; 73.20; S7.13; S7.15
(a) Vol.115,No.1, 173-178

E. TARANKO, G.M. GAVRILENKO, R. TARANKO, and V.K. FEDYANIN
On Hydrogen Chemisorption on Simple Metals
68.20 and 68.45; 73.20
(b) Vol.155,No.2, 481-490

EN-GE WANG, LI-YUAN ZHANG, and HUAI-YU WANG
Electronic Properties of Vacancy-Induced Structures in AlAs/GaAs Superlattices
71.55; 71.20; 73.20; S7.12
(b) Vol.155,No.2, 587-596

E.P. POKATILOV, S.I. BERIL, V.M. FOMIN, and G.YU. RIABUKHIN
Surface Polarons at the Contact of Two Polar Crystals
71.38 and 73.20
(b) Vol.156,No.1, 225-234

M. TOMÁŠEK, Š. PICK, G.V. GADIYAK, and YU.N. MOROKOV
Simple Analysis of Atomic Forces Acting on Silicon Surfaces
68.25 and 68.40; 73.20; S5.11
(b) Vol.156,No.1, 243-252

V.K. GOYAL, D. KUMAR, and K. SEN
Effect of Grain Boundary States Density in the Polycrystalline MIS Solar Cells
73.20 and 85; 61.70; 72.40
(a) Vol.116,No.1, 425-432

H.K. SY
Plasma Oscillations at the Surface of a Type II Superlattice
71.45 and 73.20
(b) Vol.156,No.2, 455-460

F. BECHSTEDT and D. REICHARDT
Calculation of Surface-Induced Core-Level Shifts for Covalent Semiconductors C, Si, Ge, and alpha-Sn. I.(111)2x1 Surfaces
68.20 and 73.20; S5; S5.11; S5.12
(b) Vol.156,No.2, 471-486

73.25. Surface conductivity

B. MALLIK
Electrical Conduction in Ferrocene with Adsorbed Vapours
72.80; 73.25; S12
(a) Vol.111,No.1, K61-K66

A. POGGI and E. SUSI
Effect of Passivating Oxides on the Surface Recombination Velocity in Silicon
73.25; S5.11
(a) Vol.113,No.1, K61-K66

M. CHYBICKI, J. LIWO, and K. TRZEBIATOWSKI
Electrical Surface Conductivity Modifications of Lead-Silicate Glasses
73.25; S10.15
(a) Vol.115,No.2, K185-K190

73.30. Surface double layers, Schottky barriers, and work functions

S.L. VINOGRADOV, O.K. MALINOVSKAYA, and V.A. TOLOKONNIKOV
Relaxation of Trapping Sites in ZnSe Films of Au-ZnSe-SiO_2-Si and Au-ZnSe-Ge Structures
71.55 and 73.30; S8.12
(a) Vol.111,No.1, 201-210

G.A. ADEGBOYEGA
Schottky Barrier Height Modification on n-Type Silicon by Wet Chemical Etching
73.30; S5.11
(a) Vol.111,No.1, K31-K36

A. SAKALAS and S. ZHUKAUSKAS
Unified Inverted Defect Model for GaAs Schottky Barrier Formation
73.30; S7.12
(a) Vol.111,No.1, K43-K48

L. SZARO
Steady-State Surface Photovoltage in Large-Gap Semiconductors under Sub-Bandgap Illumination
73.20 and 73.30; 72.40
(a) Vol.113,No.1, 97-106

T. HARA, A. SUGA, and R. ICHIKAWA
Properties of CVD WSi_x Films and WSi_x/GaAs Schottky Barriers
73.30 and 73.60; 73.20; S1.61; S7.12
(a) Vol.113,No.2, 459-466

L. HERNANDEZ and C. PELOSI
InGaAs/InP Schottky Barrier Diode
73.30 and 73.40; S7.11; S7.15
(a) Vol.113,No.2, 677-684

M. JAYACHANDRAN, M.J. CHOCKALINGAM, and V.K. VENKATESAN
Characteristics of Electrodeposited SnO_2-CdS/Au Schottky Devices
73.30; 85; S8.11
(a) Vol.113,No.2, K217-K222

B.M.S. LEWERENTZ and B.G. SVENSSON
Capacitance-Voltage Characteristics of Au-Si Schottky Diodes: Influence of Electron Irradiation
61.80 and 73.30; S5.11
(a) Vol.114,No.2, K147-K152

J.L. VICENTE, A. PAOLA, A. RAZZITTE, E.E. MOLA, and S.B. TRICKEY
Static Quantum Size Effects in Ultra-Thin Beryllium Films
73.30; S4
(b) Vol.155,No.2, K93-K98

73.40. Interfaces

J.-P. ZOELLNER and G. PAASCH
Conditions for the Occurence of the Electric Quantum Limit in Inversion Layers
71.25 and 73.40; S5.11; S5.12; S7.1
(b) Vol.151,No.1, 145-158

C.D. LOKHANDE and S.H. PAWAR
Electrodeposition of Thin Film Semiconductors (Review Article)
68.55; 73.40; 85; S8
(a) Vol.111,No.1, 17-40

S.V. KOVESHNIKOV, E.B. YAKIMOV, N.A. YARYKIN, and V.A. YUNKIN
Plasma Stimulated Impurity Redistribution in Silicon
61.80 and 71.55; 73.40; S5.11
(a) Vol.111,No.1, 81-88

J. ŠIKULA, P. VAŠINA, and I. HUETTEL
The Noise of the GaAlAs Super-LED Radiation
72.70 and 73.40; 78.60; S7.15
(a) Vol.111,No.1, 367-370

I. ZÓLOMY
Approximate Calculation of the Switching Voltage of MISS
73.40 and 85
(a) Vol.111,No.1, 371-376

H.-J. WUENSCHE and H. WENZEL
Influence of Surface Recombination on the Rate of Electron Leakage in Ridge-Waveguide Injection Lasers
73.40 and 85; S7.11; S7.15
(a) Vol.111,No.1, 377-386

H. PAGNIA and N. SOTNIK
Metal Influence on Switching MIM Diodes
73.40 and 85; S1.3; S6; S10
(a) Vol.111,No.1, 387-394

D.D. ASKEROV, T.D. DZHAFAROV, A.M. AGAEV, S.G. ABDINOVA, and M.S. SADYGOV
Photoelectric Properties of $CuInSe_2$-GaAs Heterojunctions
72.40 and 73.40; S8.16
(a) Vol.111,No.1, K73-K76

P.P. BASU, A. BHATTACHARYYA, and D. CHATTOPADHYAY
AC Mobility of One-Dimensional Hot Electrons in Quantum-Well Wires
72.20 and 73.40
(b) Vol.151,No.2, K129-K134

K.P. GHATAK and A. GHOSAL
On the Effective Electron Mass in n-Channel Inversion Layers on Ternary Semiconductors in the Presence of a Parallel ...
71.25 and 73.40; S8.15
(b) Vol.151,No.2, K135-K140

P. MORFOULI and G. PANANAKAKIS
Electrical Study of Thin SiO_2 Layers Obtained by Anodic Oxidation of p-Type Silicon in Pure Water Using I-U, C-U,and ...
73.60; 73.40; S5.11; S10.1
(a) Vol.111,No.2, 529-540

R. FLAGMEYER, G. OELGART, K. KREHER, R. SCHWABE, F. BUGGE, L. LEHMANN, and K. JACOBS
Characterization of MOCVD Grown (Al,Ga)As/GaAs Single Quantum Well Structures by Rutherford Backscattering and Photoluminescence
73.40 and 85; 78.55; S7.12
(a) Vol.111,No.2, 541-550

I. LUKEŠ, J. HUMLÍČEK, V. VORLÍČEK, and M. ZÁVĚTOVÁ
Infrared Reflectance of the GaAlAs/GaAs System
78.30; 73.40; S7.12
(a) Vol.111,No.2, 655-666

M.V.S. LAKSHMI and K. RAMKUMAR
I-U Characteristics of MOS Structures on Polycrystalline Silicon
73.40 and 85; S5.11
(a) Vol.111,No.2, 667-674

J. SUÑÉ, I. PLACENCIA, N. BARNIOL, E. FARRÉS; and X. AYMERICH
Degradation and Breakdown of Gate Oxides in VLSI Devices
73.40 and 77.50; 73.60; 85; S10
(a) Vol.111,No.2, 675-686

K. NATARAJAN, K. RAMKUMAR, and M. SATYAM
Capacitance of p-n Junctions under Electrical Breakdown
73.40 and 85; S5.11
(a) Vol.111,No.2, K269-K272

C. GHEZZI, D. MARTIN, A. PARISINI, J.L. STAEHLI, and L. TARRICONE
Investigation of (AlGa)As/GaAs Quantum Wells by Optical and Phototransport Measurements
73.40; 71.35; 72.40; 78.55; S7.15
(b) Vol.152,No.1, 101-110

P. STREITENBERGER, P. ZIESCHE, and N. NAFARI
The Force Constant of a Bijellic Interface
71.10 and 73.40
(b) Vol.152,No.1, 111-116

W.I. KHAN
A New Model for the Negative Resistances in Semiconductor Devices
73.40 and 85; S7.12
(a) Vol.112,No.1, 221-228

W. THIELEMANN, B. RHEINLAENDER, P. NEVERMANN, and G. OELGART
Determination of the Internal Quantum Yield and of the Injection Effiency of Red-Light Emitting (Al,Ga)As-DHS Diodes
73.40 and 78.60; 68.55; 78.55; S7.15
(a) Vol.112,No.1, 229-236

A.V. SIMASHKEVICH, L.V. GORCHAK, E.V. GILAN, and K.D. SUSHKEVICH
Study of p-n Junctions Fabricated by Phosphorus-Ion Implantation into CdTe
61.80 and 73.40; S8.13
(a) Vol.112,No.1, 305-310

V.F. KOROTOV, V.I. KHITKO, V.A. YURCHENKO, and V.S. PIVOVAR
On the Specific Features of Gallium Arsenide Doping with Donor by Ion Implantation
61.80 and 73.40; S7.12
(a) Vol.112,No.1, 331-334

J.TH. ZETTLER, J. BOLLMANN, K.H. HECKNER, and H.A. KLOSE
Radiation-Induced Defects in Silicon Due to Low-Energetic Ion-Assisted Deposition of Indium Tin Oxide
61.80 and 68.55; 73.40; S10.15
(a) Vol.112,No.1, 395-404

V.S. LYSENKO, A.N. NAZAROV, G.A. NAUMOVETS, V.B. POPOV, and A.S. TKACHENKO
Manifestation of Hydrogen in Al-SiO_2-Si Structures Subjected to a RF Plasma Annealing
61.70; 71.55; 73.40; S5.11
(a) Vol.112,No.1, K9-K12

W. KRAAK, S. TROTT, and R. HERRMANN
Weak Localization Effects in n-Type Inversion Layers of InSb Grain Boundaries under High Hydrostatic Pressure
73.40; S7.13
(b) Vol.152,No.2, 481-488

HONG CHEN and XIANG WU
Effective Mass Approximation for Tunneling States with Dissipation
71.25; 73.40
(b) Vol.152,No.2, 499-504

J. PIPREK
Numerical Analysis of TVS Ionic Current Applied to the Na -Motion in Implanted MOS Oxide Layers
66.30 and 73.40; S10.1
(a) Vol.112,No.2, 551-560

K. ŽĎÁNSKÝ
Threshold Current Optimization of InGaAsP/InP Lasers with Waveguiding Layers
73.40 and 85; S7.15
(a) Vol.112,No.2, 639-650

J.J. LIOU
Theoretical Study of Forward-Voltage Heterojunction Space-Charge-Region Capacitance Including Interface-State Effects
73.40 and 85; S7.12
(a) Vol.112,No.2, 651-660

R. BANISCH, B. TILLACK, H.H. RICHTER, B. HUNGER, P. BARNA, V. SCHILLER, A. ADAM, and J. GYULAI
On the Characteristics of CMOS-Transistors in Thick SOI Films
73.40; 68.55; S5.11
(a) Vol.112,No.2, 721-726

T. SKRABKA, M.J. SŁABY, and B. MIZERA
Ohmic Contacts to n-Type GaAs Using RTA
73.40; 85; S7.12
(a) Vol.112,No.2, K105-K110

N.A. GUSHCHINA, V.K. NIKULIN, and YU.N. TZAREV
On the Band Structure of Ultrathin Layer Semiconductor Superlattices
73.40; 71.25; S7.15
(b) Vol.153,No.1, 161-166

E. FARRÉS, J. SUÑÉ, I. PLACENCIA, N. BARNIOL, and X. AYMERICH
Si-SiO_2 Interfacial Atomic Scale Roughness Caused by Inhomogeneous Thermal Oxidation
68.55 and 73.40; 68.48; S10.1
(a) Vol.113,No.1, 83-96

F.F. SIZOV and S.D. DARCHUK
Electrical Characteristics of Narrow-Gap Multilayer Semiconductor Structures
73.40; S8.15
(a) Vol.113,No.1, 107-112

V.YU. RUD, YU.V. RUD, M. SERGINOV, and M.A. TAIROV
Production of and Physical Processes in n-p-$CdSiAs_2$ Structures
72.40 and 73.40; 85; S7.16
(a) Vol.113,No.1,, 207-214

R.V. KONAKOVA, YU.S. MELNIKOVA, E.V. MOZDOR, and V.I. FINEBERG
Breakdown of Silicon p^+-n-n^+ Diodes
73.40 and 85; S5.11
(a) Vol.113,No.1, 215-222

G. GHIBAUDO
Analytical Modelling of the MOS Transistor
73.40 and 85
(a) Vol.113,No.1, 223-240

K. BHATTACHARYYA
Alloy Scattering Limited Mobility of Two-Dimensional Electron Gas in SD-$Al_{0.3}Ga_{0.7}As/Ga_{0.87}In_{0.13}As/GaAs$ Structures
72.20 and 73.40; S7.12; S7.15
(b) Vol.153,No.2, K135-K140

B. MITRA, A. GHOSHAL, and K.P. GHATAK
On a Simplified Analysis of the Gate Capacitance of MOS Field-Effect Devices in the Presence of a Tilted Quantizing Magnetic Field
73.40 and 85
(b) Vol.153,No.2, K209-K212

Z.G. IVANOV
Superconducting Transport Properties of Ohmic Contacts
73.40 and 74.50; S7.12
(a) Vol.113,No.2, 439-446

CZ. PAWLACZYK and A. PAWLOWSKI
Electric Capacitance Anomaly near the Superionic Phase Transition in $(NH_4)_4H_2(SeO_4)_3$ Single Crystals
73.40 and 77.40; S11
(a) Vol.113,No.2, 447-452

XIN-FAN HUANG, XI-MAO BAO, and JIAN CHEN
Hydrogen Passivation and CO_2 Laser Annealing of Interface between Ar^+ Laser Recrystallized Poly-Si and SiO_2 Substrate
68.48 and 73.40; S5.11
(a) Vol.113,No.2, 453-458

V.S. LYSENKO, A.N. NAZAROV, S.A. VALIEV, I.M. ZARITSKII, T.E. RUDENKO, and A.S. TKACHENKO
EPR and TSCR Investigations of Implanted Al-SiO_2-Si Systems Treated with RF Plasma Discharge
73.40 and 85; 76.30; S5.11; S10.1
(a) Vol.113,No.2, 653-666

V.D. USIKOV
Self-Stabilzed Avalanche Process in MIS Structures under Nonstationary Conditions. I. Generation of Carriers Due to Illumination
73.40; 72.40
(a) Vol.113,No.2, 667-676

L. HERNANDEZ and C. PELOSI
InGaAs/InP Schottky Barrier Diode
73.30 and 73.40; S7.11; S7.15
(a) Vol.113,No.2, 677-684

J.J. LIOU, A. WHITTAKER, and J.S. YUAN
Modeling the Two-Dimensional Emitter-Base and Base-Collector Junction Capacitances of Bipolar Junction Transistors
73.40 and 85
(a) Vol.113,No.2, K267-K272

H. YAMAMOTO, Y. KANIE, and K. TANIGUCHI
Transmission Coefficient and Resonance Condition in Rectangular N-Fold Barrier Structures
73.20 and 73.40
(b) Vol.154,No.1, 195-200

W. LAI and P.N. BUTCHER
A Weak-Coupling Calculation of the Binding Energy and Effective Mass of a Polaron in a Multiple Quantum Well
71.38; 73.40
(b) Vol.154,No.1, 229-236

D. CHATTOPADHYAY and A. KABASI
Warm Electron Transport in GaAs Quantum Wells
72.20 and 73.40; S7.12
(b) Vol.154,No.1, 281-286

CHR. PRATSCH and M. SUHRKE
Electron Transport in Quasi-Two-Dimensional Systems. Influence of Intersubband Scattering
73.40
(b) Vol.154,No.1, 315-320

M. GASGNIER
Rare Earth Compounds (Oxides, Sulfides ...)as Thin Films and Thin Crystals. Physico-Chemical Properties and Applications(Review Article)
68.55 and 73.40; 75.70; 78.65; S1.4; S1.63; S10
(a) Vol.114,No.1, 11-72

H. WIECZOREK and W. FUHS
Deep Trapping of Carriers in a-Si:H Solar Cells Studied by Transient Photocurrents
72.40 and 73.40; 71.55; S5.11
(a) Vol.114,No.1, 413-418

G. OELGART, V. GOTTSCHALCH, H. HAEFNER, R. HEILMANN, W. KLEIN, and B. RHEINLAENDER
Characterization of GaInAsP/InP Double Heterostructure Laser-Wafers
73.40 and 85; 78.60; S7.11; S7.15
(a) Vol.114,No.1, 419-430

V. PARKHUTIK, V. SOKOL, V. SHERSHULSKII, E. UTKINA, and A. VOROBEVA
Hopping Electrical Conductivity in the Al-Anodic Al_2O_3-Al System
73.40; S4; S10.1
(a) Vol.114,No.1, K33-K36

F. BECHSTEDT and H. GERECKE
Long-Wavelength Optical Phonons in GaAs-AlAs Superlattices
63.20 and 68.30; 73.40; S7.12
(b) Vol.154,No.2, 565-582

A. SCHENK, M. STAHL, and H.-J. WUENSCHE
Calculation of Interband Tunneling in Inhomogeneous Fields
72.10 and 73.40; 77.50; S8.15
(b) Vol.154,No.2, 815-826

K.P. GHATAK, B. MITRA, and A. GHOSHAL
On the Effective Electron Mass in Superlattices of III-V Semiconductors in the Presence of a Quantizing Magnetic Field
73.40; S7.15
(b) Vol.154,No.2, K121-K126

L. TESAŘ and R. LENHARD
Evaluation of Silicon Wafers with Different Concentrations of Interstitial Oxygen
61.70 and 71.55; 73.40; S5.11
(a) Vol.114,No.2, 475-480

K. HEILIG, E. KOLBIG, and J. REINEKE
Electrical Properties of Silicon/Lead Boro-Alumino-Silicate Glass Interfaces: Insulator Charge and Distribution of Interface States
73.20 and 73.40; S5.11; S10
(a) Vol.114,No.2, 579-586

K. SOMOGYI, D.N. KORBUTYAK, L.N. LASHKEVICH, and I. POZSGAI
Cu Diffusion into GaAs from Water Based Solution
66.30 and 73.40; 78.65; S7.12
(a) Vol.114,No.2, 635-642

H. WEINERT, K. JARAŠIŪNAS, and B. SCHUMANN
Carrier Diffusion and Recombination in Type II GaAs/$Ca_xSr_{1-x}F_2$ Quantum Well Heterostructures Studied by the Transient Gating Technique
73.40; 68.55; S7.12; S9.15
(a) Vol.114,No.2, K183-K188

H.V. ANANDA and M. SATYAM
Effect of Field Plate on the Electrical Breakdown Characteristics of Bipolar Transistors
73.40 and 85; S5.11
(a) Vol.114,No.2, K259-K262

V.S. LYSENKO, A.N. NAZAROV, I.M. ZARITSKII, G. SERFŐZŐ, G. BATTISTIG, J. GYULAI, and L. DOZSA
RF Plasma Modification of Heavy Destroyed Ion Implanted Subsurface Silicon Layers
61.80; 73.40; S5.11
(a) Vol.115,No.1, 75-80

M. MEBARKI
GaSb(n)-$Ga_{0.83}Al_{0.17}Sb$(p) Heterojunction by Capacitance-Voltage Measurements
73.40; 73.20; S7.13; S7.15
(a) Vol.115,No.1, 173-178

A. FELIMBAN and M. ABDEL NABY
Effect of Annealing and Doping on the Performance of CdS/ZnTe Heterojunctions
73.40; S8.11; S8.13
(a) Vol.115,No.1, K27-K30

S.I. DRAPAK, V.N. KATERINCHUK, Z.D. KOVALYUK, and V.A. MANASSON
The Injection Peculiarities of Minority Charge Carriers in the Anisotype SIS Structure
73.40 and 78.60; S8; S10.1
(a) Vol.115,No.1, K35-K38

I.G. MARCHENKO
The Effect of Radiation Defects on the Differential Resistance of p-n Junctions during Avalanche Breakdown
61.80 and 73.40; S5.11
(a) Vol.115,No.1, K39-K42

B.C. CHAKRAVARTY, S.K. SHARMA, S.N. SINGH, B.K. DAS, RAVI KUMAR, and B.R. CHAKRABORTY
Effect of the Growth of Thin SiO Interlayers on the Performance of Si Solar Cell
68.55 and 85; 73.40; S5.11
(a) Vol.115,No.1, K125-K128

I. ZÓLOMY
Theory of the Negative Resistance Avalanche MIS Tunnel Diode
73.40 and 85; S5.11
(a) Vol.115,No.1, K129-K132

A.F. KOVALENKO, E.N. SOVYAK, and M.F. GOLOVKO
Electronic States of a Hydrogen Atom near a Hard Wall
71.55 and 73.40
(b) Vol.155,No.2, 549-558

D.A. ARONOV, B.R. MAMATKULOV, and R. MAMATKULOV
The Influence of Diffusion on the Carrier Exclusion Effect in High-Purity (Intrinsic) Semiconductors with Non-Injecting Contacts
73.40; S5.12
(a) Vol.115,No.2, K167-K170

E. NEBAUER and U. LOHSE
Lamp Furnace-Annealed Au/Ni/AuGe Ohmic Contacts for GaAs MESFETs
73.40; 85; S7.12
(a) Vol.115,No.2, K171-K174

K. NATARAJAN, K. RAMKUMAR, and M. SATYAM
Reverse Characteristics of Phosphorous Passivated Polysilicon p-n Junction Diodes
73.40 and 85; 61.70; S5.11
(a) Vol.115,No.2, K265-K268

V.V. KARYAGIN and I.I. LYAPILIN
Thermo-Galvanomagnetic Phenomena in Spin Quantum Wells
71.10 and 73.40; 72.20; S8.15
(b) Vol.156,No.1, 235-242

G.I. URUSHADZE
Quantum Voltage Fluctuations in Normal Tunnel Junctions and Departure of Conductivity from the Ohmic Law
72.20 and 73.40
(b) Vol.156,No.1, K63-K66

A. ZEHE and A. JUAREZ
Double-Peak Formation as a Doping-Superlattice Effect in Donor-Acceptor Pair-Luminescence Spectra
71.55 and 78.55; 73.40; S7.12
(b) Vol.156,No.1, K89-K92

SH.M. EFENDIEV, A.A. AGASIEV, V.E. BAGIEV, and YA.YU. GUSEINOV
Nonlinear Current-Voltage Characteristics in Me-Bi_2O_3-Me Structure
73.40; 72.20; S10
(a) Vol.116,No.1, 305-312

N. MAZUMDER, J.P. BANERJEE, and S.K. ROY
Variation of High Frequency Negative Resistance of Silicon n^+pp^+ and GaAs p^+nn^+ IMPATT Diodes with Enhancement of Reverse Saturation ...
73.40 and 85; S5.11; S7.12
(a) Vol.116,No.1, 415-424

P.C. EUTHYMIOU, P.C. BANBURY, G.J. PAPAIOANNOU, and G.E. ZARDAS
Dependence of I-V Characteristics of GaAs MESFET on Temperature and Eletron Irradiation
73.40 and 85; 61.80; S7.12
(a) Vol.116,No.1, K133-K136

M. ASCHE, L. EAVES, K. FRIEDLAND, H. KOSTIAL, and O.G. SARBEY
Transverse Transport in $n^+GaAs/Al_xGa_{1-x}As/n$-GaAs Structures
73.40; S7.12; S7.15
(b) Vol.156,No.2, 461-470

A.K. RAY, C.A. HOGARTH, and R.S. PANK
Parametric Dependence of the Conduction in Electroformed Insulating Films
73.60; 73.40
(a) Vol.116,No.2, 703-708

B. MAJKUSIAK
A Modified Method for Determination of Surface Mobility of Carriers in MOS Transistors
73.40 and 85
(a) Vol.116,No.2, 833-836

R. AGRAWAL, R. DWIVEDI, and S.K. SRIVASTAVA
Analysis of an MOS Inverter with Non-Uniform Impurity Profile
73.40 and 85
(a) Vol.116,No.2, 837-844

I. ZÓLOMY
Erratum to Theory of the Negative Resistance Avalanche MIS Tunnel Diode (phys. stat. sol. (a) 115, K129(1989))
73.40 and 85; S5.11
(a) Vol.116,No.2, K223-K224

73.60. Electronic properties of thin films

B.I. KHUDIK, V.Z. LOZOVSKII, and A.F. ZHURAVLEV
Optical Theorem in Ultra-Thin Film Electrodynamics
73.60 and 78.65
(b) Vol.151,No.1, 111-120

J.C. BERNÈDE, A. MALLOUKY, and J. POUZET
Etude de la résistivité de couches minces polycristallines de $MoSe_2$
73.60; S8
(a) Vol.111,No.1, 181-192

M.S. MURALI SASTRY
Large Scale Coalescence in Terms of a New Agglomeration Rate
68.55; 73.60; S1.3; S4
(a) Vol.111,No.1, K37-K42

F. LAGNEL, B. POUMELLEC, and C. PICARD
Electron Transport in Amorphous (Ti,Nb)O_2 Solid Solutions
73.60; S10.15
(b) Vol.151,No.2, 531-544

P. MORFOULI and G. PANANAKAKIS
Electrical Study of Thin SiO_2 Layers Obtained by Anodic Oxidation of p-Type Silicon in Pure Water Using I-U, C-U,and ...
73.60; 73.40; S5.11; S10.1
(a) Vol.111,No.2, 529-540

H. DINTNER and F. THRUM
The Homogeneous Distribution of Non-Metallic Bonds in As-Deposited NiCr-O Resistive Films
72.15 and 73.60; S1.2; S10.1
(a) Vol.111,No.2, 551-558

J. SUÑÉ, I. PLACENCIA, N. BARNIOL, E. FARRÉS, and X. AYMERICH
Degradation and Breakdown of Gate Oxides in VLSI Devices
73.40 and 77.50; 73.60; 85; S10
(a) Vol.111,No.2, 675-686

L. MAGAFAS, A.N. ANAGNOSTOPOULOS, and J.G. ANTONOPOULOS
On the Conductivity of Amorphous CdS Films
73.60; S8.11
(a) Vol.111,No.2, K175-K178

J. BÁNKUTI and G. HORVÁTH.
Thickness-Dependent Electrical Resistivity of Thin Rough Metal Films
73.60
(a) Vol.111,No.2, K185-K190

A. SINGH
Two-Step Annealing for Cobalt Disilicide Formation with Lowest Sheet Resistance
68.55 and 73.60; S1.61
(a) Vol.111,No.2, K191-K194

J. MAIER
Thin Film Conductance: Evaluation and Correlation with Defect Formation Parameters
73.60; S9
(a) Vol.112,No.1, 115-122

P. ALTANZOG, D. BATSUUR, S. BOLD, B. BURMAA, SH. CHADRAABAL, and A.G. KASANSKII
Temperature Dependence of Dark Conductivity and Steady-State Photoconductivity of a-Si:H Samples
72.40 and 73.60; S5.11
(a) Vol.112,No.1, K23-K28

N.C. HALDER
AC Conduction in Metal-Oxide Cermets. I. Ta-Oxide Resistors
73.60; S10.1
(a) Vol.112,No.2, 589-600

W. SCHARFF, K. HAMMER, B. EIBISCH, O. STENZEL, S. ROTH, and T. FRAUENHEIM
Deposition of Hyrdogenated Amorphous Carbon Films by a ECR Microwave Plasma
61.80 and 68.55; 73.60; 78.65; S5
(a) Vol.112,No.2, 667-678

N. DHIREN SINGH and H.N.K. SARMA
In situ Electrical Behaviour of Thin Films of Antimony-Tellurium Alloy
73.60; S4
(a) Vol.112,No.2, K101-K104

B.S. ACHARYA, S.K. SINGH, and B.B. NAYAK
X-Ray Emission Spectroscopy of the Electronic States in $Pb_{1-x}Hg_xS$ Thin Films
71.20 and 73.60; 78.70; S8.15
(b) Vol.153,No.1, 153-160

R. SIGNERSKI, J. KALINOWSKI, P. DI MARCO, and G. GIRO
Photoconductivity of Polycrystalline Thionaphtenindol
72.40 and 73.60; S12
(b) Vol.153,No.1, 281-294

M. ZAŁUŻNY and M. PIŁAT
On the Energy Spectrum of Size-Quantized Films in an Oriented Magnetic Field
71.25 and 73.60
(b) Vol.153,No.1, K21-K24

G. GOVINDARAJ
Electron Transport in Metallic Thin Films. A Comparative Study
72.15 and 73.60
(a) Vol.113,No.1, 113-122

K. INOUE, K. IKEZAKI, and M. KATO
Thermally Stimulated Currents in Plasma Treated FEP Films
73.60; S12
(a) Vol.113,No.1, K67-K70

H. RUECKER, R. ENDERLEIN, and F. BECHSTEDT
Strain Effects on the Band Structure of $Si/Si_{1-x}Ge_x(001)$ Superlattices
71.25 and 73.60; S5.11; S5.12
(b) Vol.153,No.2, 595-610

M. ZAŁUŹNY and W. OKULSKI
On Electron States of Quantum Wells with Anisotropic and Tilted Valleys in Magnetic Field
71.25 and 73.60; S8
(b) Vol.153,No.2, K115-K118

T. DRUESEDAU, M. ECKLER, R. BINDEMANN, and H. FIEDLER
Enhancement of Photoconductivity in Magnetron-Sputtered a-Si:H Doped with Nitrogen
68.55 and 73.60; 72.40; 78.65; S5.11
(b) Vol.153,No.2, K119-K124

T. HARA, A. SUGA, and R. ICHIKAWA
Properties of CVD WSi_x Films and WSi_x/GaAs Schottky Barriers
73.30 and 73.60; 73.20; S1.61; S7.12
(a) Vol.113,No.2, 459-466

M.J. MAŁACHOWSKI, J. PIOTROWSKI, and A. ROGALSKI
Influence of Dislocaions on the Performance of 3 to 5 μm $Hg_{1-x}Cd_xTe$ Graded Gap Photoresistors
72.40 and 73.60; S8.15
(a) Vol.113,No.2, 467-476

N.C. HALDER and L.C. NAGARAJAN
AC Conduction in Metal-Oxide Cermets. II. Sn-Oxide Resistors
73.60; S10
(a) Vol.113,No.2, 511-518

WANLU WANG and KEJUN LIAO
Persistent Photoconductivity in a-Si:H/a-SiN_x:H Multilayer Films at Low Temperature
72.40 and 73.60; S5.11; S6
(a) Vol.113,No.2, K223-K226

B.N. NARAHARI ACHAR and J.R. FERGUSON
Quantized Helicon Damping in a Semiconductor Superlattice in the Integral Quantum Hall Regime
72.20 and 73.60
(b) Vol.154,No.1, K59-K62

O.I. VASIN, A.E. KLIMOV, I.G. NEIZVESTNYI, and V.N. SHUMSKII
The Three-Layer Structure of $Pb_{0.8}Sn_{0.2}Te$ Epitaxial Films Prepared by MBE Technique on BaF_2
73.60; 68.55; S8.15
(a) Vol.114,No.1, 161-166

M.A. EL HITI and M.A. AHMED
Effect of Thermal Expansion on the Temperature Coefficient of Resistance of Double-Layer Thin Metallic Film
73.60
(a) Vol.114,No.1, 185-190

P. KASHYAP, M. JAIN, O.P. AGNIHOTRI, and H.K. SEHGAL
Effect of Composition on the Optical and Electrical Characteristics of Flash Evaporated $Cu_{2-x}In_xSe_2$ Thin Films
73.60 and 78.65; S8.16
(a) Vol.114,No.1, K29-K32

A. URBANIAK-KUCHARCZYK, T. BALCERZAK, J. MIELNICKI, and G. WIATROWSKI
The Magnetic Contribution to the Electrical Resistivity in the Ising Ferromagnet
73.60
(b) Vol.154,No.2, 583-588

J. LEDZION and J. KIERUL
Thickness Dependence of Thin Film Conductivity
72.15 and 73.60; S4
(a) Vol.114,No.2, 545-550

E.M. ZUB and S.V. SUKHVALO
On the Mechanism of Charge Transport in the $Cu_xFe_{1-x}Cr_2S_4$ Solid Solution Films
73.60; S1.63
(a) Vol.114,No.2, 573-578

M.S. MURALI SASTRY and MANJUNATHA PATTABI
Variation of the Tunnelling Barrier in Island Cu Films
73.60; 68.55; S1.3
(a) Vol.114,No.2, K179-K182

M.A. EL HITI
Dependence of the Temperature Coefficient of the Strain Coefficients of Resistance of Double-Layer Thin Metallic Films on Thermal Strains
73.60; S1.3
(a) Vol.115,No.1, 185-190

B.I. BAYRACHNYI, L.V. LIASHOK, V.P. GOMOZOV, and L.I. SKATKOV
Changes in Structure and Properties of Nb_2O_5 Anodic Films Caused by Generating Anion Defects on Their Surface
68.55; 73.60; S10
(a) Vol.115,No.1, 207-212

M.A. MOHAMMED and D.V. MORGAN
Reliability of Tantalum Oxide Film Capacitors
73.60; S10.1
(a) Vol.115,No.1, 213-222

B.B. ISMAIL and R.D. GOULD
Structural and Electronic Properties of Evaporated Thin Films of Cadmium Telluride
68.55 and 73.60; S8.13
(a) Vol.115,No.1, 237-246

O.A. GOLIKOVA and A.M. GREKHOV
Structural Network and Some Photoelectrical and Optical Parameters of Amorphous Hydrogenated Silicon
71.20; 73.60; 78.65; S5.11
(b) Vol.155,No.2, 475-480

A.A. ZAHAB, M. ABD-LEFDIL, and M. CADENE
Effect of Annealing on Structure and Electrical Conductivity of Evaporated Indium Telluride
68.55 and 73.60; S8
(a) Vol.115,No.2, 491-496

W. SCHNELLE and U. DILLNER
Electrical and Galvanomagnetic Properties of Undoped and Doped Polycrystalline Bismuth Films. I. Preparation and Experimental ...
73.60; S4
(a) Vol.115,No.2, 505-514

Y.SATYANARAYANA MURTHY, B.SRINIVASULU NAIDU, and P.JAYARAMA REDDY
Electrical Conductivity of Single Phase $AgGaSe_2$ Thin Films
73.60; S8.16
(a) Vol.115,No.2, K175-K180

PREETI KASHYAP, MUKESH JAIN, O.P. AGNIHOTRI, and H.K. SEHGAL
Effect of Composition on the Optical and Electrical Parameters of Flash Evaporated $Cu_{2-y}In_yTe_2$ Thin Films
73.60 and 78.65; S8.16
(a) Vol.115,No.2, K181-K184

U. DILLNER and W. SCHNELLE
Electrical and Galvanomagnetic Properties of Undoped and Doped Polycrystalline Bismuth Films. II. Analysis in an Anisotropic ...
73.60; 72.10; S4
(a) Vol.116,No.1, 337-342

M. LIBRA, L. BÁRDOŠ, and H. BARÁNKOVÁ
Influence of Post-Deposition Annealing on the Electrical Conductivity of Indium-Tin-Oxide Films
73.60; S10.1
(a) Vol.116,No.1, K65-K68

D.A. PAVLOV, A.F. KHOKHLOV, R.V. KUDRYAVTSEVA, and A.V. ERSHOV
Structure and Properties of Amorphous Silicon Doped with Isovalent Impurities
68.55 and 73.60; S5.11
(a) Vol.116,No.2, 697-702

A.K. RAY, C.A. HOGARTH, and R.S. PANK
Parametric Dependence of the Conduction in Electroformed Insulating Films
73.60; 73.40
(a) Vol.116,No.2, 703-708

XIANG-FENG JIN, BING MAO, SHU-QIN ZHOU, REN-YUAN QIAN, J.S. PARK, and P. EHRLICH
Carrier Transport Properties of Trans-Polyphenylacetylene
73.60; 71.55; S12
(a) Vol.116,No.2, 709-714

U. DILLNER and W. SCHNELLE
Electrical and Galvanomagnetic Properties of Undoped and Doped Polycrystalline Bismuth Films. III. Analysis in an Anisotropic ...
73.60; 72.10; S4
(a) Vol.116,No.2, 715-724

A. MITTAS, N. GEORGOULAS, D. GIRGINOUDI, and A. THANAILAKIS
Properties of Amorphous Non-Hydrogenated Silicon-Tin Alloys
73.60 and 78.65; S6
(a) Vol.116,No.2, 725-728

73.90. Other topics in electrical properties of surfaces, interfaces, and thin films

M. TROTT, S. TROTT, and CH. SCHNITTLER
Normalization, Orthogonality, and Completeness for the Finite Step Potential
71.25 and 73.90
(b) Vol.151,No.2, K123-K128

74. Superconductivity

74.10. Occurrence, critical temperature

A.E. BOCQUET, J.F. DOBSON, P.C. HEALY, S. MYHRA, and J.G. THOMPSON
X-Ray Photoelectron Spectra of High Tc Bismuth Compounds: General Trends for Layer Structures
71.45 and 79.60; 74.10; S11
(b) Vol.152,No.2, 519-532

J. STEUER
Comparison of A-15 Nb_3Sn Films Deposited on Hot and Cold Substrates
74.10 and 74.50; S1
(a) Vol.114,No.2, 665-678

S.L. KAKANI and R.K. PALIWAL
Theoretical Study of Density of States and Specific Heat of Itinerant Ferromagnetic Superconductors
74.10 and 75.40; S10
(b) Vol.155,No.1, 241-248

74.20. Theory

V.A. KOVARSKII
Bielectron (Bihole) Mechanism of Superconductivity
74.20
(b) Vol.151,No.1, K35-K40

NGUYEN VAN HIEU and NGUYEN AI VIET
On the RVB Theory of High-Tc Superconductivity
74.20
(b) Vol.151,No.2, K171-K176

A.G. SAIF and J. CHELA-FLORES
Nucleation of Superconductivity in Ceramic Oxides
74.20; 74.70; S10.15
(b) Vol.152,No.2, 617-624

K. FUKUSHIMA and H. SATO
Possible Effect of Two-Dimensionality on the Superconducting Pairing Interaction
74.20; S10.15
(b) Vol.153,No.2, K141-K146

P. WRÓBEL
Coexistence of Superconductivity and Antiferromagnetism in the Two-Dimensional Extended Hubbard Model with Inter-Site...
74.20 and 75.10
(b) Vol.153,No.2, K177-K182

A.I. ILINIKH and B.YA. SHAPIRO
Josephson Properties of a Homogeneous Superconducting Film
74.20 and 74.50
(b) Vol.154,No.2, 679-690

R. GONCZAREK
Pauli Paramagnetism in Superconducting Systems with Discontinuous Order Parameter
74.20
(b) Vol.154,No.2, K157-K162

R. MAŃKA and A. MOLAK
The Soliton Generalization of Bipolaronic Model Proposed for the High-Tc Superconductivity in Lanthanum and Yttrium Cupric Oxides
74.20; 71.38; S10
(b) Vol.155,No.1, 231-240

R.I. SHEKHMAMETEV and N.V. STAROSTIN
One-Sign Charge Carrier Coupling in Crystals with Opposite Sign Curvature Bands
74.20
(b) Vol.155,No.1, K51-K54

E.R. KORUTCHEVA
Mean-Field Theory of an Extended Hubbard Model
74.20; S10.15
(b) Vol.155,No.2, K119-K122

E.V. KHOLOPOV
Hole Polarisation Nature of Boson Modes Forming High-Tc Superconductivity
74.20; S10.15
(b) Vol.155,No.2, K131-K136

S.P. IONOV and V.S. LUBIMOV
About the Hamiltonian of Interaction of Local Pairs with Electronic Metallic Band in Superconductors
74.20
(b) Vol.155,No.2, K137-K142

V.E. ZUBKUS, S. LAPINSKAS, and E.E. TORNAU
Calculation of Phase Diagrams in the Lattice-Gas Model by a Modified Cluster Field Method. Application to High-Temperature Superconductors
64.60 and 74.20
(b) Vol.156,No.1, 93-104

V.N. ERMAKOV and S.P. KRUCHININ
One-Particle Excitation Spectrum in a Bisoliton Model of Superconductivity
71.45 and 74.20
(b) Vol.156,No.1, 333-338

E. KRUEGER
One- and Two-Dimensional Sublattices as Preconditions for High-T_c Superconductivity
74.20
(b) Vol.156,No.1, 345-354

GANG SU
The Elementary Excitations and the Quasi-Fermi Surface in the Baskaran-Zou-Anderson Mean-Field Theory
74.20
(b) Vol.156,No.1, K51-K56

A.A. KISELEV and A.V. LIAPZEV
Effects of Non-Adiabatic Electron-Phonon Interactions in Conducting Systems
74.20
(b) Vol.156,No.1, K57-K62

M. FUKUHARA and G.W. SMITH II
New Approach to the Theory of High-Temperature Superconductivity
74.20
(b) Vol.156,No.2, 583-596

J.M. WESSELINOWA
On the Theory of Superconductivity in the Extended Hubbard Model
74.20
(b) Vol.156,No.2, K129-K134

N. PRUSS, W. KRAAK, H.-U. MUELLER, A. JACOBI, H. DWELK, and R. HERRMANN
Influence of the Sintering Temperature on the Physical Properties of $YBa_2Cu_3O_{7-\delta}$ Ceramic Samples
72.20 and 74.20; S10.15
(a) Vol.116,No.2, 793-802

74.30. General properties

A.B. NAZARENKO
On the Analysis of Thermodynamical Relations for Superconductors
74.30
(b) Vol.151,No.2, K167-K170

74.40. Fluctuations and critical effects

V.I. GREBENNIKOV and O.B. SOKOLOV
The Local Electron-Spin-Density Fluctuation Mechanism of the Superconductivity in Narrow-Band Metals
74.40
(b) Vol.151,No.2, 623-632

74.50. Tunneling phenomena, Josephson effect, and proximity effects

Z.G. IVANOV
Superconducting Transport Properties of Ohmic Contacts
73.40 and 74.50; S7.12
(a) Vol.113,No.2, 439-446

C. GIOVANELLA
The Superconductor Glasses (SuG). A Fractal Description
74.50; 64.70; S10.15
(b) Vol.154,No.1, 273-280

A.I. ILINIKH and B.YA. SHAPIRO
Josephson Properties of a Homogeneous Superconducting Film
74.20 and 74.50
(b) Vol.154,No.2, 679-690

J. STEUER
Comparison of A-15 Nb_3Sn Films Deposited on Hot and Cold Substrates
74.10 and 74.50; S1
(a) Vol.114,No.2, 665-678

P. SEIDEL and M. TURTENWALD
Multi-Gap Model for Tunneling in High-T_c Superconductors
74.70; 74.50
(a) Vol.115,No.1, 273-278

74.60. Type-II superconductivity

A.G. SAIF
Magnetostatics of Anisotropic Superconducting Ellipsoid
74.60
(a) Vol.111,No.2, 573-582

W. ANDRAE, H. DANAN, and R. HERGT
Anisotropy of Superconducting Properties of a $YBa_2Cu_3O_{7-x}$ Crystal
Investigated by Torque Measurements
74.60 and 74.70; S10.15
(a) Vol.111,No.2, 583-590

A.G. SAIF
The Behaviour of the Vortex State in Magnetic Superconductors
74.60
(b) Vol.152,No.1, 261-266

74.70. Superconducting materials

J. ZIELIŃSKI and M. MATLAK
Role of Interlayer Interactions in High-T_c Superconductors
74.70
(b) Vol.151,No.1, 203-210

A.I. GUSEV and A.A. REMPEL
Superconductivity in Disordered and Ordered Niobium Carbide
74.70; S1.61
(b) Vol.151,No.1, 211-224

V. DVOŘÁK
Landau Theory of the Structural Phase Transition in $YBa_2Cu_3O_{7-x}$
64.70; 74.70; S10.15
(b) Vol.151,No.1, K23-K28

O. SMRČKOVÁ, D. SÝKOROVÁ, L. SMRČKA, and P. VAŠEK
Preparation of Pb-Doped Bi-Sr-Ca-Cu-O Superconductors
74.70; S10.15
(a) Vol.111,No.1, K77-K82

B.B. BOIKO, A.I. AKIMOV, V.I. GATALSKAYA, S.E. DEMYANENKOV, L.A. KUROCHKIN, M.L. PETROVSKII, E.K. STRIBUK, and V.M. FINSKAYA
Anomalous Electrical Resistance of Y-Ba-Cu-O System Ceramics
74.70; S10.15
(a) Vol.111,No.1, K89-K92

K. BENTE, E. SCHWARZMANN, and D. SIEBERT
EPR Studies of Cr^{2+} from Impurity Phases in the High Temperature Superconducting $YBa_2Cu_3O_{7-x}$
74.70 and 76.30; S10.15
(a) Vol.111,No.1, K93-K98

A.A. BLINKIN, V.N. GOLOVIN, V.V. DEREVYANKO, G.F. TIKHINSKII, and V.A. FINKEL
Substitution Dependence of the Crystal Structure and Superconductivity in High-T_c Superconductors
74.70; S10.15
(b) Vol.151,No.2, K161-K166

W. ANDRAE, H. DANAN, and R. HERGT
Anisotropy of Superconducting Properties of a $YBa_2Cu_3O_{7-x}$ Crystal Investigated by Torque Measurements
74.60 and 74.70; S10.15
(a) Vol.111,No.2, 583-590

B.B. BOIKO, A.I. AKIMOV, V.I. GATALSKAYA, S.E. DEMYANOV, A.L. KARPEI, L.A. KUROCHKIN, YU.N. LEONOVICH, A.K. LETKO, M.N. MURAYA et al.
High T_c Superconductivity in Tl-Ba-Ca-Cu-O Ceramics
74.70; S10.15
(a) Vol.111,No.2, K209-K214

M. PĘKAŁA, K. PĘKAŁA, and A. PEJĄCZKOWSKA
Normal State Thermoelectric Power of Bi-Sr-Ca-Cu-O Superconductors
72.20 and 74.70; S10.15
(b) Vol.152,No.1, K1-K4

R. HERRMANN, U. PREPPERNAU, CH. RELLIN, A. KRAPF, and H. DWELK
Normal State Thermoelectric Power of Bi-Sr-Ca-Cu-O High-T_c Superconductor
72.20 and 74.70; S10.15
(b) Vol.152,No.1, K5-K8

N. GUSKOS, TH. LEVENTOURI, CH. TRIKALINOS, M. CALAMIOTOU, S.M. PARASKEVAS, J. KURIATA, and L. SADLOWSKI
On the Possibility of the Excistence of Cu^{3+} Ions in High-T_c Superconducting Samples of $YBa_2Cu_3O_x$
74.70 and 76.30; S10.15
(b) Vol.152,No.1, K9-K12

J. BASZYŃSKI
Superconducting $YBa_2Cu_3O_{7-\delta}$ Textured Films on MgO Substrates
68.55 and 74.70; S10.15
(a) Vol.112,No.1, K39-K42

A.G. SAIF and J. CHELA-FLORES
Nucleation of Superconductivity in Ceramic Oxides
74.20; 74.70; S10.15
(b) Vol.152,No.2, 617-624

H. GRUBER, E. KRAUTZ, H.P. FRITZER, K. GATTERER, and A. POPITSCH
Electrical Conductivity, Magnetic Susceptibility, and Infrared Spectra in the Superconducting Systems Bi-Sr-Ca-Cu-O and Tl-Ba-Cu-O
74.70; 75.30; 78.30; S10.15
(b) Vol.153,No.1, 307-314

N. PRUSS, W. KRAAK, N. KUBICKI, H. DWELK, and R. HERRMANN
The Influence of Boron on the Electrical Properties of the Ceramic Superconductor $YBa_2Cu_3O_{7-\delta}$
74.70; S10.15
(b) Vol.153,No.1, K63-K68

Z.C. KANG, D. MORIN, L. PIERRE, O. MONNEREAU, and C. BOULESTEIX
AEM Study of the High Temperature Superconductor $Bi_2Sr_2CaCu_2O_8$
61.60 and 74.70; 61.16; S10
(a) Vol.113,No.1, 23-34

J. PIECHOTA, Z. KOZIOŁ, A. PAJĄCZKOWSKA, and H. SZYMCZAK
AC Susceptibility in Ceramic and Powdered $YBa_2Cu_3O_{7-\delta}$
74.70; S10
(a) Vol.113,No.1, 151-158

H. DWELK, D. FREUDE, R. HERRMANN, H. PFEIFFER, and N. PRUSS
Substitution of Yttrium for Boron in the Structure of $YBa_2Cu_3O_{7-\delta}$
72.20 and 74.70; 76.60; S10.15
(a) Vol.113,No.1, 159-162

XIAO HUA CHEN, YE NING WANG, and HUI MING SHEN
Elastic Behaviour Related to the Superconductivity in $YBa_2Cu_3O_x$
62.20 and 74.70; S10.1
(a) Vol.113,No.1, K85-K88

B.B. BOIKO, F.P. KORSHUNOV, G.V. GATALSKII, A.I. AKIMOV, V.I. GATALSKAYA, and E.K. STRIBUK
Effect of Irradiation by Fast Electrons and ^{60}Co-gamma Rays on Superconducting and Electrical Properties of Thallium ...
61.80 and 74.70; S10.15
(a) Vol.113,No.2, K235-K240

JIANSHENG XIA, LIEZHAO CAO, SHIFANG SUN, QIRUI ZHANG, JIAN CHEN, ZUYAO CHEN, XIANHUI CHEN, and YITAI QIAN
Superconducting Behavior of $Bi_{1.5}Pb_{0.5}Ca_2Sr_2Cu_3O_y$ at Low Magnetic Field
74.70; S10.15
(b) Vol.154,No.1, K51-K54

C.D. MUKHERJEE, R. RANGANATHAN, A.K. RAYCHAUDHURI, and N. CHATTERJEE
Thermoelectric Power of the Ni and Cd Substituted YBCO System
72.15 and 74.70; S10.15
(b) Vol.154,No.1, K55-K58

S.J. WANG, X.H. LI, Y.L. CHEN, S.Q. LI, G.H. FONG, Z. WANG, A. CHEN, and B.R. LI
Structural Changes between 100 and 300 K of the High-T_c Superconductor Y-Ba-Cu-O Studied by Positron Annihilation
74.70 and 78.70; S10.15
(a) Vol.114,No.1, 273-276

A. BOHDZIEWICZ, B. CENDLEWSKA, M. KAZIMIERSKI, J. SZYMASZEK, and J.J. WNUK
Superconducting Properties of $LuRh_{1.2}Sn_4$ Thin Films
74.70; S1.4
(a) Vol.114,No.1, K49-K52

Z. WOKULSKI and C. SULKOWSKI
Electrical Properties and Superconductivity of $TiN_{1-x}C_x$ Films
74.70; S1.62
(a) Vol.114,No.1, K53-K56

E.M. GOLOLOBOV, N.A. PRYTKOVA, ZH.M. TOMILO, D.M. TURTSEVICH, and N.M. SHIMANSKAYA
High T_c Superconductivity of Bi-Sr-Ca-Cu Based Metal Oxides Doped by Magnesium
74.70; S10.15
(a) Vol.114,No.1, K57-K62

S.P. MUNTYAN, G.A. KIOSSE, and F.M. MUNTYANU
Superconductivity in the Ba-Y-Cu-Pb-O System
74.70; S10.15
(a) Vol.114,No.1, K63-K66

T. KOPTE and E. HEGENBARTH
Behaviour of the Thermal Conductivity of High Temperature Superconductors around the Critical Temperature
66.70 and 74.70; S10.15
(a) Vol.114,No.1, K67-K70

Š. CHROMIK, H. TOLLET, C. CREMERS, D. MACHAJDÍK, V. ŠMATKO, Y. BRUYNSERAEDE, and Š. BEŇAČKA
The Composition of High T_c Superconducting Y-Ba-Cu-O Thin Films Prepared by Vacuum Coevaporation
68.55 and 74.70; S10.15
(a) Vol.114,No.2, 567-572

JINGKUI LIANG, XIAOLONG CHEN, YULING ZHANG, SISHEN XIE, JINQI HUANG, and YONGMIN NI
The Crystal Structure and Superconductivity of the Superconducting Phase $TlSr_2(Ca, La)Cu_2O_7$
61.60 and 74.70; S10.15
(a) Vol.114,No.2, 651-658

JIRONG JIN, WUMING CHEN, XIN JIN, YITONG ZHANG, MU LU, YUBAO SHENG, JIABEN JI, GUOSHENG HAN, YI TANG, and XIXIAN YAO
Effect of Neutron Irradiation on the Superconductivity of Y-Ba-Cu-O
74.70; S10.15
(a) Vol.114,No.2, K189-K192

H.-J. DIETZE and S. BECKER
Laser-Induced Plasma Deposition of Superconducting Bi-Sr-Ca-Cu-O Thin Films on Silicon
68.55 and 74.70; S10.15
(a) Vol.114,No.2, K193-K196

I.M. CHAPNIK
Superconductivity and the Periodic Table
74.70
(b) Vol.155,No.1, K39-K44

A.S. SIDORENKO, L.A. KONOPKO, A.E. KOSHELEV, and A.YU. SIMONOV
Resistive Transition Broadening of Metal-Oxide Superconductors in Magnetic Fields
74.70; S10.15
(b) Vol.155,No.1, K45-K50

G.V. BAZUEV, K.R. KRYLOV, A.I. PONOMAREV, V.I. TSIDILKOVSKII, I.M. TSIDILKOVSKII, and T.B. CHARIKOVA
Peculiarities of Galvanomagnetic and Thermoelectric Properties of $YBa_{2-x}La_xCu_3O_{7-\delta}$ Solid Solutions
74.70; 72.20; S10.15
(a) Vol.115,No.1, 267-272

P. SEIDEL and M. TURTENWALD
Multi-Gap Model for Tunneling in High-T_c Superconductors
74.70; 74.50
(a) Vol.115,No.1, 273-278

S.X. WANG, Z.H. HE, X.H. CHEN, M.R. JI, Y.T. QIAN, Z.Y. CHEN, and Q.R. ZHANG
The Effect of Substitution of Cd for Ca in the Bi-Sr-Ca-Cd-Cu-O System Studied by X-Ray Photoemission Spectroscopy
74.70 and 79.60; S10.15
(a) Vol.115,No.1, K71-K74

FANCHUN ZHENG, JIANYI JIANG, YUPIN SUN, HUAQING YIN, and JIAJU DU
Effects of Cooling Rate on the Properties and the Oxygen Content of the Superconductor $Bi_2Sr_2CaCu_2O_y$
74.70; 62.40; S10.15
(a) Vol.115,No.1, K75-K78

E.M. GOLOLOBOV, N.A. PRYTKOVA, ZH.M. TOMILO, D.M. TURTSEVICH, M.S. TSELUEVSKII, and N.M. SHIMANSKAYA
Influence of Aluminium and Indium Addition on the Properties of Superconducting $Bi_2Sr_2CaCu_2O_y$ Metal Oxide
61.70 and 74.70; S10.15
(a) Vol.115,No.1, K79-K82

M. PĘKAŁA, R. RATAJ, A. PAJĄCZKOWSKA, and B. GEGENHEIMER
Normal State Transport Study of the Ba-K-Bi-O Superconductor
72.80 and 74.70; 72.20; S10.15
(b) Vol.155,No.2, K123-K126

V.E. BORISENKO, S.YU. NIKITIN, A.B. FILONOV, and A.L. PUSHKARCHUK
Electronic Property Simulation of the $YBa_2Cu_3O_7$ Films
71.25 and 74.70; S10.15
(b) Vol.155,No.2, K127-K130

E.N. IZAKOVICH, N.A. ZHELTOVA, YU.M. SHULGA, I.N. IVLEVA,
ZH.D. SOKOLOVSKAYA, A.V. BALUEV, V.V. GANIN, O.S. ROSHUPKINA et al.
Study of Ba and Cu(III) Based Inorganic Oxide Compounds
74.70; 78.30; 79.20; 79.60
(a) Vol.115,No.2, K195-K200

N.YU. ARUTYUNOV and V.YU. TRASHCHAKOV
1D-Angular Distributions of Annihilation Photons in RE-Ba-Cu-O Oxides (RE=Y,Nd,Lu)
74.70 and 78.70; S10.15
(a) Vol.115,No.2, K201-K204

A.I. AKIMOV, B.B. BOIKO, S.I. BOROVITSKII, V.I. GATALSKAYA,
V.D. GELIKONOVA, S.E. DEMYANOV, A.M. KLUSHIN, L.A. KUROCHKIN et al.
Observation on the Josephson Effect in TlBaCaCuO-TlBaCaCuO Junctions at Temperatures above 100 K
74.70; S10.15
(a) Vol.115,No.2, K205-K206

G. GŁADYSZEWSKI, A. SMAL, Z. MITURA, and M. SUBOTOWICZ
Formation of Superconducting Bi-Sr-Ca-Cu-O Thin Films by Flash Evaporation
68.55 and 74.70; S10.15
(a) Vol.115,No.2, K207-K210

R. KIEMEL, W. WISCHERT, and S. KEMMLER-SACK
Dependence of T_c on the Hole Concentration in the High-T_c Superconductors $Bi_{1.5}Pb_{0.5}Sr_2Ca_{1-x}Gd_xCu_2O_{8+z}$
74.70; 71.30; 72.20; S10.15
(b) Vol.156,No.1, 339-344

I.S. SMIRNOVA
Twins and Dislocations in $RBa_2Cu_3O_7$ Crystals
61.70; 74.70; S10.15
(a) Vol.116,No.1, 237-254

WENHUI SU, CHANGQING JIN, LANYING HE, YIFENG WANG, HONGJIAN LIU, and XIANG LONG
Study on Oxygen Hole and Superconductivity in $Y_1Ba_2Cu_3O_{7-\delta}$ Material
64.70 and 74.70; 61.10; S10.15
(a) Vol.116,No.1, 267-274

Z. DAMM, J. KLAMUT, T. MORAWSKA-KOWAL, J. OLEJNICZAK, J. STĘPIEŃ-DAMM, CZ. SUŁKOWSKI, and E. TROJNAR
Some Properties of Superconducting (RE)$Ba_2Cu_3O_7$ Compounds with Different Bulk Densities
74.70; S10.15
(a) Vol.116,No.1, 367-370

CHUN-SHENG JIA and YAN-JUN YANG
Research on Heat Treatment of Tl-Ba-Ca-Cu-O Superconducting Ceramics
74.70; S10.15
(a) Vol.116,No.1, 371-376

HONG ZHANG, XIAO-GANG WANG, LE LUO, and SHU-HUI HU
A Study on the Positron Annihilation of Glassy $Bi_1Sr_1Ca_1Cu_2O_x$ Quenched from the Molten Liquid
61.40 and 78.70; 74.70; S10.15
(a) Vol.116,No.1, K7-K12

E.N. IZAKOVICH, N.A. ZHELTOVA, YU.M. KOROLEV, ZH.D. SOKOLOVSKAYA, V.N. SPECTOR, and M.L. KHIDEKEL
Study of Isostructural Compounds Y_2BaCuO_5 and $(Y_{1.4}Cu^{3+}_{0.6})BaCuO_5$
61.50; 74.70; S10.15
(a) Vol.116,No.1, K13-K18

B.P. GORSHUNOV, M.G. KAPLUNOV, N.D. KUSHCH, and O.I. SYROTYNSKII
Temperature Dependence of the Optical Properties of the Organic Superconductor $(ET)_2Cu(NCS)_2$ in the Submillimeter Range
78.20; 74.70; S12
(a) Vol.116,No.1, K129-K132

H. BOERNER, M. WURLITZER, H.-CH. SEMMELHACK, B. LIPPOLD, R.-M. BOETTCHER, and M. KROETZSCH
A Method for the Preparation of Single-Phase $YBa_2Cu_3O_x$ Ceramics
61.50; 74.70; S10.15
(a) Vol.116,No.2, 537-544

A. MEHDAOUI, D. BOLMONT, and B. LOEGEL
Dynamical Critical Fields Aging Effects and Restoring in Some $YBa_2Cu_3O_x$ Compounds Studied by Complex Susceptibility Measurements
74.70; 75.30; S10.15
(a) Vol.116,No.2, 777-786

Y. UMAKOSHI, W. TAKAHARA, T. YAMANE, and K.A. TANAKA
Effect of Laser Irradiation on the Superconductive Properties of $(Y_{0.95}Sm_{0.05})Ba_2Cu_3O_x$
74.70; 61.80; S10.15
(a) Vol.116,No.2, 787-792

74.90. Other topics in superconductivity

V.A. KULBACHINSKII
Galvanomagnetic Properties of Graphite Fibers
72.60 and 74.90; S5
(b) Vol.151,No.1, 185-192

75. Magnetic Properties and Materials

75.10. General theory and models of magnetic ordering
(see also 71.25. Nonlocalized single-particle electronic states 71.70. Level splitting and interactions)

R.A. EVARESTOV, V.P. SMIRNOV, and S.A. EGOROV
Band Corepresentations of Magnetic Space Groups
75.10; 71.25; 71.50; S11
(b) Vol.151,No.1, 275-282

G.S. CHADDHA and G.S. KALSI
Anisotropic Ferromagnet with Biquadratic Exchange
75.10 and 75.40
(b) Vol.151,No.1, 283-288

L.S. CAMPANA, A.CARAMICO D'AURIA, and U. ESPOSITO
Critical Field of the Alternated and Bond-Mixed Transverse Ising Chains
75.10
(b) Vol.151,No.2, K177-K180

S. FAKIOGLU
Eigenvalues and Eigenfunctions of Renormalization Group Equations for Susceptibilities and Magnetizations
75.10
(b) Vol.151,No.2, K205-K208

N. RAJ and D.R. TILLEY
Green Functions for Polariton Modes in Magnetic Superlattices
71.36; 71.45; 75.10
(b) Vol.152,No.1, 135-144

A. DOROBA and K. SOKALSKI
The Ising Model of the Two-Dimensional Quasiperiodic Lattice
75.10 and 75.30
(b) Vol.152,No.1, 275-288

T. IWASHITA and N. URYÛ
Ising Ferromagnet with Biquadratic Exchange Interaction of Negative Sign
75.10 and 75.30
(b) Vol.152,No.1, 289-302

R.A. EVARESTOV and V.P. SMIRNOV
Tensor Fields in Crystals and Band Representations of Space Groups
75.10; 71.25; 71.50; S11
(b) Vol.152,No.2, 633-638

CHUAN-ZHANG YANG and JIA-LIN ZHONG
The Site-Bond Diluted Transverse Ising Model with S=1
75.10
(b) Vol.153,No.1, 323-330

J. ULNER and K. DURCZEWSKI
Lattice Distortion-Induced Quadrupol Interactions, Phases, and Excitations in J=1 Tetragonal Paramagnets
75.10 and 75.30
(b) Vol.153,No.1, 331-342

L. WANG and R.F. O'CONNELL
Ideal Two-Dimensional Electron Gas in a Magnetic Field and at Non-Zero Temperatures. An Alternative Approach
71.25 and 75.10
(b) Vol.153,No.1, 343-360

J.L. CADORIN and W. FIGUEIREDO
Biquadratic Exchange in a Two-Spin System
75.10
(b) Vol.153,No.1, K73-K78

S.S. APLESNIN
Monte-Carlo Study of Two-Dimensional Quantum Antiferromagnets with Random Anisotropies and Spin S=1
75.10
(b) Vol.153,No.1, K79-K84

M.A. YURISHCHEV
Ising Model Consisting of Central and Three Side Chains
75.10; S12.1
(b) Vol.153,No.2, 703-710

J. RICHTER and R. SKOMSKI
Antiferromagnets with Random Anisotropy:Influence of the Dimensionality
75.10; 75.30
(b) Vol.153,No.2, 711-720

J. POZO
Solitons in Anisotropic Heisenberg Ferromagnetic Chains
75.10
(b) Vol.153,No.2, K163-K166

P. WRÓBEL
Coexistence of Superconductivity and Antiferromagnetism in the Two-Dimensional Extended Hubbard Model with Inter-Site...
74.20 and 75.10
(b) Vol.153,No.2, K177-K182

ZIDAN WANG and CHANGDE GONG
Migdal-Kadanoff Method for the Ising Model on a Periodic Nonbranching Koch-Curve Lattice
75.10
(b) Vol.154,No.1, K63-K66

T. IWASHITA and N. URYŪ
Ground State of Ising Ferromagnet with Biquadratic Exchange of Negative Sign. The Case of Large Spin S=5/2
75.10; 75.30
(b) Vol.154,No.2, 747-762

U. LINDNER and W. SCHUMACHER
To Magnetic Solutions of the Hubbard Model
75.10; 75.30
(b) Vol.155,No.1, 263-274

J.R. FALEIRO FERREIRA
On the Thermodynamical-Statistical Consistency of a Variational Approximation for the Free Energy
75.10
(b) Vol.155,No.1, 275-280

D.V. KAPOR and M.J. ŠKRINJAR
On the Goldhirsch Boson Expansion for Spin Operators
75.10
(b) Vol.155,No.1, K77-K80

SIMING ZHENG
Magnetic Properties of Disordered Ising Ferrimagnets
75.10
(b) Vol.155,No.2, 653-662

B.V. COSTA
On the Phase Shift and Bound States in One-Dimensional Antiferromagnetic Chain
75.10; 75.50
(b) Vol.155,No.2, 663-668

A. BOBÁK, J. KARABA, and Ľ. TÓTH
The Phase Transition of the Spin-1/2 Ising Model with a Random Field
75.10
(b) Vol.155,No.2, K143-K148

J. MIELNICKI, T. BALCERZAK, and G. WIATROWSKI
Ising Model with Random Local Fields in a Two-Dimensional Ferromagnet
75.10 and 75.30
(b) Vol.156,No.1, 355-362

G.Z. WEI, H.Q. NIE, and K.Y. ZHANG
Correlation Effects of the Antiferromagnetic Phase in the Hubbard Model
75.10 and 75.30
(b) Vol.156,No.1, 363-370

J.R. FALEIRO FERREIRA
An Improved Linear Chain Approximation for the Free Energy
75.10
(b) Vol.156,No.2, 647-656

J.W. TUCKER
The Dynamics of a Heisenberg Paramagnet with Crystal Field Anisotropy
75.10
(b) Vol.156,No.2, K139-K142

D. KAPOR and M. ŠKRINJAR
On Solitons in Anisotropic Heisenberg Ferromagnetic Chains
75.10
(b) Vol.156,No.2, K143-K144

S. TABOR and R. MICNAS
Erratum to Dynamic Properties of the Heisenberg Model with Crystal Field Anisotropy. $T=\infty$ Limit (phys. stat. sol. (b) 137, K49 (1986))
75.10
(b) Vol.156,No.2, K165-K166

75.20. Diamagnetism and paramagnetism

R. ALLUB
Kondo Effect and Direct Interaction: Renormalization-Group Approach
75.20; 75.30
(b) Vol.154,No.1, 347-352

75.25. Spin arrangements in magnetically ordered mateials (neutron studies, etc.)

U. KNOOP and G. HEBER
Approximate Solution for a Two-Dimensional Anisotropic Ferromagnetic Spin Lattice with Bloch Wall and Bloch Line. I. Analytical ...
75.25 and 75.50
(b) Vol.151,No.2, 693-702

NGUYEN HOANG LUONG, NGUYEN PHU THUY, LƯU TUAN TAI, and THAN DUC HIEN
Rare Earth Magnetocrystalline Anisotropy in $R_2Fe_{14}B$ (R=Sm,Er,Tm,Yb)
75.30; 75.25; S1.1; S1.4
(a) Vol.111,No.2, 591-596

U. KNOOP and G. HEBER
Approximate Solution for a Two-Dimensional Anisotropic Ferromagnetic Spin Lattice with Bloch Wall and Bloch Line. II. Numeric Computation
75.25 and 75.50
(b) Vol.152,No.1, 303-310

R.S. GEKHT, N.V. FEDOSEEVA, V.A. DOLINA, and A.D. BALAEV
Incommensurate Magnetic Structures in Triangular Antiferromagnet $CsCuCl_3$
75.25; 75.30
(b) Vol.155,No.2, 639-646

M.M. SAVOSTA, V.D. DOROSHEV, P. NOVÁK, and I. VELTRUSKÝ
NMR of ^{167}Er and Magnetic Structure of $Er_3Fe_5O_{12}$ Garnet
76.60; 75.25; S11.2
(b) Vol.155,No.2, 669-678

M. VAZQUEZ, W. FERNENGEL, and H. KRONMUELLER
Approach to Magnetic Saturation in Rapidly Quenched Amorphous Alloys
75.60; 75.25; S1.1; S1.2; S1.61; S1.62
(a) Vol.115,No.2, 547-554

75.30. Magnetically ordered materials, other intrinsic properties
(for critical-points effects, see 75.40)

V.N. BERZHANSKII and V.I. IVANOV
Critical Relaxation of Uniform Precession of Magnetization in Cubic Ferrimagnets
75.30 and 75.50; 76.50; S11.2
(b) Vol.151,No.1, 259-268

O.A. BAYUKOV, V.P. IKONNIKOV, and A.F. SAVITSKII
Magnetic and Moessbauer Studies of Anion-Substituted Nickel Ferrites
75.30; 76.80; S11.2
(b) Vol.151,No.1, 269-274

O. GOEDE, D. BACKS, W. HEIMBRODT, and M. KANIS
EPR Study of the Antiferromagnetic Phase Transition in (Cd,Mn)S
76.30; 75.30; S8.15
(b) Vol.151,No.1, 311-318

E.F. SARMENTO, H.R. DA CRUZ, and T. KANEYOSHI
A New Type of Bond-Dilution Problem in a Ferromagnetic Insulator with a Fractional Random Coordination Number
75.30 and 75.50
(b) Vol.151,No.2, 675-686

P. STEFANSKI and A. WRZECIONO
The Rare Earth Contribution to the Magnetocrystalline Anisotropy in $RFe_{10}V_2$
75.30; S1.1
(b) Vol.151,No.2, K195-K200

NGUYEN HOANG LUONG, NGUYEN PHU THUY, LUU TUAN TAI, and THAN DUC HIEN
Rare Earth Magnetocrystalline Anisotropy in $R_2Fe_{14}B$ (R=Sm,Er,Tm,Yb)
75.30; 75.25; S1.1; S1.4
(a) Vol.111,No.2, 591-596

YU.G. OKLADNOY
Anomalous Magnetic Field Dependence of Electron Mobility in Anisotropic Ferromagnetic Semiconductors
72.20; 75.30
(b) Vol.152,No.1, 217-224

A. DOROBA and K. SOKALSKI
The Ising Model of the Two-Dimensional Quasiperiodic Lattice
75.10 and 75.30
(b) Vol.152,No.1, 275-288

T. IWASHITA and N. URYÛ
Ising Ferromagnet with Biquadratic Exchange Interaction of Negative Sign
75.10 and 75.30
(b) Vol.152,No.1, 289-302

J. SCHREIBER, J. BLAESING, and S. FLACH
On the Relation between Structural Phase Transitions and Spin Reorientation Transition in Anisotropic Ferromagnets
75.30 and 75.50; 75.40
(b) Vol.152,No.1, 311-316

L. BIEGAŁA and J. MICHALSKI
Consistent Spin Wave Theory for Easy-Plane Ferromagnets
75.30
(b) Vol.152,No.1, K17-K22

I.O. TROYANCHUK, A.S. CHERNYI, and E.F. SHAPOVALOVA
Magnetic Transitions in Perovskites $Cu(Mn_{3-x}Cu_x)Mn_4O_{12}$
75.30; S10.15
(a) Vol.112,No.1, 155-160

O.A. BAYUKOV and A.F. SAVITSKII
Application of the Superexchange Model to the Evaluation Method of Exchange and 3d-Electron Transfer Parameters in Magnetodielectrics
75.30; S11.2
(b) Vol.152,No.2, K51-K54

C. DJEGA-MARIADASSOU, T. MERCERON, J.L. DORMANN, M. PORTE, and P. RENAUDIN
Verwey Transition in Two Series of Substituted Magnetites. $Li_xFe_{3-x}O_4$ and $Mg_xFe_{3-x}O_4$
75.30 and 75.50; S11.2
(a) Vol.112,No.2, 601-616

H. GRUBER, E. KRAUTZ, H.P. FRITZER, K. GATTERER, and A. POPITSCH
Electrical Conductivity Magnetic Susceptibility and Infrared Spectra in the Superconducting Systems Bi-Sr-Ca-Cu-O and Tl-Ba-Cu-O
74.70; 75.30; 78.30; S10.15
(b) Vol.153,No.1, 307-314

Ą. LIPOWSKI and Z. ONYSZKIEWICZ
Elementary Excitations in Two-Sublattice Spin Model with Three-Ion Interaction
75.30; S9
(b) Vol.153,No.1, 315-322

J. ULNER and K. DURCZEWSKI
Lattice Distortion-Induced Quadrupol Interactions, Phases, and Excitations in J=1 Tetragonal Paramagnets
75.10 and 75.30
(b) Vol.153,No.1, 331-342

I.O. TROYANCHUK, A.A. SHEMYAKOV, and V.K. PROKOPENKO
Magnetic Structure of $NiMnO_3$ and $CoMnO_3$ Ilmenites
75.30 and 76.60; S11.1
(a) Vol.113,No.1, K107-K110

JIFAN HU, MINGXI LIN, RUFENG XU, SHOUGONG ZHANG, and ZHENXI WANG
Formation of the Magnetic Phases in Y-Zr-Fe Compounds
75.30; S1.1; S1.4
(a) Vol.113,No.1, K127-K130

J. RICHTER and R. SKOMSKI
Antiferromagnets with Random Anisotropy:Influence of the Dimensionality
75.10; 75.30
(b) Vol.153,No.2, 711-720

T. TSUBOI, M. CHIBA, and R. LAIHO
Magnetic Susceptibility of One-Dimensional Ferromagnetic $CsFeCl_3$ Crystals
75.30 and 75.50; S9.16
(b) Vol.153,No.2, K159-K162

YU.M. SEIDOV and A.M. SULEIMANOV
On the Theory of Raman Scattering by Surface Collective Excitations in Singlet-Triplet Ferromagnets
78.30; 71.70; 75.30; S1.62
(b) Vol.153,No.2, K203-K208

V.V. VOROBEV, M.YA. KRUPOTKIN, and V.A. FINKEL
Anomalous Behaviour of Elastic Moduli of Terbium and Dysprosium in Magneto-Ordered State
62.20 and 64.60; 75.30; S1.4
(a) Vol.113,No.2, 375-382

G.A. GOVOR, K. BAERNER, and J.-W. SCHUENEMANN
Phase Transitions in the $MnAs_{1-x}Sb_x$ Alloy System
64.70; 75.30; S1.2
(a) Vol.113,No.2, 403-420

A. URBANIAK-KUCHARCZYK
The Hyperfine Field Distribution in Thin Films of Binary Diluted Alloys
75.30 and 75.70
(b) Vol.154,No.1, 327-332

R. ALLUB
Kondo Effect and Direct Interaction: Renormalization-Group Approach
75.20; 75.30
(b) Vol.154,No.1, 347-352

N.N. EFIMOVA, YU.A. POPKOV, and N.V. TKACHENKO
The Irreversibility and Viscosity Effects in Dilute Ferrimagnets
75.30 and 75.50
(b) Vol.154,No.1, 353-362

W. HEIMBRODT, C. BENECKE, O. GOEDE, and H.E. GUMLICH
Optical Effects Due to Antiferromagnetic Spin Ordering in (Cd,Mn)S at Highest Mn Concentrations
78.20; 75.30; S8.15
(b) Vol.154,No.1, 405-416

A. KOWALCZYK, P. STEFAŃSKI, A. WRZECIONO, and A. SZLAFEREK
Structure and Magnetic Properties of $R_2Fe_{14-x}Ni_xB$ Compounds (R = Nd and Gd)
75.30; 61.55; S1.1; S1.2
(a) Vol.114,No.1, 355-360

N.H. CHAU, N.H. DUC, T.D. HIEN, K. KROP, and J. ZUKROWSKI
Magnetic and Magnetostrictive Properties of the $Er(Fe_xCu_{1-x})_2$ Compounds
75.30 and 75.80; S1.1; S1.2; S1.4
(a) Vol.114,No.1, 361-368

G.I. KATAEV, M.R. SATTAROV, and A.M. TISHIN
Influence of Commensurability Effects on the Magnetic Phase Diagram of Terbium Single Crystals
75.30; S1.4
(a) Vol.114,No.1, K79-K82

T. TARNÓCZI and A. SIEMKO
Anomalous Crossover Effect
75.30; S1.1
(a) Vol.114,No.1, K83-K86

R. HOEHNE
On the Instability of Relaxation Processes of Induced Magnetic Anisotropy in Gallium-Iron Ferrites
75.30 and 75.50; S11.2
(a) Vol.114,No.1, K93-K98

S.A. NIKITIN, A.M. TISHIN, and S.E. BYKHOVER
Investigation of Magnetic Phase Transitions in Terbium Using the Magnetocaloric Effect
75.30; S1.4
(a) Vol.114,No.1, K99-K102

T. IWASHITA and N. URYÛ
Ground State of Ising Ferromagnet with Biquadratic Exchange of Negative Sign. The Case of Large Spin S=5/2
75.10; 75.30
(b) Vol.154,No.2, 747-762

M. AHMAD
Spin Waves with l=1 in a Degenerate Fermi Liquid of Alkali Metals in DC Magnetic Fields
75.30; S2
(b) Vol.154,No.2, K163-K166

M. MIHALIK, J. KOVÁČ, A. ZENTKO, and A. LOVAS
Some Magnetic Properties of Neutron Irradiated $Fe_{85-x}Cr_xB_{15}$ Metallic Glasses
61.80 and 75.30; 75.40; 75.60; S1.1; S1.2
(a) Vol.114,No.2, 679-684

O.A. BAYUKOV and A.F. SAVITSKII
The Prognostication Possibility of Some Magnetic Properties for Dielectrics on the Basis of Covalency Parameters of Ligand-Cation ...
75.30; S11.2
(b) Vol.155,No.1, 249-256

O. HUDÁK
Reentrant Paramagnetic-AFM-Paramagnetic Phase Transitions in La_2CuO_4
75.30 and 75.50; S11.1
(b) Vol.155,No.1, 257-262

U. LINDNER and W. SCHUMACHER
To Magnetic Solutions of the Hubbard Model
75.10; 75.30
(b) Vol.155,No.1, 263-274

E.L. NAGAEV
Local Levels in Ferromagnetic Semiconductors near and above the Curie Point
71.55 and 75.30
(b) Vol.155,No.1, K71-K76

R.I. ZAINULLINA, K.B. VLASOV, M.A. MILYAEV, E.V. USTELEMOVA, and V.N. SYROMYATNIKOV
Oscillatory Temperature Dependence of Crystal Structure and Magnetic Parameters in $FeGe_2$
64.60; 75.30; S1.61
(b) Vol.155,No.2, 419-426

R.S. GEKHT, N.V. FEDOSEEVA, V.A. DOLINA, and A.D. BALAEV
Incommensurate Magnetic Structures in Triangular Antiferromagnet $CsCuCl_3$
75.25; 75.30
(b) Vol.155,No.2, 639-646

X.K. SUN, X.C. KOU, Y.C. CHUANG, R.W. ZHAO, J.Y. LI, F.M. YANG, R. GROESSINGER, and H.R. KIRCHMAYR
3d Anisotropy and Preferential Site Occupation in $R_2Fe_{13}M_1B$ Compounds (R=Y,Gd, M=Ga,Si,Al)
75.50; 75.30; S1.1
(a) Vol.115,No.2, K219-K224

A.I. GUSEV
Short-Range Order in Nonstoichiometric Transition Metal Carbides Nitrides and Oxides (Review Article)
61.60 and 75.30; 76.60; S1.61; S1.62; S10
(b) Vol.156,No.1, 11-40

J. MIELNICKI, T. BALCERZAK, and G. WIATROWSKI
Ising Model with Random Local Fields in a Two-Dimensional Ferromagnet
75.10 and 75.30
(b) Vol.156,No.1, 355-362

G.Z. WEI, H.Q. NIE, and K.Y. ZHANG
Correlation Effects of the Antiferromagnetic Phase in the Hubbard Model
75.10 and 75.30
(b) Vol.156,No.1, 363-370

A.YA. FISHMAN, N.M. KOVTUN, V.YA. MITROFANOV, V.K. PROKOPENKO, and A.A. SHEMYAKOV
Supertransferred Hyperfine Fields at Nuclei of Diamagnatic Copper Ions in Chalcogenide Spinels Fe Cu Cr S
75.30; 76.60; S11.2
(b) Vol.156,No.2, 597-604

V.E. ADAMIAN, A.A. ARTSRUNI, A. BENAISSA, A.N. KOCHARIAN, M.A. MELIKIAN, and A.G. TONEIAN
Magnetic Properties of Model Multicomponent Compounds of CsCl Structure
64.75 and 75.30; S1.4
(b) Vol.156,No.2, 633-640

T. HAI and Z.Y. LI
A Study for Critical Behaviour of Ferromagnetic Thin Films
75.30 and 75.70
(b) Vol.156,No.2, 641-646

P. STEFANSKI, A. SZLAFEREK, and A. WRZECIONO
The Intra- and Intersublattice Exchange Interactions in $RFe_{12-x}T_x$ Intermetallics
75.30; S1.1; S1.2; S1.4
(b) Vol.156,No.2, 657-662

A. MEHDAOUI, D. BOLMONT, and B. LOEGEL
Dynamical Critical Fields Aging Effects and Restoring in Some $YBa_2Cu_3O_x$ Compounds Studied by Complex Susceptibility Measurements
74.70; 75.30; S10.15
(a) Vol.116,No.2, 777-786

75.40. Critical-point effects, specific heats, short-range order
(see also 65.40. Heat capacities)

G.S. CHADDHA and G.S. KALSI
Anisotropic Ferromagnet with Biquadratic Exchange
75.10 and 75.40
(b) Vol.151,No.1, 283-288

ZIDAN WANG and CHANGDE GONG
The Density of States and the Thermal Properties of a Ferromagnetic Heisenberg Model on a Periodic Fractal Lattice
75.40
(b) Vol.151,No.1, K77-K80

J. SCHREIBER, J. BLAESING, and S. FLACH
On the Relation between Structural Phase Transitions and Spin Reorientation Transition in Anisotropic Ferromagnets
75.30 and 75.50; 75.40
(b) Vol.152,No.1, 311-316

S.A. NIKITIN, A.M. TISHIN, and P.I. LEONTEV
Magnetocaloric Effect and Refringerant Capacity of Tb-Dy Alloys
75.40; S4
(a) Vol.113,No.1, K117-K122

S.J. GŁADYSZ
Curie Temperature of Magnetic Semiconductors in Nonlocal s-d(f) Model
75.40
(b) Vol.153,No.2, K171-K176

A.A. POVZNER, S.V. KORTOV, and P.V. GELD
Magneto-Volume Effects in the Weak Itinerant Ferromagnets on the Basis of Solid Solutions of $Fe_{1-y}Mn_ySi$ and $Fe_{1-x}Co_xSi$
65.70 and 75.40; S1.61
(a) Vol.114,No.1, 315-326

O. HUDAK
Thermodynamics of Localized Magnetic Singlets and 1/2-Spins in the Plane: Paramagnetic Phase of La_2CuO_4
75.40; S11.1
(b) Vol.154,No.2, 743-746

M. MIHALIK, J. KOVÁČ, A. ZENTKO, and A. LOVAS
Some Magnetic Properties of Neutron Irradiated $Fe_{85-x}Cr_xB_{15}$ Metallic Glasses
61.80 and 75.30; 75.40; 75.60; S1.1; S1.2
(a) Vol.114,No.2, 679-684

J. BISCHOF, M. DIVIŠ, P. SVOBODA, and Z. SMETANA
The Specific Heat of $HoCu_2$ in Magnetic Fields
75.40; S1.4
(a) Vol.114,No.2, K229-K232

S.L. KAKANI and R.K. PALIWAL
Theoretical Study of Density of States and Specific Heat of Itinerant Ferromagnetic Superconductors
74.10 and 75.40; S10
(b) Vol.155,No.1, 241-248

A.S.T. PIRES
Dynamically Convergent Calculations of the Response Function of a Ferromagnet in the Linear Chain Approximation
75.40
(b) Vol.155,No.1, K67-K70

R. KUENTZLER, E. MURRAY, Y. DOSSMANN, and D. AST
Electronic Properties of AlMn and AlMnSi Quasicrystals
75.40; 65.40; 71.20; S1.2
(a) Vol.116,No.1, 399-404

75.50. Studies of specific magnetic materials

V.N. BERZHANSKII and V.I. IVANOV
Critical Relaxation of Uniform Precession of Magnetization in Cubic Ferrimagnets
75.30 and 75.50; 76.50; S11.2
(b) Vol.151,No.1, 259-268

D. VLADIKOVA, L. ILKOV, and S. KARBANOV
Influence of the Microstructure on Some Microwave Properties of Substituted Nickel Ferrites
64.80; 72.20; 75.50; S11.2
(a) Vol.111,No.1, 145-154

T. REININGER, N. MOSER, A. HOFMANN, and H. KRONMUELLER
The Kinetics of the Magnetically Induced Anisotropy in Amorphous Ferromagnetic $Co_{71}Fe_1Mo_1Mn_4Si_{14}B_9$ and $Fe_{41}Ni_{41}B_{18}$ Alloy
75.50; S1.1; S1.2
(a) Vol.111,No.1, 243-256

P. SVOBODA, Z. SMETANA, and A.V. ANDREEV
Magnetic Behaviour of $TmCu_2$ below T_N
75.50; S1.2; S1.4
(a) Vol.111,No.1, 285-288

F.M. KAMARLY, N.G. GUSEINOV, and A.YU. YANGIROV
Magnetic Susceptibility and Electrophysical Properties of SiTe Single Crystals
61.70; 72.80; 75.50; S8
(a) Vol.111,No.1, K17-K20

V.I. GARASIM, D.M. ZAYACHUK, R.D. IVANCHUK, V.B. LOTOTSKII, P.M. STARIK, and V.A. SHENDEROVSKII
Peculiarities of Physical Properties of Lead and Tin Selenide Solid Solutions
72.80 and 75.50; S8
(a) Vol.111,No.1, K99-K104

K. TABATA and H. KIDO
Preparation and Magnetic and Electrical Properties of $La_{1-x}Ce_xCoO_3$ Compounds
72.80 and 75.50; S10.15
(a) Vol.111,No.1, K105-K108

E.F. SARMENTO, H.R. DA CRUZ, and T. KANEYOSHI
A New Type of Bond-Dilution Problem in a Ferromagnetic Insulator with a Fractional Random Coordination Number
75.30 and 75.50
(b) Vol.151,No.2, 675-686

U. KNOOP and G. HEBER
Approximate Solution for a Two-Dimensional Anisotropic Ferromagnetic Spin Lattice with Bloch Wall and Bloch Line. I. Analytical ...
75.25 and 75.50
(b) Vol.151,No.2, 693-702

M. MATLAK
Curie Temperature of EuO under Pressure
75.50; S10.1
(b) Vol.151,No.2, K187-K190

L. POTOCKÝ, P. KOLLÁR, É. KISDI-KOSZÓ, Z. JURÁNEK, and J. DEGRO
Surface Coercitive Force of Some Metallic Glasses
75.50; S1.1
(a) Vol.111,No.2, 603-610

J. RIVAS, M.A. LOPEZ-QUINTELA, F. WALZ, and H. KRONMUELLER
Fractal Analysis of Magnetic After-Effects in the Amorphous Alloy $Fe_{40}Ni_{40}P_{14}B_6$ at 500 K
75.60; 75.50; S1.1; S1.2
(a) Vol.111,No.2, 611-616

U. KNOOP and G. HEBER
Approximate Solution for a Two-Dimensional Anisotropic Ferromagnetic Spin Lattice with Bloch Wall and Bloch Line. II. Numeric Computation
75.25 and 75.50
(b) Vol.152,No.1, 303-310

J. SCHREIBER, J. BLAESING, and S. FLACH
On the Relation between Structural Phase Transitions and Spin Reorientation Transition in Anisotropic Ferromagnets
75.30 and 75.50; 75.40
(b) Vol.152,No.1, 311-316

E.A. LITTLE, D.J. BUTTLE, and C.B. SCRUBY
Radation Damage Studies in Model Ferritic Alloys Using Micromagnetic Techniques
61.80; 75.50; S1.1
(a) Vol.112,No.1, 55-72

KH.YA. MULYUKOV, R.Z. VALIEV, G.F. KORZNIKOVA, and V.V. STOLYAROV
The Amorphous $Fe_{83}Nd_{13}B_4$ Alloy Crystallization Kinetics and High Coercivity State Formation
75.50; 61.40; S1.1
(a) Vol.112,No.1, 137-144

YU.G. CHUKALKIN, V.R. SHTIRZ, and B.N. GOSHCHITSKII
The Structure and Magnetism of Amorphous $Y_3Fe_5O_{12}$
61.40 and 75.50; S11.2
(a) Vol.112,No.1, 161-174

X.C. KOU, X.K. SUN, Y.C. CHUANG, T.S. ZHAO, R. GROESSINGER, R. EIBLER, and H.R. KIRCHMAYR
Anisotropy Energy of Nd Ion in $Nd_2Fe_{12-x}Co_2Al_xB$ Compounds
75.50; S1.1; S1.2; S1.4
(a) Vol.112,No.1, 175-180

M. BOURROUS, N. MOSER, and H. KRONMUELLER
Reversibility and Annealing of Magnetic After-Effect in Ferromagnetic Amorphous Alloys
75.60; 75.50; S1.2
(a) Vol.112,No.1, 181-188

M. FAEHNLE, R. PAWELLEK, and H. KRONMUELLER
The Magnetic Field Effect on the Linear Magnetostriction of Amorphous Ferromagnets
75.80; 75.50
(a) Vol.112,No.1, 189-196

A.I. BAZHIN, V.V. KOZEL, V.P. SHESTOV, M. SUCHANSKA, and S.V. TEPLOV
Temperature Dependence of Energy Distribution of Excited Atoms Sputtered from an Ion-Bombarded Ni(100) Surface
79.20; 75.50; S1.2
(a) Vol.112,No.1, 437-440

A.D. BALAEV, O.A. BAYUKOV, and A.F. SAVITSKII
Magnetic and Moessbauer Studies of Magnesium-Chromium Ferrites
75.50 and 76.80; S11.2
(b) Vol.152,No.2, 639-644

C. DJEGA-MARIADASSOU, T. MERCERON, J.L. DORMANN, M. PORTE, and P. RENAUDIN
Verwey Transition in Two Series of Substituted Magnetites. $Li_xFe_{3-x}O_4$ and $Mg_xFe_{3-x}O_4$
75.30 and 75.50; S11.2
(a) Vol.112,No.2, 601-616

M. JURCZYK, G.S. BURCHANOV, and O.D. CHISTYAKOV
Magnetic and Structural Properties of $Y_2Fe_{14-x}Nb_xB$ Alloys
75.50; S8.16
(a) Vol.112,No.2, K121-K126

M. BOURROUS and H. KRONMUELLER
Analysis of the Stabilization Kinetics of Amorphous $Co_{58}Ni_{10}Fe_5Si_{11}B_{16}$
75.60; 75.50; S1.2
(a) Vol.113,No.1, 169-180

J. GONZÁLEZ, M. VÁZQUEZ, J.M. BARANDIARÁN, M.A. ILLARRAMENDI, A. SALAZAR, and A. HERNADO
Magnetic Anisotropies Induced by Current Annealing and Their Dependence on Annealing Temperature
75.50; S1.1; S1.2
(a) Vol.113,No.1, 187-192

J. HORN and H.-J. ZACHARIAS
The Lines of Constant Vacancy Content in the Phase Diagram of the Al-Fe Ferrite System
61.70 and 75.50; S11.2
(a) Vol.113,No.1, K103-K106

A.N. CHUVYROV, L.A. KOSYKH, G.V. LEPLYANIN, V.N. SALIMGAREEVA, and N.S. SANNIKOVA
Magnetically Ordered Phase in Quasi-One-Dimensional Systems Based on Polyacetylene
75.50; S12.1
(a) Vol.113,No.1, K111-K116

S.A. BELOKON, S.O. KLIMONSKII, and V.D. KUZNETSOV
The Influence of $Pb_{0.99}Mn_{0.01}Te$ Solid Solution Ageing and Annealing on Its Paramagnetic Curie Temperature
75.50; S8
(a) Vol.113,No.1, K123-K126

G. BUSIELLO, L. DE CESARE, K. LUKIERSKA-WALASEK, and K. WALASEK
Exotic Glassy Phase in the Site-Diluted Transverse Ising Model at Zero Temperature?
63.75 and 64.60; 75.50
(b) Vol.153,No.2, 561-572

T. TSUBOI, M. CHIBA, and R. LAIHO
Magnetic Susceptibility of One-Dimensional Ferromagnetic $CsFeCl_3$ Crystals
75.30 and 75.50; S9.16
(b) Vol.153,No.2, K159-K162

K. HISATAKE, I. MATSUBARA, K. MAEDA, T. FUJIHARA, S. SUZUKI, H. WAKAO, T. HIGASHI, and K. UEMATSU
Gamma-Ray Induced Change of the Initial Permeability in Yttrium Iron Garnet at 77 K
75.50; S11.2
(a) Vol.113,No.2, K245-K248

E. BURZO, N. PLUGARU, V. POP, and I. CREANGA
Magnetic Properties of $YCo_{4-x}Fe_xB$ Compounds
61.55 and 75.50; S1.1
(a) Vol.113,No.2, K253-K256

D. BAHADUR, RAM BILAS, and B.S.R. MURTY
Electronic Properties of the Defect Perovskite System $La_{1-x}Li_xMO_3$
61.60 and 72.20; 75.50; S11
(b) Vol.154,No.1, 297-304

J.N. DAOU, J.P. BURGER, and P. VAJDA
Anomalous Behavior of the Low-Temperature Spin Disorder Resistivity in PrH_{2+x} Compounds
72.15; 75.50; S1.4
(b) Vol.154,No.1, 321-326

N.N. EFIMOVA, YU.A. POPKOV, and N.V. TKACHENKO
The Irreversibility and Viscosity Effects in Dilute Ferrimagnets
75.30 and 75.50
(b) Vol.154,No.1, 353-362

Z. FRAIT and D. FRAITOVÁ
Surface Spin-Wave Mode Excitation in Bulk Amorphous FeNiB Alloys
75.50 and 76.50; S1.1; S1.2
(b) Vol.154,No.1, 363-366

J. RICHTER
On the Stability of the Magnetic Long-Range Order in Three-Dimensional Antiferromagnets with Random Anisotropy
75.50
(b) Vol.154,No.1, K79-K84

M.M. KORSUKOVA, V.N. GURIN, YU.B. KUZMA, N.F. CHABAN, S.I. CHYKHRII, V.V. MOSHCHALKOV, N.B. BRANDT, A.A. GIPPIUS, and KHO KHYU NYAN
Crystal Structure, Electrical, and Magnetic Properties of the New Ternary Compounds $LnAlB_{14}$
61.60 and 72.80; 75.50; S1.4
(a) Vol.114,No.1, 265-272

R. HOEHNE
On the Instability of Relaxation Processes of Induced Magnetic Anisotropy in Gallium-Iron Ferrites
75.30 and 75.50; S11.2
(a) Vol.114,No.1, K93-K98

L.M. SANDRATSKII and P.G. GULETSKII
Comparative Analysis of Tendencies in the Temperature-Induced Changes in Electronic Structure of Iron and Nickel
71.25; 75.50; S1.1; S1.2
(b) Vol.154,No.2, 623-636

K. SUMIYAMA, Y. HIROSE, and Y. NAKAMURA
Magnetic and Electrical Properties of Nonequilibrium Ni-Al Alloys Produced by Vapor Quenching
72.15 and 75.50; 61.55; S1.2
(a) Vol.114,No.2, 693-704

M. JURCZYK and O.D. CHISTYAKOV
Structure and Magnetic Properties of $Y_2Fe_{14-x}Re_xB$
75.50; S1.1
(a) Vol.114,No.2, K219-K222

O. HUDÁK
Reentrant Paramagnetic-AFM-Paramagnetic Phase Transitions in La_2CuO_4
75.30 and 75.50; S11.1
(b) Vol.155,No.1, 257-262

M.A. VASILIEV, S.D. GORODETSKII, and A.G. BLASHUK
Distribution of the Effective Debye Temperature in the Surface Layers of Single Crystals of the Fe-Ni System
75.50; S1.1; S1.2
(a) Vol.115,No.1, K95-K100

SHOUYE ZHANG, GUANGYAO FONG, and MING MAO
Spin-Canting Effect and Magnetic Moment of Li-Zn Ferrite
75.50; S11.2
(b) Vol.155,No.2, 647-652

B.V. COSTA
On the Phase Shift and Bound States in One-Dimensional Antiferromagnetic Chain
75.10; 75.50
(b) Vol.155,No.2, 663-668

S. ICHIKAWA, S. OHTA, K. HATAKEYAMA, S. ANZAI, and A. TAKASE
Curie Temperatures of Monoclinic $(Cr_{1-x}M_x)_3Te_4$ with M:Ni, Co, and Vacancy
75.50; S1.63
(a) Vol.115,No.2, K215-K218

X.K. SUN, X.C. KOU, Y.C. CHUANG, R.W. ZHAO, J.Y. LI, F.M. YANG, R. GROESSINGER, and H.R. KIRCHMAYR
3d Anisotropy and Preferential Site Occupation in $R_2Fe_{13}M_1B$ Compounds (R=Y,Gd, M=Ga,Si,Al)
75.50; 75.30; S1.1
(a) Vol.115,No.2, K219-K224

I.O. TROYANCHUK and S.N. PASTUSHONOK
Antiferromagnetic-Ferromagnetic Transition in Orthomanganites
75.50; S10.15
(a) Vol.115,No.2, K225-K228

M. JURCZYK and O.D. CHISTYAKOV
Note on the Crystallographic and Magnetic Properties of $YCo_{10}V_2$
75.50; S1.2
(a) Vol.115,No.2, K229-K232

J. GONZALEZ ESTEVEZ and E. DU TREMOLET DE LACHEISSERIE
Influence of Annealing on the Magnetostriction of an Amorphous $Co_{45}Ni_{30}Si_{15}B_{10}$ Alloy
75.50 and 75.80; S1
(a) Vol.115,No.2, K233-K236

75.60. Domain effects, magnetization curves, and hysteresis

U. HARTMANN
Internal Structure of Two-Dimensional Domain Walls in Finite Bulk Crystals
75.60
(b) Vol.151,No.1, 289-310

A.H. WAFIK
Effect of Rolling and Heat Treatment on the Irreversible Magnetization in Nickel-Chromium Wire
75.60; S1.2
(a) Vol.111,No.1, 265-270

K. ELK and V. CHRISTOPH
Influence of the Magnetostatic Interaction on the Outer Hysteresis Loop
75.60
(b) Vol.151,No.2, 687-692

J. RIVAS, M.A. LOPEZ-QUINTELA, F. WALZ, and H. KRONMUELLER
Fractal Analysis of Magnetic After-Effects in the Amorphous Alloy $Fe_{40}Ni_{40}P_{14}B_6$ at 500 K
75.60; 75.50; S1.1; S1.2
(a) Vol.111,No.2, 611-616

G.E. ABROSIMOVA, A.P. ZHUKOV, and B.K. PONOMAREV
Effect of Heat Treatment on Some Properties of Fe- and Co-Based Amorphous Alloys
75.60; S1.1
(a) Vol.111,No.2, K237-K242

P.P. SHATSKII and D.A. YABLONSKII
Dynamic Inhomogeneous Magnetoelectric Effect in Magnetics with Domain Structure. Resonance Excitation of Domain Boundaries with Electric ...
75.60 and 75.80
(b) Vol.152,No.1, K23-K28

E.A. ZAVADSKII and V.A. ZABLOTSKII
Phase Transition in Magnetic Bubble Lattice
75.60
(a) Vol.112,No.1, 145-154

M. BOURROUS, N. MOSER, and H. KRONMUELLER
Reversibility and Annealing of Magnetic After-Effect in Ferromagnetic Amorphous Alloys
75.60; 75.50; S1.2
(a) Vol.112,No.1, 181-188

C. DE FRANCISCO, J.M. MUÑOZ, J. AYALA, and J.I. IÑIGUEZ
An Investigation of the Magnetic Disaccommodation in Zinc Ferrites Sintered under Very Low Oxidation Pressures
75.60; S11.2
(a) Vol.112,No.1, K43-K48

R. RIESENBERG, A. UNKROTH, and D. DIETRICH
Stress Controlled Adhesion and Magnetic Properties of Ion Implanted Co Films
62.20; 61.80; 75.60; S12
(a) Vol.112,No.2, K85-K90

A.P. ZHUKOV and B.K. PONOMAREV
Start Field Distribution in a Bistable Amorphous Ribbon
75.60; S1.1
(a) Vol.112,No.2, K127-K130

S. NIEBER and H. KRONMUELLER
Nucleation Fields in Periodic Multilayers
75.60; 75.70
(b) Vol.153,No.1, 367-376

M. DIVIŠ, P. SVOBODA, Z. SMETANA, V. NEKVASIL, and A.V. ANDREEV
The Influence of Crystal-Field Splitting on the Paramagnetic Magnetisation of $TmCu_2$
71.70 and 75.60; S1.2
(b) Vol.153,No.1, K69-K72

M. BOURROUS and H. KRONMUELLER
Analysis of the Stabilization Kinetics of Amorphous $Co_{58}Ni_{10}Fe_5Si_{11}B_{16}$
75.60; 75.50; S1.2
(a) Vol.113,No.1, 169-180

I. TOMÁŠ, D. TOMÁŠ, and T. DOLEŽAL
A Quick Estimate of Saturation Magnetization Value from the Bitter Powder Picture of Open-Flux Domain Structures
75.60; S11.2
(a) Vol.113,No.1, K95-K98

J. EHLERT and W. SPERBER
A Numerical Investigation of Nonuniformities in Micromagnetics
75.60
(a) Vol.113,No.1, K99-K102

M. BOURROUS and H. KRONMUELLER
Structural Transition from the As-Quenched to the Relaxed State in Metallic Glasses
61.40 and 64.60; 75.60; S1.2
(a) Vol.113,No.2, 383-392

A.I. MITSEK and V.N. PUSHKAR
The Topology of Magnetic Inhomogeneities in Thin Ferromagnetic Layers. The Model of the Magnetic Vortex
75.60 and 75.70; S1.1; S1.2; S11.2
(a) Vol.113,No.2, 529-540

J. CASTRO, J. RIVAS, H.J. BLYTHE, J. INIGUEZ, and C. DE FRANCISCO
Orientation and Diffusion Magnetic After-Effects in Vacancy-Doped Polycrystalline Magnetite in the Temperature Range 300 to 700 K
75.60; S10.1
(a) Vol.113,No.2, 541-550

A. BOETTCHER and H. SCHAEFER
Mapping of Magnetic Domain Structures inside of Ferromagnetic Materials by X-Ray Topography
75.60; S1.1
(a) Vol.113,No.2, 551-560

TH. KLUPSCH
Magnetization Curve and Magnetic Characterization of a System of Fine Single Domain Particles with Oblate Ellipsoidal Shape and ...
75.60
(a) Vol.113,No.2, 561-572

M. JIRSA, G. VÉRTESY, and J. PAČES
Temperature Dependence of the FMR in Garnet Films with Bubble-Domain Lattice
75.70 and 76.50; 75.60; S11.2
(a) Vol.113,No.2, 573-578

I.O. TROYANCHUK, V.N. DERKACHENKO, and E.F. SHAPOVALOVA
Magnetic Properties of $A_2Mn_2O_7$ Pyrochlores
75.60; S10
(a) Vol.113,No.2, K249-K252

TH. KLUPSCH, W. SCHUEPPEL, and M. ROESLER
Magnetization Curve and Magnetic Characterization of Planar Hexaferrite Powders Prepared by Glass Crystallization
75.60; S11.2
(a) Vol.114,No.1, 327-340

G.D. KHATTAK, V. KEITH, and PH. MARTIN
Magnetization of $Cd_{1-x}Mn_xSe$
75.60; S8.15
(a) Vol.114,No.1, K71-K74

M. MIHALIK, J. KOVÁČ, A. ZENTKO, and A. LOVAS
Some Magnetic Properties of Neutron Irradiated $Fe_{85-x}Cr_xB_{15}$ Metallic Glasses
61.80 and 75.30; 75.40; 75.60; S1.1; S1.2
(a) Vol.114,No.2, 679-684

L.F. KISS, G. MARTINEK, A. FORKL, and H. KRONMUELLER
Correlation between Microstructure and Magnetic Properties in FeNdB Permanent Magnets Doped with Al_2O_3
75.60; S1.1; S1.4
(a) Vol.114,No.2, 685-692

V.M. FEDOSYUK, M.U. SHELEG, O.I. KASYUTICH, and M.M. MALYUSH
Columnar Structure and Texture [001] in Co-Ni-W Films
75.60; 75.70; S1.2
(a) Vol.115,No.1, 279-284

U. HARTMANN
Bit Analysis of Magnetic Recording Media by Force Microscopy
75.60
(a) Vol.115,No.1, 285-292

A.V. CHZHAN, YU.M. FEDOROV, and T.I. ISAEVA
Features of the Temperature Dependence of the Initial Magnetic Susceptibility in $FeBO_3$
75.60; S10
(a) Vol.115,No.1, K101-K104

H. DRAGANOVSKÁ, J. SLÁMA, P. BUTVIN, and B. BUTVINOVÁ
Domain Patterns and Magnetic Properties of Some Amorphous Metallic Ribbons
75.60; S1.1; S1.2
(a) Vol.115,No.2, 533-538

K. SUN, M.X. LIN, T. HE, and H.L. LUO
The Moessbauer Study of alpha-Fe_2O_3 Fine Particles with and without Adsorbed Cobalt
75.60; 76.80; S10
(a) Vol.115,No.2, 539-546

M. VAZQUEZ, W. FERNENGEL, and H. KRONMUELLER
Approach to Magnetic Saturation in Rapidly Quenched Amorphous Alloys
75.60; 75.25; S1.1; S1.2; S1.61; S1.62
(a) Vol.115,No.2, 547-554

DECHENG TIAN, KAIYANG LI, and DIHUA DING
Magnetic Aftereffect and Mossbauer Studies of Structural Relaxation in $(Fe_{1-x}Ce_x)_{84}B_{16}$ Amorphous Alloys
61.40; 75.60; 76.80; S1.1
(a) Vol.116,No.1, 203-212

A.H. WAFIK, N.A. RAZIK, and M.H. ZAYAN
Barkhausen Jumps and Microstructure in Low Carbon Steel
75.60; S1.1
(a) Vol.116,No.1, 383-388

J. EHLERT, F.K. HUEBNER, and W. SPERBER
A Numerical Method for Solving Three-Dimensional Micromagnetic Problems
75.60
(a) Vol.116,No.1, 389-398

R. SKOMSKI and V. CHRISTOPH
Power Law Behaviour of Magnetic Viscosity
75.60 and 75.90
(b) Vol.156,No.2, K149-K152

K. MAKI, A. FORKL, T. DRAGON, and H. KRONMUELLER
Investigation of the Domain Structure of Sintered FeNdB-Al Magnets
75.60; S1.1
(a) Vol.116,No.2, 803-812

M. CELASCO, A. MASOERO, P. MAZZETTI, and A. STEPANESCU
Study of the Propagation of a Perturbation along a Bloch Wall
75.60
(a) Vol.116,No.2, 813-818

L. JAHN, J.S. PASTUSCHENKOV, and V. CHRISTOPH
On the Angular Dependence of the Coercivity of NdFeB Hard Magnets
75.70; 75.60; S4
(a) Vol.116,No.2, K179-K184

75.70. Magnetic films and plates

J. BARNAS
Magnetostatic Modes in Disordered Superstructures
75.70; 76.50
(b) Vol.151,No.1, K71-K76

J. ULNER
Remarks on the Theory of Slightly Deformed Thin Ferromagnetic Plates
75.70; 75.80
(b) Vol.151,No.2, K181-K186

W. MACIEJEWSKI
On the Study of Interfacial Exchange Coupling between Ferromagnetic Films by the Spin Wave Resonance Method
75.70; 76.50
(b) Vol.151,No.2, K191-K194

A.V. BOLTUSHKIN, V.G. SHADROV, and V.M. FEDOSYUK
Hard-Magnetic Co-W, Co-Ni-W Films with Perpendicular Magnetic Anisotropy
75.70; S1.2
(a) Vol.111,No.2, K225-K228

E.Z. KATSNELSON, A.G. KAROZA, and M.M. CHERVINSKII
Photomagnetization of Ferrite Films and Its Effect on Optical Spectra
75.70 and 78.20; S11.2
(a) Vol.11.,No.2, K233-K236

A. URBANIAK-KUCHARCZYK and Z. OSETEK
Moessbauer Linshapes in the Case of Diluted Ferromagnetic Alloy Thin Films
75.70 and 76.80
(b) Vol.152,No.1, 267-274

S. NIEBER and H. KRONMUELLER
Nucleation Fields in Periodic Multilayers
75.60; 75.70
(b) Vol.153,No.1, 367-376

R. SAKAI and Y. SHIMADA
Coercive Forces and Magnetostriction of Fe-C Films Deposited on ZnO Substrates
75.70; S1.1
(a) Vol.113,No.1, K131-K134

A.I. MITSEK and V.N. PUSHKAR
The Topology of Magnetic Inhomogeneities in Thin Ferromagnetic Layers. The Model of the Magnetic Vortex
75.60 and 75.70; S1.1; S1.2; S11.2
(a) Vol.113,No.2, 529-540

M. JIRSA, G. VÉRTESY, and J. PAČES
Temperature Dependence of the FMR in Garnet Films with Bubble-Domain Lattice
75.70 and 76.50; 75.60; S11.2
(a) Vol.113,No.2, 573-578

A. URBANIAK-KUCHARCZYK
The Hyperfine Field Distribution in Thin Films of Binary Diluted Alloys
75.30 and 75.70
(b) Vol.154,No.1, 327-332

M. GASGNIER
Rare Earth Compounds (Oxides, Sulfides ...)as Thin Films and Thin Crystals. Physico-Chemical Properties and Applications(Review Article)
68.55 and 73.40; 75.70; 78.65; S1.4; S1.63; S10
(a) Vol.114,No.1, 11-72

V.I. MALYUTIN, V.E. OSUKHOVSKII, YU.D. VOROBIEV, and L.A. CHEBOTKEVICH
On the Effect of Annealing on the Coercive Force of Polycrystalline Magnetic Films
75.70; S1.2
(a) Vol.114,No.1, K75-K78

V.M. FEDOSYUK, M.U. SHELEG, O.I. KASYUTICH, and M.M. MALYUSH
Columnar Structure and Texture [001] in Co-Ni-W Films
75.60; 75.70; S1.2
(a) Vol.115,No.1, 279-284

Z. ONYSZKIEWICZ and A. WIERZBICKI
Influence of the Ferromagnetic Bulk Substrate on the Thermodynamical Properties of Metamagnetic Thin Films
75.70
(b) Vol.156,No.1, K71-K76

T. HAI and Z.Y. LI
A Study for Critical Behaviour of Ferromagnetic Thin Films
75.30 and 75.70
(b) Vol.156,No.2, 641-646

L. JAHN, J.S. PASTUSCHENKOV, and V. CHRISTOPH
On the Angular Dependence of the Coercivity of NdFeB Hard Magnets
75.70; 75.60; S4
(a) Vol.116,No.2, K179-K184

75.80. Magnetomechanical and magnetoelectric effects, magnetostriction
For galvanomagnetic effects, see 72.15 and 72.20
For magneto-optical effects, see 78.20

J. GUTIERREZ, J.M. BARANDIARÁN, and O.V. NIELSEN
Magnetoelastic Properties of Some Fe-Rich Fe-Co-Si-B Metallic Glasses
75.80; S1.1; S1.2
(a) Vol.111,No.1, 279-284

J. ULNER
Remarks on the Theory of Slightly Deformed Thin Ferromagnetic Plates
75.70; 75.80
(b) Vol.151,No.2, K181-K186

M. JERGEL, G. VLASÁK, and P. DUHAJ
Magnetostriction Study of Some Co-Based and Fe-Based Amorphous Alloys
75.80; S1.1; S1.2
(a) Vol.111,No.2, 597-602

R. PAWELLEK and M. FAEHNLE
The Magnetic Field Effect on the Linear Magnetostriction of Amorphous Ferromagnets. I. Basic Formalism
75.80; S1.2
(a) Vol.111,No.2, 617-624

G. BALASUBRAMANIAN, A.N. TIWARI, and C.M. SRIVASTAVA
Annealing Effects in the Metallic Glass $Fe_{65}Cr_8B_{27}$
75.80; 76.50; S1.1
(a) Vol.111,No.2, K229-K232

P.P. SHATSKII and D.A. YABLONSKII
Dynamic Inhomogeneous Magnetoelectric Effect in Magnetics with Domain Structure. Resonance Excitation of Domain Boundaries with Electric ...
75.60 and 75.80
(b) Vol.152,No.1, K23-K28

M. FAEHNLE, R. PAWELLEK, and H. KRONMUELLER
The Magnetic Field Effect on the Linear Magnetostriction of Amorphous Ferromagnets
75.80; 75.50
(a) Vol.112,No.1, 189-196

75.80. Magnetomechanical and magnetoelectric effects, magnetostriction

A. SIEMKO, H.K. LACHOWICZ, N. MOSER, A. FORKL, and H. KRONMUELLER
Stress-Induced Sign Change of λ_s Low-Magnetostrictive Metallic Glass as Seen by Its Domain Structure
75.80; S1.1; S1.2
(a) Vol.113,No.1, 181-186

N.H. CHAU, N.H. DUC, T.D. HIEN, K. KROP, and J. ZUKROWSKI
Magnetic and Magnetostrictive Properties of the $Er(Fe_xCu_{1-x})_2$ Compounds
75.30 and 75.80; S1.1; S1.2; S1.4
(a) Vol.114,No.1, 361-368

E.L. ALBUQUERQUE and J. MARTINS S. FILHO
Dielectric Function in Magnetoelectric Crystals
75.80 and 78.20
(b) Vol.154,No.2, K179-K182

M. HRABČÁK, M. MIHÁLIK, J. KOVÁČ, and A. LOVAS
Saturation Magnetostriction and Curie Temperature of Neutron Irradiated $Fe_{75}Cr_{10}B_{15}$ and $Fe_{77}Cr_8B_{15}$ Amorphous Alloys
75.80; 61.80; S1.1
(a) Vol.115,No.1, K83-K86

N.P. KOLMAKOVA, R.Z. LEVITIN, and V.N. ORLOV
Anomalies of Crystal Lattice Parameters in Paramagnetic Rare-Earth Garnets at Low Temperatures
65.70 and 75.80; S11.2
(a) Vol.115,No.1, K87-K90

J. GONZALEZ ESTEVEZ and E. DU TREMOLET DE LACHEISSERIE
Influence of Annealing on the Magnetostriction of an Amorphous $Co_{45}Ni_{30}Si_{15}B_{10}$ Alloy
75.50 and 75.80; S1
(a) Vol.115,No.2, K233-K236

A.I. MITSEK
Collective Excitations of Localized Electrons and Their Effect on Magnetic-Material Kinetics
71.45; 72.15; 75.80; S1.2; S1.4; S1.5
(b) Vol.156,No.2, 507-514

I.E. CHUPIS
Temperature Evolution of the Magnetoelectric Effect in Ferroelectrics-Ferromagnetics
75.80
(b) Vol.156,No.2, K135-K138

M. FAEHNLE and J. FURTHMUELLER
The Effect of External Stress on the Saturation Magnetostriction Constant of Metallic Glasses
75.80; S1.2
(a) Vol.116,No.2, 819-826

75.90. Other topics in magnetic properties and materials

R. SKOMSKI and V. CHRISTOPH
Power Law Behaviour of Magnetic Viscosity
75.60 and 75.90
(b) Vol.156,No.2, K149-K152

76. Magnetic Resonances and Relaxation in Condensed Matter; Moessbauer Effect

76.20. General theory of resonances and relaxation

R.KH. SABIROV
On the Calculation of Thermodynamic Quantities of the Paramagnetic Spin System at Low Temperature
76.20
(b) Vol.151,No.2, K201-K204

D.A. TAYURSKII
Spin Kinetics in Paramagnets at Low Temperature
76.20; 76.30; 76.60; 76.70
(b) Vol.152,No.2, 645-656

76.30. Electron paramagnetic resonance and relaxation

R. CYWIŃSKI, E. MUGEŃSKI, W. NOWY-WIECHUŁA, and J. WIECHUŁA
Suzuki Phase Precipitates in Doubly Doped KCl:Pb^{2+},Mn^{2+} Crystals
61.70; 76.30; 78.50; S9.11
(b) Vol.151,No.1, 47-52

A.A. BUGAI, I.N. KRAVTSOVA, V.M. MAKSIMENKO, B.D. SHANINA, and V.I. SHAKHOVTSOV
Properties of the Si-S1 Center in $Si_{1-x}Ge_x$ Solid Solutions
71.55; 76.30; S5.11; S5.12
(b) Vol.151,No.1, 159-166

O. GOEDE, D. BACKS, W. HEIMBRODT, and M. KANIS
EPR Study of the Antiferromagnetic Phase Transition in (Cd,Mn)S
76.30; 75.30; S8.15
(b) Vol.151,No.1, 311-318

WEN-CHEN ZHENG
A Study of Distortion of the Immediate Environment of the Chromium Ion in Alums
61.70; 62.20; 76.30; S11.1
(b) Vol.151,No.1, K11-K16

WEN-CHEN ZHENG
A Further Study of Tetragonal Distortion of the Cr^{3+}-V_c Centre in MgO from Optical and EPR Spectra
61.70 and 71.70; 76.30; S10.1
(b) Vol.151,No.1, K17-K22

V.A. IVANSHIN, I.N. KURKIN, W. BRUNNER, and G. VOELKEL
Electron Spin-Lattice Relaxation of the $(SeO_2)^-$ Radical in X-Irradiated $KH_3(SeO_3)_2$
76.30 and 77.80; S11.1
(b) Vol.151,No.1, K53-K58

A.A. ROMANYUKHA, YU.N. SHVACHKO, A.V. SKRIPOV, and V.V. USTINOV
Conduction Electron Spin Resonance in the New Organic Superconductor $(ET)_2Cu(SCN)_2$
76.30; S12
(b) Vol.151,No.1, K59-K64

YU.N. SHVACHKO, A.A. ROMANYUKHA, A.V. SKRIPOV, and V.V. USTINOV
Effects of Electron Irradiation on ESR Properties of β-$(ET)_2I_3$
76.30; S12
(b) Vol.151,No.1, K65-K70

A.A. KARANOVICH and A.V. DVURECHENSKII
The EPR Study of Inhomogeneous Deformations in Neutron-Irradiated Silicon
61.70; 76.30; S5.11
(a) Vol.111,No.1, 257-264

S. WAPLAK and T. KRAJEWSKI
EPR Studies of $LiNaSO_4$ Chromium Cr^{6+} Doped Monocrystals after X-Ray Irradiation
76.30; 61.80; 66.70; S11
(a) Vol.111,No.1, 271-278

K. BENTE, E. SCHWARZMANN, and D. SIEBERT
EPR Studies of Cr^{2+} from Impurity Phases in the High Temperature Superconducting $YBa_2Cu_3O_{7-x}$
74.70 and 76.30; S10.15
(a) Vol.111,No.1, K93-K98

P. STEGGER and G. LEHMANN
Dynamic Effects in a New Substitutional Center of Trivalent Iron in Quartz
61.70; 76.30; S10.1
(b) Vol.151,No.2, 463-468

R.M. KRISHNA, J.L. RAO, PREM CHAND, and S.V.J. LAKSHMAN
Electron Paramagnetic Resonance Study of Copper(II) Ions in Single Crystals of Thallium Magnesium Sullphate Hexahydrate
71.10 and 76.30; S11
(b) Vol.151,No.2, 615-622

F.S. DZHEPAROV and E.K. HENNER
Magnetic Resonance in Magnetically Diluted Solids at Low Temperatures
76.30
(b) Vol.151,No.2, 663-674

WEN-CHEN ZHENG
A Study of the Spin-Hamiltonian Parameters of Isoelectronic Impurity Ions in the Same Host Crystal
71.70; 76.30; S9.12; S10; S10.1
(b) Vol.151,No.2, K145-K150

V.K. JAIN, N.V. VUGMAN, and M.L.N. GRILLO
Electron Paramagnetic Resonance Studies of Irradiated Single Crystals of Cesium Chloride Doped with Nickel Cyanide Complexes
76.30; S9.11
(a) Vol.111,No.2, 625-628

L.K. AMINOV, A.V. VINOKUROV, I.N. KURKIN, YU.YU. KOSTETSKII, S.L. KORABLEVA, I.KH. SALIKHOV, S.I. SAKHAEVA, and M.A. TEPLOV
Distortion of a Crystal Field in the van Vleck Paramagnet $LiTmF_4$ under Activation by Isovalent Impurities
71.70; 76.30; 78.50; S9.16
(b) Vol.152,No.1, 191-202

N. GUSKOS, TH. LEVENTOURI, CH. TRIKALINOS, M. CALAMIOTOU, S.M. PARASKEVAS, J. KURIATA, and L. SADLOWSKI
On the Possibility of the Excistence of Cu^{3+} Ions in High-T_c Superconducting Samples of $YBa_2Cu_3O_x$
74.70 and 76.30; S10.15
(b) Vol.152,No.1, K9-K12

D. GOURIER, T. GBEHI, R. VISOCEKAS, J. THERY, and D. VIVIEN
Color Centres in Barium Hexaaluminate (Phase I). Study by EPR, Optical Absorption, and Thermally Stimulated Luminescence
61.70; 76.30; 78.50; 78.60; S11
(b) Vol.152,No.2, 415-430

KE-QIN LU, SHENG-LI GUO, and PING-FENG WU
An Interpretation of the Pressure Dependence of EPR Zero-Field Splitting of Mn^{2+} in $Na_2ZnCl_4 \cdot 3\ H_2O$
71.70 and 76.30; S9.16
(b) Vol.152,No.2, 577-582

R.S. EACHUS, M.T. OLM, R. JANES, and M.C.R. SYMONS
Observation by ESR of Electrons Localised at Intrinsic Shallow Traps in Silver Halide Systems
71.50 and 76.30; 71.38; 76.40; S9
(b) Vol.152,No.2, 583-592

D.A. TAYURSKII
Spin Kinetics in Paramagnets at Low Temperature
76.20; 76.30; 76.60; 76.70
(b) Vol.152,No.2, 645-656

N. GUSKOS
Ground State of the Gd^{3+} Ion in MF_2 Single Crystals under the Influence of External Static Electric Field
71.70; 76.30; S9.12
(b) Vol.153,No.1, 243-248

MAO-LU DU and MIN-GUANG ZHAO
Gyromagnetic Factor and Spin-Orbit Coupling of Ligands in $CsVX_3$ (X=Cl,Br,I)
71.70; 76.30; S1.2
(b) Vol.153,No.1, 249-256

J.L. RAO, G.L. NARENDRA, B. SREEDHAR, and S.V.J. LAKSHMAN
Electron Spin Resonance and Optical Absorption Spectra of Mn^{2+} Ions in K_2SO_4-$ZnSO_4$ Glasses
71.70; 76.30; 78.50; S11
(b) Vol.153,No.1, 257-262

S.Z. KORCZAK, W. KORCZAK, M. SUBOTOWICZ, and H. WASIEWICZ
EPR Linewidth in $Pb_{1-x}Mn_xTe$
76.30; S8.15
(b) Vol.153,No.1, 361-366

T. KAWAKUBO and M. OKADA
Optical and Electron Paramagnetic Resonance Studies on Neutron-Irradiated GaP Crystals
71.55; 76.30; 78.50; S7.11
(b) Vol.153,No.1, K93-K96

WEN-CHEN ZHENG
Local Thermal Expansion around the Impurity in Chromium-doped alpha-$LiIO_3$ Crystals
65.70; 71.70; 76.30; S11.1
(b) Vol.153,No.2, K103-K108

SHENG-LI GUO, KE-QIN LIU, LIN-RUI PING, and GAO-CHAO LIU
Analysis of the EPR of Cr^{3+}-Vacancy Center MgO
71.70 and 76.30; S10.1
(b) Vol.153,No.2, K167-K170

M. ANWAR, C.A. HOGARTH, and K.A.K. LOTT
ESR Studies of MoO_3/In_2O_3 Amorphous Thin Films Deposited by Co-Evaporation
68.55 and 76.30; S10.15
(a) Vol.113,No.2, 483-488

V.S. LYSENKO, A.N. NAZAROV, S.A. VALIEV, I.M. ZARITSKII, T.E. RUDENKO, and A.S. TKACHENKO
EPR and TSCR Investigations of Implanted Al-SiO_2-Si Systems Treated with RF Plasma Discharge
73.40 and 85; 76.30; S5.11; S10.1
(a) Vol.113,No.2, 653-666

L.L. KULYUK, N.F. PERELMAN, S.I. RADAUTSAN, A.V. SIMINEL, and E.E. STRUMBAN
About the Possibility of Double Centre Formation in InP:Fe:O
71.55 and 72.40; 76.30; 78.55; S7.11
(a) Vol.113,No.2, K227-K230

L.E. MISIAK, S.K. MISRA, and U. ORHUN
Study of Temperature Variation of EPR Linewidth of Gd^{3+}-Doped $LiYb_xY_{1-x}F_4$ Single Crystals. Estimation of Yb^{3+} Spin-Lattice
71.70 and 76.30; S9.16
(b) Vol.154,No.1, 249-258

S.K. MISRA and CHUNZHENG WANG
EPR and Optical-Absorption Studies of Cu^{2+}-Doped $Mg(CH_3COO)_2 \cdot 4\ H_2O$ Single Crystal. Core Polarization and Molecular-Orbital Coefficients
71.70; 76.30; 78.50; S12
(b) Vol.154,No.1, 259-272

H. ISOTALO, J. LAAKSO, P. KUIVALAINEN, H. STUBB, J.-E. OESTERHOLM, and P. YLI-LAHTI
Charge Transport in $FeCl_3$-Doped Polyethylenevinylacetate-Polyoctylthiophene Blends
72.20; 72.60; 76.30; S12
(b) Vol.154,No.1, 305-314

L. FIRLEJ, A. GRAJA, J. WOLAK, R.N. LYUBOVSKAYA, and L.M. GOLDENBERG
ESR Studies of Single Crystals of Bis(ethylenedithio)tetrahiafulvalene Based Conductor with Octaiodinemercurate Anion $(BEDT\text{-}TTF)_4Hg_3I_8$
64.70 and 76.30; 72.20; S12.1
(b) Vol.154,No.1, 333-340

SHENG-LI GUO and KE-QIN LU
EPR Analysis of Fe^{3+} in CsCl and NH_4Cl
76.30; S9; S9.12
(b) Vol.154,No.1, K67-K72

WEN-CHEN ZHENG
Investigation of the Pressure and the Stress Dependence of Low-Temperature EPR Spectra for $ZnSiF_6 \cdot 6\ H_2O{:}V^{2+}$ Crystals
76.30; 71.70; S9.15
(b) Vol.154,No.1, K73-K78

K. SWIATEK, M. GODLEWSKI, D. HOMMEL, and H. HARTMANN
On the Incorporation of Rare Earth Ions into II-VI Compounds: ZnS:Eu
61.70; 76.30; S8.11
(a) Vol.114,No.1, 127-134

M. ANWAR, C.A. HOGARTH, and K.A.K. LOTT
Electron Spin Resonance Study of Thermally Evaporated Amorphous Thin Films of MoO_3/SiO
68.55 and 76.30; S10.15
(a) Vol.114,No.1, 225-232

A. BALS, S.I. GORBACHUK, J. KLIAVA, and N.G. KAKAZEI
EPR Study of the Structure of Disordered Polycrystals. Mn^{2+} in Lithographic Stone
76.30; S10.15
(a) Vol.114,No.1, 305-314

A.D. STEVENS, M.C.R. SYMONS, and R.S. EACHUS
ESR Study of Silver Clusters in AgCl and AgBr. An Attempt to Detect Photographic Latent Images
61.70; 76.30; S9
(b) Vol.154,No.2, 483-496

WEN-CHEN ZHENG
Theoretical Investigation of the Spin-Lattice Coupling Coefficients G_{11} and G_{44} for Cr^{3+} in Cubic $SrTiO_3$ Crystals
71.70 and 76.30; S11.1
(b) Vol.154,No.2, K167-K170

GUANGHUA CHEN, FANGQING ZHANG, WEI JIA, and WEI CHEN
Study on Defects in Reactively Sputtered a-GeN_x:H Films
68.60 and 76.30; S6
(a) Vol.114,No.2, K171-K174

R.S. DE BIASI
Comments on Concentration Peculiarities of a Set of Impurity Centres in Fluorite Crystals Activated by Gadolinium
61.70; 76.30; 78.60; S9.15
(b) Vol.155,No.1, K3-K4

F. MAES, F. CALLENS, P. MATTHYS, and E. BOESMAN
Electron Paramagnetic Resonance of an O_3^- Centre in KBr
76.30; 61.70; S9.11
(b) Vol.155,No.1, K55-K58

A.B. VASSILIKOU-DOVA, H. MATTHIES, K. RECKER, F. WALLRAFEN, and G. LEHMANN
EPR of Mn^{2+} in Ba_2ZnF_6 Single Crystals
76.30; 71.55; S9.16
(b) Vol.155,No.1, K59-K62

A. POEPPL and G. VOELKEL
ESR Investigation of the Oxygen Vacancy in Pure and Bi_2O_3-Doped ZnO Ceramics
71.55; 76.30; S10.1
(a) Vol.115,No.1, 247-256

A.V. EGRANOV, A.I. NEPOMNYACHIKH, V.V. OTROSHOK, and B.P. CHERNYAGO
H_2^--Centre in LiF-H^-
61.70; 76.30; 78.40; S9.11
(b) Vol.155,No.2, 365-372

J. KREISSL, W. ULRICI, and W. GEHLHOFF
Electron Paramagnetic Resonance Study of the Dynamical Jahn-Teller Effect on the $3d^1$ Impurities Ti^{3+} and V^{4+} in GaP
71.55; 76.30; S7.11
(b) Vol.155,No.2, 597-608

WEN-CHEN ZHENG
Theoretical Considerations on the Axial Crystal-Field Splitting and Its Dependence on the Bonding Angle Distortion in Tetragonal ...
71.70 and 76.30; S9.16
(b) Vol.155,No.2, K149-K152

M. HOEHNE
EPR of Lithium-Induced Silicon-$5d^n$ Pairs in Silicon
71.55 and 76.30; S5.11
(b) Vol.156,No.1, 325-332

A.A. ALYBAKOV, V.A. GUBANOVA, M.ZH. ZHANIBEKOV, and N. TOICHEV
EPR Spectra of X-Irradiated KBr:CrO_4^{2-}, Ba^{2+} Single Crystals
76.30; 61.80; S9.11
(b) Vol.156,No.1, K77-K80

V.A. NADOLINNYI, V.V. VASILEV, and I.P. MIKHAILOVSKII
Photo-Induced Metastable States in Silicon Nitride Amorphous Layers
76.30 and 78.55; S6
(a) Vol.116,No.1, K105-K110

E.K. SADYKOV and A.I. SKVORTSOV
Propagation of Gamma-Radiation through a Resonance Medium with RF Modulated Magnetic Hyperfine Interaction
76.80; 76.30
(b) Vol.156,No.2, 605-614

C.K. SARKAR and P. BANERJI
Estimation of the Alloy Scattering Potential in $Ga_{0.47}In_{0.53}As$ Using Cyclotron Resonance Measurements
76.30; 72.20; S7.15
(b) Vol.156,No.2, K145-K148

M. KUMEDA, M. KIWAKI, and T. SHIMIZU
Erratum to Temperature Dependence of the E′ Center Creation in Silica Glasses (phys. stat. sol. (b) 147, K1 (1988))
61.70 and 76.30; S10.1
(b) Vol.156,No.2, K163-K164

W. ZAPART, M.B. ZAPART, and A.I. OTKO
EPR Detection of the Ferroelastic Phase Transition in $KSc(WO_4)_2$
76.30; 64.70; S11.1
(a) Vol.116,No.2, K191-K194

76.40. Diamagnetic and cyclotron resonances

R.S. EACHUS, M.T. OLM, R. JANES, and M.C.R. SYMONS
Observation by ESR of Electrons Localised at Intrinsic Shallow Traps in Silver Halide Systems
71.50 and 76.30; 71.38; 76.40; S9
(b) Vol.152,No.2, 583-592

76.50. Ferromagnetic, antiferromagnetic, and ferrimagnetic resonances; spin-wave resonance
(see also 75.30. Spin waves)

V.N. BERZHANSKII and V.I. IVANOV
Critical Relaxation of Uniform Precession of Magnetization in Cubic Ferrimagnets
75.30 and 75.50; 76.50; S11.2
(b) Vol.151,No.1, 259-268

J. BARNAS
Magnetostatic Modes in Disordered Superstructures
75.70; 76.50
(b) Vol.151,No.1, K71-K76

W. MACIEJEWSKI
On the Study of Interfacial Exchange Coupling between Ferromagnetic Films by the Spin Wave Resonance Method
75.70; 76.50
(b) Vol.151,No.2, K191-K194

G. BALASUBRAMANIAN, A.N. TIWARI, and C.M. SRIVASTAVA
Annealing Effects in the Metallic Glass $Fe_{65}Cr_{8}B_{27}$
75.80; 76.50; S1.1
(a) Vol.111,No.2, K229-K232

M. JIRSA, G. VÉRTESY, and J. PAČES
Temperature Dependence of the FMR in Garnet Films with Bubble-Domain Lattice
75.70 and 76.50; 75.60; S11.2
(a) Vol.113,No.2, 573-578

Z. FRAIT and D. FRAITOVÁ
Surface Spin-Wave Mode Excitation in Bulk Amorphous FeNiB Alloys
75.50 and 76.50; S1.1; S1.2
(b) Vol.154,No.1, 363-366

D. FRAITOVÁ
A Theoretical Study of the FMR Surface Mode Excitation in Metals (Parallel Configuration)
76.50
(b) Vol.155,No.1, 281-288

K.G. NIKIFOROV, A.G. GUREVICH, L.YA. PASENKO, S.I. RADAUTSAN, and L.M. EMIRYAN
Ferromagnetic Resonance and Cr^{4+} Ions in the Magnetic Semiconductor $HgCr_2Se_4$
76.50; S8.16
(a) Vol.116,No.2, K185-K190

76.60. Nuclear magnetic resonance and relaxation

A.A. SUKHOVSKII, I.P. ALEKSANDROVA, and YU. N. MOSKVICH
NMR Confirmation of the Existence of Superstructure in the Ferroelectric Phase of NH_4HSeO_4
64.70; 76.60; S11.1
(a) Vol.111,No.2, 477-484

D.A. TAYURSKII
Spin Kinetics in Paramagnets at Low Temperature
76.20; 76.30; 76.60; 76.70
(b) Vol.152,No.2, 645-656

K. KOKKO, E. OJALA, and K. MANSIKKA
Fermi Level Density of States in Ag-Au Alloys. Nuclear Magnetic Spin-Lattice Relaxation Rate and Low-Temperature Specific Heat
71.20; 65.40; 76.60; S1.3
(b) Vol.153,No.1, 235-242

H. DWELK, D. FREUDE, R. HERRMANN, H. PFEIFFER, and N. PRUSS
Substitution of Yttrium for Boron in the Structure of $YBa_2Cu_3O_{7-\delta}$
72.20 and 74.70; 76.60; S10.15
(a) Vol.113,No.1, 159-162

J. CZAPLICKI, N. WEIDEN, and A. WEISS
Proton Reorientations and Tunneling in $Cd(NH_3)_6Cl_2$ from NMR
76.60; S9.16
(a) Vol.113,No.1, 163-168

I.O. TROYANCHUK, A.A. SHEMYAKOV, and V.K. PROKOPENKO
Magnetic Structure of $NiMnO_3$ and $CoMnO_3$ Ilmenites.
75.30 and 76.60; S11.1
(a) Vol.113,No.1, K107-K110

K. TOMPA, R. PIRLOT, L. GRENACS, and J. LEHMANN
Hyperfine Fields at Implanted ^{12}Boron in F.C.C. Metals
76.60; 71.50; S1; S1.3; S4
(b) Vol.153,No.2, 721-726

N.T. BAGRAEV, I.S. POLOVTSEV, and K. SCHMALZ
An ONP Investigation of Deep Level Defects Produced by Heat Treatment at 600 and 800 °C in Cz-Si
61.70 and 71.55; 76.60; S5.11
(a) Vol.114,No.1, 233-246

L.F. MARTYNOVA, Z.B. CHACHKHIANI, and V.I. CHECHERNIKOV
Knight Shift on ^{45}Sc Nuclei in Scandium Compounds and Alloys
76.60; S1.2
(a) Vol.114,No.1, K87-K92

TH. ERGE, D. MICHEL, J. PETERSSON, and W. WINDSCH
^{35}Cl NMR Study in the Paraelectric Phase of Tris-Sarcosine Calcium Chloride
76.60; 77.80; S12
(a) Vol.114,No.2, 705-712

T. TANABE, R. IKEDA, and D. NAKAMURA
The Highest-Temperature Crystalline Phase of Tetramethylammonium Thiocyanate with the Cations Performing Three-Dimensional ...
66.30 and 76.60; S12
(a) Vol.114,No.2, K143-K146

A.V. ZALESSKII, V.G. KRIVENKO, V.S. LUTOVINOV, V.A. MURASHOV, and T.A. KHIMICH
NMR Study of the Supertransferred Hyperfine Fields at ^{57}Fe Nuclei and Morin Transition in Al-Substituted Hematite Crystals
76.60; S10.1
(b) Vol.155,No.1, K63-K66

M.M. SAVOSTA, V.D. DOROSHEV, P. NOVAK, and I. VELTRUSKY
NMR of ^{167}Er and Magnetic Structure of $Er_3Fe_5O_{12}$ Garnet
76.60; 75.25; S11.2
(b) Vol.155,No.2, 669-678

N.S. SULLIVAN and Y. LIN
Low-Energy Excitations in the Orientational Glass State of Solid Hydrogen
71.20; 76.60; S5.2
(b) Vol.155,No.2, K109-K112

G.G. BOKUCHAVA, L.L. BUISHVILI, and A.I. TUGUSHI
Spin-Lattice Relaxation of Nuclei in Amorphous Dielectrics Containing Paramagnetic Impurities
76.60
(b) Vol.155,No.2, K153-K156

A.I. GUSEV
Short-Range Order in Nonstoichiometric Transition Metal Carbides Nitrides and Oxides (Review Article)
61.60 and 75.30; 76.60; S1.61; S1.62; S10
(b) Vol.156,No.1, 11-40

T.SH. ABESADZE, A.I. TUGUSHI, and V.N. KHELASHVILI
The Influence of Ultrasound on the Nuclear Magnetic Resonance in Disordered Materials Containing Paramagnetic Impurities with ...
76.60
(b) Vol.156,No.1, K67-K70

A.YA. FISHMAN, N.M. KOVTUN, V.YA. MITROFANOV, V.K. PROKOPENKO, and A.A. SHEMYAKOV
Supertransferred Hyperfine Fields at Nuclei of Diamagnatic Copper Ions in Chalcogenide Spinels $Fe_{1-x}Cu_xCr_2S_4$
75.30; 76.60; S11.2
(b) Vol.156,No.2, 597-604

YU.N. MOSKVICH, B.I. CHERKASOV, A.M. POLYAKOV, A.A. SUKHOVSKII, and R.L. DAVIDOVICH
NMR Study of Internal Motions in a New Family of Ionic Conductors M_2SbF_5
76.60; 66.30; 72.20; S9.16
(b) Vol.156,No.2, 615-632

76.70. Magnetic double resonances and cross effets

M. KIKNADZE, R. MIRIANASHVILI, and T. SANADZE
RFDS Studies of Thermally Non-Stable Trigonal Centers of Tm^{2+} in SrF_2 and BaF_2 Single Crystals
76.70; S9.12
(b) Vol.151,No.1, K49-K52

M. KIKNADZE, R MIRIANASHVILI, and T. SANADZE
Ligand Hyperfine Interaction and Lattice Distortion for Cubic Er^{3+} Centers in BaF_2
76.70; S9.12
(b) Vol.152,No.1, K13-K16

D.A. TAYURSKII
Spin Kinetics in Paramagnets at Low Temperature
76.20; 76.30; 76.60; 76.70
(b) Vol.152,No.2, 645-656

M.M. KAZANIN and V.V. KUTASOV
Fe Impurity in alpha- and beta-Boron and in Some Boron Compounds
71.55; 72.20; 76.70; S5
(a) Vol.113,No.1, 143-150

76.80. Moessbauer effect; other gamma-ray spectroscopy

YAN SHI, HANMIN JIN, and MINGZHI JIN
Interpretation of ^{57}Fe Moessbauer Measurements on $Nd_2Fe_{14}B$ at 83 and 295 K
76.80; S1.1; S1.4
(b) Vol.151,No.1, 249-258

O.A. BAYUKOV, V.P. IKONNIKOV, and A.F. SAVITSKII
Magnetic and Moessbauer Studies of Anion-Substituted Nickel Ferrites
75.30; 76.80; S11.2
(b) Vol.151,No.1, 269-274

J.J. BARA and B.F. BOGACZ
Investigation of ^{57}Fe-Doped theta-$Al_xV_2O_5$ Bronzes
61.60; 76.80; S10
(a) Vol.111,No.1, 53-58

R.G. GABRIELYAN and KH.V. KOTANDJIAN
Moessbauer Radiation Transfer in the Plane Layer. Anisotropic Case
76.80
(b) Vol.151,No.2, 655-662

R. FOLLATH and H. JEX
Temperature Dependent Frequency Shifts in the Resonance Absorption and Resonance Fluorescence Scattering of Moessbauer gamma-Quanta
76.80; S9.11
(b) Vol.151,No.2, 703-710

E.L. ALBUQUERQUE and P.W. MAURIZ
Debye-Waller Factor for Moessbauer Interface Atoms: Elastic Approximation
63.20 and 76.80
(b) Vol.151,No.2, K109-K114

A. URBANIAK-KUCHARCZYK and Z. OSETEK
Moessbauer Linshapes in the Case of Diluted Ferromagnetic Alloy Thin Films
75.70 and 76.80
(b) Vol.152,No.1, 267-274

J. PAVLOVSKY and T. ZEMCIK
Comparison of Surface and Bulk Crystallization of the Amorphous $Fe_{70}Co_{10}B_{20}$ Alloy
61.60; 76.80; S1.1
(a) Vol.112,No.1, K1-K4

A.D. BALAEV, O.A. BAYUKOV, and A.F. SAVITSKII
Magnetic and Moessbauer Studies of Magnesium-Chromium Ferrites
75.50 and 76.80; S11.2
(b) Vol.152,No.2, 639-644

J. UTZIG
Moessbauer Spectroscopy of Cobalt in Germanium
76.80; S1.2; S5.12
(b) Vol.153,No.1. 377-382

S. BALUCH, M. MIGLIERINI, R. GROENE, and J. SITEK
Positron Lifetime and Moessbauer Study of $Fe_{80-x}Ni_xB_{20}$ Metallic Glasses
76.80 and 78.70; S1.1
(a) Vol.113,No.1, K143-K146

K.N. SHRIVASTAVA
Physical Properties of Fractals. Moessbauer Recoilless Fraction, Raman Scattering, and Spin-Lattice Relaxation
63.90; 76.80; 78.30
(b) Vol.153,No.2, 547-560

V.I. KUZNETSOV, E.N. YURCHENKO, M.T. PROTASOVA, E.A. TARABAN, O.P. KRIVORUCHKO, and R.A. BUYANOV
Moessbauer Investigation of the Influence of Al(III) Ions on the Properties of Mixed Fe(III)-Al(III) Hydroxides and Oxides
64.75; 76.80; S10
(a) Vol.113,No.2, 359-364

S. MOTIEJŪNAS, D. BALTRŪNAS, N. GEČIAUSKAITĖ, and K. MAKARIUNAS
Correlation between the Moessbauer and Semiconductor Parameters of Tin Telluride
76.80; 61.55; S1.63
(b) Vol.154,No.1, 341-346

L. T. TSANKOV
Time Behaviour of Recoilless gamma-Radiation Scattered by RF-Induced Substrates in Moessbauer Crystals
76.80
(a) Vol.114,No.1, 341-354

E.L. SAW, M.H. KUOK, C.T. YAP, and J.B. WARD
Temperature Dependence of the Moessbauer Spectra of $CsSnCl_3$
76.80; S11
(a) Vol.114,No.2, K223-K228

Y.L. CHEN, P. MAO, DECHENG TIAN, X.Z. LI, and J.J. ZHU
Moessbauer Study of Amorphous $(Fe_{1-x}Ga_x)_{84}B_{16}$
76.80; S1.1
(a) Vol.115,No.1, K91-K94

K. SUN, M.X. LIN, T. HE, and H.L. LUO
The Moessbauer Study of alpha-Fe_2O_3 Fine Particles with and without Adsorbed Cobalt
75.60; 76.80; S10
(a) Vol.115,No.2, 539-546

DECHENG TIAN, KAIYANG LI, and DIHUA DING
Magnetic Aftereffect and Mossbauer Studies of Structural Relaxation in $(Fe_{1-x}Ce_x)_{84}B_{16}$ Amorphous Alloys
61.40; 75.60; 76.80; S1.1
(a) Vol.116,No.1, 203-212

E.K. SADYKOV and A.I. SKVORTSOV
Propagation of Gamma-Radiation through a Resonance Medium with RF Modulated Magnetic Hyperfine Interaction
76.80; 76.30
(b) Vol.156,No.2, 605-614

F.G. REQUEJO, A.G. BIBILONI, C.P. MASSOLO, J. DESIMONI, and M. RENTERIA
TDPAC Study of Stoichiometric and Reduced MoO_3. Localization of Indium Trapped Oxygen Single Vacancies
61.60; 61.70; 76.80; S10.1
(a) Vol.116,No.2, 503-512

V. PIERRON-BOHNES, C. LEROUX, J.P. AMBROISE, A. MENELLE, and P. BASTIE
Measurement of the Mean Static Displacement and of the Debye Waller Factor between 5 and 1464 K in Fe-19.5at%Al
61.55; 61.12; 76.80; S1.1
(a) Vol.116,No.2, 529-536

76.90. Other topics in magnetic resonances and relaxation

C.P. HILTON, J.J. DAVIES, J.E. NICHOLLS, and O. GOEDE
Optically-Detected Magnetic Resonance of Excitons Bound at Te_2 Isoelectronic Centres in $CdS_{1-x}Te_x$
71.35 and 71.55; 76.90; 78.55; S8.11; S8.15
(b) Vol.151,No.1, 175-184

M. SOUIRI, A. GOLTZENÉ, and C. SCHWAB
Extension of the Crystal Chemistry Approach for the Location of Muonium in Crystals
61.70; 76.90
(b) Vol.156,No.1, 71-76

77. **Dielectric Properties and Materials**
(for conductivity phenomena, see 72.20 and 72.80)

77.20. Permittivity

R. JAKUBAS, P.E. TOMASZEWSKI, and L. SOBCZYK
Phase Transition in $(CH_3NH_3)_3Bi_2Cl_9$
64.70; 77.20; S12
(a) Vol.111,No.1, K27-K30

H. KRESSE, H. STETTIN, F. GOUDA, and G. ANDERSSON
Dielectric Relaxation of a Mixture of a Nematic Liquid Crystal and a Non-Liquid Crystalline Compound
61.25 and 77.20; S12
(a) Vol.111,No.2, K265-K268

R. NAVIL KUMAR and C.P.G. VALLABHAN
Effect of Deuteration on the Electrical Conductivity, Dielectric Constant, and Phase Transition in $(NH_4)_3H(SO_4)_2$
72.80 and 77.20; 64.70; S11.1
(a) Vol.112,No.2, K113-K118

R. CACH and Z. CZAPLA
Dielectric Nonlinearity of Diglycine Nitrate (DGN) Crystal in the Paraelectric Phase
77.20; 77.80; S12
(a) Vol.113,No.1, K139-K142

R. MIZERIS, J. GRIGAS, R.N.P. CHOUDHARY, and T.V. NARASAIAH
Microwave Dielectric Investigations of TlH_2PO_4
77.20; 77.80; S11.1
(a) Vol.113,No.2, 597-602

A.G. FOKIN and T.D. SHERMERGOR
Electromagnetic Wave Scattering in Inhomogeneous Media with Spatial Dispersion
77.20 and 78.90
(b) Vol.154,No.1, 367-376

M.A. AHMED, M.K. EL-NIMR, A. TAWFIK, and A.M. ABOELATA
Dielectric Behaviour in Co-Zn Ferrites
77.20 and 77.40; S11.2
(a) Vol.114,No.1, 377-380

T. GOLDMANN, E. GERDES, and J. SIMANOWSKI
Dielectric Properties of Passivation Glasses in the System $PbO-B_2O_3-SiO_2-Al_2O_3-CeO_2$ Depending on Glass Composition
77.20 and 77.40; S10.15
(a) Vol.114,No.1, 387-396

S. MISHRA, A.V. KRISHNA RAO, and K.V. RAO
Dielectric Properties of Tourmaline under Different Conditions
77.20 and 77.40; 77.70; S11
(a) Vol.114,No.1, K115-K118

J. MONECKE
Microstructure Dependence of Material Properties of Composites
77.20; 64.80
(b) Vol.154,No.2, 805-814

M. POLOMSKA and F. SMUTNÝ
$Rb_4LiH_3(SeO_4)_4$: A New Tetragonal Crystal with Low-Temperature Ferroelastic Phase
64.70; 77.20; S11
(b) Vol.154,No.2, K103-K106

A.K. PACHAURI, A.K. SHARMA, and H.C. SAXENA
Volume Dependence of Short Range Force Constant and Dielectric Properties of Alkali Halides
63.90; 77.20; S9.11
(b) Vol.155,No.1, 73-76

H. ZANGAR, J.L. MIANE, C. COURSEILLE, N.B. CHANH, M. COUZI, and Y. MLIK
Structural Phase Transition in the Perovskite-Type Layer Compound $(C_3H_7NH_3)_2PbCl_4$
64.70; 61.14; 77.20; S11
(a) Vol.115,No.1, 107-118

B.K. CHOUDHARY, R.N.P. CHOUDHARY, and K.V. RAO
Dielectric Properties of $TbAsO_4$ Single Crystals
77.20 and 77.40; S11.1
(a) Vol.115,No.1, 301-306

K. KUMAZAKI, L. VIÑA, C. UMBACH, and M. CARDONA
Interband Critical Point Parameters Determined by Ellipsometry in $Cd_xHg_{1-x}Se$
71.25 and 78.20; 77.20; S8.15
(b) Vol.156,No.1, 371-376

V.V. LOPATIN and A.V. KABYSHEV
On the Connection between the Building of Aluminium Nitride Ceramics and Its Dielectric Properties
61.40; 72.80; 77.20; S1.62
(a) Vol.116,No.1, 221-230

S.K. ARORA and T. MATHEW
Dielectric Studies of $CuWO_4$ Crystals
77.20 and 77.40; S11.1
(a) Vol.116,No.1, 405-414

R. CACH, S. DACKO, and Z. CZAPLA
Dielectric Properties and State Equation Parameters of $(CH_3)_2NH_2Al(SO_4)_2 \cdot 6H_2O$ Ferroelectric Crystals
77.80; 77.20; S12
(a) Vol.116,No.2, 827-832

F. BALLER, B. GATHER, B. HELLERMANN, H. HESSE, and E. KRAETZIG
The Influence of Stress on the Dielectric Properties of KTN Crystals
77.20; 77.80; S11.1
(a) Vol.116,No.2, K195-K198

77.30. Polarization and depolarization effects

C. CHRISTODOULIDES, L. APEKIS, P. PISSIS, and D. DAOUKAKI-DIAMANTI
The Determination of the Distibutions of the Parameters of Thermally Stimulated Depolarization Current Peaks. Theory
77.30
(a) Vol.111,No.1, 325-334

Z. ZIKMUND and J. FOUSEK
Stabilization of Polarization Vector in Uniaxial Ferroelectrics by Means of Chiral Dopants. Macroscopic Symmetry Conditions
77.30 and 77.80; S12
(a) Vol.112,No.2, 625-634

M. OLSZOWY
Electret State in the Polyethylene Terephthalate Films Implanted by Oxygen and Bismuth Ions
77.30; 61.80; S12
(a) Vol.113,No.2, 609-614

S.S. BAWA, K. SAXENA, and S. CHANDRA
Surface and Bulk Switching Dynamics in Surface Stabilized Ferroelectric Liquid Crystals
77.30 and 77.80; S12
(a) Vol.114,No.1, 369-376

S.N. FEDOSOV
Voltage Recovery in Open-Circuit Polar Electret
77.30; S12
(a) Vol.114,No.1, K103-K106

D. TRIANTIS, C. ANASTASIADIS, and S. ORFANAKOS
Evaluation of Activation Energy from the Ionic Thermocurrent Spectrum
61.70; 77.30
(a) Vol.114,No.2, 491-496

S.N. FEDOSOV
Deep Trapping of Injected Carriers in Ferroelectric Polymer
77.80; 77.30; S12
(a) Vol.115,No.1, 293-300

77.40. Dielectric loss and relaxation

CZ. PAWLACZYK and A. PAWŁOWSKI
Electric Capacitance Anomaly near the Superionic Phase Transition in $(NH_4)_4H_2(SeO_4)_3$ Single Crystals
73.40 and 77.40; S11
(a) Vol.113,No.2, 447-452

M.A. AHMED, M.K. EL-NIMR, A. TAWFIK, and A.M. ABOELATA
Dielectric Behaviour in Co-Zn Ferrites
77.20 and 77.40; S11.2
(a) Vol.114,No.1, 377-380

T. GOLDMANN, E. GERDES, and J. SIMANOWSKI
Dielectric Properties of Passivation Glasses in the System $PbO-B_2O_3-SiO_2-Al_2O_3-CeO_2$ Depending on Glass Composition
77.20 and 77.40; S10.15
(a) Vol.114,No.1, 387-396

S. MISHRA, A.V. KRISHNA RAO, and K.V. RAO
Dielectric Properties of Tourmaline under Different Conditions
77.20 and 77.40; 77.70; S11
(a) Vol.114,No.1, K115-K118

H. KLIEM and G. ARLT
The Temperature Dependence of Dielectric Response in Relation to Distribution Functions
77.40
(b) Vol.155,No.1, 309-316

B.K. CHOUDHARY, R.N.P. CHOUDHARY, and K.V. RAO
Dielectric Properties of $TbAsO_4$ Single Crystals
77.20 and 77.40; S11.1
(a) Vol.115,No.1, 301-306

R. ARORA and A. KUMAR
Dielectric Relaxation in Glassy Se and $Se_{100-x}Te_x$ Alloys
77.40; S6
(a) Vol.115,No.1, 307-314

S.K. ARORA and T. MATHEW
Dielectric Studies of $CuWO_4$ Crystals
77.20 and 77.40; S11.1
(a) Vol.116,No.1, 405-414

77.50. Dielectric breakdown and space-charge effects

J. SUNÉ, I. PLACENCIA, N. BARNIOL, E. FARRÉS, and X. AYMERICH
Degradation and Breakdown of Gate Oxides in VLSI Devices
73.40 and 77.50; 73.60; 85; S10
(a) Vol.111,No.2, 675-686

C. DERVOS, P.D. BOURKAS, and E.A. KAYAFAS
Potential Profile of Dielectrics during High-Voltage Pulse Application
72.20 and 77.50; S12
(a) Vol.112,No.1, 123-130

A. SCHENK, M. STAHL, and H.-J. WUENSCHE
Calculation of Interband Tunneling in Inhomogeneous Fields
72.10 and 73.40; 77.50; S8.15
(b) Vol.154,No.2, 815-826

77.55. Dielectric thin films

S.M.J. AKHTAR, D. RISTAU, J. EBERT, and H. WELLING
Characterization of Dielectric Films and Damage Threshold at 1.064 μm
78.65; 77.55; S9; S10
(a) Vol.115,No.1, 191-198

A. THOMAS, A. ZEHE, B. SCHRECKENBACH, J.-W. ERBEN, and R. GROETZSCHEL
Ion Beam Lithography of Polycrystalline CaF_2 Films Deposited on Silicon
61.80 and 68.55; 77.55; S9.12
(a) Vol.116,No.2, 735-744

77.60. Piezoelectricity and electrostriction
(for piezo-optical effects, see 78.20)

A. NEOGI and S. GHOSH
Parametric Amplification in Magnetised Piezoelectric Semiconducting Plasmas
78.20; 77.60; S7.13
(b) Vol.152,No.2, 691-700

P. CORDIER, B. BOULOGNE, N. DOUKHAN, and J.C. DOUKHAN
Lattice Defects in Lithium Tantalate
61.70; 77.60; S11.1
(a) Vol.112,No.2, 493-510

B. MRÓZ and R. LAIHO
Elastic Properties of $K_4LiH_3(SO_4)_4$ Crystals
77.60; 62.20; S11.1
(a) Vol.115,No.2, 575-580

77.70. Pyroelectric and electrocaloric effects

P.E. TOMASZEWSKI
On the Effect of Doping on the Mechanism of Phase Transition of Some Inorganic Compounds (Comment)
64.70 and 77.80; 77.70; S11.1
(a) Vol.112,No.1, K13-K14

S. MISHRA, A.V. KRISHNA RAO, and K.V. RAO
Dielectric Properties of Tourmaline under Different Conditions
77.20 and 77.40; 77.70; S11
(a) Vol.114,No.1, K115-K118

77.80. Ferroelectricity and antiferroelectricity

V.M. BURLAKOV and M.R. YAKHEEV
Raman Scattering in Ferroelectric $TlGaSe_2$. Soft Mode and Order Parameter
78.30; 77.80; S8.16
(b) Vol.151,No.1, 337-346

V.A. IVANSHIN, I.N. KURKIN, W. BRUNNER, and G. VOELKEL
Electron Spin-Lattice Relaxation of the $(SeO_2)^-$ Radical in X-Irradiated $KH_3(SeO_3)_2$
76.30 and 77.80; S11.1
(b) Vol.151,No.1, K53-K58

S.S. BAWA, K. SAXENA, and S. CHANDRA
Polarization Reversal Current Behaviour for Parallel and Antiparallel Rubbed Surface Stabilized Ferroelectric Liquid Crystals
77.80; S12
(a) Vol.111,No.1, 309-314

J.M. WESSELINOWA
Relaxation Time for KDP-Type Ferroelectrics in an External Electric Field
77.80
(b) Vol.151,No.2, K209-K212

P.E. TOMASZEWSKI
On the Effect of Doping on the Mechanism of Phase Transition of Some Inorganic Compounds (Comment)
64.70 and 77.80; 77.70; S11.1
(a) Vol.112,No.1, K13-K14

S.A. RÓŻAŃSKI, W. KUCZYŃSKI, and J. MAŁECKI
Biaxiality of the Ferroelectric Liquid Crystal DOBAMBC in Electric Field
77.80 and 78.20; 61.30; S12
(a) Vol.112,No.2, 617-624

Z. ZIKMUND and J. FOUSEK
Stabilization of Polarization Vector in Uniaxial Ferroelectrics by Means of Chiral Dopants. Macroscopic Symmetry Conditions
77.30 and 77.80; S12
(a) Vol.112,No.2, 625-634

P.K. BAJPAI
Ferroelectric Phase Transition in $(NH_4)_2BeF_4$. Possibility of Molecular Distortion Type of Transition
64.60; 77.80; S11
(b) Vol.153,No.1, 101-106

C.N.W. DARLINGTON
On the Changes in Structure of PLZT(8.7/65/35) between 80 and 750K
64.70; 61.10; 77.80
(a) Vol.113,No.1, 63-70

R. CACH and Z. CZAPLA
Dielectric Nonlinearity of Diglycine Nitrate (DGN) Crystal in the Paraelectric Phase
77.20; 77.80; S12
(a) Vol.113,No.1, K139-K142

I.R. YUKHNOVSKII and N.A. KORYNEVSKII
The Investigation of the Ferroelectric Phase Transition in Cluster Systems of Order-Disorder Type. I.Partition Function Functional
64.60; 77.80
(b) Vol.153,No.2, 583-594

V.M. BURLAKOV, A.P. RYABOV, M.R. YAKHEEV, E.A. VINOGRADOV, N.N. MELNIK, and N.M. GASANLY
Raman Spectroscopy of Soft and Rigid Modes in Ferroelectric $TlInS_2$
78.30; 77.80; S8.16
(b) Vol.153,No.2, 727-740

R. MIZERIS, J. GRIGAS, R.N.P. CHOUDHARY, and T.V. NARASAIAH
Microwave Dielectric Investigations of TlH_2PO_4
77.20; 77.80; S11.1
(a) Vol.113,No.2, 597-602

R.A. RUPP, A.E. KRUMINS, and K. KERPELIN
Dielectric, Electrooptic, and Holographic Properties Close to the Diffuse Phase Transition of PLZT 10/65/35
78.20; 64.60; 77.80; S10
(a) Vol.113,No.2, 637-652

M. TAYA, T. ASAJI, and D. NAKAMURA
A Low-Temperature Phase Transition in Ferroelectric $(CH_3NH_3)HgCl_3$ Definitely Located by Pyroelectric and Dielectric Measurements
64.70 and 77.80; S12
(a) Vol.114,No.1, 157-160

S.S. BAWA, K. SAXENA, and S. CHANDRA
Surface and Bulk Switching Dynamics in Surface Stabilized Ferroelectric Liquid Crystals
77.30 and 77.80; S12
(a) Vol.114,No.1, 369-376

A.K. ABIEV, E.Z. ALIEV, B.A. NATIG, and R.M. SARDARLY
Electron Irradiation Effect on Dielectric Properties of $TlInS_2$ Single Crystals
77.80; 61.80; S8.16
(a) Vol.114,No.1, K119-K122

I.R. YUKHNOVSKII and N.A. KORYNEVSKII
The Investigation of the Ferroelectric Phase Transition in Cluster Systems of Order-Disorder Type. II. Two-Particle Cluster System
64.60; 77.80
(b) Vol.154,No.2, 519-534

V.I. TORGASHEV, YU.I. YUZYUK, F. SMUTNÝ, P. VANĚK, and B. BŘEZINA
Raman Spectra of the Incommensurate and Ferroelectric Phases of Rb_2CoCl_4 Single Crystals
78.30; 77.80; S9.16
(b) Vol.154,No.2, 777-788

I.N. ZAKHARCHENKO, E.S. NIKITIN, V.M. MUKHORTOV, YU.I. GOLOVKO, M.G. RADCHENKO, and V.P. DUDKEVICH
Peculiarities of Ferroelectric Phase Transition in $(Ba,Sr)TiO_3$ Heteroepitaxial Films Obtained by HF Cathode Sputtering
64.70 and 68.55; 77.80; S10.15
(a) Vol.114,No.2, 559-566

TH. ERGE, D. MICHEL, J. PETERSSON, and W. WINDSCH
^{35}Cl NMR Study in the Paraelectric Phase of Tris-Sarcosine Calcium Chloride
76.60; 77.80; S12
(a) Vol.114,No.2, 705-712

C.A. KLEINT, U. STOEPEL, and A. ROST
X-Ray Diffraction and Conductivity Investigations of Lanthanum-Doped Barium Titanate Ceramics
64.70 and 77.80; 64.80; S10.15
(a) Vol.115,No.1, 165-172

S.N. FEDOSOV
Deep Trapping of Injected Carriers in Ferroelectric Polymer
77.80; 77.30; S12
(a) Vol.115,No.1, 293-300

S.K. GUPTA, V.A. KIRIKOV, V.V. GLADKII, and B. BŘEZINA
Effect of Mechanical Stresses on Anomalies of Dielectric Properties of the Incommensurate Phase in Rb_2CoCl_4 Crystals
77.80; S9.16
(a) Vol.115,No.2, 569-574

X. GERBAUX, A. HADNI, and A. KITADE
Far IR Spectra of Ferroelectric Rochelle Salt and Sodium Ammonium Tartrate
78.30; 77.80; S11
(a) Vol.115,No.2, 587-598

S. AHMED, A.M. SHAMAH, A. IBRAHIM, and F. HANNA
X-Ray Studies of the High Temperature Phase Transition of Ammonium Sulphate Crystals
64.70; 77.80; S11.1
(a) Vol.115,No.2, K149-K154

P. KONSIN and T. OERD
Influence of Hydrostatic Pressure on the Ferroelectric Phase Transition Temperature in (Pb,Sn)Te
62.50 and 63.75; 63.70; 77.80; S8
(b) Vol.156,No.1, K37-K40

JUN CHEN and DUAN FENG
Crystallographic Shear Planes and Stacking Disorders in Perovskites $KNbO_3$ and $NaNbO_3$
61.70; 77.80; S10
(a) Vol.116,No.1, 231-236

H. OTOP
Some Experimental Observations of Exoemission from Ammonium Hydrogen Selenates
77.80 and 79.75; S12
(a) Vol.116,No.1, K115-K118

R. CACH, S. DACKO, and Z. CZAPLA
Dielectric Properties and State Equation Parameters of $(CH_3)_2NH_2Al(SO_4)_2\cdot 6H_2O$ Ferroelectric Crystals
77.80; 77.20; S12
(a) Vol.116,No.2, 827-832

F. BALLER, B. GATHER, B. HELLERMANN, H. HESSE, and E. KRAETZIG
The Influence of Stress on the Dielectric Properties of KTN Crystals
77.20; 77.80; S11.1
(a) Vol.116,No.2, K195-K198

ZHONGMING LIU, XIAOHUA CHEN, HUIMIN SHEN, YENING WANG, HUIFANG YOUNG, and P.C.W. FUNG
Ferroelectric Domain Wall Motion and Related Internal Friction in TGS Crystal
62.20 and 77.80; S12
(a) Vol.116,No.2, K199-K204

78. Optical Properties and Condensed-Matter Spectroscopy and Other Interactions of Matter with Particles and Radiation

(for phonon spectra, see 63)

78.20. Optical properties and materials

M. CASALBONI, R. FRANCINI, F.J. LOHMEIER, and U.M. GRASSANO
Two-Photon Absorption of Bismuth Germanate
78.20; S11
(b) Vol.151,No.1, 347-352

G.A. MEDVEDKIN, YU.V. RUD, and M.A. TAIROV
Fundamental Optical Absorption Edge in $MnIn_2Te_4$ Single Crystals
78.20; S1.63; S8.16
(a) Vol.111,No.1, 289-300

A.N. GEORGOBIANI, A.N. GRUZINTSEV, A.V. SPITSYN, and I.M. TIGINYANU
Change of the Optical Properties of CdS Single Crystals upon High Dose Indium and Gallium Implantation
78.20 and 78.55; 78.60; S8.11
(a) Vol.111,No.1, 301-308

G. MIZUTANI and N. NAGASAWA
Exciton-Grating Induced Light Mixing at the Giant Two-Photon Resonances of Excitonic Molecules in CuBr
71.35 and 78.20; S9
(b) Vol.151,No.2, 591-598

E. HEINER
General Properties of Highly Excited Direct Semiconductors in the Athermal Stage
72.10; 71.10; 78.20
(b) Vol.151,No.2, 645-654

G.A. MEDVEDKIN, V.D. PROCHUKHAN, YU.V. RUD, and M.A. TAIROV
Anisotropy of the Optical Band Edge Absorption in $CuInTe_2$ Single Crystals
78.20; S8.16
(b) Vol.151,No.2, 711-720

V.A. BIBIK and N.A. DAVYDOVA
Transformation of Excitonic Spectra of PbI_2 Layer Crystals under Order-Disorder-Order Structural Transitions
71.35 and 78.20; S9
(b) Vol.151,No.2, 733-738

A.D. PRINS, K.R. ALLAKHVERDIEV, S.S. BABAEV, S.G. GUSEINOV, E. MEKHTIEV, M.M. SHIRINOV, M.M. TAGYEV, and D.J. DUNSTAN
Effect of High Pressure on the Optical Transmission Spectra of $A^{III}B^{III}C_2^{IV}$ Crystals
78.20; S8.16
(b) Vol.151,No.2, 759-764

R. NIES and F.R. KESSLER
Coherent and Incoherent Magnetooptical Reflection and Transmission Effects of Multilayer Structures
78.20; 78.50; S5.11
(a) Vol.111,No.2, 639-654

E.Z. KATSNELSON, A.G. KAROZA, and M.M. CHERVINSKII
Photomagnetization of Ferrite Films and Its Effect on Optical Spectra
75.70 and 78.20; S11.2
(a) Vol.111,No.2, K233-K236

M. GAAFAR, M.A. LASHEEN, A.A. ALY, and A.M. AHMED
Optical Properties of $C_6H_{6-y}X_y$ (y=2,3 and X=Cl,Br)
78.20; S12
(a) Vol.111,No.2, K259-K264

F.M. GASHIMZADE and E.V. TAGIROV
Free-Carrier Absorption for Intervalley Phonon Scattering
78.20; S8
(b) Vol.152,No.1, 329-334

TZU-YU HUANG and JUH-TZENG LUE
Theoretical Calculation of the Optical Constants of Transition Metal Silicides
78.20; S1.61
(b) Vol.152,No.1, 335-346

A.A. KAMINSKII, A.V. BUTASHIN, A.A. DEMIDOVICH, V.G. KOPTEV, B.V. MILL, and A.P. SHKADAREVICH
Broad-Band Tunable Stimulated Emission from Octahedral Cr^{3+} Ions in New Acentric Crystals with Ca-Gallogermanate Structure
78.20 and 78.45; 78.55; S10
(a) Vol.112,No.1, 197-206

R. CYWIŃSKI and M. MANFREDI
Aggregation Processes in the NaCl:Eu^{2+} System Induced by the High-Temperature Annealing
78.20; 71.55; S9.11
(a) Vol.112,No.1, 215-220

L. JONIKAS, K. JARAŠIŪNAS, and J. VAITKUS
Study of Carrier Dynamics and Radiation Defects in Ion-Implanted Silicon by Transient Grating Technique
71.55 and 78.20; S5.11
(a) Vol.112,No.1, 375-380

A.A. KAMINSKII, A.V. BUTASHIN, I.A. MASLYANITSIN, and V.D. SHIGORIN
Nonlinear Optical Properties of Acentric Crystals with Ca-Gallogermanate Structure
78.20; S10
(a) Vol.112,No.1, K49-K52

G. MIZUTANI and N. NAGASAWA
Exciton-Grating Induced Metamorphoses of the Degenerate Four-Wave Mixing Spectra in the Giant Two-Photon Resonances of the Excitonic ...
71.35 and 78.20; S9
(b) Vol.152,No.2, 489-498

V.G. KISELEV
On Kink Dynamics in Media with Increasing Absorption Optical Bistability
78.20
(b) Vol.152,No.2, 667-674

V.YA. REZNICHENKO, M.I. STRASHNIKOVA, and V.V. CHERNY
Dispersion Properties of Submicron-Thin CdS Crystals
78.20; S8.11
(b) Vol.152,No.2, 675-684

M.V. MAGNITSKAYA, N.I. KULIKOV, and A.L. SHELEKHOV
Frequency-Modulated Optical Spectroscopy of Niobium: Theory and Experiment
78.20; S1
(b) Vol.152,No.2, 685-690

A. NEOGI and S. GHOSH
Parametric Amplification in Magnetised Piezoelectric Semiconducting Plasmas
78.20; 77.60; S7.13
(b) Vol.152,No.2, 691-700

V.YA. KOSTYUCHENKO, E.M. SKOK, and S.A. STUDENIKIN
Magneto-Resonant Absorption from the Local Center in Degenerate n-InSb
78.20; 71.55; S7.13
(b) Vol.152,No.2, K59-K62

S.A. RÓŻAŃSKI, W. KUCZYŃSKI, and J. MAŁECKI
Biaxiality of the Ferroelectric Liquid Crystal DOBAMBC in Electric Field
77.80 and 78.20; 61.30; S12
(a) Vol.112,No.2, 617-624

M. ABD-LEFDIL and M. CADÈNE
Optical Studies of Oriented Undoped Polyace'ylene
78.20; S12
(a) Vol.112,No.2, 635-638

M. KULIK and K. ŻUKOWSKA
The Multilayer Model of Implanted GaAs. Ellipsometric Investigation
61.80; 78.20; S7.12
(a) Vol.112,No.2, 785-788

M. KULIK and J. ŻUK
Ellipsometric Investigation of the Kinetics of Native Oxide Growth on Aluminium and Argon Implanted GaAs
61.80 and 68.55; 78.20; S7.12
(a) Vol.112,No.2, 789-792

M. KULIK, D. JAWORSKA, and J. TRZCIANKA
Determination of the Damage Profiles in Implanted GaAs from the Change of (n,k) Optical Parameters
61.80; 78.20; S7.12
(a) Vol.112,No.2, 793-798

V.I. GAVRILENKO, N.I. KLYUI, V.G. LITOVCHENKO, and B.N. ROMANYUK
Investigation of the Ion Implanted Silicon by Electroreflectance Spectroscopy
61.70; 78.20; S5.11
(a) Vol.112,No.2, 805-810

J. TATARKIEWICZ
Optical Effects of High Energy Implantations in Semiconductors
61.80 and 78.20; 78.30; 78.50; 78.55; S5.11; S7.12
(b) Vol.153,No.1, 11-48

M.J. PONNAMBALAM
Stark Effect and the F-Center Excited States
61.70 and 71.55; 78.20
(b) Vol.153,No.1, 79-88

H.N. NAZARENO and J.C. GALLARDO
Bloch Oscillations of an Electron in a Crystal under the Action of an Electric Field
71.25; 78.20
(b) Vol.153,No.1, 179-184

P. GILLIOT, J.C. MERLE, R. LEVY, M. ROBINO, and S. HOENERLAGE
Laser Induced Absorption of CuCl Microcrystallites in a Glass Matrix
78.20; S9
(b) Vol.153,No.1, 403-410

C. ALBERS and A. JACOBI
Temperature Correction of x-Value Determination in $Hg_{1-x}Cd_xTe$ by Optical Transmission Measurements
78.20; S8.15
(a) Vol.113,No.1, K135-K138

N.I. GRIGORCHUK and L.G. GRECHKO
Lineshape of Light Absorption by Thermalized Frenkel Excitons under Band-to-Band Transitions
71.35 and 78.20
(b) Vol.153,No.2, 633-640

G. BABONAS, R. DAGYS, and G. PUKINSKAS
Electronic Structure and Magnetooptical Effects in gamma-Ln_2S_3
78.20; S8
(b) Vol.153,No.2, 741-750

J. SCHLOESSER and A. STAHL
A Symmetry-Adapted RPA: Dynamics of the Electronic Density Matrix in a Semiconductor
71.25 and 78.20; S8.11
(b) Vol.153,No.2, 773-786

R.A. RUPP, A.E. KRUMINS, and K. KERPELIN
Dielectric, Electrooptic, and Holographic Properties Close to the Diffuse Phase Transition of PLZT 10/65/35
78.20; 64.60; 77.80; S10
(a) Vol.113,No.2, 637-652

M.I. SHMIGLYUK, P.I. BARDETSKII, and S.D. TIRON
Nonlinear Optical Nutation on Interexciton Transition in Cu_2O. An Analogue of Equidistant Three-Level System
71.35; 78.20; S10.1
(b) Vol.154,No.1, 209-218

G. BABONAS, R.R. GAŁĄZKA, G. PUKINSKAS, and A. ŠILEIKA
Crystal Structure and Magnetooptical Properties of $Cd_{1-x}Mn_xTe$ with High Mn Concentrations
78.20; 61.60; S8.15
(b) Vol.154,No.1, 389-396

W. HEIMBRODT, C. BENECKE, O. GOEDE, and H.E. GUMLICH
Optical Effects Due to Antiferromagnetic Spin Ordering in (Cd,Mn)S at Highest Mn Concentrations
78.20; 75.30; S8.15
(b) Vol.154,No.1, 405-416

P.S. GHOSH, D.K. GHOSH, and L.K. SAMANTA
Studies of Optical Properties and Applications of Some Mixed Ternary Semiconductors
78.20; 85; S7.15
(a) Vol.114,No.1, 381-386

B. KOSTUREK and R. JAKUBAS
Optical Birefringence of $(CH_3NH_3)_5Bi_2Br_{11}$
78.20; S12
(a) Vol.114,No.1, K111-K114

M. SINGH and S.Y. TANG
Magneto-Optics of Holes in the Presence of Internal Strains and Temperature in Single Quantum Well Structures
71.25 and 72.10; 78.20
(b) Vol.154,No.2, 589-602

R. GOLDHAHN and H. ROEPPISCHER
Electroreflectance Lineshape of Step-Like Doped GaAs Structures
78.20; S7.12
(b) Vol.154,No.2, 769-776

E.L. ALBUQUERQUE and J. MARTINS S. FILHO
Dielectric Function in Magnetoelectric Crystals
75.80 and 78.20
(b) Vol.154,No.2, K179-K182

Y.S. TANG
Photoreflectance of Multiple Quantum Wells: Modulation Intensity Dependence
78.20 and 78.65
(b) Vol.154,No.2, K187-K190

V.A. GANSHIN, YU.N KORKISHKO, T.V. MOROZOVA, and V.V. SARAIKIN
The Study of Proton Exchange in Lithium Tantalate Crystals
61.70; 78.20; S11.1
(a) Vol.114,No.2, 457-466

S. LAKSHMI RAGHAVAN, N. RAMESH, and K. SRINIVASAN
Dispersion of the Quadratic Electro-Optical (Kerr) Constants in $KCl:Cd^{2+}$ and $KCl:Sr^{2+}$
78.20; 61.70; S9.11
(a) Vol.114,No.2, 749-756

JIPING YANG, S. RENTSCH, H. WABNITZ, and R. ERIKASHVILI
Transient Behaviour of Ce^{3+}- and Ce^{3+}/Nd^{3+}-Doped Silica Glasses after Excitation with Picosecond UV Pulses
78.20; S10
(a) Vol.114,No.2, 757-764

R. RICHTER, T. BREMER, P. HERTEL, and E. KRAETZIG
Refractive Index and Concentration Profiles of Proton-Exchanged $LiNbO_3$ Waveguides
78.20; 61.70; S10.15
(a) Vol.114,No.2, 765-774

D. LUPU
Hydrogen Diffusion and Radiationless Electron Transfer in Metal Hydrides
66.30; 71.38; 78.20; S1.2; S1.4; S3
(b) Vol.155,No.1, 65-72

V.S. STEPANYUK, A. SZÁSZ, B.L. GRIGORENKO, O.V. FARBEROVICH, and A.A. KATSNELSON
Electronic Structure and Optical Properties of MgO: Band Structure Calculation and Cluster Model
71.25; 78.20; S10.1
(b) Vol.155,No.1, 179-184

V.S. STEPANYUK, A. SZÁSZ, O.V. FARBEROVICH, A.A. GRIGORENKO, A.V. KOZLOV, and V.V. MIKHAILIN
An Electronic Band Structure Calculation and the Optical Properties of Alkaline-Earth Sulphides
71.25; 78.20; S8
(b) Vol.155,No.1, 215-220

C.N. MISHRA, R.P. GOYAL, and B.R.K. GUPTA
Photoelastic Behaviour of Alkaline Earth Fluoride Crystals
62.20 and 78.20; S9.12
(b) Vol.155,No.1, K81-K84

G.A. MEDVEDKIN, YU.V. RUD, and M.A. TAIROV
Photoelectric Anisotropy of II-IV-V Ternary Semiconductors (Review Article)
72.40; 78.20; S7.16
(a) Vol.115,No.1, 11-50

Z. ŠIMŠA and P. GOERNERT
Preparation and Faraday Rotation Spectra of YIG:Pb,Pt Garnet Films
68.55 and 78.20; S11.2
(a) Vol.115,No.1, 229-236

D. TALAPATRA and R.K. MUKHERJEE
Optical Studies of Electronic Transitions of Mn^{2+} Ion in $Na_2Mn_3Cl_8$ Single Crystal
71.70; 78.20; S9.16
(b) Vol.155,No.2, 541-548

S.S. KOTSUR and I.V. STASYUK
The Quantum Theory of the Optical Activity of Ionic Dielectric Crystals
78.20
(b) Vol.155,No.2, 697-708

V.K. LUKYANYUK, S.P. VORONYUK, and Z.D. KOVALYUK
Alkali-Metal-Intercalated Indium and Gallium Selenides: Non-Monotonous Shift of Exciton Lines
78.20; S8.12
(b) Vol.155,No.2, 717-722

V.I. GAVRILENKO, J. HUMLIČEK, N.I. KLYUI, and V.G. LITOVCHENKO
Electron Energy Structure and Optical Properties of Microcrystalline Silicon
78.20; 71.25; S5.11
(b) Vol.155,No.2, 723-732

L.M. SANDRATSKII
On a Peculiarity of the Optical Characteristics of Rare-Earth Metals
71.25; 78.20; S1.4
(b) Vol.155,No.2, K105-K108

W. KUCHARCZYK and P. GÓRSKI
Bond Polarizability Calculation of the Quadratic Electrooptic Effect in KDP
78.20; S11.1
(b) Vol.155,No.2, K157-K160

B.M. ANGHELOV
Energy Levels and Intensity Parameters of $LiTbP_4O_{12}$
71.30 and 78.20; S10
(b) Vol.155,No.2, K165-K168

L. LEJČEK, A. FOUSKOVÁ, and M. PAVEL
Note on Light Scattering by Dislocations in TGS
61.70 and 78.20; S12
(a) Vol.115,No.2, 445-450

I. ILIEV, T. DIMOV, D. RIBAROV, and H. LANGE
Composition Determination of CdS_xSe_{1-x} Mixed Crystals by Optical Dispersion
78.20; S8.15
(a) Vol.115,No.2, K243-K246

D. BRACH, P. GLAREMIN, and H.-J. WEBER
Optical Investigations of the Phase Transition in Pure and Cr^{3+}-Doped Potassium Alum
64.60; 78.20; 78.55; S11.1
(b) Vol.156,No.1, 195-204

K. KUMAZAKI, L. VIÑA, C. UMBACH, and M. CARDONA
Interband Critical Point Parameters Determined by Ellipsometry in $Cd_xHg_{1-x}Se$
71.25 and 78.20; 77.20; S8.15
(b) Vol.156,No.1, 371-376

A.M. SHVAIKA, I.V. STASYUK, and S.M. MOKHNYAK
On the Microscopic Theory of Optical Properties of Crystals with Incommensurate Phases
78.20
(b) Vol.156,No.1, 377-382

O.G. VLOKH, A.V. KITYK, O.M. MOKRY, and V.G. GRYBYK
Pressure-Temperature Phase Diagram of $\{N(CH_3)_4\}_2CuCl_4$ Crystals. Birefringent and Elastic Properties
64.60; 62.20; 78.20; S12
(a) Vol.116,No.1, 287-294

B.P. GORSHUNOV, M.G. KAPLUNOV, N.D. KUSHCH, and O.I. SYROTYNSKII
Temperature Dependence of the Optical Properties of the Organic Superconductor $(ET)_2Cu(NCS)_2$ in the Submillimeter Range
78.20; 74.70; S12
(a) Vol.116,No.1, K129-K132

D. VIGNAUD and J.L. FARVACQUE
Free-Carrier Optical Absorption Induced by Dislocation Scattering Mechanisms in Semiconductors: Application to GaAs
78.20; 61.70; S7.12
(b) Vol.156,No.2, 717-724

78.30. Infrared and Raman spectra and scattering

A.A. KLOCHIKHIN and S.G. OGLOBLIN
Effect of a Random Potential on the Resonant One-LO-Phonon Raman Process
78.30
(b) Vol.151,No.1, 319-330

V.M. BURLAKOV and M.R. YAKHEEV
Raman Scattering in Ferroelectric $TlGaSe_2$. Soft Mode and Order Parameter
78.30; 77.80; S8.16
(b) Vol.151,No.1, 337-346

K. KUMAZAKI
Resonant Raman Scattering near $E_1+\Delta_1$ in $Cd_xHg_{1-x}Se$
78.30; S8.15
(b) Vol.151,No.1, 353-362

J. OTTO
Electric Field Induced Splitting of the Stretching Vibration Lines SH^-, SeH^-, and TeH^- Centres in Alkali Halides
78.30 and 78.50; S9.11
(b) Vol.151,No.1, 363-374

J. PASTRŇÁK, J. OSWALD, I. GREGORA, V. VORLÍČEK, and M. BABINSKÝ
Lateral Distribution of Si Related Defects and Carrier Density in GaAs Crystals
78.55; 71.55; 78.30; S7.12
(a) Vol.111,No.1, 345-360

I. LUKEŠ, J. HUMLÍČEK, V. VORLÍČEK, and M. ZÁVĚTOVÁ
Infrared Reflectance of the GaAlAs/GaAs System
78.30; 73.40; S7.12
(a) Vol.111,No.2, 655-666

GONG YU, GUANGHUA CHEN, and FANGQING ZHANG
Theoretical Study of Raman Spectrum in $a\text{-}Si_{1-x}N_x:H$ Films
61.40 and 68.30; 78.30; S5.11
(b) Vol.152,No.1, 73-78

G.A. KOUROUKLIS and E. ANASTASSAKIS
Pressure-Induced Phase Transitions and Anharmonicity Study of Alkaline-Earth Fluorides
64.60; 63.20; 78.30; S9.12
(b) Vol.152,No.1, 89-100

K.R. ALLAKHVERDIEV, S.S. BABAEV, M.M. TAGIEV, and M.M. SHIRINOV
Low-Temperature IR and Raman Scattering Spectra of $TlInS_2$ Layered Crystal
64.60 and 78.30; S8.16
(b) Vol.152,No.1, 317-328

C. TRALLERO-GINER and R. RIERA
Multiple LO-Phonon Raman Scattering and Hot Excitons
71.35 and 78.30; S9
(b) Vol.152,No.1, 357-368

G.G. BOGACHEV, M.N. ILIEV, and V. PETROV
Raman Scattering of $ScBO_3$
78.30; S11.1
(b) Vol.152,No.1, K29-K32

K. ISHII, N. MORITA, H. NAKAYAMA, S. TSUNEKAWA, and T. FUKUDA
Raman Spectra of $LaNbO_4$ in the Ferroelastic Phase and the Relaxation after the State Shift
78.30; S11.1
(a) Vol.112,No.1, 207-214

E.Z. KATSNELSON, A.G. KAROZA, L.A. MELESHCHENKO, and L.A. BASHKIROV
IR-Reflection Spectra of Manganese-Zinc Ferrites
78.30; S11.2
(b) Vol.152,No.2, 657-666

M. ISHII, K. SHIBATA, and H. WADA
Raman Study of $M_xMo_6S_{8-y}$ (M = Yb, Fe, Cr)
78.30; 63.20; S1.63
(b) Vol.152,No.2, K63-K68

A.K. PRADHAN and R.N.P. CHOUDHARY
Raman and Infrared Effect of Sm^{3+} in $YAsO_4$ Crystals
78.30; 63.20; S11.1
(b) Vol.152,No.2, K69-K72

V.V. PETROV and V.S. PROSOLOVICH
Thermal Defects in Silicon Doped with Rare-Earth Elements
61.70 and 71.55; 72.20; 78.30; S5.11
(a) Vol.112,No.2, 561-568

V. RIEDE, H. NEUMANN, N. SHARIF, F.S. HASOON, and H. SOBOTTA
Structural and Infrared Optical Properties of $(CuInSe_2)_{1-x}(2InAs)_x$ Solid Solutions
61.60 and 78.30; S7.15; S8.15
(a) Vol.112,No.2, K139-K144

J. TATARKIEWICZ
Optical Effects of High Energy Implantations in Semiconductors
61.80 and 78.20; 78.30; 78.50; 78.55; S5.11; S7.12
(b) Vol.153,No.1, 11-48

J.M. CHEN, L.M. XIE, G.R. BAI, and Y.G. ZHAO
Temperature Effects of the Several Intense IR Absorption Peaks in NTD Si:H
63.20 and 78.30; S5.11
(b) Vol.153,No.1, 107-114

H. GRUBER, E. KRAUTZ, H.P. FRITZER, K. GATTERER, and A. POPITSCH
Electrical Conductivity Magnetic Susceptibility and Infrared Spectra in the Superconducting Systems Bi-Sr-Ca-Cu-O and Tl-Ba-Cu-O
74.70; 75.30; 78.30; S10.15
(b) Vol.153,No.1, 307-314

A. LEHMANN
Theory of Infrared Spectra of Thin Inhomogeneous Insulating Films
78.30 and 78.65
(b) Vol.153,No.1, K85-K88

N.M. GASANLY, B.A. NATIG, A.E. BAKHYSHOV, and K.E. SHIRINOV
Long-Wave Optical Phonons in $InSbSe_3$ Layer Single Crystals
63.20 and 78.30; S8.16
(b) Vol.153,No.1, K89-K92

YU.I. DURNEV, B.S. KULBUZHEV, A.U. MALSAGOV, L.M. RABKIN, V.I. TORGASHEV, and YU.I. YUZYK
Vibrational Spectra and Phase Transitions in Layered Semiconducting Ferroelectrics with $TlGaSe_2$ Structure. I. Temperatur Dependence ...
64.60 and 78.30; S8.16
(b) Vol.153,No.2, 517-528

YU.G. GONCHAROV, G.V. KOZLOV, B.S. KULBUZHEV, V.B. SHIROKOV, V.I. TORGASHEV, A.A. VOLKOV, and YU.I. YUZYK
Vibration Spectra and Phase Transitions in Layered Semiconducting Ferroelectrics with $TlGaSe_2$ Structure. II. Thermodynamic ...
64.60 and 78.30; S8.16
(b) Vol.153,No.2, 529-538

K.N. SHRIVASTAVA
Physical Properties of Fractals. Moessbauer Recoilless Fraction, Raman Scattering, and Spin-Lattice Relaxation
63.90; 76.80; 78.30
(b) Vol.153,No.2, 547-560

V.M. BURLAKOV, A.P. RYABOV, M.R. YAKHEEV, E.A. VINOGRADOV, N.N. MELNIK, and N.M. GASANLY
Raman Spectroscopy of Soft and Rigid Modes in Ferroelectric $TlInS_2$
78.30; 77.80; S8.16
(b) Vol.153,No.2, 727-740

K. KUMAZAKI
Resonant Raman Scattering in $Zn_xHg_{1-x}Se$
78.30; S8.15
(b) Vol.153,No.2, 751-756

H. WEINERT, A. ZUKAUSKAS, G. TAMULAITIS, and V. STEPANKEVICIUS
Some Hints on the Influence of the Barrier on Excitonic Optical Properties of GaAs-GaAlAs Superlattices Obtained by ...
78.55; 71.35; 78.30; S7.12; S7.15
(b) Vol.153,No.2, K193-K198

YU.M. SEIDOV and A.M. SULEIMANOV
On the Theory of Raman Scattering by Surface Collective Excitations in Singlet-Triplet Ferromagnets
78.30; 71.70; 75.30; S1.62
(b) Vol.153,No.2, K203-K208

S.E. SARKISOV, A.A. KAMINSKII, and A.V. BUTASHIN
Optical Phonons and Lattice-Activator Interactions of Li^{3+} Ions in the Disordered Oxide Single Crystals with Ca-Gallogermanate ...
78.30; 71.70; S11.2
(a) Vol.113,No.2, 585-596

P. LOŠTÁK, R. NOVOTNÝ, E. URBANOVA, and J. HORÁK
Optical Properties of $Bi_{2-x}In_xSe_3$ Single Crystals
78.30; S8.15
(a) Vol.113,No.2, 615-622

M. PALPACUER, M. ROLLAND, and C. BENOIT
Polarized Spectra of Oriented Polyacetylene Doped with Iodine
78.30; 78.65; S12
(a) Vol.113,No.2, 623-630

M.O. MANARESH, D.W. FISCHER, and W.C. MITCHEL
The EL2 Defect in GaAs: Some Recent Developments (Review Article)
61.70 and 71.55; 72.40; 78.30; 78.50; 78.55; S7.12
(b) Vol.154,No.1, 11-42

C.P. Chang, K.F. PAI, C.S. FANG, and W.S. TSE
Dispersion Curves of Phonon-Plasmon Coupled Polaritons in n-Type $Al_xGa_{1-x}As$
63.20; 71.36; 78.30; S7.15
(b) Vol.154,No.1, 135-142

G.A. ADEGBOYEGA, A. POGGI, E. SUSI, and M. REICHE
The Effects of High Temperature Anneal on the Electrical Activity of Iron in B-Doped Cz Silicon
71.55; 72.20; 78.30; S5.11
(a) Vol.114,No.1, 247-252

M.N. KHAN
The Electronic and Spectroscopic Studies of Semiconducting Germanium-Tellurite Glasses Containing Iron Oxide
72.20 and 78.30; S10.15
(a) Vol.114,No.1, 285-292

XIAODING QI, ZUNDU LUO, QINGZHEN HUANG, JINGKUI LIANG, JIMING CHENG, YICHUAN HUANG, MINWANG QIU, and HONG ZHANG
A New Laser Host Crystal. Lathanum Tungstate $La_2(WO_4)_3$
78.45; 78.30; 78.55; S11.1
(a) Vol.114,No.1, K127-K130

R.K. RAM, S.S. KUSHWAHA, and A. SHUKLA
Phonon Assignments in II-VI and III-V Semiconductor Compounds Having Zincblende-Type Structure
63.20; 78.30; S7.1; S8.1
(b) Vol.154,No.2, 553-564

V.I. TORGASHEV, YU.I. YUZYUK, F. SMUTNY, P. VANĚK, and B. BŘEZINA
Raman Spectra of the Incommensurate and Ferroelectric Phases of Rb_2CoCl_4 Single Crystals
78.30; 77.80; S9.16
(b) Vol.154,No.2, 777-788

P.P. LOTTICI
Breathing-Like A-Symmetry Vibrations in $ZnGa_2(S_xSe_{1-x})_4$ and $CdGa_2(S_xSe_{1-x})_4$
63.20; 78.30; S8.16
(b) Vol.154,No.2, K117-K120

H. BOUDRIOT and H.A. SCHNEIDER
Influence of Dipole Scattering on the Free Carrier Absorption
78.30; 72.20; S7.1
(b) Vol.154,No.2, K175-K178

A. BECKMANN, K. DETTMER, and F.R. KESSLER
Thermoabsorption Studies of the Indirect Exciton Bandgap of Si and GaP
78.30 and 78.40; 71.35; S5.11; S7.11
(b) Vol.154,No.2, K191-K196

V.G. MALESH, YU.M. AZHNYUK, D.B. GOYER, and A.V. GOMONNAI
Disorder-Activated First-Order Raman Scattering by Acoustic Phonons in Electron-Irradiated GaP Crystals
61.80 and 78.30; S7.11
(b) Vol.154,No.2, K197-K200

I.P. ALEKSANDROVA, PH. COLOMBAN, F. DENOYER, N. LE CALVE, A. NOVAK, B. PASQUIER, and A. ROZYCKI
Phase Transitions in Ammonium Hydrogen Selenates NH_4HSeO_4 and ND_4DSeO_4
64.70; 78.30; S11.1
(a) Vol.114,No.2, 531-544

K.H. HERRMANN, U. JANK, and K.-P. MOELLMANN
Bipolar Generation within the Exponential Tail of the Interband Absorption Edge in HgCdTe
72.40 and 78.30; S8.15
(a) Vol.114,No.2, K249-K254

K. KNEIPP, H. BOCK, and W. SALEWSKI
Raman Study of Thorium Tungsten Cathodes
78.30; S1
(a) Vol.114,No.2, K255-K258

J. TATARKIEWICZ, M. CARDONA, A. BREITSCHWERDT, C.-T. LIN, R. GROETZSCHEL, and W. WITTHUHN
In Search of Hydrogen Local Vibrational Modes in CuCl, CuBr, CuI, ZnS, and AlSb
63.20; 78.30; S7.13; S8.11; S9
(b) Vol.155,No.1, 77-88

T.D. IBRAGIMOV, G.G. KURBANOVA, and V.S. GORELIK
Raman Scattering in Gallium Selenide and Sulfide Films
63.20 and 68.30; 78.30; S8
(b) Vol.155,No.1, 113-116

XING-KUI ZHANG, HUA XIA, and AN HU
Raman Scattering in Nb-Cu Metallic Multilayers
68.30; 78.30; S1.2; S1.3
(b) Vol.155,No.1, 137-146

T. KUZUBA and M. ISHII
Elastic Constant c_{44} of Uniaxial Layered Crystals
62.20; 63.20; 78.30; S1.63
(b) Vol.155,No.1, K13-K16

J. MONECKE
Collective Excitations in Microcrystal Aggregates and Composites
68.30 and 78.30; 63.10
(b) Vol.155,No.2, 437-444

W. OKULSKI and M. ZALUZNY
Intersubband Raman Scattering in Size-Quantized Narrow-Gap Semiconductor Films
71.25 and 78.30
(b) Vol.155,No.2, 709-716

F.M. GASHIMZADE, S.T. PAVLOV, R.S. NADIRZADE, T.G. ISMAILOV, and B.I. BELITSKII
Resonance Interband Raman Scattering in Semimagnetic Semiconductors
78.30
(b) Vol.155,No.2, K161-K164

X. GERBAUX, A. HADNI, and A. KITADE
Far IR Spectra of Ferroelectric Rochelle Salt and Sodium Ammonium Tartrate
78.30; 77.80; S11
(a) Vol.115,No.2, 587-598

E.N. IZAKOVICH, N.A. ZHELTOVA, YU.M. SHULGA, I.N. IVLEVA, ZH.D. SOKOLOVSKAYA, A.V. BALUEV, V.V. GANIN, O.S. ROSHUPKINA et al.
Study of Ba and Cu(III) Based Inorganic Oxide Compounds
74.70; 78.30; 79.20; 79.60
(a) Vol.115,No.2, K195-K200

Y. TOKUMITSU, H. KITAYAMA, A. KAWABUCHI, T. IMURA, and Y. OSAKA
Raman Spectra of $ZnS_xTe_{1-x}/GaAs(100)$
78.30; S8.15
(a) Vol.115,No.2, K261-K264

YI-JIAN JIANG, LI-JI LIAO, GANG CHEN, and PENG-XIANG ZHANG
Generalized Formulae of Frequency Dispersion and Raman Scattering Cross Section of Phonon-Polaritons
63.20 and 78.30; 71.36
(b) Vol.156,No.1, 145-150

P.S.R. PRASAD and H.D. BIST
Vibrational Spectroscopic Study of Order-Disorder Transition in Ethylammonium Tetrachloro Cadmate
64.70 and 78.30; S12
(a) Vol.116,No.1, 275-286

A.M. POLUBOTKO
About the SERS Phenomenon. I. The Cross-Section in Adiabatic Perturbation Theory
78.30
(b) Vol.156,No.2, 677-686

T. KUZUBA and M. ISHII
Erratum to Elastic Constant c_{44} of Uniaxial Layered Crystals (phys. stat. sol. (b) 155, K13 (1989))
62.20; 63.20; 78.30; S1.63
(b) Vol.156,No.2, K167-K168

V. RIEDE, H. SOBOTTA, H. NEUMANN, C. ASCHERON, C. NEELMEIJER, and A. SCHINDLER
Depth Dependence of Localized Mode Absorptions in Proton-Implanted InP
61.80 and 78.30; 63.20; S7.11
(a) Vol.116,No.2, K147-K152

78.35. Brillouin and Rayleigh scattering

P. AGHAMKAR and P.K. SEN
Stimulated Brillouin Scattering in Doped Centrosymmetric Semiconductors
78.35; S7.13
(b) Vol.151,No.2, 739-746

HUA XIA, X.K. ZHANG, A. HU, W. ZHANG, S.N. ZHU, and C. TIEN
Brillouin Scattering from Surface Phonons in Fibonacci Metallic Superlattices
63.20; 78.35; S1.2
(a) Vol.112,No.1, 99-104

A. NEOGI and S. GHOSH
Stimulated Brillouin Scattering in a Transversely Magnetised Doped Centrosymmetric Semiconductor
78.35; S7.13
(b) Vol.156,No.2, 705-716

78.40. Visible and ultraviolet spectra

SH.M. EFENDIEV, A.M. MAMEDOV, and N.G. DARVISHOV
VUV Reflection Spectra of $PbWO_4$ Single Crystals
78.40; 71.25; S11.1
(b) Vol.151,No.1, K87-K90

T. TSUBOI
Vibronic Structure in the Absorption Spectra of $CaF_2:Sm^{2+}$
63.20 and 78.40; S9.12
(b) Vol.151,No.1, K91-K96

RENYUAN QIAN, SONGQING SHI, TONGMING LIU, and DAOBEN ZHU
Carrier Transport Properties of Poly(1,6-Bis-9-Carbazolyl-2,4-Hexadinyne)
72.80; 72.40; 78.40; S12.1
(a) Vol.111,No.1, K67-K72

I.A. AKSENOV, A.I. LUKOMSKII, and L.A. MAKOVETSKAYA
Edge Absorption and Cathodoluminescence Study of $CuAl_xIn_{1-x}S_2$ Solid Solutions
78.40 and 78.60; S8.16
(a) Vol.111,No.1, K109-K112

A. MANAR, M.J.M. GOMES, B. HOENERLAGE, and R. LEVY
Time Evolution of the Refractive Index of CuCl under Picosecond Pulsed Excitation
71.35 and 78.40; S9
(b) Vol.151,No.2, 747-758

QIGUANG LI, SHAN JIANG, WEI LU, XIAOMING FANG, S.C. CHEN, and W. GIRIRAT
Optical Absorption Experiments of Cycling through the Structural Transition under Hydrostatic Pressure in $Cd_{1-x}Mn_xSe$
62.50 and 78.40; S8.15
(b) Vol.152,No.1, K33-K36

K. JEZIERSKI, J. MISIEWICZ, and F. KROLICKI
Influence of the Surface Quality on the Reflectance for Zn_3P_2
78.40; S7
(a) Vol.112,No.2, K135-K138

E.M. JABOORI and A.K. ABASS
Optical Properties of Chemically Deposited Fe-Cr Mixed Oxide Coatings
78.40; S10.15
(a) Vol.113,No.2, 579-584

V.V. SOBOLEV, A.I. KOZLOV, YU.I. POLYGALOV, V.E. TUPITSYN, and A.S. POPLAVNOI
Reflectivity Spectra and Band Structure of the Zinc and Cadmium Diphosphides
71.25 and 78.40; 71.35; S7
(b) Vol.154,No.1, 377-388

M.YA. BAKIROV, B.A. NAYAFOV, V.S. MAMEDOV, and R.S. MADATOV
Electrical and Optical Properties of Amorphous $Ge_{1-x}Si_x$ Semiconductors
72.80 and 78.40; S5.11; S5.12
(a) Vol.114,No.1, K45-K48

A.G. ARESHKIN, L.G. SUSLINA, and D.L. FEDOROV
On Concentration Dependences of the Band Parameters in $A^{II}B^{VI}$ Solid Solutions with Structural Phase Transitions
64.70 and 71.25; 71.35; 78.40; S8.15
(b) Vol.154,No.2, 609-622

A.A. O'CONNEL-BRONIN and A.B. TRESHCHALOV
Dynamics of Exciton Reflection Spectra on the Change of the Incidence Angle in NaI Crystals
78.40; 71.35; S9.11
(b) Vol.154,No.2, 763-768

A. BECKMANN, K. DETTMER, and F.R. KESSLER
Thermoabsorption Studies of the Indirect Exciton Bandgap of Si and GaP
78.30 and 78.40; 71.35; S5.11; S7.11
(b) Vol.154,No.2, K191-K196

E. FLACH, H.-E. GUMLICH, CH. JUNG, and M. KRAUSE
Reflectance Spectroscopy on $Zn_{1-x}Mn_xS$ in the Energy Range from 11 to 20 eV by Use of Synchrotron Radiation
78.40; 71.25; 71.35; S8.15
(b) Vol.155,No.1, 317-328

M. OŽVOLDOVÁ
Optical Properties of Uranium in Potassium-Alumino-Phosphate Glasses
78.40 and 78.60; S10.15
(a) Vol.115,No.1, 315-324

A.V. EGRANOV, A.I. NEPOMNYACHIKH, V.V. OTROSHOK, and B.P. CHERNYAGO
H_2^--Centre in LiF-H^-
61.70; 76.30; 78.40; S9.11
(b) Vol.155,No.2, 365-372

N.A. MIRONOVA, V.N. SKVORTSOVA, A.S. SMIRNOV, D.V. RIEKSTINYA, G.A. GRINVALDS, and L.A. LITVINOV
Anomalous Behaviour of $MgO \cdot 2.5Al_2O_3$ under Radiation
61.80 and 78.40; 78.55; S11.2
(a) Vol.116,No.1, K125-K128

SH. EFENDIEV, V. GAVRYUSHIN, G. RACIUKAITIS, G. PUZONAS, A. KAZLAUSKAS, N. DARVISHOV, and S. SHAKHDGAN
Two-Photon Spectroscopy of $PbMoO_4$ Single Crystals. Indirect Interband Transitions
78.40; 71.25; S11.1
(b) Vol.156,No.2, 697-704

78.45. Stimulated emission

A.A. KAMINSKII, A.V. BUTASHIN, A.A. DEMIDOVICH, V.G. KOPTEV, B.V. MILL, and A.P. SHKADAREVICH
Broad-Band Tunable Stimulated Emission from Octahedral Cr^{3+} Ions in New Acentric Crystals with Ca-Gallogermanate Structure
78.20 and 78.45; 78.55; S10
(a) Vol.112,No.1, 197-206

A.A. KAMINSKII, M.I. DEMCHUK, N.V. ZHAVORONKOV, and V.P. MICHAILOV
Ultrashort Pulses of Stimulated Emission by Nd^{3+}-Doped ($^4F_{3/2}-^4I_{11/2}$ Channel) Anisotropic Fluorides, Acentric Dispersed Silicates, ...
78.45; S9.16; S11.2
(a) Vol.113,No.2, K257-K260

XIAODING QI, ZUNDU LUO, QINGZHEN HUANG, JINGKUI LIANG, JIMING CHENG, YICHUAN HUANG, MINWANG QIU, and HONG ZHANG
A New Laser Host Crystal. Lathanum Tungstate $La_2(WO_4)_3$
78.45; 78.30; 78.55; S11.1
(a) Vol.114,No.1, K127-K130

A.A. KAMINSKII, A. KHOLOV, P.V. KLEVTSOV, and S.KH. KHAFIZOV
Spectroscopy and Stimulated Emission of Nd^{3+}-Doped Tetragonal $NaBi(MoO_4)_2$ and $NaBi(WO_4)_2$ Disordered Crystals
78.45 and 78.55; 71.55; 78.50; S10
(a) Vol.114,No.2, 713-720

O. ZIEMANN, B. RHEINLAENDER, G. SCHWARZ, and R. SCHULZ
Influence of Time-Dependent Lateral Temperature Profile on Power-Current Characteristics of (Al,Ga)As Oxide Stripe Lasers
78.45; S7.15
(a) Vol.114,No.2, K237-K240

M. AIDARALİEV, N.V. ZOTOVA, S.A. KARANDASHOV, B.A. MATVEEV, N.M. STUS, and G.N. TALALAKIN
Spontaneous and Stimulated Emission from InAsSbP/InAs Heterostructures
78.45 and 78.55; S7.12; S7.15
(a) Vol.115,No.1, K117-K120

78.50. Impurity and defect absorption in solids

R. CYWIŃSKI, E. MUGEŃSKI, W. NOWY-WIECHUŁA, and J. WIECHUŁA
Suzuki Phase Precipitates in Doubly Doped $KCl:Pb^{2+},Mn^{2+}$ Crystals
61.70; 76.30; 78.50; S9.11
(b) Vol.151,No.1, 47-52

J. OTTO
Electric Field Induced Splitting of the Stretching Vibration Lines SH^-, SeH^-, and TeH^- Centres in Alkali Halides
78.30 and 78.50; S9.11
(b) Vol.151,No.1, 363-374

W.P. AMBROSE and A.J. SIEVERS
Determination of the Orientation of NO_2^- in KI by Persistent IR Spectral Hole Burning
63.20 and 78.50; S9.11
(b) Vol.151,No.1, K97-K100

R. SUBRAHMANYAM and Y.V.G.S. MURTI
Laser Induced $F-F^-$ Conversions in KCl Crystals
61.70; 71.55; 78.50; S9.11
(b) Vol.151,No.2, 445-452

F. DE MATTEIS, A. SCACCO, F. SOMMA, G. BALDACCHINI, E. GIOVENALE, and U.M. GRASSANO
Complex Centres in Additively Coloured $RbCl:Li^+$
61.70; 78.50; 78.55; S9.11
(b) Vol.151,No.2, 453-462

J.D. COLLINS, G.A. GLEDHILL, R. MURRAY, P.S. NANDHRA, and R.C. NEWMAN
The Selective Trapping of Arsenic Interstitial Atoms by Impurities in Gallium Arsenide
61.70; 78.50; S7.12
(b) Vol.151,No.2, 469-478

J.L. RAO, M.R. REDDY, and S.V.J. LAKSHMAN
Electronic Absorption Spectrum in Ni^{2+} Ions Doped in Zinc Thallium Sulphate Hexahydrate
71.55 and 78.50; S11
(b) Vol.151,No.2, 599-604

R. BALTRAMIEJŪNAS, V. GAVRYUSHIN, G. RAČIUKAITIS, V. KUBERTAVIČIUS, and A. KAZLAUSKAS
Two-Photon Absorption via Deep Local Levels as Virtual Intermediate States in Semiconductors
71.55 and 78.50; S8.12; S10.1
(b) Vol.151,No.2, 721-732

R. NIES and F.R. KESSLER
Coherent and Incoherent Magnetooptical Reflection and Transmission Effects of Multilayer Structures
78.20; 78.50; S5.11
(a) Vol.111,No.2, 639-654

V.V. PETROV and V.V. CHORNYI
Recharging and Negative Annealing of Divacancies in Neutron-Irradiated Silicon
61.70 and 78.50; S5.11
(a) Vol.111,No.2, K141-K144

L. BOSI and M. NIMIS
On the Decay Properties of the M′-Center in LiF Crystals
61.70; 78.50; S9.11
(b) Vol.152,No.1, 67-72

P.S.S. PRASAD and S. RADHAKRISHNA
Chromium Centres in $3\ CdSO_4 \cdot 8\ H_2O$ Single Crystals
61.80 and 71.55; 78.50; S11.1
(b) Vol.152,No.1, 171-180

SANG-BO ZHAO and KANG-WEI ZHOU
Strong-Field Scheme for $C^{*}_{3v}(d^3)$ and the Spectra of Ruby and Cr_2O_3
71.70; 78.50; S10.1
(b) Vol.152,No.1, 181-190

L.K. AMINOV, A.V. VINOKUROV, I.N. KURKIN, YU.YU. KOSTETSKII, S.L. KORABLEVA, I.KH. SALIKHOV, S.I. SAKHAEVA, and M.A. TEPLOV
Distortion of a Crystal Field in the van Vleck Paramagnet LiTmF under Activation by Isovalent Impurities
71.70; 76.30; 78.50; S9.16
(b) Vol.152,No.1, 191-202

E. WENDLER, W. WESCH, and G. GOETZ
Defects in Weakly Damaged Ion Implanted GaAs and Other III-V Semiconductors
61.80; 78.50; S7.11; S7.12
(a) Vol.112,No.1, 289-300

D. GOURIER, T. GBEHI, R. VISOCEKAS, J. THERY, and D. VIVIEN
Color Centres in Barium Hexaaluminate (Phase I). Study by EPR, Optical Absorption, and Thermally Stimulated Luminescence
61.70; 76.30; 78.50; 78.60; S11
(b) Vol.152,No.2, 415-430

U. BECKER, H. ZIMMERMANN, P. RUDOLPH, and R. BOYN
Optical Study of Impurity Distribution in Vertical-Bridgman Grown CdTe Crystals
71.55; 64.75; 78.50; 78.55; S8.13
(a) Vol.112,No.2, 569-578

I.D. KOZMIK, I.I. GRIGORCHAK, Z.D. KOVALYUK, and I.Z. MARCHUK
Kinetic and Optical Properties of H_xInSe and H_xGaSe Proton Intercalates and Their Thermodynamic Model
61.70; 72.80; 78.50; S8
(a) Vol.112,No.2, K61-K64

J. TATARKIEWICZ
Optical Effects of High Energy Implantations in Semiconductors
61.80 and 78.20; 78.30; 78.50; 78.55; S5.11; S7.12
(b) Vol.153,No.1, 11-48

J.L. RAO, G.L. NARENDRA, B. SREEDHAR, and S.V.J. LAKSHMAN
Electron Spin Resonance and Optical Absorption Spectra of Mn^{2+} Ions in K_2SO_4-$ZnSO_4$ Glasses
71.70; 76.30; 78.50; S11
(b) Vol.153,No.1, 257-262

SHENG-LI GUO and KE-QIN LU
Analysis of the Spectra of Mn^{2+} in Rhodonite
71.55 and 78.50; S11.1
(b) Vol.153,No.1, K45-K50

T. KAWAKUBO and M. OKADA
Optical and Electron Paramagnetic Resonance Studies on Neutron-Irradiated GaP Crystals
71.55; 76.30; 78.50; S7.11
(b) Vol.153,No.1, K93-K96

S. NOWAK, T. KACZOR, and J. ZMIJA
Photochromic Phenomenon Kinetics
61.70 and 78.50; 71.55; S10.15
(a) Vol.113,No.1, 199-206

L. KOVÁCS, R. SOMMERFELDT, YE MING, and E. KRAETZIG
Light-Induced Charge Transport in $LiNbO_3$:Mg,Cr Crystals
61.70 and 72.40; 78.50; S11.1
(a) Vol.113,No.1, K75-K78

S. EMURA and M. ISHIGURO
Effects of Hydrostatic Pressure on the Cu^+ Center in Potassium Halides
71.55; 78.50; 78.55; S9.11
(b) Vol.153,No.2, 683-694

A. MARTÍN, F.J. LÓPEZ, and H. AREND
X-Ray Induced Optical Absorption in Cs_2CdBr_4
78.50; S9.16
(b) Vol.153,No.2, K199-K202

M.O. MANARESH, D.W. FISCHER, and W.C. MITCHEL
The EL2 Defect in GaAs: Some Recent Developments (Review Article)
61.70 and 71.55; 72.40; 78.30; 78.50; 78.55; S7.12
(b) Vol.154,No.1, 11-42

S.K. MISRA and CHUNZHENG WANG
EPR and Optical-Absorption Studies of Cu^{2+}-Doped $Mg(CH_3COO)_2 \cdot 4\,H_2O$ Single Crystal. Core Polarization and Molecular-Orbital Coefficients
71.70; 76.30; 78.50; S12
(b) Vol.154,No.1, 259-272

E. GOLDYS
Photoionisation and Inter (Conduction) Band Absorption in S and Te Doped Gallium Phosphide
78.50; S7.11
(b) Vol.154,No.1, 397-404

A.S. KROCHUK, O.R. ONUFRIV, and Z.P. CHORNIY
Characteristic Properties of the Radiation Colouring Mechanism in MeFX Compounds (Me = Sr,Ba, X = Cl,Br)
61.70; 78.50; S9.16
(b) Vol.154,No.1, K9-K12

C. ZALDO, L. CONTRERAS, L. ARIZMENDI, and E. DIEGUEZ
Optical Properties of $Bi_2Ge_3O_9$ Single Crystals. X-Ray Irradiation Effects
78.50; 78.55; S10
(a) Vol.114,No.1, 397-406

K.P. OBOTH, F.J. LOHMEIER, and F. FISCHER
VUV and UV Spectroscopy of Pb^{2+} and Bi^{2+} Centres in Alkaline-Earth Fluorides
78.50; 78.55; S9.12
(b) Vol.154,No.2, 789-804

E.R. LÓPEZ-TÉLLEZ, D. CÁRDENAS-GARCÍA, and C. RUÍZ-MEJÍA
F_A-Center Optical Absorption in CsF
71.55 and 78.50; S9.11
(b) Vol.154,No.2, K171-K174

A.A. KAMINSKII, A. KHOLOV, P.V. KLEVTSOV, and S.KH. KHAFIZOV
Spectroscopy and Stimulated Emission of Nd^{3+}-Doped Tetragonal $NaBi(MoO_4)_2$ and $NaBi(WO_4)_2$ Disordered Crystals
78.45 and 78.55; 71.55; 78.50; S10
(a) Vol.114,No.2, 713-720

G.J. PAPAIOANNOU, E. ANAGNOSTAKIS, P.C. EUTHYMIOU, C.D. KOURKOUTAS, and G. ZARDAS
On the Impurity Photoconductivity of Semi-Insulating GaP:Cr
72.40; 78.50; S7.11
(a) Vol.114,No.2, K215-K218

V.S. KISHAN KUMAR, S.B.S. SASTRY, and B.S.V.S.R. ACHARYULU
Optical Studies on Europium Doped K_2SO_4
61.70 and 78.50; 78.55; S11.1
(b) Vol.155,No.2, 679-684

M. TAIBI, J. ARIDE, E. ANTIC-FIDANCEV, M. LEMAITRE-BLAISE, and P. PORCHER
Crystal Field Parameters of Eu^{3+} and Nd^{3+} in Ln_2BaZnO_5 Phases (Ln=La,Nd,Y)
71.70; 78.50; 78.55; S10.15
(a) Vol.115,No.2, 523-532

K.-E. PEIPONEN, P. SILFSTEN, P. KETOLAINEN, and T. TSUBOI
A Remark on the Thermo-Optical Properties of the $F_A(II)$ Centre in KCl:Li Crystals
61.70 and 78.50; S9.11
(b) Vol.156,No.1, K81-K84

H. DONKER, W.M.A. SMIT, and G. BLASSE
The Luminescence of $CaCO_3:Sn^{2+}$
78.55; 78.50; S11.1
(b) Vol.156,No.1, K93-K96

W. HOERIG, V.F. ORLENKO, H. NEUMANN, and C. ASCHERON
Near-Edge Optical Absorption Spectra of Proton-Implanted GaAs
61.70 and 78.50; S7.12
(a) Vol.116,No.1, K119-K124

GH. STOICESCU, S.V. NISTOR, and C.D. MATEESCU
Aggregation of Bismuth in NaCl Crystals
61.70; 66.30; 78.50; S9.11
(b) Vol.156,No.2, 411-418

V.A. KOVARSKII and M.A. CHECHELNITSKII
A Microscopical Model of Optical Spectra of Local Crystal Centres Caused by Order-Disorder Phase Transition
64.70 and 78.50
(b) Vol.156,No.2, K157-K162

78.55. Photoluminescence

C.P. HILTON, J.J. DAVIES, J.E. NICHOLLS, and O. GOEDE
Optically-Detected Magnetic Resonance of Excitons Bound at Te_2 Isoelectronic Centres in $CdS_{1-x}Te_x$
71.35 and 71.55; 76.90; 78.55; S8.11; S8.15
(b) Vol.151,No.1, 175-184

A. MARTÍN, F.J. LÓPEZ, and H. AREND
Luminescence Properties of $\{(CH_3)_4N\}_2ZnCl_4$ Single Crystals
78.55 and 78.60; S9.16
(b) Vol.151,No.1, 331-336

A.N. GEORGOBIANI, A.N. GRUZINTSEV, A.V. SPITSYN, and I.M. TIGINYANU
Change of the Optical Properties of CdS Single Crystals upon High Dose Indium and Gallium Implantation
78.20 and 78.55; 78.60; S8.11
(a) Vol.111,No.1, 301-308

J. WIEHL, R. HIRRLE, W. WISCHERT, and S. KEMMLER-SACK
Cathodo- and Photoluminescence in the Mn^{2+} Activated Garnets $Cd_{3-x}Ca_xAl_2Ge_3O_{12}$
78.55 and 78.60; 71.55; S11.2
(a) Vol.111,No.1, 315-324

J. PASTRNAK, J. OSWALD, I. GREGORA, V. VORLÍČEK, and M. BABINSKY
Lateral Distribution of Si Related Defects and Carrier Density in GaAs Crystals
78.55; 71.55; 78.30; S7.12
(a) Vol.111,No.1, 345-360

F. DE MATTEIS, A. SCACCO, F. SOMMA, G. BALDACCHINI, E. GIOVENALE, and U.M. GRASSANO
Complex Centres in Additively Coloured $RbCl:Li^+$
61.70; 78.50; 78.55; S9.11
(b) Vol.151,No.2, 453-462

R. FLAGMEYER, G. OELGART, K. KREHER, R. SCHWABE, F. BUGGE, L. LEHMANN, and K. JACOBS
Characterization of MOCVD Grown (Al,Ga)As/GaAs Single Quantum Well Structures by Rutherford Backscattering and Photoluminescence
73.40 and 85; 78.55; S7.12
(a) Vol.111,No.2, 541-550

R. HIRRLE, J. WIEHL, W. WISCHERT, and S. KEMMLER-SACK
Cathodo- and Photoluminescence in the Tb^{3+} Activated Garnets of Type $Cd_{3-x}Ca_xAl_2Ge_3O_{12}$
78.55 and 78.60; 71.55; S11.2
(a) Vol.111,No.2, 629-638

J. MISIEWICZ, Z. GUMIENNY, R. CYWINSKI, and E. MUGENSKI
The Luminescence of Zn_3P_2 in the 0.84 to 1.1 μm Wavelength Range
78.55; 71.55; S7.1
(a) Vol.111,No.2, K249-K252

C. GHEZZI, D. MARTIN, A. PARISINI, J.L. STAEHLI, and L. TARRICONE
Investigation of (AlGa)As/GaAs Quantum Wells by Optical and Phototransport Measurements
73.40; 71.35; 72.40; 78.55; S7.15
(b) Vol.152,No.1, 101-110

A.A. KAMINSKII, A.V. BUTASHIN, A.A. DEMIDOVICH, V.G. KOPTEV, B.V. MILL, and A.P. SHKADAREVICH
Broad-Band Tunable Stimulated Emission from Octahedral Cr^{3+} Ions in New Acentric Crystals with Ca-Gallogermanate Structure
78.20 and 78.45; 78.55; S10
(a) Vol.112,No.1, 197-206

W. THIELEMANN, B. RHEINLAENDER, P. NEVERMANN, and G. OELGART
Determination of the Internal Quantum Yield and of the Injection Effiency of Red-Light Emitting (Al,Ga)As-DHS Diodes
73.40 and 78.60; 68.55; 78.55; S7.15
(a) Vol.112,No.1, 229-236

G. LIIDJA, V. NAGIRNII, T. SOOVIK, and S. ZAZUBOVICH
New Off-Centre System: Jahn-Teller Mercury-Like Impurities in Alkali Halides
71.55 and 78.55; S9.11
(b) Vol.152,No.2, 563-576

U. BECKER, H. ZIMMERMANN, P. RUDOLPH, and R. BOYN
Optical Study of Impurity Distribution in Vertical-Bridgman Grown CdTe Crystals
71.55; 64.75; 78.50; 78.55; S8.13
(a) Vol.112,No.2, 569-578

A. KOZANECKI
Backscattering and Optical Studies of Ytterbium Implanted into A^{III}-B^{V} Compounds
61.80; 78.55; S7.11; S7.15
(a) Vol.112,No.2, 777-780

W. DZIURDA, L. GABLA, and M. TULETA
Bombardment-Induced Photon Emission from Modified Surface of Glass
61.80 and 78.55; S10
(a) Vol.112,No.2, 811-814

J. TATARKIEWICZ
Optical Effects of High Energy Implantations in Semiconductors
61.80 and 78.20; 78.30; 78.50; 78.55; S5.11; S7.12
(b) Vol.153,No.1, 11-48

J.S. NKOMA
Radiation Induced Luminescence from Dipoles Immersed in a Solid with a Rough Surface
68.20 and 78.55; S7.11
(b) Vol.153,No.1, 383-390

C. BENECKE, W. BUSSE, H.-E. GUMLICH, H. HOFFMANN, S. HOFFMANN, A. KROST, and H. WALDMANN
On the Excitation and Emission of Infrared Luminescence of Wide Band Gap Mn II-VI Semimagnetic Semiconductors
78.55; 71.70; S8.15
(b) Vol.153,No.1, 391-402

S.E. SARKISOV, K.K. PUKHOV, A.A. KAMINSKII, A.G. PETROSYAN, and T.I. BUTAEVA
Manifestation of Electron-Phonon Interaction in Insulating Crystals Doped with Pr^{3+} Ions
63.20 and 78.55; 71.55; S10.15
(a) Vol.113,No.1, 193-198

I.YA. FUGOL, A.M. RATNER, YU.I. RYBALKO, and YU.L. STESHENKO
On the Mechanism of Vibrational Relaxation of Two-Centre Excitations in Solid Neon
63.90; 71.45; 78.55; S5.2
(b) Vol.153,No.2, 539-546

C. GOURDON and P. LAVALLARD
Exiton Transfer Between Localized States in $CdS_{1-x}Se_x$ Alloys
71.35 and 78.55; S8.15
(b) Vol.153,No.2, 641-652

P.E. MOZOL, N.A. SKUBENKO, P.A. SKUBENKO, YU.P. GNATENKO, E.A. SALKOV, and Z.D. KOVALYUK
Effect of Laser Irradiation on Low-Temperature Photoconductivity and Photoluminescence Spectra of Gallium Selenide
71.35 and 71.55; 72.40; 78.55; S8
(b) Vol.153,No.2, 667-674

S. EMURA and M ISHIGURO
Effects of Hydrostatic Pressure on the Cu^+ Center in Potassium Halides
71.55; 78.50; 78.55; S9.11
(b) Vol.153,No.2, 683-694

I. YU, H.L. PARK, H.K. KIM, S.K. CHANG, and C.H. CHUNG
Sodium Related Luminescence Centers in $CaS:Bi^{3+}$ Phosphor
78.55; S8
(b) Vol.153,No.2, K183-K186

J. E. MUNOZ SANTIUSTE and J. GARCIA-SOLĖ
Lifetime of Eu^{2+} Luminescence in Alkali Halides. Dependence on the Host Matrix
78.55; S9.11
(b) Vol.153,No.2, K187-K192

H. WEINERT, A. ŻUKAUSKAS, G. TAMULAITIS, and V. STEPANKEVICIUS
Some Hints on the Influence of the Barrier on Excitonic Optical Properties of GaAs-GaAlAs Superlattices Obtained by ...
78.55; 71.35; 78.30; S7.12; S7.15
(b) Vol.153,No.2, K193-K198

K.D. GLINCHUK, N.S. ZAYATS, and A.V. PROKHOROVICH
Effect of Proton Bombardment on GaAs Luminescence
71.55 and 78.55; 61.80; S7.12
(a) Vol.113,No.2, 497-502

L.L. KULYUK, N.F. PERELMAN, S.I. RADAUTSAN, A.V. SIMINEL, and E.E. STRUMBAN
About the Possibility of Double Centre Formation in InP:Fe:O
71.55 and 72.40; 76.30; 78.55; S7.11
(a) Vol.113,No.2, K227-K230

S.L. PINJARE, N. BALASUBRAMANYAM, and VIKRAM KUMAR
Photoluminescence at 0.944 eV from Heat Treated n-Type Silicon
78.55; 61.70; 71.55; S5.11
(a) Vol.113,No.2, K261-K264

YU.O. DERID, E.I. KOZHACHENKO, S.I. RADAUTSAN, and I.M. TIGINYANU
Radiative Recombination in $ZnGa_2S_4$ Single Crystals
78.55; S8.16
(a) Vol.113,No.2, K265-K266

M.O. MANARESH, D.W. FISCHER, and W.C. MITCHEL
The EL2 Defect in GaAs: Some Recent Developments (Review Article)
61.70 and 71.55; 72.40; 78.30; 78.50; 78.55; S7.12
(b) Vol.154,No.1, 11-42

M. PAWŁOWSKA, W. STREK, and I. TRABJERG
Temperature Dependence of Luminescence of $RbMnCl_3{:}Sm^{3+}$ Crystal
78.55; S9.16
(b) Vol.154,No.1, K89-K92

S.I. RADAUTSAN, I.M. TIGINYANU, V.N. FULGA, and YU.O. DERID
Quasi-Continuously Distributed Traps and Photoluminescence in $ZnGa_2Se_4$ Single Crystals
71.55; 78.55; S8.16
(a) Vol.114,No.1, 259-264

C. ZALDO, L. CONTRERAS, L. ARIZMENDI, and E. DIEGUEZ
Optical Properties of $Bi_2Ge_3O_9$ Single Crystals. X-Ray Irradiation Effects
78.50; 78.55; S10
(a) Vol.114,No.1, 397-406

XIAODING QI, ZUNDU LUO, QINGZHEN HUANG, JINGKUI LIANG, JIMING CHENG, YICHUAN HUANG, MINWANG QIU, and HONG ZHANG
A New Laser Host Crystal. Lathanum Tungstate $La_2(WO_4)_3$
78.45; 78.30; 78.55; S11.1
(a) Vol.114,No.1, K127-K130

K.P. OBOTH, F.J. LOHMEIER, and F. FISCHER
VUV and UV Spectroscopy of Pb^{2+} and Bi^{2+} Centres in Alkaline-Earth Fluorides
78.50; 78.55; S9.12
(b) Vol.154,No.2, 789-804

E. MIHÓKOVÁ
On Negative Probabilities from the Agranovich and Galanin Model of Luminescence and Energy Transfer
78.55; 66.90
(b) Vol.154,No.2, K183-K186

J.W. TOMM, J. GRIESCHE, H. SCHMIDT, D. GENZOW, and J. SIEMON
Strain in Lead Salt Layers. A Luminescence Study
68.55; 78.55; S8
(a) Vol.114,No.2, 621-628

A.A. KAMINSKII, A. KHOLOV, P.V. KLEVTSOV, and S.KH. KHAFIZOV
Spectroscopy and Stimulated Emission of Nd^{3+}-Doped Tetragonal $NaBi(MoO_4)_2$ and $NaBi(WO_4)_2$ Disordered Crystals
78.45 and 78.55; 71.55; 78.50; S10
(a) Vol.114,No.2, 713-720

L.N. SKUJA
Photoluminescence of Intrinsic Defects in Glassy GeO_2. Twofold Coordinated Ge and Nonbridging Oxygen
78.55; S10.1
(a) Vol.114,No.2, 731-738

A.A. KLYUKANOV, E.A. SENOKOSOV, and L.E. CHIBOTARU
The Shape of Multi-Plasmon Radiation Band in Direct Gap Semiconductors
71.45 and 78.55; S8.12
(b) Vol.155,No.1, 295-308

G. OELGART, B. LIPPOLD, R. HEILMANN, H. NEUMANN, and B. JACOBS
Determination of the Germanium Acceptor Ionisation Energy of $Al_xGa_{1-x}As$ ($0<x<0.40$) by Hall Effect and Luminescence
71.55; 72.80; 78.55; S7.15
(a) Vol.115,No.1, 257-266

M. AIDARALIEV, N.V. ZOTOVA, S.A. KARANDASHOV, B.A. MATVEEV, N.M. STUS, and G.N. TALALAKIN
Spontaneous and Stimulated Emission from InAsSbP/InAs Heterostructures
78.45 and 78.55; S7.12; S7.15
(a) Vol.115,No.1, K117-K120

J. DEXPERT-GHYS, Y. CHARREIRE, P. ESTEBAN-PUGES, L. ALBERT, and H. DEXPERT
Optical and EXAFS Investigations of ZnS:Tm^{3+} Powders
71.55; 61.10; 78.55; S8.11
(b) Vol.155,No.2, 571-580

V.S. KISHAN KUMAR, S.B.S. SASTRY, and B.S.V.S.R. ACHARYULU
Optical Studies on Europium Doped K_2SO_4
61.70 and 78.50; 78.55; S11.1
(b) Vol.155,No.2, 679-684

M. TAIBI, J. ARIDE, E. ANTIC-FIDANCEV, M. LEMAITRE-BLAISE, and P. PORCHER
Crystal Field Parameters of Eu^{3+} and Nd^{3+} in Ln_2BaZnO_5 Phases (Ln=La,Nd,Y)
71.70; 78.50; 78.55; S10.15
(a) Vol.115,No.2, 523-532

E. GRILLI, M. GUZZI, and M. PASTORI
Time Resolved Photoluminescence of $CdIn_2S_4$ Single Crystals
78.55; 71.55; S8.16
(a) Vol.115,No.2, 599-606

D. BRACH, P. GLAREMIN, and H.-J. WEBER
Optical Investigations of the Phase Transition in Pure and Cr^{3+}-Doped Potassium Alum
64.60; 78.20; 78.55; S11.1
(b) Vol.156,No.1, 195-204

A. ZEHE and A. JUAREZ
Double-Peak Formation as a Doping-Superlattice Effect in Donor-Acceptor Pair-Luminescence Spectra
71.55 and 78.55; 73.40; S7.12
(b) Vol.156,No.1, K89-K92

H. DONKER, W.M.A. SMIT, and G. BLASSE
The Luminescence of $CaCO_3$:Sn^{2+}
78.55; 78.50; S11.1
(b) Vol.156,No.1, K93-K96

G. SCHRAMM
Quantitative Analysis of Sb Doped Silicon by Photoluminescence
78.55; S5.11
(a) Vol.116,No.1, K97-K100

V. TABAKOVA
Emission Spectra of Tallium Centres in Barium Chloride Single Crystals
78.55; S9.11
(a) Vol.116,No.1, K101-K104

V.A. NADOLINNYI, V.V. VASILEV, and I.P. MIKHAILOVSKII
Photo-Induced Metastable States in Silicon Nitride Amorphous Layers
76.30 and 78.55; S6
(a) Vol.116,No.1, K105-K110

N.A. MIRONOVA, V.N. SKVORTSOVA, A.S. SMIRNOV, D.V. RIEKSTINYA, G.A. GRINVALDS, and L.A. LITVINOV
Anomalous Behaviour of $MgO\cdot2.5Al_2O_3$ under Radiation
61.80 and 78.40; 78.55; S11.2
(a) Vol.116,No.1, K125-K128

H. DONKER, N. YAMASHITA, W.M.A. SMIT, and G. BLASSE
Luminescence Decay Times of the Sb^{3+}, Pb^{2+}, and Bi^{3+} Ions in Alkaline-Earth Sulfides
71.70; 78.55; S8.11
(b) Vol.156,No.2, 537-544

U. STUTENBAEUMER, H.-E. GUMLICH, and H. ZUBER
The Energy Transfer between Free Excitons, Bound Excitons, and the 3d-Shell of Mn^{2+} in $Zn_{1-x}Mn_xSe$ Mixed Crystals and s-d Type Exchange
71.35 and 71.70; 78.55; S8.15
(b) Vol.156,No.2, 561-568

P.V. CHERNOV, E.M. DIANOV, V.N. KARPECHEV, L.S. KORNIENKO, I.O. MOROZOVA, A.O. RYBALTOVSKII, V.O. SOKOLOV, and V.B. SULIMOV
Spectroscopic Manifestations of Self-Trapped Holes in Silica: Theory and Experiment
78.55; 71.55; S10.1
(b) Vol.156,No.2, 663-676

Y.S. TANG
A Lineshape Analysis of the near 1.54 μm Emission Band of Er^{3+} in Silicon
78.55; S5.11
(b) Vol.156,No.2, K153-K156

G. SCHRAMM
Determination of Free Exciton Capture Cross Section of Si:Sb by Photoluminescence (PL)
78.55; S5.11
(a) Vol.116,No.2, K209-K212

78.60. Other luminescence and radiative recombination

For photoconduction and photovoltaic effects, see 72.40

A. MARTIN, F.J. LOPEZ, and H. AREND
Luminescence Properties of $\{(CH_3)_4N\}_2ZnCl_4$ Single Crystals
78.55 and 78.60; S9.16
(b) Vol.151,No.1, 331-336

N. NISHIMAKI, S. SHIMANUKI, and Y. NASU
Copper Centers Induced by Optical Bleaching of Cu^- Center in KCl and KBr Crystals
78.60; S9.11
(b) Vol.151,No.1, K81-K86

A.N. GEORGOBIANI, A.N. GRUZINTSEV, A.V. SPITSYN, and I.M. TIGINYANU
Change of the Optical Properties of CdS Single Crystals upon High Dose Indium and Gallium Implantation
78.20 and 78.55; 78.60; S8.11
(a) Vol.111,No.1, 301-308

J. WIEHL, R. HIRRLE, W. WISCHERT, and S. KEMMLER-SACK
Cathodo- and Photoluminescence in the Mn^{2+} Activated Garnets $Cd_{3-x}Ca_xAl_2Ge_3O_{12}$
78.55 and 78.60; 71.55; S11.2
(a) Vol.111,No.1, 315-324

I. GODMANIS and W. HOHENAU
On the Nature of Afterglow of the X-Ray Induced Luminescence in Crystalline and Glassy SiO_2
78.60; S10.1
(a) Vol.111,No.1, 335-344

J. ŠIKULA, P. VASINA, and I. HUETTEL
The Noise of the GaAlAs Super-LED Radiation
72.70 and 73.40; 78.60; S7.15
(a) Vol.111,No.1, 367-370

I.A. AKSENOV, A.I. LUKOMSKII, and L.A. MAKOVETSKAYA
Edge Absorption and Cathodoluminescence Study of $CuAl_xIn_{1-x}S_2$ Solid Solutions
78.40 and 78.60; S8.16
(a) Vol.111,No.1, K109-K112

F. FIRSZT and H.L. OCZKOWSKI
Distribution of Traps in ZnSe:Cu Crystals Obtained by Fractional Glow Technique
71.55 and 78.60; S8.12
(a) Vol.111,No.1, K113-K118

P.L. MUTHAL, S.V. MOHARIL, and B.T. DESHMUKH
Radiation-Induced Sensitization in Polyethylene
78.60; S12
(a) Vol.111,No.1, K119-K122

K. LOEBE, R. BOYN, and R. REETZ
Stability of Sm Centers in Electroluminescent ZnS:SmF_3 MISIM Structures during Ageing
78.60; S8.11
(a) Vol.111,No.1, K129-K132

R. HIRRLE, J. WIEHL, W. WISCHERT, and S. KEMMLER-SACK
Cathodo- and Photoluminescence in the Tb^{3+} Activated Garnets of Type $Cd_{3-x}Ca_xAl_2Ge_3O_{12}$
78.55 and 78.60; 71.55; S11.2
(a) Vol.111,No.2, 629-638

W. HERGERT
The Information Radius in Electron Beam or Light Beam Probing of Semiconductors
78.60
(a) Vol.111,No.2, K253-K258

W. THIELEMANN, B. RHEINLAENDER, P. NEVERMANN, and G. OELGART
Determination of the Internal Quantum Yield and of the Injection Effiency of Red-Light Emitting (Al,Ga)As-DHS Diodes
73.40 and 78.60; 68.55; 78.55; S7.15
(a) Vol.112,No.1, 229-236

D.F. MARIANI, F. GONZALES, J. VIGNOLO, M. JIMENEZ DE CASTRO, and A. IBARRA
Activation Energies for the Thermoluminescence Processes of Al_2O_3 Gamma-Irradiated at Room Temperature
78.60; S10.1
(a) Vol.112,No.1, K53-K56

D. GOURIER, T. GBEHI, R. VISOCEKAS, J. THERY, and D. VIVIEN
Color Centres in Barium Hexaaluminate (Phase I). Study by EPR, Optical Absorption, and Thermally Stimulated Luminescence
61.70; 76.30; 78.50; 78.60; S11
(b) Vol.152,No.2, 415-430

R. KUDUK, M. KULIK, D. MĄCZKA, K. PYSZNIAK, and J. ŻUK
Depth Profile Measurements of Ions Implanted in Semiconductor Materials by Particle Induced Photon Emission
61.70; 78.60; 79.20
(a) Vol.112,No.2, 799-804

S.S. CAMARGO,JR. and J.P. VON DER WEID
Triplet Molecular Excitations in Potassium Halide Cyanide Crystals. I. The CN- Impurity Centre
71.35 and 78.60; S9.11
(b) Vol.153,No.2, 757-764

S.S. CAMARGO,JR. and J.P. VON DER WEID
Triplet Molecular Excitations in Potassium Halide Cyanide Crystals. II. The $K(CN)_xCl_{1-x}$ Mixed System
71.35 and 78.60; S9.15
(b) Vol.153,No.2, 765-772

Y. KIRSH, S. SHOVAL, S. YARIV, and N. KRISTIANPOLLER
Natural and Beta Induced Thermoluminescence in Microcline
78.60; S11
(a) Vol.113,No.2, 631-636

R.K. GARTIA, S.J. SINGH, and P.S. MAZUMDAR
The Determination of the Activation Energy of Thermally Stimulated Luminescence Peaks Obeying the General Order Kinetics
78.60
(a) Vol.114,No.1, 407-412

G. OELGART, V. GOTTSCHALCH, H. HAEFNER, R. HEILMANN, W. KLEIN, and B. RHEINLAENDER
Characterization of GaInAsP/InP Double Heterostructure Laser-Wafers
73.40 and 85; 78.60; S7.11; S7.15
(a) Vol.114,No.1, 419-430

I. GŁOWACKI and J. ULAŃSKI
Precipitation of NIPC Dopant in Polycarbonate Matrix as Detected by Thermoluminescence
78.60; S12
(a) Vol.114,No.1, K107-K110

B.P. CHANDRA and D.P. BISEN
Theoretical Approach to the Mechanoluminescence of Thermoluminescent Crystals
78.60; S9.11
(a) Vol.114,No.1, K123-K126

Y. KIRSH, J.E. TOWNSEND, and P.D. TOWNSEND
Kinetic Analysis of the New Sensitization Effect in the TL of Silica Optical Fibres
78.60; S10.1
(a) Vol.114,No.2, 739-748

M.K. JAYARAJ and C.P.G. VALLABHAN
Low Voltage Driven White Light Emitting Thin Film EL Device
78.60; S8.11
(a) Vol.114,No.2, K233-K236

V. BICHEVIN, G. ALSEITOV, M. GAPANOV, and H. KAAMBRE
Excimer-Laser-Induced Exoemission and Thermoluminescence of Some Alkali Halides
78.60 and 79.75; 61.80; S9.11
(a) Vol.114,No.2, K241-K244

Y. FUKUDA, A. TOMITA, and N. TAKEUCHI
Thermoluminescence and Thermally Stimulated Exoelectron Emission of Sintered MgB_4O_7 Doped with $DyCl_3$ and Tm_2O_3
78.60 and 79.75; S10
(a) Vol.114,No.2, K245-K248

R.S. DE BIASI
Comments on Concentration Peculiarities of a Set of Impurity Centres in Fluorite Crystals Activated by Gadolinium
61.70; 76.30; 78.60; S9.15
(b) Vol.155,No.1, K3-K4

M. OZVOLDOVA
Optical Properties of Uranium in Potassium-Alumino-Phosphate Glasses
78.40 and 78.60; S10.15
(a) Vol.115,No.1, 315-324

F. FERMI, A. NAGORNYI, and C. PARACCHINI
Defect Annealing in Pure CdF_2 Crystals
61.70 and 78.60; S9.12
(a) Vol.115,No.1, 325-334

R. DUBOURG, M. SCHVOERER, R. BERGER, and B. DUMERCQ
Mécanismes de thermoluminescence dans des fluorines CaF_2 naturelles et de synthese
78.60; 61.70; S9.12
(a) Vol.115,No.1, 335-346

S.I. DRAPAK, V.N. KATERINCHUK, Z.D. KOVALYUK, and V.A. MANASSON
The Injection Peculiarities of Minority Charge Carriers in the Anisotype SIS Structure
73.40 and 78.60; S8; S10.1
(a) Vol.115,No.1, K35-K38

I.A. AKSENOV, I.R. GULAKOV, V.I. LIPNITSKII, A.I. LUKOMSKII, and L.A. MAKOVETSKAYA
Cathodoluminescence and Electrical Properties of $CuAlS_2$
72.80 and 78.60; S8.16
(a) Vol.115,No.1, K113-K116

A.F. BURENKOV, V.S. VARYCHENKO, A.M. ZAITSEV, F.F. KOMAROV, G.G. KONOPLYANIK, A.A. MELNIKOV, V.F. STELMAKH, and V.V. TKACHEV
Depth Distribution of Defects and Impurities in Diamond and Cubic Boron Nitride after High-Energy Ion Implantation
61.80; 61.70; 78.60; S5; S7.1
(a) Vol.115,No.2, 427-436

I.S. GORBAN, N.S. KORETS, TS.A. KRYSKOV, and M.V. CHUKICHEV
Cathodoluminescence in Doped CdP_2 and $CdSiP_2$ Crystals
78.60; 71.55; S7
(a) Vol.115,No.2, 555-560

N.S. KORETS, N.V. STUCHINSKAYA, and M.V. CHUKICHEV
Cathodoluminescence of Cadmium Diphosphide Crystals under Uniaxial Deformation
78.60; S7
(a) Vol.115,No.2, 561-568

E. CHIMCZAK and M. BERTRANDT-ZYTKOWIAK
Blocking Effect in Electroluminescence of ZnS:Mn Driven by Short Voltage Pulses
78.60; 78.65; S8.11
(a) Vol.115,No.2, 581-586

E.M. SHISHONOK, V.B. SHIPILO, A.I. LUKOMSKII, and T.V. RAPINCHUK
The Nature of Some Luminescence Centers in Cubic Boron Nitride
78.60; S7.1
(a) Vol.115,No.2, K237-K242

Y. FUKUDA, A. TOMITA, and N. TAKEUCHI
Thermoluminescence and Thermally Stimulated Exoelectron Emission of $Ca_2B_5O_8Cl$ Activated with Divalent Europium or Lead
78.60 and 79.75; S9.16
(a) Vol.115,No.2, K251-K254

E.M. IBRAGIMOVA and V.N. STEPANOV
Surface Defect States at Silica-Polymer Interfaces in Fiber-Optic Waveguides
61.80 and 78.60; S10.1; S12
(a) Vol.115,No.2, K255-K260

A. OPANOWICZ
On the Kinetics Order of Thermoluminescence in Insulating Crystals
71.55 and 78.60
(a) Vol.116,No.1, 343-348

M.V. RAMANA and G.S. SASTRY
Thermoluminescence in Oxy-Fluoro-Borate Glasses
78.60; S10.15
(a) Vol.116,No.2, K205-K208

78.65. Optical properties of thin films

B.I. KHUDIK, V.Z. LOZOVSKII, and A.F. ZHURAVLEV
Optical Theorem in Ultra-Thin Film Electrodynamics
73.60 and 78.65
(b) Vol.151,No.1, 111-120

W. HENRION and H. LANGE
Optical Properties of High-Refractory Disilicide Thin Films
78.65; S1.61
(b) Vol.151,No.1, 375-382

U. TRUTSCHEL, M. GOLZ, and M. ABRAHAM
Generalized Transfer-Matrix Description of the Optical Properties of Spatially Dispersive Periodic Metal Multilayers
78.65
(b) Vol.151,No.1, 383-398

S.A. AL-KUHAIMI
Optical and Compositional Properties of $CuInSe_2$ Thin Films at High Temperature
68.55 and 78.65; S8.16
(a) Vol.111,No.2, 523-528

S.N. SAHU and A. BOURDILON
First Non-Aqueous Electrocodeposition in InP, In, and P Films
68.55; 78.65; S4; S5; S7.11
(a) Vol.111,No.2, K179-K184

W. SCHARFF, K. HAMMER, B. EIBISCH, O. STENZEL, S. ROTH, and T. FRAUENHEIM
Deposition of Hyrdogenated Amorphous Carbon Films by a ECR Microwave Plasma
61.80 and 68.55; 73.60; 78.65; S5
(a) Vol.112,No.2, 667-678

B.I. KHUDIK, V.Z. LOZOVSKII, and I.V. NAZARENKO-BARYAKHTAR
Macroscopic Electrodynamics of Ultra-Thin Films
78.65
(b) Vol.153,No.1, 167-178

A. LEHMANN
Theory of Infrared Spectra of Thin Inhomogeneous Insulating Films
78.30 and 78.65
(b) Vol.153,No.1, K85-K88

M.E. ABU-ZEID, Y.A. YOUSSEF, A.E. RAKHSHANI, and J. VARGHESE
Investigation of Cu_2O Thin Film Using Thermal Interferometry
68.55; 78.65; S10
(a) Vol.113,No.1, 123-130

K. SCHMIDT, L. BAUCH, G. KLUGE, and P. SUEPTITZ
Preparation and Properties of Spin Coated Amorphous $GeSe_2$ Layers
68.55; 78.65; S8
(a) Vol.113,No.1, K55-K60

T. DRUESEDAU, M. ECKLER, R. BINDEMANN, and H. FIEDLER
Enhancement of Photoconductivity in Magnetron-Sputtered a-Si:H Doped with Nitrogen
68.55 and 73.60; 72.40; 78.65; S5.11
(b) Vol.153,No.2, K119-K124

K. PONNURAJU, P.R. VAYA, and J. SOBHANADRI
The Effect of Substrate Temperature on the Optical Properties of $Pb_{1-x}Sn_xTe$ Films Formed by Molecular Beam Epitaxy
78.65; 68.55; S8.15
(a) Vol.113,No.2, 477-482

M. PALPACUER, M. ROLLAND, and C. BENOIT
Polarized Spectra of Oriented Polyacetylene Doped with Iodine
78.30; 78.65; S12
(a) Vol.113,No.2, 623-630

M. GASGNIER
Rare Earth Compounds (Oxides, Sulfides ...)as Thin Films and Thin Crystals. Physico-Chemical Properties and Applications(Review Article)
68.55 and 73.40; 75.70; 78.65; S1.4; S1.63; S10
(a) Vol.114,No.1, 11-72

M. ANWAR and C.A. HOGARTH
A Study of Optical Absorption in MoO_3/SiO Amorphous Thin Films
78.65; 68.55; S10.15
(a) Vol.114,No.1, 215-224

P. KASHYAP, M. JAIN, O.P. AGNIHOTRI, and H.K. SEHGAL
Effect of Composition on the Optical and Electrical Characteristics of Flash Evaporated $Cu_{2-x}In_xSe_2$ Thin Films
73.60 and 78.65; S8.16
(a) Vol.114,No.1, K29-K32

Y.S. TANG
Photoreflectance of Multiple Quantum Wells: Modulation Intensity Dependence
78.20 and 78.65
(b) Vol.154,No.2, K187-K190

E. HAJTO, P.J.S. EWEN, P.G. HILL, and A.E. OWEN
Spectroscopic Investigation of Spin-Deposited As-S Amorphous Films
78.65; 68.55; S8
(a) Vol.114,No.2, 587-598

K. SOMOGYI, D.N. KORBUTYAK, L.N. LASHKEVICH, and I. POZSGAI
Cu Diffusion into GaAs from Water Based Solution
66.30 and 73.40; 78.65; S7.12
(a) Vol.114,No.2, 635-642

S. SAHA, U. PAL, A.K. CHAUDHURI, V.V. RAO, and H.D. BANERJEE
Optical Properties of CdTe Thin Films
78.65; S8.13
(a) Vol.114,No.2, 721-730

S.M.J. AKHTAR, D. RISTAU, J. EBERT, and H. WELLING
Characterization of Dielectric Films and Damage Threshold at 1.064 μm
78.65; 77.55; S9; S10
(a) Vol.115,No.1, 191-198

L. MALACKY, J. KUZMIK, and V. CAMBEL
Optical Properties of Metal/GaInAs Contacts for Schottky Photodiodes
78.65 and 85; S7.15
(a) Vol.115,No.1, K121-K124

O.A. GOLIKOVA and A.M. GREKHOV
Structural Network and Some Photoelectrical and Optical Parameters of Amorphous Hydrogenated Silicon
71.20; 73.60; 78.65; S5.11
(b) Vol.155,No.2, 475-480

E. CHIMCZAK and M. BERTRANDT-ŻYTKOWIAK
Blocking Effect in Electroluminescence of ZnS:Mn Driven by Short Voltage Pulses
78.60; 78.65; S8.11
(a) Vol.115,No.2, 581-586

N.N. DINH, V.T. BICH, L.Q. MINH, and N.T. TOAN
Preparation of Lu-Diphthalocyanine Thin Films by Electron Beam Deposition and Electrochromic Effect
68.55 and 78.65; S12
(a) Vol.115,No.2, K163-K166

PREETI KASHYAP, MUKESH JAIN, O.P. AGNIHOTRI, and H.K. SEHGAL
Effect of Composition on the Optical and Electrical Parameters of Flash Evaporated $Cu_{2-y}In_yTe_2$ Thin Films
73.60 and 78.65; S8.16
(a) Vol.115,No.2, K181-K184

O. STENZEL, B. RAU, and G. SCHAARSCHMIDT
Optical Properties of Amorphous Carbon Deposited by an Ion-Plating Technique
78.65; S5
(a) Vol.115,No.2, K247-K250

I.YU. GOLINEY and V.N. RUDKO
Surface Electromagnetic Waves in Quasiperiodic Superlattices
78.65; S7.12; S7.15
(b) Vol.156,No.1, 211-224

A.K. ABASS and E.M. JABOORI
Optical Properties of Cr_2O_3 Coatings in the Fundamental Absorption Region
78.65; S10.1
(a) Vol.116,No.1, K111-K114

V.P. GRIBKOVSKII, L.G. ZIMIN, S.V. GAPONENKO, I.E. MALINOVSKII, P.I. KUZNETSOV, and G.G. YAKUSHCHEVA
Nonlinear Laser Spectroscopy of Exciton Absorption in MOCVD Grown ZnSe Monocrystalline Films
71.35 and 78.65; S8.12
(b) Vol.156,No.2, 449-454

A. MITTAS, N. GEORGOULAS, D. GIRGINOUDI, and A. THANAILAKIS
Properties of Amorphous Non-Hydrogenated Silicon-Tin Alloys
73.60 and 78.65; S6
(a) Vol.116,No.2, 725-728

78.70. Other interactions of matter with particles and radiation

E. SCHULZ, G. DRAEGER, W. CZOLBE, and O. BRUEMMER
X-Ray Spectroscopic Investigations of Titanium Oxides and Basic Materials for Ferroelectric Ceramics
78.70; 71.25; S10
(a) Vol.111,No.1, 361-366

LING CHEN, MINGKANG TENG, GUANGHOU WANG, XIAOYUN LI, and TIANCHANG LU
Positron Annihilation Lifetime Study of Positive Temperature Coefficient $BaTiO_3$
78.70; S11.1
(a) Vol.111,No.1, K123-K128

I.I. GEGUSIN, V.N. DATSYUK, A.D. YUMATOFF, and R.V. VEDRINSKII
Atomic and Solid State Effects in X-Ray Absorption Spectra of Ionic Crystals
71.25 and 78.70; S9.11
(b) Vol.151,No.2, 581-590

R.M. DE LA CRUZ and R. PAREJA
Porous Silicon Investigated by Positron Annihilation
61.70; 78.70; S5.11
(a) Vol.111,No.2, 463-468

K.P. AREFIEV, O.V. BOEV, and YU.P. SUROV
Positron Annihilation Processes in Mordenite Samples
78.70; S11
(a) Vol.111,No.2, K243-K248

G. ANTONIOLI, P.P. LOTTICI, and C. RAZZETTI
The Structure of the Defect Chalcopyrite $ZnGa_2Se_4$ Studied by EXAFS
61.60 and 78.70; S8.16
(b) Vol.152,No.1, 39-50

F.M. MOHAMMAD, B.K. SHARMA, H. SINGH, and B.L. AHUJA
Study of Electron Distribution in Rhodium by Compton Scattering Technique
71.25 and 78.70; S1
(b) Vol.152,No.1, 145-152

P.A. APANASEVICH, V.D. ASIMOVA, V.A. ZAPOROZHCHENKO, R.G. ZAPOROZHCHENKO, and I.V. PILIPOVICH
Energetic and Temporal Characteristics of Picosecond Pulse Self-Diffraction in Silicon
78.70; 71.25; S5.11
(b) Vol.152,No.1, 347-356

G.B. SMITH, D.R. MCKENZIE, and P.J. MARTIN
An XPS Study of Chemical Order in Hydrogenated Amorphous Silicon-Carbon Alloy Films
61.40 and 68.55; 78.70; S5; S5.11
(b) Vol.152,No.2, 475-480

V.V. NEMOSHKALENKO, P.K. NIKOLYUK, P.V. GEL, L.I. NIKOLAEV, V.KH. KASIYANENKO, A.V. KHRUSHCHAK, and B.P. MAMKO
X-Ray Spectroscopic Study of Electronic Structure of the Ce-Co-Si System
78.70; S1.61
(b) Vol.152,No.2, K55-K58

B.S. ACHARYA, S.K. SINGH, and B.B. NAYAK
X-Ray Emission Spectroscopy of the Electronic States in $Pb_{1-x}Hg_xS$ Thin Films
71.20 and 73.60; 78.70; S8.15
(b) Vol.153,No.1, 153-160

N.R. LOKHANDE, V.B. SAPRE, and C. MANDE
Energy Gap and Chemical Shift of X-Ray K Absorption Discontinuity of Arsenic in Its Compounds
71.25 and 78.70; S8; S10.1
(b) Vol.153,No.1, K35-K38

S. BALUCH, M. MIGLIERINI, R. GROENE, and J. SITEK
Positron Lifetime and Moessbauer Study of $Fe_{80-x}Ni_xB_{20}$ Metallic Glasses
76.80 and 78.70; S1.1
(a) Vol.113,No.1, K143-K146

S. SATPATHY and A.P. PATHAK
Planar Channeling of Electrons and Positrons in Crystals. Local-Density Calculation of Planar Averaged Potential and ..
61.80 and 78.70; S5; S5.11; S5.12
(b) Vol.153,No.2, 455-464

J. BODZENTA and Z. KLESZCZEWSKI
Theoretical Desciption of Light Diffraction on Acoustic Pulses in Braggs Region
62.65 and 78.70; 62.60
(b) Vol.153,No.2, 573-578

B.D. SHRIVASTAVA, S.K. JOSHI, and K.B. PANDEYA
X-Ray K-Absorption Studies of Cobalt(II) Complexes in Solid State
78.70; S12
(a) Vol.113,No.2, 603-608

C. Q. TANG and X. CHEN
Positron Lifetimes in Deformed Two-Phase Brass
61.70 and 78.70; S1.2
(a) Vol.113,No.2, K161-K164

C. NAGESWARA RAO and D.S.R. MURTY
Compton Cross-Sections of Molybdenum
78.70; S1.2
(b) Vol.154,No.1, K85-K88

S.J. WANG, X.H. LI, Y.L. CHEN, S.Q. LI, G.H. FONG, Z. WANG, A. CHEN, and B.R. LI
Structural Changes between 100 and 300 K of the High-T_c Superconductor Y-Ba-Cu-O Studied by Positron Annihilation
74.70 and 78.70; S10.15
(a) Vol.114,No.1, 273-276

G. ANTONIOLI, P.P. LOTTICI, and C. RAZZETTI
A Study of the Second and Third Neighbour Structure in Chalcopyrite $CuGaSe_2$ by EXAFS
61.60; 78.70; S8.16
(b) Vol.154,No.2, 461-468

T.-M. WANG, B.-Y. WANG, B.-X. FENG, C.-L. LIU, B.-H. JIANG, and Z.-Y. XU
The Recovery Behaviour of Quenched-in Vacancies in CuZnAl Alloy
61.70; 78.70; S1.2
(a) Vol.114,No.2, 451-456

V.N. BRUDNYI, A.D. POGREBNYAK, YU.P. SUROV, and A.S. RUDNEV
Positron Trapping at Point Defects in Electron-Irradiated GaAs
61.70; 61.80; 78.70; S7.12
(a) Vol.114,No.2, 481-490

DECHENG TIAN, ZHU-XIN YU, CAI-GUO ZHAN, and SHAO-LING WANG
Positron Annihilation Study of Ageing of an Al-Mg-Si-RE Alloy
61.70 and 78.70; 64.70; S4
(a) Vol.114,No.2, K135-K138

H. AOURAG, A. BELAIDI, T. KOBAYASI, R.N. WEST, and B. KHELIFA
Positron Annihilation in Si and Ge
71.10 and 78.70; S5.11; S5.12
(b) Vol.155,No.1, 191-200

YU.V. ALEKSANDROV, V.G. TSIRELSON, I.M. REZNIK, and R.P. OZEROV
The Crystal Electron Energy and Compton Profile Calculations from X-Ray Diffraction Data
71.45; 78.70
(b) Vol.155,No.1, 201-208

L.L. BALASHOVA, O.V. CHUMANOVA, and N.M. KABACHNIK
Z_2-Oscillations of the Proton Stopping Maximum in Gases and Solids
71.45 and 78.70; 79.20
(b) Vol.155,No.1, 289-294

R.K. SHUKLA, S. SWARUP, A. KUMAR, and A.N. NIGAM
Structural Studies in Glassy Semiconducting Se-Te Alloys
61.40 and 78.70; S4
(a) Vol.115,No.1, K105-K108

TRUONG BA HA and I.YA. DUBOVSKAYA
Diffraction X-Ray Radiation under Multiwave Diffraction
78.70
(b) Vol.155,No.2, 685-696

J. SERNA, J.CH. ABBÉ, and G. DUPLÂTRE
Size of Voids in Polyethylene
61.40 and 61.70; 78.70; S12
(a) Vol.115,No.2, 389-398

N.YU. ARUTYUNOV and V.YU. TRASHCHAKOV
1D-Angular Distributions of Annihilation Photons in RE-Ba-Cu-O Oxides (RE=Y,Nd,Lu)
74.70 and 78.70; S10.15
(a) Vol.115,No.2, K201-K204

V.H. ARAKELIAN and L.A. BAGDASSARIAN
Theory of Surface Polariton Excitation by Charged Particles in a Layer with Account of Retardation
71.36 and 78.70
(b) Vol.156,No.1, 275-286

E. DĘBOWSKA and J. RUDZINSKA-GIRULSKA
Densities of Interstitial Electrons in Pd(Ni)-Noble Metal Alloys
71.25; 78.70; S1.2; S1.3
(b) Vol.156,No.1, 303-308

A.M. AFANASEV, R.M. IMAMOV, E.KH. MUKHAMEDZHANOV, and A.A. BZHEUMIKHOV
Double-Channel X-Ray Standing Wave Technique
61.10; 78.70; S7.12
(a) Vol.116,No.1, 197-202

HONG ZHANG, XIAO-GANG WANG, LE LUO, and SHU-HUI HU
A Study on the Positron Annihilation of Glassy $Bi_1Sr_1Ca_1Cu_2O_x$ Quenched from the Molten Liquid
61.40 and 78.70; 74.70; S10.15
(a) Vol.116,No.1, K7-K12

I.YA. DUBOVSKAYA and G.K. SAVCHUK
The Parametric X-Ray Radiation in Case of Three-Wave Asymmetric Diffraction
78.70; S5.11
(b) Vol.156,No.2, 687-696

A.H. ROSTOMYAN, P.H. BEZIRGANYAN, and A.M. ROSTOMYAN
X-Ray Resonators. I.Theory
61.10 and 78.70
(a) Vol.116,No.2, 483-492

A.M. ROSTOMYAN, A.H. ROSTOMYAN, and P.H. BEZIRGANYAN
X-Ray Resonators. II. Calculation and Experiment
61.10 and 78.70
(a) Vol.116,No.2, 493-502

J.J. DONAGHY and K.E. VAN NESS
A Trapping Model Analysis of the Positron Lifetime Spectra of Polymers
78.70; S12
(a) Vol.116,No.2, K213-K218

78.90. Other topics in optical properties of condensed matter and other interactions of matter with particles and radiation

A.G. FOKIN and T.D. SHERMERGOR
Electromagnetic Wave Scattering in Inhomogeneous Media with Spatial Dispersion
77.20 and 78.90
(b) Vol.154,No.1, 367-376

79. Electron and Ion Emission by Liquids and Solids; Impact Phenomena

79.20. Impact phenomena (including electron spectra and sputtering)

L.B. GUOBA, A.A. VITKAUSKAS, J.V. KAMENECKAS, V.R. SARGUNAS, and A.P. SALALAS
The Influence of Ion-Irradiation on Interdiffusion and Annealing Behavior of Ohmic Contacts to GaAs
68.48; 79.20; S7.12
(a) Vol.111,No.2, 507-514

K. PAPROCKI, I. BRYŁOWSKA, W. ANDRAE, and J. KRYNICKI
Recoil Implantation from Sb Thin Films under Ion Bombardment
61.80 and 79.20; S5.11
(a) Vol.112,No.1, 301-304

W. SZYSZKO and J. SIELANKO
Monte Carlo Calculation of Depth Distribution of Implanted Ions and Sputtering Yield. Comparison for Some of the ...
61.80 and 79.20; S5; S5.11
(a) Vol.112,No.1, 315-318

J. SIELANKO
Influence of the Primary Ion Beam on the Depth Distribution of Implanted Atoms Measured by SIMS
61.80; 79.20; S5.11
(a) Vol.112,No.1, 319-322

W. SOSZKA
Krypton Saturation of the Metal Surface during Kr^+ Ion Bombardment Determined Due to Autoionization Electron Analysis
61.80; 68.20; 79.20; S1
(a) Vol.112,No.1, 425-430

P. CZUBA, A. PORADZISZ, Z. POSTAWA, T. DOHNALIK, and M. SZYMONSKI
Electronic and Cascade Processes in Ion-Induced Sputtering of Alkali Halides
79.20; S9.11
(a) Vol.112,No.1, 431-436

A.I. BAZHIN, V.V. KOZEL, V.P. SHESTOV, M. SUCHAŃSKA, and S.V. TEPLOV
Temperature Dependence of Energy Distribution of Excited Atoms Sputtered from an Ion-Bombarded Ni(100) Surface
79.20; 75.50; S1.2
(a) Vol.112,No.1, 437-440

P.W. ZUKOWSKI, S.B. KANTOROW, D. MĄCZKA, and W.F. STELMAKH
Procecces of Radiation Defect Interaction and Amorphisation of Silicon at Large Implantation Doses
61.70; 79.20; S5.11
(a) Vol.112,No.2, 695-698

YA.V. FATTACHOV, I.B. KHAIBULLIN, R.M. BAYAZITOV, and E.M. MISYURYOV
Anisotropic Local Melting of Semiconductors by Incoherent Light Pulses
79.20; 64.70; S5.11; S5.12
(a) Vol.112,No.2, 699-702

K. GAERTNER, G. GOETZ, C. KASCHNER, I. KASKO, A. WITZMANN, and A. ZENTGRAF
Laser Induced Silicide Formation
68.55 and 79.20; 64.60; S1.61
(a) Vol.112,No.2, 747-752

R. KUDUK, M. KULIK, D. MĄCZKA, K. PYSZNIAK, and J. ŻUK
Depth Profile Measurements of Ions Implanted in Semiconductor Materials by Particle Induced Photon Emission
61.70; 78.60; 79.20
(a) Vol.112,No.2, 799-804

A.YU. BASAI, V.V. KAPLIN, E.I. ROZUM, A.M. SLUPSKII, and S.A. VOROBIEV
Autodeflection of Electron Beams from Dielectric Solids
79.20
(a) Vol.112,No.2, K131-K134

A.R. URMANOV, V.G. GRIGORIEV, and A.A. PUZANOV
A Kinetic Theory for the Motion of Accelerated Particles in Solids. II. Dechanneling in Complex Crystals
61.80 and 79.20; S10
(b) Vol.153,No.1, 65-78

YU.V. LILENKO, K.V. SHASTOV, A.S. PETROV, A.V. VOITSEKHOVSKII, V.S. KULIKAUSKAS, V.N. KUZNETSOV, and A.P. MAMONTOV
Radiation-Induced Defects in Implanted $Hg_{1-x}Cd_xTe$ Crystals
61.80; 72.20; 79.20; S8.15
(a) Vol.113,No.2, 285-294

G. KLUGE, A. THOMAS, R. KLABES, and P. SUEPTITZ
Applicability of Silver Implanted Amorphous $GeSe_2$ as Inorganic Negative Resist
61.80; 66.30; 79.20; S8
(a) Vol.113,No.2, K171-K176

V.G. BESHENKOV, CH.V. KOPETSKII, and YU.A. SHIYANOV
Phase Composition of Thin Films as Revealed by Auger Depth Profiling
64.80 and 68.55; 79.20; S1.61
(a) Vol.114,No.1, 191-198

D. CZEKAJ, E.K. HOLLMANN, A.B. KOZIREV, V.A. VOLPYAS, and A.G. ZAITSEV
Initial Energy Distribution of Sputtered Atoms in DC Diode Systems
79.20; S1.2
(a) Vol.114,No.1, K21-K24

R. HEYNER and G. MARX
Depth Profiles and Surface Characterization by Glow Discharge Optical Emission Spectroscopy
68.20; 79.20; S1.2
(a) Vol.114,No.2, 609-620

H.-J. AUGUST and J. WERNISCH
Analytical Expressions for the Electron Backscattering Coefficient
79.20; S1; S5; S5.11
(a) Vol.114,No.2, 629-634

H. BUDINOV, V. STAVROV, and R. BURKOVA
Flash Lamp Annealing of Phosphorus-Implanted Silicon
61.70; 79.20; S5.11
(a) Vol.114,No.2, K131-K134

L.L. BALASHOVA, O.V. CHUMANOVA, and N.M. KABACHNIK
Z_2-Oscillations of the Proton Stopping Maximum in Gases and Solids
71.45 and 78.70; 79.20
(b) Vol.155,No.1, 289-294

S. BAUNACK and A. ZEHE
A Study of UV/Ozone Cleaning Procedure for Silicon Surfaces
68.20; 68.55; 79.20; S5.11
(a) Vol.115,No.1, 223-228

L.N. ALEKSANDROV, V.YU. BALANDIN, A.V. DVURECHENSKII, and O.A. KULYASOVA
A Two-Dimensional Model of SOI Structure Crystallization by Pulse Nanosecond Heating
68.55 and 79.20; S5.11; S10.1
(a) Vol.115,No.1, K23-K26

H. LANGE, H. REIS, and F. FENSKE
AES Study of the Interaction of Ni and Al Overlayers with W-Ti and W-Re Diffusion Barriers
68.55 and 79.20; S1.2
(a) Vol.115,No.2, 497-504

E.N. IZAKOVICH, N.A. ZHELTOVA, YU.M. SHULGA, I.N. IVLEVA, ZH.D. SOKOLOVSKAYA, A.V. BALUEV, V.V. GANIN, O.S. ROSHUPKINA et al.
Study of Ba and Cu(III) Based Inorganic Oxide Compounds
74.70; 78.30; 79.20; 79.60
(a) Vol.115,No.2, K195-K200

M. ROESLER and W. BRAUER
Influence of Dynamic Screening on the Electron Yield-Stopping Power Relation for Proton Impact in Aluminum
79.20; S4
(b) Vol.156,No.1, K85-K88

C.-M. LIEGENER
Calculation of the Auger Spectra of Clusters Modelling Solid Hydrogen Fluoride
79.20; S9
(b) Vol.156,No.2, 441-448

79.40. Thermionic emission

J. DĄBEK and B. ADAMCZYK
Al^{+} Beam Generation by Means of the Pencil Type Thermoemission Ion Source
79.40
(a) Vol.112,No.2, 829-834

79.60. Photoemission and photoelectron spectra

A. HUPFER, D. HIRSCH, and S. SCHULZE
Photoemission on $A_3^{II}B_2^{V}$ Semiconductor Material
71.20 and 79.60; S7
(b) Vol.152,No.2, 505-518

A.E. BOCQUET, J.F. DOBSON, P.C. HEALY, S. MYHRA, and J.G. THOMPSON
X-Ray Photoelectron Spectra of High T_c Bismuth Compounds: General Trends for Layer Structures
71.45 and 79.60; 74.10; S11
(b) Vol.152,No.2, 519-532

H. SOMMER, D.W. BULLETT, H. NEUMANN, and G. KUEHN
Valence Band Density of States in $LiInSe_2$
71.20; 79.60; S8.16
(b) Vol.152,No.2, K47-K50

W. UNGER and W. POSSART
An XPS Investigation of the Interface between Thin Polymethylmethacrylate (PMMA) Films and Natively Oxidized Al
68.48 and 79.60; S4; S12
(a) Vol.114,No.2, K175-K178

S.X. WANG, Z.H. HE, X.H. CHEN, M.R. JI, Y.T. QIAN, Z.Y. CHEN, and Q.R. ZHANG
The Effect of Substitution of Cd for Ca in the Bi-Sr-Ca-Cd-Cu-O System Studied by X-Ray Photoemission Spectroscopy
74.70 and 79.60; S10.15
(a) Vol.115,No.1, K71-K74

G.P. NIZHNIKOVA and O.V. FARBEROVICH
Theoretical Investigation of XPS, UPS, and BIS in Rare-Earth Compounds - Thulium Monochalcogenides
71.20; 71.50; 79.60; S1.63
(b) Vol.155,No.2, 491-500

V.K. ADAMCHUK, A.E. KUCHMA, and S.L. MOLODTSOV
Effects of Low-Energy Electron Transport in Insulators
71.25; 79.60; S9.11
(b) Vol.155,No.2, 525-534

E.N. IZAKOVICH, N.A. ZHELTOVA, YU.M. SHULGA, I.N. IVLEVA, ZH.D. SOKOLOVSKAYA, A.V. BALUEV, V.V. GANIN, O.S. ROSHUPKINA et al.
Study of Ba and Cu(III) Based Inorganic Oxide Compounds
74.70; 78.30; 79.20; 79.60
(a) Vol.115,No.2, K195-K200

R. MAMY and X. ZAOUI
Growth of Au on MoS_2 Studied by Photoemission
68.55 and 79.60; S8
(b) Vol.156,No.2, K109-K112

79.70. Field emission and field ionization

K. KORNARZYNSKI, K. KISZCZAK, A. LATUSZYŃSKI, D. MĄCZKA, J. PARTYKA, and YU.V. YUSHKIEVICH
The Generation of Molecular and Muliply-Charged Ions in the Hollow-Cathode Ion Source
79.70
(a) Vol.112,No.2, 815-820

J. MITTENBACHER and R. MUEHLE
A Scanning Ion Microbeam System with a Liquid Metal Ion Source
61.80 and 79.70; 61.16; S1.3; S4
(a) Vol.112,No.2, 825-828

79.75. Exoelectron emission

T. KUBOZOE and Y. WATANABE
Thermally Stimulated Exoelectron Emission from LiF Single Crystals Excited with Electrons and X-Rays and Subsequently with UV Light
71.55 and 79.75; S9.11
(a) Vol.111,No.2, K201-K204

V. BICHEVIN, G. ALSEITOV, M. GAPANOV, and H. KAAMBRE
Excimer-Laser-Induced Exoemission and Thermoluminescence of Some Alkali Halides
78.60 and 79.75; 61.80; S9.11
(a) Vol.114,No.2, K241-K244

Y. FUKUDA, A. TOMITA, and N. TAKEUCHI
Thermoluminescence and Thermally Stimulated Exoelectron Emission of Sintered MgB_4O_7 Doped with $DyCl_3$ and Tm_2O_3
78.60 and 79.75; S10
(a) Vol.114,No.2, K245-K248

V. BICHEVIN and H. KÄÄMBRE
Low-Temperature Thermo- and Photostimulated Exoemission (TSEE and PSEE) of KBr
79.75; S9.11
(a) Vol.115,No.1, K109-K112

Y. FUKUDA, A. TOMITA, and N. TAKEUCHI
Thermoluminescence and Thermally Stimulated Exoelectron Emission of $Ca_2B_5O_9Cl$ Activated with Divalent Europium or Lead
78.60 and 79.75; S9.16
(a) Vol.115,No.2, K251-K254

H. OTOP
Some Experimental Observations of Exoemission from Ammonium Hydrogen Selenates
77.80 and 79.75; S12
(a) Vol.116,No.1, K115-K118

79.90. Other topics in emission and impact phenomena of condensed matter

A. HORODENSKI
Evaluation of Pulse Shape of Ion Beams Produced by the Ionotron-Type Ion Sources
79.90
(a) Vol.112,No.2, 821-824

85. Devices

C.D. LOKHANDE and S.H. PAWAR
Electrodeposition of Thin Film Semiconductors (Review Article)
68.55; 73.40; 85; S8
(a) Vol.111,No.1, 17-40

I. ZÓLOMY
Approximate Calculation of the Switching Voltage of MISS
73.40 and 85
(a) Vol.111,No.1, 371-376

H.-J. WUENSCHE and H. WENZEL
Influence of Surface Recombination on the Rate of Electron Leakage in Ridge-Waveguide Injection Lasers
73.40 and 85; S7.11; S7.15
(a) Vol.111,No.1, 377-386

H. PAGNIA and N. SOTNIK
Metal Influence on Switching MIM Diodes
73.40 and 85; S1.3; S6; S10
(a) Vol.111,No.1, 387-394

R. FLAGMEYER, G. OELGART, K. KREHER, R. SCHWABE, F. BUGGE, L. LEHMANN, and K. JACOBS
Characterization of MOCVD Grown (Al,Ga)As/GaAs Single Quantum Well Structures by Rutherford Backscattering and Photoluminescence
73.40 and 85; 78.55; S7.12
(a) Vol.111,No.2, 541-550

M.V.S. LAKSHMI and K. RAMKUMAR
I-U Characteristics of MOS Structures on Polycrystalline Silicon
73.40 and 85; S5.11
(a) Vol.111,No.2, 667-674

J. SUÑÉ, I. PLACENCIA, N. BARNIOL, E. FARRÉS, and X. AYMERICH
Degradation and Breakdown of Gate Oxides in VLSI Devices
73.40 and 77.50; 73.60; 85; S10
(a) Vol.111,No.2, 675-686

K. NATARAJAN, K. RAMKUMAR, and M. SATYAM
Capacitance of p-n Junctions under Electrical Breakdown
73.40 and 85; S5.11
(a) Vol.111,No.2, K269-K272

W.I. KHAN
A New Model for the Negative Resistances in Semiconductor Devices
73.40 and 85; S7.12
(a) Vol.112,No.1, 221-228

S.R. BLIGHT and H. THOMAS
Observation of Bulk Traps in Ion-Implanted GaAs MESFETs
71.55; 85; S7.12
(a) Vol.112,No.1, K29-K34

K. ZDANSKY
Threshold Current Optimization of InGaAsP/InP Lasers with Waveguiding Layers
73.40 and 85; S7.15
(a) Vol.112,No.2, 639-650

J.J. LIOU
Theoretical Study of Forward-Voltage Heterojunction Space-Charge-Region Capacitance Including Interface-State Effects
73.40 and 85; S7.12
(a) Vol.112,No.2, 651-660

T. SKRABKA, M.J. SŁABY, and B. MIZERA
Ohmic Contacts to n-Type GaAs Using RTA
73.40; 85; S7.12
(a) Vol.112,No.2, K105-K110

J. BELAN and V. MIKOLAJ
Voltage Noise of the Breakdown Region of Subsurface Zener Diodes
72.70; 85; S5.11
(a) Vol.112,No.2, K119-K120

V.YU. RUD, YU.V. RUD, M. SERGINOV, and M.A. TAIROV
Production of and Physical Processes in n-p-$CdSiAs_2$ Structures
72.40 and 73.40; 85; S7.16
(a) Vol.113,No.1, 207-214

R.V. KONAKOVA, YU.S. MELNIKOVA, E.V. MOZDOR, and V.I. FINEBERG
Breakdown of Silicon p^+-n-n^+ Diodes
73.40 and 85; S5.11
(a) Vol.113,No.1, 215-222

G. GHIBAUDO
Analytical Modelling of the MOS Transistor
73.40 and 85
(a) Vol.113,No.1, 223-240

N.M. FICZA and A.L. TOTH
The Effect of Lead Impurity on Amorphous Si Thin Film Transistors
61.70 and 85; 68.55; S5.11; S10.1
(a) Vol.113,No.1, K147-K148

B. MITRA, A. GHOSHAL, and K.P. GHATAK
On a Simplified Analysis of the Gate Capacitance of MOS Field-Effect Devices in the Presence of a Tilted Quantizing Magnetic Field
73.40 and 85
(b) Vol.153,No.2, K209-K212

V.S. LYSENKO, A.N. NAZAROV, S.A. VALIEV, I.M. ZARITSKII, T.E. RUDENKO, and A.S. TKACHENKO
EPR and TSCR Investigations of Implanted $Al-SiO_2-Si$ Systems Treated with RF Plasma Discharge
73.40 and 85; 76.30; S5.11; S10.1
(a) Vol.113,No.2, 653-666

M. JAYACHANDRAN, M.J. CHOCKALINGAM, and V.K. VENKATESAN
Characteristics of Electrodeposited SnO_2-CdS/Au Schottky Devices
73.30; 85; S8.11
(a) Vol.113,No.2, K217-K222

J.J. LIOU, A. WHITTAKER, and J.S. YUAN
Modeling the Two-Dimensional Emitter-Base and Base-Collector Junction Capacitances of Bipolar Junction Transistors
73.40 and 85
(a) Vol.113,No.2, K267-K272

P.S. GHOSH, D.K. GHOSH, and L.K. SAMANTA
Studies of Optical Properties and Applications of Some Mixed Ternary Semiconductors
78.20; 85; S7.15
(a) Vol.114,No.1, 381-386

G. OELGART, V. GOTTSCHALCH, H. HAEFNER, R. HEILMANN, W. KLEIN, and B. RHEINLAENDER
Characterization of GaInAsP/InP Double Heterostructure Laser-Wafers
73.40 and 85; 78.60; S7.11; S7.15
(a) Vol.114,No.1, 419-430

P. BROGOWSKI, H. MUCHA, and J. PIOTROWSKI
Modification of Mercury Cadmium Telluride, Mercury Manganese Telluride, and Mercury Zinc Telluride by Ion Etching
61.80 and 72.80; 85; S8.15
(a) Vol.114,No.1, K37-K40

H.V. ANANDA and M. SATYAM
Effect of Field Plate on the Electrical Breakdown Characteristics of Bipolar Transistors
73.40 and 85; S5.11
(a) Vol.114,No.2, K259-K262

Ľ. MALACKY, J. KUZMIK, and V. CAMBEL
Optical Properties of Metal/GaInAs Contacts for Schottky Photodiodes
78.65 and 85; S7.15
(a) Vol.115,No.1, K121-K124

B.C. CHAKRAVARTY, S.K. SHARMA, S.N. SINGH, B.K. DAS, RAVI KUMAR, and B.R. CHAKRABORTY
Effect of the Growth of Thin SiO Interlayers on the Performance of Si Solar Cell
68.55 and 85; 73.40; S5.11
(a) Vol.115,No.1, K125-K128

I. ZOLOMY
Theory of the Negative Resistance Avalanche MIS Tunnel Diode
73.40 and 85; S5.11
(a) Vol.115,No.1, K129-K132

H. BERGNER and T. DAMM
Binary States in Integrated Circuits Investigated by Laser Scanning Microscopy with Optical Beam Induced Current
72.40 and 85; S5.11
(a) Vol.115,No.2, 607-616

E. NEBAUER and U. LOHSE
Lamp Furnace-Annealed Au/Ni/AuGe Ohmic Contacts for GaAs MESFETs
73.40; 85; S7.12
(a) Vol.115,No.2, K171-K174

K. NATARAJAN, K. RAMKUMAR, and M. SATYAM
Reverse Characteristics of Phosphorous Passivated Polysilicon p-n Junction Diodes
73.40 and 85; 61.70; S5.11
(a) Vol.115,No.2, K265-K268

N. MAZUMDER, J.P. BANERJEE, and S.K. ROY
Variation of High Frequency Negative Resistance of Silicon n^+pp^+ and GaAs p^+nn^+ IMPATT Diodes with Enhancement of Reverse Saturation ...
73.40 and 85; S5.11; S7.12
(a) Vol.116,No.1, 415-424

V.K. GOYAL, D. KUMAR, and K. SEN
Effect of Grain Boundary States Density in the Polycrystalline MIS Solar Cells
73.20 and 85; 61.70; 72.40
(a) Vol.116,No.1, 425-432

P.C. EUTHYMIOU, P.C. BANBURY, G.J. PAPAIOANNOU, and G.E. ZARDAS
Dependence of I-V Characteristics of GaAs MESFET on Temperature and Eletron Irradiation
73.40 and 85; 61.80; S7.12
(a) Vol.116,No.1, K133-K136

V.A. GURTOV and P.A. RAIKERUS
Radiation Charge in Silicon Dioxide Considering Diffusion Currents
72.20; 71.55; 85; S5.11
(a) Vol.116,No.2, 769-776

B. MAJKUSIAK
A Modified Method for Determination of Surface Mobility of Carriers in MOS Transistors
73.40 and 85
(a) Vol.116,No.2, 833-836

R. AGRAWAL, R. DWIVEDI, and S.K. SRIVASTAVA
Analysis of an MOS Inverter with Non-Uniform Impurity Profile
73.40 and 85
(a) Vol.116,No.2, 837-844

U. BRAATZ, J. KUNSCH, E. KRAUSE, and E. JAEGER
Sweep Drive Circuit for Streak Cameras Based on ZnS and ZnSe Optoelectronic Switches
72.40 and 85; S8.11; S8.12
(a) Vol.116,No.2, K219-K222

I. ZOLOMY
Erratum to Theory of the Negative Resistance Avalanche MIS Tunnel Diode (phys. stat. sol. (a) 115, K129(1989))
73.40 and 85; S5.11
(a) Vol.116,No.2, K223-K224

S. Substances
(classification according to physica status solidi)

S1. Transition metals and their alloys and compounds
(for oxides, see S.10)

LANPING YUE
Effect of Heat Treatment on the Internal Friction Peaks of a-PdNiSi near T_g and T_x
62.40; S1
(a) Vol.111,No.2, K155-K160

A. SZASZ and D.J. FABIAN
Subsurface Hydrogen Exciton-Like States in Solid Catalysts
68.45 and 73.20; 71.35; S1
(b) Vol.152,No.1, 117-122

F.M. MOHAMMAD, B.K. SHARMA, H. SINGH, and B.L. AHUJA
Study of Electron Distribution in Rhodium by Compton Scattering Technique
71.25 and 78.70; S1
(b) Vol.152,No.1, 145-152

M. BUCKI, A. GEMPERLE, S.A. KONSTANTINOV, M.M. KANTOR, F. CERNY, K. KULT, G.S. BURCHANOV, and E.V. OTTENBERG
Modification of Surface Layers of Molybdenum by B^+ and N^+ Ion Implantation
61.80 and 68.20; S1
(a) Vol.112,No.1, 335-338

W. SOSZKA
Krypton Saturation of the Metal Surface during Kr^+ Ion Bombardment Determined Due to Autoionization Electron Analysis
61.80; 68.20; 79.20; S1
(a) Vol.112,No.1, 425-430

M.V. MAGNITSKAYA, N.I. KULIKOV, and A.L. SHELEKHOV
Frequency-Modulated Optical Spectroscopy of Niobium: Theory and Experiment
78.20; S1
(b) Vol.152,No.2, 685-690

S.N. ROMANOV, A.I. MELKER, and D.B. MIZANDRONTSEV
Athermal Clusters of Vacancies in B.C.C. and H.C.P. Metals
61.70; 61.80; S1; S1.1
(b) Vol.153,No.1, K5-K10

S. SUZUKI
Hydrogen Cold-Work Peak in High Purity Molybdenum
61.70 and 62.40; S1
(a) Vol.113,No.1, 51-56

K. TOMPA, R. PIRLOT, L. GRENACS, and J. LEHMANN
Hyperfine Fields at Implanted ^{12}Boron in F.C.C. Metals
76.60; 71.50; S1; S1.3; S4
(b) Vol.153,No.2, 721-726

A. MORAWIEC
Calculation of Polycrystal Elastic Constants from Single-Crystal Data
62.20; S1; S10
(b) Vol.154,No.2, 535-542

H.-J. AUGUST and J. WERNISCH
Analytical Expressions for the Electron Backscattering Coefficient
79.20; S1; S5; S5.11
(a) Vol.114,No.2, 629-634

J. STEUER
Comparison of A-15 Nb_3Sn Films Deposited on Hot and Cold Substrates
74.10 and 74.50; S1
(a) Vol.114,No.2, 665-678

K. KNEIPP, H. BOCK, and W. SALEWSKI
Raman Study of Thorium Tungsten Cathodes
78.30; S1
(a) Vol.114,No.2, K255-K258

J. GONZALEZ ESTEVEZ and E. DU TREMOLET DE LACHEISSERIE
Influence of Annealing on the Magnetostriction of an Amorphous
$Co_{45}Ni_{30}Si_{15}B_{10}$ Alloy
75.50 and 75.80; S1
(a) Vol.115,No.2, K233-K236

M. OKUMURA, H. MIZUBAYASHI, S. OKUDA, and Y. TAGISHI
Internal Friction Study on Anomalous Self Interstitial
Atom-Dislocation Interaction in Nb after Low Temperature Irradiation
61.80 and 62.40; S1
(a) Vol.116,No.2, 583-594

S1.1. Fe

YAN SHI, HANMIN JIN, and MINGZHI JIN
Interpretation of ^{57}Fe Moessbauer Measurements on $Nd_2Fe_{14}B$ at 83 and
295 K
76.80; S1.1; S1.4
(b) Vol.151,No.1, 249-258

M. KLOSEK, J.W. MORON, and T. POLOCZEK
Activation Energy of Linear Heating Crystallization in
$Ni_{70.3}Fe_{4.7}Si_{15}B_{10}$ Glassy Alloy Determined Using the Internal ...
61.40 and 64.70; 62.40; S1.1; S1.2
(a) Vol.111,No.1, 127-136

R. PASIANOT, G. GALAMBOS, E.J. SAVINO, L. AMARAL, M. BEHAR, D. FARKAS, and S. RAO
Argon Irradiation of Sn Thin Layers Deposited on Fe Substrates
61.80 and 68.20; S1.1; S4
(a) Vol.111,No.1, 173-180

T. REININGER, N. MOSER, A. HOFMANN, and H. KRONMUELLER
The Kinetics of the Magnetically Induced Anisotropy in Amorphous Ferromagnetic $Co_{71}Fe_1Mo_1Mn_4Si_{14}B_9$ and $Fe_{41}Ni_{41}B_{18}$ Alloy
75.50; S1.1; S1.2
(a) Vol.111,No.1, 243-256

J. GUTIÈRREZ, J.M. BARANDIARAN, and O.V. NIELSEN
Magnetoelastic Properties of Some Fe-Rich Fe-Co-Si-B Metallic Glasses
75.80; S1.1; S1.2
(a) Vol.111,No.1, 279-284

P. STEFANSKI and A. WRZECIONO
The Rare Earth Contribution to the Magnetocrystalline Anisotropy in $RFe_{10}V_2$
75.30; S1.1
(b) Vol.151,No.2, K195-K200

NGUYEN HOANG LUONG, NGUYEN PHU THUY, LUU TUAN TAI, and THAN DUC HIEN
Rare Earth Magnetocrystalline Anisotropy in $R_2Fe_{14}B$ (R=Sm,Er,Tm,Yb)
75.30; 75.25; S1.1; S1.4
(a) Vol.111,No.2, 591-596

M. JERGEL, G. VLASĀK, and P. DUHAJ
Magnetostriction Study of Some Co-Based and Fe-Based Amorphous Alloys
75.80; S1.1; S1.2
(a) Vol.111,No.2, 597-602

L. POTOCKY, P. KOLLAR, E. KISDI-KOSZO, Z. JURANEK, and J. DEGRO
Surface Coercitive Force of Some Metallic Glasses
75.50; S1.1
(a) Vol.111,No.2, 603-610

J. RIVAS, M.A. LOPEZ-QUINTELA, F. WALZ, and H. KRONMUELLER
Fractal Analysis of Magnetic After-Effects in the Amorphous Alloy $Fe_{40}Ni_{40}P_{14}B_6$ at 500 K
75.60; 75.50; S1.1; S1.2
(a) Vol.111,No.2, 611-616

A. SCHUBERT and B. MICHEL
Simultaneous Line Broadening and Texture Analysis
61.10 and 68.20; S1.1
(a) Vol.111,No.2, K137-K140

G. BALASUBRAMANIAN, A.N. TIWARI, and C.M. SRIVASTAVA
Annealing Effects in the Metallic Glass $Fe_{65}Cr_8B_{27}$
75.80; 76.50; S1.1
(a) Vol.111,No.2, K229-K232

G.E. ABROSIMOVA, A.P. ZHUKOV, and B.K. PONOMAREV
Effect of Heat Treatment on Some Properties of Fe- and Co-Based Amorphous Alloys
75.60; S1.1
(a) Vol.111,No.2, K237-K242

J. VAN DER STEL, H.B. VAN VELDHUIZEN, G.W. KOEBRUGGE, J. SIETSMA, and A. VAN DEN BEUKEL
Change of Electrical Resistivity and Youngs Modulus during Crystallization of Amorphous $Fe_{40}Ni_{40}B_{20}$
61.40 and 64.70; 62.20; 72.15; S1.1; S1.2
(a) Vol.112,No.1, 35-40

E.A. LITTLE, D.J. BUTTLE, and C.B. SCRUBY
Radation Damage Studies in Model Ferritic Alloys Using Micromagnetic Techniques
61.80; 75.50; S1.1
(a) Vol.112,No.1, 55-72

M.A.Z. VASCONCELLOS, J.A.T. BORGES DA COSTA, W.H. SCHREINER, and I.J.R. BAUMVOL
A Comparison of Thermal Annealing and Ion Beam Mixing in the Fe/Al Bilayer
68.48 and 68.55; S1.1; S4
(a) Vol.112,No.1, 105-114

KH.YA. MULYUKOV, R.Z. VALIEV, G.F. KORZNIKOVA, and V.V. STOLYAROV
The Amorphous $Fe_{83}Nd_{13}B_4$ Alloy Crystallization Kinetics and High Coercivity State Formation
75.50; 61.40; S1.1
(a) Vol.112,No.1, 137-144

X.C. KOU, X.K. SUN, Y.C. CHUANG, T.S. ZHAO, R. GROESSINGER, R. EIBLER, and H.R. KIRCHMAYR
Anisotropy Energy of Nd Ion in $Nd_2Fe_{12-x}Co_2Al_xB$ Compounds
75.50; S1.1; S1.2; S1.4
(a) Vol.112,No.1, 175-180

W. ROSINSKI and J. KACZMAREK
Present Position and Further Developments of Ion Implantation in Metals
61.80 and 68.20; S1.1
(a) Vol.112,No.1, 245-252

A. TUROS
Ion Implantation Studies of Collision Cascades in Solids
61.70; S1.1; S1.2; S1.3; S1.4
(a) Vol.112,No.1, 253-262

T. OTMIANOWSKI and B. KAMIENSKA-STEFANIUK
Some Aspects of the Surface Wear Process of Steel Implanted with Nitrogen
61.80 and 68.20; S1.1
(a) Vol.112,No.1, 339-342

J. JAGIELSKI, G. GAWLIK, A. PODGORSKI, A. TUROS, A. DYGO, and N. MADI
Thermal Effects on Depth Distribution of Nitrogen Atoms Implanted in Ferrous Alloys
61.80; 61.70; S1.1
(a) Vol.112,No.1, 343-346

T. BARBASZEWSKI, M. DABROWSKI, M. DRWIEGA, J. GAWLIK, and E. LIPINSKA
Implantation Profiles of Nitrogen and Titanium in a Low Energy Bombarded High Speed Tool Steel
61.80 and 68.20; S1.1
(a) Vol.112,No.1, 347-352

P. TARKOWSKI, I. BRYLOWSKA, and M. SUBOTOWICZ
Influence of Nitrogen Implantation on Corrosion Process in Steel
61.80; 68.20; S1.1
(a) Vol.112,No.1, 353-356

J. JĘDRKOWIAK, J. MARTAN, J. MASALSKI, and D.B. BOGOMOLOV
Influence of Ar, Al, and N Ion Implantation on the Surface Properties of the OOH18M2Nb Steel
61.80; 68.20; S1.1
(a) Vol.112,No.1, 357-360

V.L. KUTUZOV, M.YU. OVSYANNIKOV, I.G. ROMANOV, A.D. POGREBNYAK, and G.E. REMNEV
Mechanical and Frictional Properties of Tool Steels Exposed to HPIB Irradiation
62.20; 61.80; S1.1
(a) Vol.112,No.1, 361-364

J. ŁUNARSKI and M. ZIELECKI
Relationship between Microhardness and Fatigue Strength after Glass Micro-Bead Peening and Ion Implantation Processes
62.20; 61.80; S1.1
(a) Vol.112,No.1, 365-374

J. PAVLOVSKÝ and T. ZEMCIK
Comparison of Surface and Bulk Crystallization of the Amorphous $Fe_{70}Co_{10}B_{20}$ Alloy
61.60; 76.80; S1.1
(a) Vol.112,No.1, K1-K4

E. HENSEL, H. SOMMER, P. KNOTHE, and E. RICHTER
Silicon Nitride Layers on Tool Steel Produced by Ion Beam Mixing and Ion Beam Assisted Deposition
68.48 and 68.55; S1.1; S7
(a) Vol.112,No.2, 533-540

WEN-CAI WANG and YU CHEN
Study on the Short-Range Order Structures of Fe-Ge Amorphous Thin Films by EXAFS
61.40 and 68.55; S1.1; S5.12
(a) Vol.112,No.2, 541-550

M. SOWA, W. SZYSZKO, L. GŁUSIEC, J. TRZCIANKA, and J. ZINKIEWICZ
Diffusion of Indium Implanted into Steel
66.30; S1.1
(a) Vol.112,No.2, 835-838

M. SOWA, W. SZYSZKO, J. SIELANKO, L. GŁUSIEC, and B. KAMIENSKA
Microhardness of Boron, Titanium, and Nitrogen Implanted Steel
62.20; 66.30; S1.1
(a) Vol.112,No.2, 839-844

A.P. ZHUKOV and B.K. PONOMAREV
Start Field Distribution in a Bistable Amorphous Ribbon
75.60; S1.1
(a) Vol.112,No.2, K127-K130

QIANG WANG, XIAO-WEN ZHANG, and BING-LIN GU
Order-Disorder Model of Phase Transition in A_xB_{1-x} Binary Systems of F.C.C. Structure
64.60; 64.75; S1.1; S1.2
(b) Vol.153,No.1, 139-144

S.N. ROMANOV, A.I. MELKER, and D.B. MIZANDRONTSEV
Athermal Clusters of Vacancies in B.C.C. and H.C.P. Metals
61.70; 61.80; S1; S1.1
(b) Vol.153,No.1, K5-K10

A.A. HOSSEINI and I.P. JONES
The Radiation Induced Segregation of Nickel in B.C.C. Iron Alloys
64.75; 61.80; S1.1
(a) Vol.113,No.1, 57-62

A. SIEMKO, H.K. LACHOWICZ, N. MOSER, A. FORKL, and H. KRONMUELLER
Stress-Induced Sign Change of λ_s Low-Magnetostrictive Metallic Glass as Seen by Its Domain Structure
75.80; S1.1; S1.2
(a) Vol.113,No.1, 181-186

J. GONZÁLEZ, M. VAZQUEZ, J.M. BARANDIARAN, M.A. ILLARRAMENDI, A. SALAZAR, and A. HERNADO
Magnetic Anisotropies Induced by Current Annealing and Their Dependence on Annealing Temperature
75.50; S1.1; S1.2
(a) Vol.113,No.1, 187-192

JIFAN HU, MINGXI LIN, RUFENG XU, SHOUGONG ZHANG, and ZHENXI WANG
Formation of the Magnetic Phases in Y-Zr-Fe Compounds
75.30; S1.1; S1.4
(a) Vol.113,No.1, K127-K130

R. SAKAI and Y. SHIMADA
Coercive Forces and Magnetostriction of Fe-C Films Deposited on ZnO Substrates
75.70; S1.1
(a) Vol.113,No.1, K131-K134

S. BALUCH, M. MIGLIERINI, R. GROENE, and J. SITEK
Positron Lifetime and Moessbauer Study of $Fe_{80-x}Ni_xB_{20}$ Metallic Glasses
76.80 and 78.70; S1.1
(a) Vol.113,No.1, K143-K146

A.I. MITSEK and V.N. PUSHKAR
The Topology of Magnetic Inhomogeneities in Thin Ferromagnetic Layers. The Model of the Magnetic Vortex
75.60 and 75.70; S1.1; S1.2; S11.2
(a) Vol.113,No.2, 529-540

A. BOETTCHER and H. SCHAEFER
Mapping of Magnetic Domain Structures inside of Ferromagnetic Materials by X-Ray Topography
75.60; S1.1
(a) Vol.113,No.2, 551-560

E. BURZO, N. PLUGARU, V. POP, and I. CREANGĂ
Magnetic Properties of $YCo_{4-x}Fe_xB$ Compounds
61.55 and 75.50; S1.1
(a) Vol.113,No.2, K253-K256

Z. FRAIT and D. FRAITOVÀ
Surface Spin-Wave Mode Excitation in Bulk Amorphous FeNiB Alloys
75.50 and 76.50; S1.1; S1.2
(b) Vol.154,No.1, 363-366

A. KOWALCZYK, P. STEFAŃSKI, A. WRZECIONO, and A. SZLAFEREK
Structure and Magnetic Properties of $R_2Fe_{14-x}Ni_xB$ Compounds (R = Nd and Gd)
75.30; 61.55; S1.1; S1.2
(a) Vol.114,No.1, 355-360

N.H. CHAU, N.H. DUC, T.D. HIEN, K. KROP, and J. ZUKROWSKI
Magnetic and Magnetostrictive Properties of the $Er(Fe_xCu_{1-x})_2$ Compounds
75.30 and 75.80; S1.1; S1.2; S1.4
(a) Vol.114,No.1, 361-368

T. TARNOCZI and A. SIEMKO
Anomalous Crossover Effect
75.30; S1.1
(a) Vol.114,No.1, K83-K86

L.M. SANDRATSKII and P.G. GULETSKII
Comparative Analysis of Tendencies in the Temperature-Induced Changes in Electronic Structure of Iron and Nickel
71.25; 75.50; S1.1; S1.2
(b) Vol.154,No.2, 623-636

M. MIHALIK, J. KOVAC, A. ZENTKO, and A. LOVAS
Some Magnetic Properties of Neutron Irradiated $Fe_{85-x}Cr_xB_{15}$ Metallic Glasses
61.80 and 75.30; 75.40; 75.60; S1.1; S1.2
(a) Vol.114,No.2, 679-684

L.F. KISS, G. MARTINEK, A. FORKL, and H. KRONMUELLER
Correlation between Microstructure and Magnetic Properties in FeNdB Permanent Magnets Doped with Al_2O_3
75.60; S1.1; S1.4
(a) Vol.114,No.2, 685-692

M.S. ANAND, B.M. PANDE, and R.P. AGARWALA
Recovery of Plastically Deformed A-203 Steel
62.20; S1.1
(a) Vol.114,No.2, K159-K162

M. JURCZYK and O.D. CHISTYAKOV
Structure and Magnetic Properties of $Y_2Fe_{14-x}Re_xB$
75.50; S1.1
(a) Vol.114,No.2, K219-K222

N. GANEV, I. KRAUS, and J. TRPCEVSKA
X-Ray Measurement of Residual Stresses Induced by Laser Surface Treatment
62.20; 68.20; S1.1
(a) Vol.115,No.1, K13-K16

M. HRABCAK, M. MIHALIK, J. KOVAC, and A. LOVAS
Saturation Magnetostriction and Curie Temperature of Neutron Irradiated $Fe_{75}Cr_{10}B_{15}$ and $Fe_{77}Cr_8B_{15}$ Amorphous Alloys
75.80; 61.80; S1.1
(a) Vol.115,No.1, K83-K86

Y.L. CHEN, P. MAO, DECHENG TIAN, X.Z. LI, and J.J. ZHU
Moessbauer Study of Amorphous $(Fe_{1-x}Ga_x)_{84}B_{16}$
76.80; S1.1
(a) Vol.115,No.1, K91-K94

M.A. VASILIEV, S.D. GORODETSKII, and A.G. BLASHUK
Distribution of the Effective Debye Temperature in the Surface Layers of Single Crystals of the Fe-Ni System
75.50; S1.1; S1.2
(a) Vol.115,No.1, K95-K100

MOHEB H. YOUSSEF, M. ROUSHDY, and N.K. GOBRAN
Origin of Periodic Fluctuation with Temperature of Ultrasonic Attenuation in Zinc, Aluminium, and Iron
62.80; S1.1; S1.2; S4
(b) Vol.155,No.2, 453-460

S. ENZO, G. COCCO, I. SOLETTA, C. ANTONIONE, and L. BATTEZZATI
Structuring Effects in FeSiB Metallic Glasses. An EXAFS Approach
61.40 and 64.60; S1.1
(a) Vol.115,No.2, 459-466

YUANSHENG WANG, MING MAO, and YIZHEN HE
Influence of a Third Metallic Component on the Crystallization of Metallic Glasses Fe-Ni-Si-B
61.40 and 64.70; S1.1; S1.2
(a) Vol.115,No.2, 467-474

V.N. NOVIKOV
Precision Dilatometric Studies of Metallic Glasses
61.40 and 64.70; S1.1; S1.2
(a) Vol.115,No.2, 475-482

H. DRAGANOVSKÁ, J. SLAMA, P. BUTVIN, and B. BUTVINOVÁ
Domain Patterns and Magnetic Properties of Some Amorphous Metallic Ribbons
75.60; S1.1; S1.2
(a) Vol.115,No.2, 533-538

M. VAZQUEZ, W. FERNENGEL, and H. KRONMUELLER
Approach to Magnetic Saturation in Rapidly Quenched Amorphous Alloys
75.60; 75.25; S1.1; S1.2; S1.61; S1.62
(a) Vol.115,No.2, 547-554

X.K. SUN, X.C. KOU, Y.C. CHUANG, R.W. ZHAO, J.Y. LI, F.M. YANG, R. GROESSINGER, and H.R. KIRCHMAYR
3d Anisotropy and Preferential Site Occupation in $R_2Fe_{13}M_1B$ Compounds (R=Y,Gd, M=Ga,Si,Al)
75.50; 75.30; S1.1
(a) Vol.115,No.2, K219-K224

N. SINGH
Lattice Dynamics of gamma-Iron at 1428 K
63.20; S1.1
(b) Vol.156,No.1, K33-K36

DECHENG TIAN, KAIYANG LI, and DIHUA DING
Magnetic Aftereffect and Mossbauer Studies of Structural Relaxation in $(Fe_{1-x}Ce_x)_{84}B_{16}$ Amorphous Alloys
61.40; 75.60; 76.80; S1.1
(a) Vol.116,No.1, 203-212

M.S.R. CHARI and R.G. SHARMA
The Electronic Lorenz Number and Magnon Thermal Conductivity of Cobalt and Iron
72.15; S1.1; S1.2
(a) Vol.116,No.1, 377-382

A.H. WAFIK, N.A. RAZIK, and M.H. ZAYAN
Barkhausen Jumps and Microstructure in Low Carbon Steel
75.60; S1.1
(a) Vol.116,No.1, 383-388

L.J. HUANG, Q.M. CHEN, B.X. LIU, Y.D. FAN, and H.D. LI
Icosahedral Incommensurate Fe-Mo Phase Formed by Solid State Transformation
64.70; S1.1
(a) Vol.116,No.1, K51-K56

A.O. MEKHRABOV, Z.M. BABAEV, and Z.A. MATYSINA
The Effect of a Third Component Impurity C on the Heat Capacity of A_3B-Type Ordered Alloys
64.60 and 65.40; S1.1; S1.2
(b) Vol.156,No.2, 419-430

P. STEFANSKI, A. SZLAFEREK, and A. WRZECIONO
The Intra- and Intersublattice Exchange Interactions in $RFe_{12-x}T_x$ Intermetallics
75.30; S1.1; S1.2; S1.4
(b) Vol.156,No.2, 657-662

V. PIERRON-BOHNES, C. LEROUX, J.P. AMBROISE, A. MENELLE, and P. BASTIE
Measurement of the Mean Static Displacement and of the Debye Waller Factor between 5 and 1464 K in Fe-19.5at%Al
61.55; 61.12; 76.80; S1.1
(a) Vol.116,No.2, 529-536

V.I. DIMITROV, YU.I. KLECHIN, and A.V. APOSTOLOV
Calculation of Elastic Moduli for Amorphous $Fe_{80}B_{20}$
62.20; 61.40; S1.1
(a) Vol.116,No.2, 627-632

K. MAKI, A. FORKL, T. DRAGON, and H. KRONMUELLER
Investigation of the Domain Structure of Sintered FeNdB-Al Magnets
75.60; S1.1
(a) Vol.116,No.2, 803-812

S1.2. Other 3d metals

C. MALINOWSKA-ADAMSKA
The Effect of Pressure on the Thermodynamic Properties of Crystals near the Lattice Instability Point
65.50; S1.2; S1.3; S2; S4
(b) Vol.151,No.1, 53-60

H.S. KIM and S.S. SHEININ
An Electron Microscope Method for Determing the Nature of Stacking Faults in F.C.C. Materials which Does not Rely on ...
61.16 and 61.70; S1.2
(a) Vol.111,No.1, 47-52

F. HAIDER
Calculation of TEM Contrast of Small Coherent Precipitates
61.16; 61.70; S1.2; S1.3
(a) Vol.111,No.1, 71-80

K. YAMAKAWA, B. HOHLER, and H. KRONMUELLER
Diffusion of Hydrogen in Nickel Based Alloys
66.30; 72.15; S1.2
(a) Vol.111,No.1, 89-98

D.M. VANDERWALKER
Fracture in Titanium Diboride
61.70 and 62.50; S1.2
(a) Vol.111,No.1, 119-126

M. KLOSEK, J.W. MORON, and T. POLOCZEK
Activation Energy of Linear Heating Crystallization in $Ni_{70.3}Fe_{4.7}Si_{15}B_{10}$ Glassy Alloy Determined Using the Internal ...
61.40 and 64.70; 62.40; S1.1; S1.2
(a) Vol.111,No.1, 127-136

T. REININGER, N. MOSER, A. HOFMANN, and H. KRONMUELLER
The Kinetics of the Magnetically Induced Anisotropy in Amorphous Ferromagnetic $Co_{71}Fe_{1}Mo_{1}Mn_{4}Si_{14}B_{9}$ and $Fe_{41}Ni_{41}B_{18}$ Alloy
75.50; S1.1; S1.2
(a) Vol.111,No.1, 243-256

A.H. WAFIK
Effect of Rolling and Heat Treatment on the Irreversible Magnetization in Nickel-Chromium Wire
75.60; S1.2
(a) Vol.111,No.1, 265-270

J. GUTIERREZ, J.M. BARANDIARÁN, and O.V. NIELSEN
Magnetoelastic Properties of Some Fe-Rich Fe-Co-Si-B Metallic Glasses
75.80; S1.1; S1.2
(a) Vol.111,No.1, 279-284

P. SVOBODA, Z. SMETANA, and A.V. ANDREEV
Magnetic Behaviour of $TmCu_2$ below T_N
75.50; S1.2; S1.4
(a) Vol.111,No.1, 285-288

V.P. GOLTSEV, V.V. KHODASEVICH, G. GOETZ, I.I. PRIKHODKO, V.M. DRAKO, and V.V. UGLOV
Study of Ion Bombardment of Double-Layer Thin Zr/Ni Films
61.80 and 68.48; S1.2
(a) Vol.111,No.1, K13-K16

H. DINTNER and F. THRUM
The Homogeneous Distribution of Non-Metallic Bonds in As-Deposited NiCr-O Resistive Films
72.15 and 73.60; S1.2; S10.1
(a) Vol.111,No.2, 551-558

M. JERGEL, G. VLASAK, and P. DUHAJ
Magnetostriction Study of Some Co-Based and Fe-Based Amorphous Alloys
75.80; S1.1; S1.2
(a) Vol.111,No.2, 597-602

J. RIVAS, M.A. LOPEZ-QUINTELA, F. WALZ, and H. KRONMUELLER
Fractal Analysis of Magnetic After-Effects in the Amorphous Alloy $Fe_{40}Ni_{40}P_{14}B_6$ at 500 K
75.60; 75.50; S1.1; S1.2
(a) Vol.111,No.2, 611-616

R. PAWELLEK and M. FAEHNLE
The Magnetic Field Effect on the Linear Magnetostriction of Amorphous Ferromagnets. I. Basic Formalism
75.80; S1.2
(a) Vol.111,No.2, 617-624

A.V. BOLTUSHKIN, V.G. SHADROV, and V.M. FEDOSYUK
Hard-Magnetic Co-W, Co-Ni-W Films with Perpendicular Magnetic Anisotropy
75.70; S1.2
(a) Vol.111,No.2, K225-K228

YIFAN CHENG and RENHUI WANG
Dynamical Theory of Electron Diffraction for Quasicrystals
61.14; 61.70; S1.2; S4
(b) Vol.152,No.1, 33-38

P. MODRAK and R. WOJNECKI
Wannier Functions for d-Band of Ferromagnetic Nickel
71.50; S1.2
(b) Vol.152,No.1, 203-208

J. VAN DER STEL, H.B. VAN VELDHUIZEN, G.W. KOEBRUGGE, J. SIETSMA, and A. VAN DEN BEUKEL
Change of Electrical Resistivity and Youngs Modulus during Crystallization of Amorphous $Fe_{40}Ni_{40}B_{20}$
61.40 and 64.70; 62.20; 72.15; S1.1; S1.2
(a) Vol.112,No.1, 35-40

D.M. VANDERWALKER
Hydride Formation on Dislocations in Titanium
61.70; 61.16; S1.2
(a) Vol.112,No.1, 73-78

HUA XIA, X.K. ZHANG, A. HU, W. ZHANG, S.N. ZHU, and C. TIEN
Brillouin Scattering from Surface Phonons in Fibonacci Metallic Superlattices
63.20; 78.35; S1.2
(a) Vol.112,No.1, 99-104

X.C. KOU, X.K. SUN, Y.C. CHUANG, T.S. ZHAO, R. GROESSINGER, R. EIBLER, and H.R. KIRCHMAYR
Anisotropy Energy of Nd Ion in $Nd_2Fe_{12-x}Co_2Al_xB$ Compounds
75.50; S1.1; S1.2; S1.4
(a) Vol.112,No.1, 175-180

M. BOURROUS, N. MOSER, and H. KRONMUELLER
Reversibility and Annealing of Magnetic After-Effect in Ferromagnetic Amorphous Alloys
75.60; 75.50; S1.2
(a) Vol.112,No.1, 181-188

A. TUROS
Ion Implantation Studies of Collision Cascades in Solids
61.70; S1.1; S1.2; S1.3; S1.4
(a) Vol.112,No.1, 253-262

G. BARDOS
An Approximation for Solving the Transport Equation of the Angular Flux of Ions in Matter
61.80; S1.2; S5.11
(a) Vol.112,No.1, 311-314

A.I. BAZHIN, V.V. KOZEL, V.P. SHESTOV, M. SUCHAŃSKA, and S.V. TEPLOV
Temperature Dependence of Energy Distribution of Excited Atoms Sputtered from an Ion-Bombarded Ni(100) Surface
79.20; 75.50; S1.2
(a) Vol.112,No.1, 437-440

S. ERKOÇ
A New Empirical Many-Body Potential Energy Function: Application to Microclusters
63.10; 61.55; S1.2; S1.3; S2; S5
(b) Vol.152,No.2, 447-454

R. RAMJI RAO and A. PADMAJA
Effective Second Order Elastic Constants and the Second Pressure Derivatives of a Hexagonal Crystal
62.20; S1.2; S3
(b) Vol.152,No.2, 455-462

M. STRÓŻAK, M. SUBOTOWICZ, and P. MIKOŁAJCZAK
Formation of Unusual Alloys by Means of Ion Beam Mixing of the Pd/V Superlattice
61.55 and 61.80; 68.48; S1.2
(a) Vol.112,No.2, 739-742

M.B. ZIKRY and K.H. GEORGY
Creep Rate Sensitivities of Copper in an Alpha Phase Copper-Zinc Solid Solution at 473 K
62.20; S1.2
(a) Vol.112,No.2, K91-K96

QIANG WANG, XIAO-WEN ZHANG, and BING-LIN GU
Order-Disorder Model of Phase Transition in A_xB_{1-x} Binary Systems of F.C.C. Structure
64.60; 64.75; S1.1; S1.2
(b) Vol.153,No.1, 139-144

M. BABIKER and M.G. COTTAM
A Quantized Screening Theory for the Surface Energy of Solids in the Hydrodynamic Jellium Model
68.40; S1.2; S3; S4
(b) Vol.153,No.1, 145-152

A. JEZIERSKI
Density of States in Transition Metal-Metalloid Amorphous Alloys
71.20; S1.2
(b) Vol.153,No.1, 185-192

MAO-LU DU and MIN-GUANG ZHAO
Gyromagnetic Factor and Spin-Orbit Coupling of Ligands in $CsVX_3$ (X=Cl,Br,I)
71.70; 76.30; S1.2
(b) Vol.153,No.1, 249-256

J. UTZIG
Moessbauer Spectroscopy of Cobalt in Germanium
76.80; S1.2; S5.12
(b) Vol.153,No.1, 377-382

M. DIVIŠ, P. SVOBODA, Z. SMETANA, V. NEKVASIL, and A.V. ANDREEV
The Influence of Crystal-Field Splitting on the Paramagnetic Magnetisation of $TmCu_2$
71.70 and 75.60; S1.2
(b) Vol.153,No.1, K69-K72

G. RAATZ and J. WOLTERSDORF
Structure of Metal Deposits on Ceramic Materials Studied in the Ni/MgO System
68.55; 68.48; S1.2; S10.1
(a) Vol.113,No.1, 131-142

M. BOURROUS and H. KRONMUELLER
Analysis of the Stabilization Kinetics of Amorphous $Co_{58}Ni_{10}Fe_5Si_{11}B_{16}$
75.60; 75.50; S1.2
(a) Vol.113,No.1, 169-180

A. SIEMKO, H.K. LACHOWICZ, N. MOSER, A. FORKL, and H. KRONMUELLER
Stress-Induced Sign Change of λ_s Low-Magnetostrictive Metallic Glass as Seen by Its Domain Structure
75.80; S1.1; S1.2
(a) Vol.113,No.1, 181-186

J. GONZALEZ, M. VAZQUEZ, J.M. BARANDIARAN, M.A. ILLARRAMENDI, A. SALAZAR, and A. HERNADO
Magnetic Anisotropies Induced by Current Annealing and Their Dependence on Annealing Temperature
75.50; S1.1; S1.2
(a) Vol.113,No.1, 187-192

S. ARAJS and T.A. JENKINS
Electrical Conduction in Some Directionally Solidified Eutectic Composites
72.15; S1.2
(a) Vol.113,No.1, K71-K74

H. LOEFFLER, C. ESCHRICH, and O. SIMMICH
Recalculation of Several Metastable Phase Lines of the Al-Zn System
61.55 and 64.75; S1.2
(a) Vol.113,No.2, 269-276

M. BOURROUS and H. KRONMUELLER
Structural Transition from the As-Quenched to the Relaxed State in Metallic Glasses
61.40 and 64.60; 75.60; S1.2
(a) Vol.113,No.2, 383-392

G.A. GOVOR, K. BAERNER, and J.-W. SCHUENEMANN
Phase Transitions in the $MnAs_{1-x}Sb_x$ Alloy System
64.70; 75.30; S1.2
(a) Vol.113,No.2, 403-420

A.I. MITSEK and V.N. PUSHKAR
The Topology of Magnetic Inhomogeneities in Thin Ferromagnetic Layers. The Model of the Magnetic Vortex
75.60 and 75.70; S1.1; S1.2; S11.2
(a) Vol.113,No.2, 529-540

C. Q. TANG and X. CHEN
Positron Lifetimes in Deformed Two-Phase Brass
61.70 and 78.70; S1.2
(a) Vol.113,No.2, K161-K164

M.Z. BUTT and M.A. KHAN
Short-Range Order in Copper-Manganese Alloy Single Crystals
62.20; S1.2
(a) Vol.113,No.2, K189-K194

D.Y. LI, X.F. WU, and T. KO
The Effect of Stress on the Lattice Instability of a TiNi Alloy
64.70; 61.10; 62.20; 72.15; S1.2
(b) Vol.154,No.1, 85-96

Z. FRAIT and D. FRAITOVA
Surface Spin-Wave Mode Excitation in Bulk Amorphous FeNiB Alloys
75.5Q and 76.50; S1.1; S1.2
(b) Vol.154,No.1, 363-366

V.E. EGORUSHKIN and V.B. LAPIN
Electronic Structure of Disordered TiNi and Pseudobinary $Ti(Ni_{0.8}Cu_{0.2})$
71.20 and 71.25; S1.2
(b) Vol.154,No.1, K43-K46

C. NAGESWARA RAO and D.S.R. MURTY
Compton Cross-Sections of Molybdenum
78.70; S1.2
(b) Vol.154,No.1, K85-K88

K.Z. BOTROS and A.S. TITI
On Assessing Deviations from Symmetrical Laue Diffraction Conditions Using TEM Images of Stacking Faults
61.16 and 61.70; S1.2
(a) Vol.114,No.1, 83-90

O.S. RAJORA and A.E. CURZON
The Use of Electron-Image Contrast to Measure the Composition and Thickness of Cu-Mo Thin Films
68.60; 68.55; S1.2
(a) Vol.114,No.1, 199-206

A. KOWALCZYK, P. STEFAŃSKI, A. WRZECIONO, and A. SZLAFEREK
Structure and Magnetic Properties of $R_2Fe_{14-x}Ni_xB$ Compounds (R = Nd and Gd)
75.30; 61.55; S1.1; S1.2
(a) Vol.114,No.1, 355-360

N.H. CHAU, N.H. DUC, T.D. HIEN, K. KROP, and J. ZUKROWSKI
Magnetic and Magnetostrictive Properties of the $Er(Fe_xCu_{1-x})_2$ Compounds
75.30 and 75.80; S1.1; S1.2; S1.4
(a) Vol.114,No.1, 361-368

D. CZEKAJ, E.K. HOLLMANN, A.B. KOZIREV, V.A. VOLPYAS, and A.G. ZAITSEV
Initial Energy Distribution of Sputtered Atoms in DC Diode Systems
79.20; S1.2
(a) Vol.114,No.1, K21-K24

V.I. MALYUTIN, V.E. OSUKHOVSKII, YU.D. VOROBIEV, and L.A. CHEBOTKEVICH
On the Effect of Annealing on the Coercive Force of Polycrystalline Magnetic Films
75.70; S1.2
(a) Vol.114,No.1, K75-K78

L.F. MARTYNOVA, Z.B. CHACHKHIANI, and V.I. CHECHERNIKOV
Knight Shift on ^{45}Sc Nuclei in Scandium Compounds and Alloys
76.60; S1.2
(a) Vol.114,No.1, K87-K92

L.M. SANDRATSKII and P.G. GULETSKII
Comparative Analysis of Tendencies in the Temperature-Induced Changes in Electronic Structure of Iron and Nickel
71.25; 75.50; S1.1; S1.2
(b) Vol.154,No.2, 623-636

T.-M. WANG, B.-Y. WANG, B.-X. FENG, C.-L. LIU, B.-H. JIANG, and Z.-Y. XU
The Recovery Behaviour of Quenched-in Vacancies in CuZnAl Alloy
61.70; 78.70; S1.2
(a) Vol.114,No.2, 451-456

S. ABIS and G. RIONTINO
A Scanning Electrical Resistivity (SER) Study of Phase Separation in an Al-22at%Zn Alloy
64.75; 72.15; S1.2; S4
(a) Vol.114,No.2, 511-518

R. HEYNER and G. MARX
Depth Profiles and Surface Characterization by Glow Discharge Optical Emission Spectroscopy
68.20; 79.20; S1.2
(a) Vol.114,No.2, 609-620

M. MIHALIK, J. KOVÁČ, A. ZENTKO, and A. LOVAS
Some Magnetic Properties of Neutron Irradiated $Fe_{85-x}Cr_xB_{15}$ Metallic Glasses
61.80 and 75.30; 75.40; 75.60; S1.1; S1.2
(a) Vol.114,No.2, 679-684

K. SUMIYAMA, Y. HIROSE, and Y. NAKAMURA
Magnetic and Electrical Properties of Nonequilibrium Ni-Al Alloys Produced by Vapor Quenching
72.15 and 75.50; 61.55; S1.2
(a) Vol.114,No.2, 693-704

D. LUPU
Hydrogen Diffusion and Radiationless Electron Transfer in Metal Hydrides
66.30; 71.38; 78.20; S1.2; S1.4; S3
(b) Vol.155,No.1, 65-72

XING-KUI ZHANG, HUA XIA, and AN HU
Raman Scattering in Nb-Cu Metallic Multilayers
68.30; 78.30; S1.2; S1.3
(b) Vol.155,No.1, 137-146

S.R. HOFFMAN and L.M. SCARFONE
A Two-Band Model of Substitutionally Disordered Ternary Metal Alloys in the Coherent Potential Approximation
71.20; 72.15; S1.2; S1.3
(b) Vol.155,No.1, K33-K38

K. SATO, P.E. DONOVAN, and W.M. STOBBS
Optimisation of the Fourier Component of Potential in Bloch Wave TEM Image Contrast Calculations
61.16; S1.2; S4
(a) Vol.115,No.1, 61-74

V.M. FEDOSYUK, M.U. SHELEG, O.I. KASYUTICH, and M.M. MALYUSH
Columnar Structure and Texture [001] in Co-Ni-W Films
75.60; 75.70; S1.2
(a) Vol.115,No.1, 279-284

M.A. VASILIEV, S.D. GORODETSKII, and A.G. BLASHUK
Distribution of the Effective Debye Temperature in the Surface Layers of Single Crystals of the Fe-Ni System
75.50; S1.1; S1.2
(a) Vol.115,No.1, K95-K100

MOHEB H. YOUSSEF, M. ROUSHDY, and N.K. GOBRAN
Origin of Periodic Fluctuation with Temperature of Ultrasonic Attenuation in Zinc, Aluminium, and Iron
62.80; S1.1; S1.2; S4
(b) Vol.155,No.2, 453-460

S. ERKOC
A New Empirical Many-Body Potential Energy Function. Application to Microclusters: Elements in B.C.C., F.C.C., and H.C.P. Structures
63.10; 61.55; S1.2; S2; S5
(b) Vol.155,No.2, 461-466

YUANSHENG WANG, MING MAO, and YIZHEN HE
Influence of a Third Metallic Component on the Crystallization of Metallic Glasses Fe-Ni-Si-B
61.40 and 64.70; S1.1; S1.2
(a) Vol.115,No.2, 467-474

V.N. NOVIKOV
Precision Dilatometric Studies of Metallic Glasses
61.40 and 64.70; S1.1; S1.2
(a) Vol.115,No.2, 475-482

H. LANGE, H. REIS, and F. FENSKE
AES Study of the Interaction of Ni and Al Overlayers with W-Ti and W-Re Diffusion Barriers
68.55 and 79.20; S1.2
(a) Vol.115,No.2, 497-504

H. DRAGANOVSKÁ, J. SLÁMA, P. BUTVIN, and B. BUTVINOVÁ
Domain Patterns and Magnetic Properties of Some Amorphous Metallic Ribbons
75.60; S1.1; S1.2
(a) Vol.115,No.2, 533-538

M. VAZQUEZ, W. FERNENGEL, and H. KRONMUELLER
Approach to Magnetic Saturation in Rapidly Quenched Amorphous Alloys
75.60; 75.25; S1.1; S1.2; S1.61; S1.62
(a) Vol.115,No.2, 547-554

M. JURCZYK and O.D. CHISTYAKOV
Note on the Crystallographic and Magnetic Properties of $YCo_{10}V_2$
75.50; S1.2
(a) Vol.115,No.2, K229-K232

E. DĘBOWSKA and J. RUDZIŃSKA-GIRULSKA
Densities of Interstitial Electrons in Pd(Ni)-Noble Metal Alloys
71.25; 78.70; S1.2; S1.3
(b) Vol.156,No.1, 303-308

N.Y. JIN, F. PHILLIPP, and A. SEEGER
Investigation of Defect Ordering in Heavily Irradiated Metals by High-Voltage Electron Microscopy
61.70; 61.16; S1.2; S1.3
(a) Vol.116,No.1, 91-112

I.V. ZOLOTUKHIN, V.I. BELYAVSKII, V.A. KHONIK, and T.N. RYABTSEVA
Internal Friction in Cold-Rolled Metallic Glasses $Cu_{50}Ti_{50}$ and $Ni_{78}Si_{8}B_{14}$
62.40; S1.2
(a) Vol.116,No.1, 255-266

M.S.R. CHARI and R.G. SHARMA
The Electronic Lorenz Number and Magnon Thermal Conductivity of Cobalt and Iron
72.15; S1.1; S1.2
(a) Vol.116,No.1, 377-382

R. KUENTZLER, E. MURRAY, Y. DOSSMANN, and D. AST
Electronic Properties of AlMn and AlMnSi Quasicrystals
75.40; 65.40; 71.20; S1.2
(a) Vol.116,No.1, 399-404

K. EFTAXIAS and V. HADJICONTIS
Analysis of Self-Diffusion Data in V and Nb
66.30; S1.2
(b) Vol.156,No.2, 393-402

A.O. MEKHRABOV, Z.M. BABAEV, and Z.A. MATYSINA
The Effect of a Third Component Impurity C on the Heat Capacity of A_3B-Type Ordered Alloys
64.60 and 65.40; S1.1; S1.2
(b) Vol.156,No.2, 419-430

A.I. MITSEK
Collective Excitations of Localized Electrons and Their Effect on Magnetic-Material Kinetics
71.45; 72.15; 75.80; S1.2; S1.4; S1.5
(b) Vol.156,No.2, 507-514

P. STEFAŃSKI, A. SZLAFEREK, and A. WRZECIONO
The Intra- and Intersublattice Exchange Interactions in $RFe_{12-x}T_x$ Intermetallics
75.30; S1.1; S1.2; S1.4
(b) Vol.156,No.2, 657-662

L. TERZIEV, J. LECOMTE-BECKERS, P. RASHEV, and T. GREDAY
Study of Transition Phases during Isothermal Ageing at 250 °C of Zn-27%Al Alloy
61.55; 61.14; 64.70; S1.2
(a) Vol.116,No.2, 521-528

H. MIZUBAYASHI, S. OKUDA, and Y. IKEDA
Inelastic Properties of Amorphous CuTi
62.40; S1.2
(a) Vol.116,No.2, 633-644

H. RUPPERSBERG, I. DETEMPLE, and J. KRIER
Evaluation of Strongly Non-Linear Surface-Stress Fields $\sigma_{xx}(z)$ and $\sigma_{yy}(z)$ from Diffraction Experiments
62.20 and 68.25; S1.2
(a) Vol.116,No.2, 681-688

C. LEXCELLENT and G. SOMÉ
Comportement en fluage à haute température de la phase γ du Cu-16,6%atSn. Tentative de corrélation avec la diffusion en volume
62.20; 66.30; S1.2
(a) Vol.116,No.2, 689-696

M. FAEHNLE and J. FURTHMUELLER
The Effect of External Stress on the Saturation Magnetostriction Constant of Metallic Glasses
75.80; S1.2
(a) Vol.116,No.2, 819-826

R.B. MCLELLAN
The Thermodynamics of Pd-(Ce,Y,Er)-H Ternary Solutions
65.50; S1.2; S1.4
(a) Vol.116,No.2, K157-K160

S1.3. Noble metals

C. MALINOWSKA-ADAMSKA
The Effect of Pressure on the Thermodynamic Properties of Crystals near the Lattice Instability Point
65.50; S1.2; S1.3; S2; S4
(b) Vol.151,No.1, 53-60

F. HAIDER
Calculation of TEM Contrast of Small Coherent Precipitates
61.16; 61.70; S1.2; S1.3
(a) Vol.111,No.1, 71-80

H. PAGNIA and N. SOTNIK
Metal Influence on Switching MIM Diodes
73.40 and 85; S1.3; S6; S10
(a) Vol.111,No.1, 387-394

M.S. MURALI SASTRY
Large Scale Coalescence in Terms of a New Agglomeration Rate
68.55; 73.60; S1.3; S4
(a) Vol.111,No.1, K37-K42

A. TUROS
Ion Implantation Studies of Collision Cascades in Solids
61.70; S1.1; S1.2; S1.3; S1.4
(a) Vol.112,No.1, 253-262

S. ERKOC
A New Empirical Many-Body Potential Energy Function: Application to Microclusters
63.10; 61.55; S1.2; S1.3; S2; S5
(b) Vol.152,No.2, 447-454

J. MITTENBACHER and R. MUEHLE
A Scanning Ion Microbeam System with a Liquid Metal Ion Source
61.80 and 79.70; 61.16; S1.3; S4
(a) Vol.112,No.2, 825-828

K. KOKKO, E. OJALA, and K. MANSIKKA
Fermi Level Density of States in Ag-Au Alloys. Nuclear Magnetic Spin-Lattice Relaxation Rate and Low-Temperature Specific Heat
71.20; 65.40; 76.60; S1.3
(b) Vol.153,No.1, 235-242

K. TOMPA, R. PIRLOT, L. GRENACS, and J. LEHMANN
Hyperfine Fields at Implanted ^{12}Boron in F.C.C. Metals
76.60; 71.50; S1; S1.3; S4
(b) Vol.153,No.2, 721-726

N.YU. LEBEDEV and K. LENKEIT
Azimuthal Angular Dependence of the Backscattering Yield in the Axial-to-Planar Channeling Transition Case
61.80; S1.3
(b) Vol.153,No.2, K97-K102

H. HAGENSCHULTE and TH. HEUMANN
On the Determination of Enhancement Factors of Solvent Diffusion
66.30; S1.3
(b) Vol.154,No.1, 71-84

R.V. GOPALA RAO and A. SATPATHY
Application of Statistical Mechanics Using Asymmetric Electrolytes. Model Evaluation of Structural Properties of Ionic Gold-Cesium Melts
61.25; S1.3; S2
(b) Vol.154,No.2 427-438

E. MALISZEWSKI, J. SOSNOWSKI, and S. BEDNARSKI
Lattice Dynamics of the Force Constant Disordered Crystal $Pd_{0.90}Ru_{0.10}$
63.20; 62.20; S1.3
(b) Vol.154,No.2, 543-552

A. HEILMANN, C. HAMANN, G. KAMPFRATH, and DO NGOC UAN
TEM Investigations on Plasma Polymer Silver Composite Films
68.55; S1.3; S12
(a) Vol.114,No.2, 551-558

M.S. MURALI SASTRY and MANJUNATHA PATTABI
Variation of the Tunnelling Barrier in Island Cu Films
73.60; 68.55; S1.3
(a) Vol.114,No.2, K179-K182

XING-KUI ZHANG, HUA XIA, and AN HU
Raman Scattering in Nb-Cu Metallic Multilayers
68.30; 78.30; S1.2; S1.3
(b) Vol.155,No.1, 137-146

T. SOMA, Y. HORIKAGO, and H.-MATSUO KAGAYA
Temperature Effect on the Elastic Coefficients of Noble Metals
62.20; 63.20; S1.3
(b) Vol.155,No.1, K5-K8

S.R. HOFFMAN and L.M. SCARFONE
A Two-Band Model of Substitutionally Disordered Ternary Metal Alloys in the Coherent Potential Approximation
71.20; 72.15; S1.2; S1.3
(b) Vol.155,No.1, K33-K38

M.A. EL HITI
Dependence of the Temperature Coefficient of the Strain Coefficients of Resistance of Double-Layer Thin Metallic Films on Thermal Strains
73.60; S1.3
(a) Vol.115,No.1, 185-190

E. DEBOWSKA and J. RUDZIŃSKA-GIRULSKA
Densities of Interstitial Electrons in Pd(Ni)-Noble Metal Alloys
71.25; 78.70; S1.2; S1.3
(b) Vol.156,No.1, 303-308

M.YA. GAMARNIK and YU.YU. SIDORIN
Change of the Unit Cell Parameters in Highly Dispersed Gold, Silver, and Copper Powders
61.10 and 68.90; S1.3
(b) Vol.156,No.1, K1-K4

K. MEINEL, O. LICHTENBERGER, and M. KLAUA
Growth, Stability, and Desorption of Ag Films on Ni(111)
68.55; 68.48; S1.3
(a) Vol.116,No.1, 47-54

P.B. BARNA, M. EL HITI, L. MALICSKÓ, and F.M. REICHA
On the Orientation Relationships of Decorating Gold Particles on Alkali Halides
68.48 and 68.55; S1.3; S9.11
(a) Vol.116,No.1, 55-66

N.Y. JIN, F. PHILLIPP, and A. SEEGER
Investigation of Defect Ordering in Heavily Irradiated Metals by High-Voltage Electron Microscopy
61.70; 61.16; S1.2; S1.3
(a) Vol.116,No.1, 91-112

G.A. HASSAN and H.H. HASSAN
Temperature Dependence of the Stress-Strain Characteristics of Technically Pure Silver Subjected to Prior Alternating Torsion
62.20; S1.3
(a) Vol.116,No.1, K43-K46

E. KESSLER
On the Measurement of the Heat of Condensation of Ag, Bi, Sb, and Pb
68.45; S1.3; S4
(a) Vol.116,No.1, K61-K64

S1.4. Lanthanides

YAN SHI, HANMIN JIN, and MINGZHI JIN
Interpretation of ^{57}Fe Moessbauer Measurements on $Nd_2Fe_{14}B$ at 83 and 295 K
76.80; S1.1; S1.4
(b) Vol.151,No.1, 249-258

R.B. PRABHU and A.F. BARBOSA
Low Temperature Resistivity of Mix-Valent $Ce_{0.76}La_{0.14}Th_{0.1}$
72.15; S1.4
(b) Vol.151,No.1, K45-K48

P. SVOBODA, Z. SMETANA, and A.V. ANDREEV
Magnetic Behaviour of $TmCu_2$ below T_N
75.50; S1.2; S1.4
(a) Vol.111,No.1, 285-288

NGUYEN HOANG LUONG, NGUYEN PHU THUY, LUU TUAN TAI, and THAN DUC HIEN
Rare Earth Magnetocrystalline Anisotropy in $R_2Fe_{14}B$ (R=Sm,Er,Tm,Yb)
75.30; 75.25; S1.1; S1.4
(a) Vol.111,No.2, 591-596

X.C. KOU, X.K. SUN, Y.C. CHUANG, T.S. ZHAO, R. GROESSINGER, R. EIBLER, and H.R. KIRCHMAYR
Anisotropy Energy of Nd Ion in $Nd_2Fe_{12-x}Co_2Al_xB$ Compounds
75.50; S1.1; S1.2; S1.4
(a) Vol.112,No.1, 175-180

A. TUROS
Ion Implantation Studies of Collision Cascades in Solids
61.70; S1.1; S1.2; S1.3; S1.4
(a) Vol.112,No.1, 253-262

F. LÓPEZ-AGUILAR, J. COSTA-QUINTANA, and S. BALLE
Hubbard Systems with Two Channels for the Localization
71.20 and 71.25; 71.50; S1.4
(b) Vol.152,No.2, 543-550

JIFAN HU, MINGXI LIN, RUFENG XU, SHOUGONG ZHANG, and ZHENXI WANG
Formation of the Magnetic Phases in Y-Zr-Fe Compounds
75.30; S1.1; S1.4
(a) Vol.113,No.1, K127-K130

V.V. VOROBEV, M.YA. KRUPOTKIN, and V.A. FINKEL
Anomalous Behaviour of Elastic Moduli of Terbium and Dysprosium in Magneto-Ordered State
62.20 and 64.60; 75.30; S1.4
(a) Vol.113,No.2, 375-382

J.N. DAOU, J.P. BURGER, and P. VAJDA
Anomalous Behavior of the Low-Temperature Spin Disorder Resistivity in PrH_{2+x} Compounds
72.15; 75.50; S1.4
(b) Vol.154,No.1, 321-326

M. GASGNIER
Rare Earth Compounds (Oxides, Sulfides ...)as Thin Films and Thin Crystals. Physico-Chemical Properties and Applications(Review Article)
68.55 and 73.40; 75.70; 78.65; S1.4; S1.63; S10
(a) Vol.114,No.1, 11-72

M.M. KORSUKOVA, V.N. GURIN, YU.B. KUZMA, N.F. CHABAN, S.I. CHYKHRII, V.V. MOSHCHALKOV, N.B. BRANDT, A.A. GIPPIUS, and KHO KHYU NYAN
Crystal Structure, Electrical, and Magnetic Properties of the New Ternary Compounds $LnAlB_{14}$
61.60 and 72.80; 75.50; S1.4
(a) Vol.114,No.1, 265-272

N.H. CHAU, N.H. DUC, T.D. HIEN, K. KROP, and J. ZUKROWSKI
Magnetic and Magnetostrictive Properties of the $Er(Fe_xCu_{1-x})_2$ Compounds
75.30 and 75.80; S1.1; S1.2; S1.4
(a) Vol.114,No.1, 361-368

A. BOHDZIEWICZ, B. CENDLEWSKA, M. KAZIMIERSKI, J. SZYMASZEK, and J.J. WNUK
Superconducting Properties of $LuRh_{1.2}Sn_4$ Thin Films
74.70; S1.4
(a) Vol.114,No.1, K49-K52

G.I. KATAEV, M.R. SATTAROV, and A.M. TISHIN
Influence of Commensurability Effects on the Magnetic Phase Diagram of Terbium Single Crystals
75.30; S1.4
(a) Vol.114,No.1, K79-K82

S.A. NIKITIN, A.M. TISHIN, and S.E. BYKHOVER
Investigation of Magnetic Phase Transitions in Terbium Using the Magnetocaloric Effect
75.30; S1.4
(a) Vol.114,No.1, K99-K102

L.F. KISS, G. MARTINEK, A. FORKL, and H. KRONMUELLER
Correlation between Microstructure and Magnetic Properties in FeNdB Permanent Magnets Doped with Al_2O_3
75.60; S1.1; S1.4
(a) Vol.114,No.2, 685-692

J. BISCHOF, M. DIVIS, P. SVOBODA, and Z. SMETANA
The Specific Heat of $HoCu_2$ in Magnetic Fields
75.40; S1.4
(a) Vol.114,No.2, K229-K232

D. LUPU
Hydrogen Diffusion and Radiationless Electron Transfer in Metal Hydrides
66.30; 71.38; 78.20; S1.2; S1.4; S3
(b) Vol.155,No.1, 65-72

L.M. SANDRATSKII
On a Peculiarity of the Optical Characteristics of Rare-Earth Metals
71.25; 78.20; S1.4
(b) Vol.155,No.2, K105-K108

A.I. MITSEK
Collective Excitations of Localized Electrons and Their Effect on Magnetic-Material Kinetics
71.45; 72.15; 75.80; S1.2; S1.4; S1.5
(b) Vol.156,No.2, 507-514

V.E. ADAMIAN, A.A. ARTSRUNI, A. BENAISSA, A.N. KOCHARIAN, M.A. MELIKIAN, and A.G. TONEIAN
Magnetic Properties of Model Multicomponent Compounds of CsCl Structure
64.75 and 75.30; S1.4
(b) Vol.156,No.2, 633-640

P. STEFAŃSKI, A. SZLAFEREK, and A. WRZECIONO
The Intra- and Intersublattice Exchange Interactions in $RFe_{12-x}T_x$ Intermetallics
75.30; S1.1; S1.2; S1.4
(b) Vol.156,No.2, 657-662

KEQIN ZHENG, SHENGYAN LIU, and DECHENG TIAN
Investigation of the Behavior of Internal Friction of Al-Mg-Si-RE Alloys
62.40; 61.70; S1.4; S3
(a) Vol.116,No.2, 621-626

R.B. MCLELLAN
The Thermodynamics of Pd-(Ce,Y,Er)-H Ternary Solutions
65.50; S1.2; S1.4
(a) Vol.116,No.2, K157-K160

S1.5. Actinides

K.P. THAKUR and S. AHMAD
Phase Transformation in F.C.C. Thorium and Its Theoretical Stability on the Path of Transformation
64.60; S1.5
(b) Vol.155,No.2, 445-452

A.I. MITSEK
Collective Excitations of Localized Electrons and Their Effect on Magnetic-Material Kinetics
71.45; 72.15; 75.80; S1.2; S1.4; S1.5
(b) Vol.156,No.2, 507-514

S1.61. Carbides, silicides, germanides

A.I. GUSEV and A.A. REMPEL
Superconductivity in Disordered and Ordered Niobium Carbide
74.70; S1.61
(b) Vol.151,No.1, 211-224

W. HENRION and H. LANGE
Optical Properties of High-Refractory Disilicide Thin Films
78.65; S1.61
(b) Vol.151,No.1, 375-382

T. KOJIMA
Semiconducting and Thermoelectric Properties of Sintered Iron Silicide
72.60; 72.20; S1.61
(a) Vol.111,No.1, 233-242

T. KANOMATA and T. GOTO
Note on the Crystal Structure of FeCoGe and FeNiGe
61.60; S1.61
(a) Vol.111,No.1, K1-K6

V.P. ZHUKOV, N.I. MEDVEDEVA, and V.A. GUBANOV
The Electronic Band Structure and Non-Empirical Calculations of Cohesive Properties of Refractory Compounds (Review Article)
71.25; S1.61; S1.62
(b) Vol.151,No.2, 407-440

D. ELEFANT, C. GLADUN, A. HEINRICH, J. SCHUMANN, and H. VINZELBERG
Comment on the Metal-Semiconductor Transition in Amorphous $Si_{1-x}Cr_x$ Films at Room Temperature
72.80; S1.61
(b) Vol.151,No.2, K155-K156

A.I. GUSEV
Structural Stability Boundaries for Nonstoichiometric Compounds
61.70; S1.61
(a) Vol.111,No.2, 443-450

V.V. TOKAREV, V.E. BORISENKO, T.M. PYATKOVA, and D.I. ZAROVSKII
Argon Ion Implantation Influence on Phase Formation and Structure of Nickel Silicides
61.70 and 68.55; S1.61
(a) Vol.111,No.2, 499-506

A. SINGH
Two-Step Annealing for Cobalt Disilicide Formation with Lowest Sheet Resistance
68.55 and 73.60; S1.61
(a) Vol.111,No.2, K191-K194

TZU-YU HUANG and JUH-TZENG LUE
Theoretical Calculation of the Optical Constants of Transition Metal Silicides
78.20; S1.61
(b) Vol.152,No.1, 335-346

V.V. NEMOSHKALENKO, P.K. NIKOLYUK, P.V. GEL, L.I. NIKOLAEV, V.KH. KASIYANENKO, A.V. KHRUSHCHAK, and B.P. MAMKO
X-Ray Spectroscopic Study of Electronic Structure of the Ce-Co-Si System
78.70; S1.61
(b) Vol.152,No.2, K55-K58

A.V. SUVOROV, K.K. BOURDELLE, N.G. CHECHENIN, and V.N. MAKAROV
Studies of Amorphous Layer Formation in SiC under Ga Bombardment
61.80; 64.70; S1.61
(a) Vol.112,No.2, 707-714

J. DANILOVICH, K. GAERTNER, G. GOETZ, A. NOVIKOV, and B. WEBER
Ion Implanted Buried Si_3N_4 Layers below Epitaxial $NiSi_2$ Layers
68.55; 61.80; S1.61
(a) Vol.112,No.2, 733-738

I. BRYLOWSKA, K. PAPROCKI, and R. ILKHAMOV
The RBS Study of Ion Induced Mixing in Sb-Si System
61.80 and 68.48; S1.61
(a) Vol.112,No.2, 743-746

K. GAERTNER, G. GOETZ, C. KASCHNER, I. KASKO, A. WITZMANN, and A. ZENTGRAF
Laser Induced Silicide Formation
68.55 and 79.20; 64.60; S1.61
(a) Vol.112,No.2, 747-752

M. VATSKICHEVA, L. VATSKICHEV, J. MUCHA, and M. KOLENDA
On Some Structural Properties of $(Ni_{80}Fe_{20})_{100-x}Ge_x$ Evaporated Thin Films
61.50 and 68.55; S1.61
(a) Vol.112,No.2, K57-K60

A.A. REMPEL and A.I. GUSEV
Heat Capacity of Niobium Carbide in Different Structural States
65.40; S1.61
(a) Vol.113,No.2, 353-358

T. HARA, A. SUGA, and R. ICHIKAWA
Properties of CVD WSi_x Films and WSi_x/GaAs Schottky Barriers
73.30 and 73.60; 73.20; S1.61; S7.12
(a) Vol.113,No.2, 459-466

V.G. BESHENKOV, CH.V. KOPETSKII, and YU.A. SHIYANOV
Phase Composition of Thin Films as Revealed by Auger Depth Profiling
64.80 and 68.55; 79.20; S1.61
(a) Vol.114,No.1, 191-198

A.A. POVZNER, S.V. KORTOV, and P.V. GELD
Magneto-Volume Effects in the Weak Itinerant Ferromagnets on the Basis of Solid Solutions of $Fe_{1-y}Mn_ySi$ and $Fe_{1-x}Co_xSi$
65.70 and 75.40; S1.61
(a) Vol.114,No.1, 315-326

A.I. GUSEV and A.A. REMPEL
Local Static and Dynamic Atomic Displacements in Disordered Niobium Carbide
61.50; 61.12; S1.61
(b) Vol.154,No.2, 453-460

B.I. CRAIG and P.V. SMITH
Determination of the Bulk Properties of Cubic SiC by the Periodic LUC-MINDO Method
71.10; 62.20; S1.61
(b) Vol.154,No.2, K127-K130

KH.R. KAZDAEV, M.T. AKCHULAKOV, E.M. BAYADILOV, V.I. ENGELKO, A.V. LAZARENKO, and E.S. CHEBUKOV
Influence of Annealing Regimes on Phase Transitions in Nitrogen and Carbon Ion Implanted Molybdenum
61.80 and 68.20; S1.61; S1.62
(a) Vol.115,No.1, 179-184

R.I. ZAINULLINA, K.B. VLASOV, M.A. MILYAEV, E.V. USTELEMOVA, and V.N. SYROMYATNIKOV
Oscillatory Temperature Dependence of Crystal Structure and Magnetic Parameters in $FeGe_2$
64.60; 75.30; S1.61
(b) Vol.155,No.2, 419-426

M. VAZQUEZ, W. FERNENGEL, and H. KRONMUELLER
Approach to Magnetic Saturation in Rapidly Quenched Amorphous Alloys
75.60; 75.25; S1.1; S1.2; S1.61; S1.62
(a) Vol.115,No.2, 547-554

A.I. GUSEV
Short-Range Order in Nonstoichiometric Transition Metal Carbides Nitrides and Oxides (Review Article)
61.60 and 75.30; 76.60; S1.61; S1.62; S10
(b) Vol.156,No.1, 11-40

I.V. BAZHIN, A.B. KOLPACHEV, and I.YA. NIKIFOROV
Electronic Energy Structure of Substoichiometric Molybdenum Carbides in Different Crystallographic Modifications
71.20; S1.61
(b) Vol.156,No.1, 309-318

D. HESSE and R. MATTHEIS
The Growth Behaviour of Epitaxial $NiSi_2$ Islands of A- and B-Types during the Reaction of Nickel Vapour with the Si(111) Surface
68.48 and 68.55; S1.61
(a) Vol.116,No.1, 67-80

P. WERNER, R. MATTHEIS, D. HESSE, R. HILLEBRAND, and J. HEYDENREICH
HREM Investigations of the $NiSi_2$/Si{111} Interfaces Bounding A-Type NiSi Islands on Si(111)
68.48 and 68.55; S1.61
(a) Vol.116,No.1, 81-90

E. PIPPEL and J. WOLTERSDORF
Electron Microscopy of TiC Containing Alumina Ceramics
61.70; 61.16; S1.61; S10
(a) Vol.116,No.1, 165-178

V.V. TOKAREV, V.E. BORISENKO, A.I. DEMCHENKO, and T.M. PYATKOVA
Growth of Epitaxial Nickel Disilicide during Rapid Thermal Processing of Argon-Implanted Nickel Films on Silicon
68.55; 61.70; S1.61; S5.11
(a) Vol.116,No.1, 331-336

G. ANDRÉ, A. OLÉS, and A. SZYTUŁA
Neutron Diffraction Study of $NdCoSi_2$
61.12; S1.61
(a) Vol.116,No.1, K29-K32

S1.62. Nitrides, phosphides, arsenides

V.P. ZHUKOV, N.I. MEDVEDEVA, and V.A. GUBANOV
The Electronic Band Structure and Non-Empirical Calculations of Cohesive Properties of Refractory Compounds (Review Article)
71.25; S1.61; S1.62
(b) Vol.151,No.2, 407-440

P. ČAPKOVÁ, M. MERISALO, P. BOUKAL, M. BLOMBERG, and W. LENGAUER
Thermal Vibrations in Vanadium Nitride Studied by X-Ray Diffraction
63.20; S1.62
(a) Vol.112,No.2, K81-K84

B.X. LIU, X. ZHOU, and H.-D. LI
Thermodynamics and Growth Kinetical Consideration of Metal-Nitride Formation by Nitrogen Implantation (Review Article)
61.80 and 64.60; 61.55; S1.62
(a) Vol.113,No.1, 11-22

YU.M. SEIDOV and A.M. SULEIMANOV
On the Theory of Raman Scattering by Surface Collective Excitations in Singlet-Triplet Ferromagnets
78.30; 71.70; 75.30; S1.62
(b) Vol.153,No.2, K203-K208

Z. WOKULSKI and C. SUŁKOWSKI
Electrical Properties and Superconductivity of $TiN_{1-x}C_x$ Films
74.70; S1.62
(a) Vol.114,No.1, K53-K56

KH.R. KAZDAEV, M.T. AKCHULAKOV, E.M. BAYADILOV, V.I. ENGELKO, A.V. LAZARENKO, and E.S. CHEBUKOV
Influence of Annealing Regimes on Phase Transitions in Nitrogen and Carbon Ion Implanted Molybdenum
61.80 and 68.20; S1.61; S1.62
(a) Vol.115,No.1, 179-184

M. VAZQUEZ, W. FERNENGEL, and H. KRONMUELLER
Approach to Magnetic Saturation in Rapidly Quenched Amorphous Alloys
75.60; 75.25; S1.1; S1.2; S1.61; S1.62
(a) Vol.115,No.2, 547-554

A.I. GUSEV
Short-Range Order in Nonstoichiometric Transition Metal Carbides Nitrides and Oxides (Review Article)
61.60 and 75.30; 76.60; S1.61; S1.62; S10
(b) Vol.156,No.1, 11-40

V.V. LOPATIN and A.V. KABYSHEV
On the Connection between the Building of Aluminium Nitride Ceramics and Its Dielectric Properties
61.40; 72.80; 77.20; S1.62
(a) Vol.116,No.1, 221-230

A.N. PILYANKEVICH, A.V. KURDYUMOV, and N.F. OSTROVSKAYA
On the Crystallographic Irreversibility of Some Phase Transformation in Boron Nitride
61.50 and 64.70; S1.62
(a) Vol.116,No.1, K1-K6

S1.63. Sulfides, selenides, tellurides

G.A. MEDVEDKIN, YU.V. RUD, and M.A. TAIROV
Fundamental Optical Absorption Edge in $MnIn_2Te_4$ Single Crystals
78.20; S1.63; S8.16
(a) Vol.111,No.1, 289-300

N. VOUROUTZIS and C. MANOLIKAS
Phase Transformations in Cuprous Telluride
64.70; S1.63
(a) Vol.111,No.2, 491-498

M. ISHII, K. SHIBATA, and H. WADA
Raman Study of $M_xMo_6S_{8-y}$ (M = Yb, Fe, Cr)
78.30; 63.20; S1.63
(b) Vol.152,No.2, K63-K68

S. MOTIEJUNAS, D. BALTRŪNAS, N. GEČIAUSKAITE, and K. MAKARIUNAS
Correlation between the Moessbauer and Semiconductor Parameters of Tin Telluride
76.80; 61.55; S1.63
(b) Vol.154,No.1, 341-346

M. GASGNIER
Rare Earth Compounds (Oxides, Sulfides ...)as Thin Films and Thin Crystals. Physico-Chemical Properties and Applications(Review Article)
68.55 and 73.40; 75.70; 78.65; S1.4; S1.63; S10
(a) Vol.114,No.1, 11-72

W. PAULUS, G.A. WIEGERS, A. MEETSMA, S. VAN SMAALEN, and J.L. DE BOER
The Structure of $Cu_{0.38}NbSe_2$. An Orthorhombic Superstructure of a Hexagonal Basic Structure Occurring in Three Orientation Variants
61.60; S1.63
(a) Vol.114,No.1, 91-98

E.M. ZUB and S.V. SUKHVALO
On the Mechanism of Charge Transport in the $Cu_xFe_{1-x}Cr_2S_4$ Solid Solution Films
73.60; S1.63
(a) Vol.114,No.2, 573-578

T. KUZUBA and M. ISHII
Elastic Constant c_{44} of Uniaxial Layered Crystals
62.20; 63.20; 78.30; S1.63
(b) Vol.155,No.1, K13-K16

G.P. NIZHNIKOVA and O.V. FARBEROVICH
Theoretical Investigation of XPS, UPS, and BIS in Rare-Earth Compounds - Thulium Monochalcogenides
71.20; 71.50; 79.60; S1.63
(b) Vol.155,No.2, 491-500

N. VOUROUTZIS and C. MANOLIKAS
Phase Transformations and Discommensurations in Hexagonal Cuprous Telluride
61.14 and 64.70; S1.63
(a) Vol.115,No.2, 399-412

S. ICHIKAWA, S. OHTA, K. HATAKEYAMA, S. ANZAI, and A. TAKASE
Curie Temperatures of Monoclinic $(Cr_{1-x}M_x)_3Te_4$ with M:Ni, Co, and Vacancy
75.50; S1.63
(a) Vol.115,No.2, K215-K218

T. KUZUBA and M. ISHII
Erratum to Elastic Constant c_{44} of Uniaxial Layered Crystals (phys. stat. sol. (b) 155, K13 (1989))
62.20; 63.20; 78.30; S1.63
(b) Vol.156,No.2, K167-K168

S2. Alkali metals and their alloys and compounds
(for alkali halides, see S9.11)

C. MALINOWSKA-ADAMSKA
The Effect of Pressure on the Thermodynamic Properties of Crystals near the Lattice Instability Point
65.50; S1.2; S1.3; S2; S4
(b) Vol.151,No.1, 53-60

K. KOKKO
Determination of Electron Momentum Density from Electron Number Density. A New Version of Semiclassical Transformation
71.10; S2
(b) Vol.151,No.1, 139-144

S. ERKOC
A New Empirical Many-Body Potential Energy Function: Application to Microclusters
63.10; 61.55; S1.2; S1.3; S2; S5
(b) Vol.152,No.2, 447-454

ZHENGJU YANG and F. LUTY
Theoretical Investigation of Ferroelastic Phase Transition of Pure and Mixed Alkali Cyanide Systems by Elastic Dipole Model (I)
64.60; S2
(b) Vol.154,No.1, 167-180

N. SINGH, N.S. BANGER, and S.P. SINGH
Lattice Dynamics of B.C.C. Sr: A Dynamical Treatment
63.20; S2
(b) Vol.154,No.1, K19-K22

M. RANI, A. PRATAP, and N.S. SAXENA
Dynamics of Liquid Potassium
63.50; S2
(b) Vol.154,No.1, K23-K28

R.V. GOPALA RAO and A. SATPATHY
Application of Statistical Mechanics Using Asymmetric Electrolytes. Model Evaluation of Structural Properties of Ionic Gold-Cesium Melts
61.25; S1.3; S2
(b) Vol.154,No.2, 427-438

A. MEYER and W.H. YOUNG
Interatomic Forces and Pseudopotential Parameterization
61.50; S2
(b) Vol.154,No.2, 469-474

M. AHMAD
Spin Waves with l=1 in a Degenerate Fermi Liquid of Alkali Metals in DC Magnetic Fields
75.30; S2
(b) Vol.154,No.2, K163-K166

S. ERKOC
A New Empirical Many-Body Potential Energy Function. Application to Microclusters: Elements in B.C.C., F.C.C., and H.C.P. Structures
63.10; 61.55; S1.2; S2; S5
(b) Vol.155,No.2, 461-466

R.N. JOARDER and S.M. MUJIBUR RAHMAN
Temperature Dependence of Entropies and Specific Heats of Liquid Na and K
65.50; 65.20; S2
(b) Vol.156,No.1, 181-194

O. AKINLADE and O. BELHACHEMI
The Electrical Resistivity of Some Lithium Based Liquid Alloys
72.15; S2
(b) Vol.156,No.2, 569-576

S3. Alkaline earth metals and their alloys and compounds
(for alkaline earth halides, see S9.12)

R. RAMJI RAO and A. PADMAJA
Effective Second Order Elastic Constants and the Second Pressure Derivatives of a Hexagonal Crystal
62.20; S1.2; S3
(b) Vol.152,No.2, 455-462

M. BABIKER and M.G. COTTAM
A Quantized Screening Theory for the Surface Energy of Solids in the Hydrodynamic Jellium Model
68.40; S1.2; S3; S4
(b) Vol.153,No.1, 145-152

D. LUPU
Hydrogen Diffusion and Radiationless Electron Transfer in Metal Hydrides
66.30; 71.38; 78.20; S1.2; S1.4; S3
(b) Vol.155,No.1, 65-72

K. ARADHANA and R.P.S. RATHORE
Two- and Three-Body Forces in Lattice Dynamics and Elastic Behaviour of B.C.C. Calcium and Strontium
63.20; 62.20; S3
(b) Vol.156,No.1, 77-82

KEQIN ZHENG, SHENGYAN LIU, and DECHENG TIAN
Investigation of the Behavior of Internal Friction of Al-Mg-Si-RE Alloys
62.40; 61.70; S1.4; S3
(a) Vol.116,No.2, 621-626

S4. Other metals and their alloys

C. MALINOWSKA-ADAMSKA
The Effect of Pressure on the Thermodynamic Properties of Crystals near the Lattice Instability Point
65.50; S1.2; S1.3; S2; S4
(b) Vol.151,No.1, 53-60

F. POVOLO and E.B. HERMIDA
Analysis of Stress Relaxation and Creep Curves for a Log-Normal Distribution of Relaxation Times
62.20; S4; S11
(b) Vol.151,No.1, 71-84

TH. BORN
Calculation of the Electrical Resistivity of Liquid Sn-Te Alloys according to the Effective-Medium Theory
72.15; S4
(b) Vol.151,No.1, K41-K44

T.H. YOUSSEF, F.A. SAADALAH, and I.K. BISHAY
Plastic Instability during Combined Torsion-Tension Deformation in Al-0.72% Si Alloy
62.20; S4
(a) Vol.111,No.1, 137-144

R. PASIANOT, G. GALAMBOS, E.J. SAVINO, L. AMARAL, M. BEHAR, D. FARKAS, and S. RAO
Argon Irradiation of Sn Thin Layers Deposited on Fe Substrates
61.80 and 68.20; S1.1; S4
(a) Vol.111,No.1, 173-180

M.S. MURALI SASTRY
Large Scale Coalescence in Terms of a New Agglomeration Rate
68.55; 73.60; S1.3; S4
(a) Vol.111,No.1, K37-K42

S.M. PATTALWAR, R.N. DIXIT, S.Y. SHETE, and B.K. BASU
Low-Temperature Heat Capacity Measurement in a Few Alloys near $PdPb_2$ Composition
65.40; S4
(b) Vol.151,No.2, K115-K118

S. POPOVIC, H. LOEFFLER, B. GRZETA, G. WENDROCK, and P. CZURRATIS
Lattice Constant of Al-Rich Matrix of Two Al-(8 and 24)at% Zn Alloys in Equilibrium with (Metastable) Guinier Preston Zones
61.55; 62.20; 64.70; S4
(a) Vol.111,No.2, 417-430

S.N. SAHU and A. BOURDILON
First Non-Aqueous Electrocodeposition in InP, In, and P Films
68.55; 78.65; S4; S5; S7.11
(a) Vol.111,No.2, K179-K184

YIFAN CHENG and RENHUI WANG
Dynamical Theory of Electron Diffraction for Quasicrystals
61.14; 61.70; S1.2; S4
(b) Vol.152,No.1, 33-38

M.A.Z. VASCONCELLOS, J.A.T. BORGES DA COSTA, W.H. SCHREINER, and I.J.R. BAUMVOL
A Comparison of Thermal Annealing and Ion Beam Mixing in the Fe/Al Bilayer
68.48 and 68.55; S1.1; S4
(a) Vol.112,No.1, 105-114

A. SAKAKIBARA, M. YAMADA, T. KANADANI, and M. OHTA
Effect of the Thickness of the Specimen on the Formation of GP Zones in Al-4wt%Zn Alloy
64.75; S4
(a) Vol.112,No.1, K15-K18

S. ERKOÇ and S. KATIRCIOĞLU
Molecular-Dynamics Simulation of Aluminum Microclusters
61.55; S4
(b) Vol.152,No.2, K37-K42

G. GŁADYSZEWSKI, P. MIKOŁAJCZAK, and M. SUBOTOWICZ
Ion Beam Mixing in Bi-Sb Superlattices
61.55 and 61.80; 68.48; S4
(a) Vol.112,No.2, 753-756

J. MITTENBACHER and R. MUEHLE
A Scanning Ion Microbeam System with a Liquid Metal Ion Source
61.80 and 79.70; 61.16; S1.3; S4
(a) Vol.112,No.2, 825-828

N. DHIREN SINGH and H.N.K. SARMA
In situ Electrical Behaviour of Thin Films of Antimony-Tellurium Alloy
73.60; S4
(a) Vol.112,No.2, K101-K104

M. BABIKER and M.G. COTTAM
A Quantized Screening Theory for the Surface Energy of Solids in the Hydrodynamic Jellium Model
68.40; S1.2; S3; S4
(b) Vol.153,No.1, 145-152

P. FANTUCCI and G. PACCHIONI
Theoretical INDO Investigation of the Electronic Structure of Aluminum Clusters
71.10; 71.25; S4
(b) Vol.153,No.1, 193-206

N.A. DOUMANIS and CH.G. TRICALINOS
Derivation of the Fokker-Planck Kinetic Equation for Planar Channeling of Relativistic Positively Charged Particles
61.80; S4
(b) Vol.153,No.1, K1-K4

S.G. PSAKHIE, S.V. ALEKSEEV, and V.E. PANIN
Influence of Shear on the Binding Energy of Complexes of Impurities and Vacancy in Aluminium
61.70; 62.20; S4
(b) Vol.153,No.1, K11-K16

S.A. NIKITIN, A.M. TISHIN, and P.I. LEONTEV
Magnetocaloric Effect and Refringerant Capacity of Tb-Dy Alloys
75.40; S4
(a) Vol.113,No.1, K117-K122

K. TOMPA, R. PIRLOT, L. GRENACS, and J. LEHMANN
Hyperfine Fields at Implanted ^{12}Boron in F.C.C. Metals
76.60; 71.50; S1; S1.3; S4
(b) Vol.153,No.2, 721-726

A.W. ZHU and T.S. KE
Characteristics of the Internal Friction Peak Associated with Bamboo Grain Boundaries
62.40; S4
(a) Vol.113,No.2, 393-402

T.S. KE and A.W. ZHU
Experimental Evidence on the Existence of the Bamboo Boundary Internal Friction Peak
62.40; S4
(a) Vol.113,No.2, K195-K198

T. KANADANI and A. SAKAKIBARA
GP Zone Formation by Nucleation and Growth in Al-4mass%Zn Alloy
62.20 and 64.75; S4
(a) Vol.113,No.2, K199-K202

XIAOYU HU and H. MARGOLIN
Dislocation Arrangements in Stress Gradients
61.70; 62.20; S4
(b) Vol.154,No.1, 61-70

WEI LIU, ZHONGGUANG WANG, and YUEBO XIA
Crystallographic Characteristics of Slip Band Continuity across Grain Boundaries in Multicrystalline Aluminum during Cyclic Deformation
68.25; S4
(a) Vol.114,No.1, 177-184

T. KANADANI and A. SAKAKIBARA
Short-Range Clustering in Dilute Al-Cu Alloy
64.75 and 72.15; S4
(a) Vol.114,No.1, K17-K20

V. PARKHUTIK, V. SOKOL, V. SHERSHULSKII, E. UTKINA, and A. VOROBEVA
Hopping Electrical Conductivity in the Al-Anodic Al_2O_3-Al System
73.40; S4; S10.1
(a) Vol.114,No.1, K33-K36

S. ABIS and G. RIONTINO
A Scanning Electrical Resistivity (SER) Study of Phase Separation in an Al-22at%Zn Alloy
64.75; 72.15; S1.2; S4
(a) Vol.114,No.2, 511-518

J. LEDZION and J. KIERUL
Thickness Dependence of Thin Film Conductivity
72.15 and 73.60; S4
(a) Vol.114,No.2, 545-550

DECHENG TIAN, ZHU-XIN YU, CAI-GUO ZHAN, and SHAO-LING WANG
Positron Annihilation Study of Ageing of an Al-Mg-Si-RE Alloy
61.70 and 78.70; 64.70; S4
(a) Vol.114,No.2, K135-K138

W. UNGER and W. POSSART
An XPS Investigation of the Interface between Thin Polymethylmethacrylate (PMMA) Films and Natively Oxidized Al
68.48 and 79.60; S4; S12
(a) Vol.114,No.2, K175-K178

T. KANADANI and A. SAKAKIBARA
Effect of the Specimen Thickness on the Aging of an Al-10mass%Zn Alloy after Quenching from Low Temperature
72.15; S4
(a) Vol.114,No.2, K197-K200

XIAOYU HU and H. MARGOLIN
The Stress Fields on Secondary Slip Systems Due to a Group of Dislocations in a Stress Gradient
61.70; 62.20; S4
(b) Vol.155,No.1, 53-64

B. MITRA, A. GHOSHAL, and K.P. GHATAK
On the Electric Susceptibility Mass in Bi under Arbitrary Magnetic Quantization
71.25; S4
(b) Vol.155,No.1, K23-K28

K. SATO, P.E. DONOVAN, and W.M. STOBBS
Optimisation of the Fourier Component of Potential in Bloch Wave TEM Image Contrast Calculations
61.16; S1.2; S4
(a) Vol.115,No.1, 61-74

T.S. KE and B.L. CHENG
Mechanical Model of the Bamboo Boundary Internal Friction Peak
62.40; S4
(a) Vol.115,No.1, 119-124

D.O. NORTHWOOD and I.O. SMITH
Steady-State Creep and Strain Transients for Stress Change Tests in an Al-0.5wt%Zn Solid Solution Alloy
62.20; S4
(a) Vol.115,No.1, 125-134

O.I. BARKALOV, I.T. BELASH, and A.F. GUROV
Thermally-Activated Amorphization of the Quenched High-Pressure Phase in the Cd-Sb System
64.70; S4
(a) Vol.115,No.1, K19-K22

R.K. SHUKLA, S. SWARUP, A. KUMAR, and A.N. NIGAM
Structural Studies in Glassy Semiconducting Se-Te Alloys
61.40 and 78.70; S4
(a) Vol.115,No.1, K105-K108

MOHEB H. YOUSSEF, M. ROUSHDY, and N.K. GOBRAN
Origin of Periodic Fluctuation with Temperature of Ultrasonic Attenuation in Zinc, Aluminium, and Iron
62.80; S1.1; S1.2; S4
(b) Vol.155,No.2, 453-460

J.L. VICENTE, A. PAOLA, A. RAZZITTE, E.E. MOLA, and S.B. TRICKEY
Static Quantum Size Effects in Ultra-Thin Beryllium Films
73.30; S4
(b) Vol.155,No.2, K93-K98

R.Z. VALIEV, R.SH. MUSALIMOV, and N.K. TSENEV
The Non-Equilibrium State of Grain Boundaries and the Grain Boundary Precipitations in Aluminium Alloys
61.70; S4
(a) Vol.115,No.2, 451-458

W. SCHNELLE and U. DILLNER
Electrical and Galvanomagnetic Properties of Undoped and Doped Polycrystalline Bismuth Films. I. Preparation and Experimental ...
73.60; S4
(a) Vol.115,No.2, 505-514

T. KANADANI
Effect of Soft Surface Layer on the Fatigue Strength of Age-Hardened Al-12mass%Zn Alloy
62.20; S4
(a) Vol.115,No.2, K147-K148

M. ROESLER and W. BRAUER
Influence of Dynamic Screening on the Electron Yield-Stopping Power Relation for Proton Impact in Aluminum
79.20; S4
(b) Vol.156,No.1, K85-K88

U. DILLNER and W. SCHNELLE
Electrical and Galvanomagnetic Properties of Undoped and Doped Polycrystalline Bismuth Films. II. Analysis in an Anisotropic ...
73.60; 72.10; S4
(a) Vol.116,No.1, 337-342

W. ŁUŻNY, S. NIZIOŁ, J. MAYER, and I. NATKANIEC
Inelastic Neutron Scattering on Solid and Liquid Gallium
61.12; S4
(a) Vol.116,No.1, K25-K28

E. KESSLER
On the Measurement of the Heat of Condensation of Ag, Bi, Sb, and Pb
68.45; S1.3; S4
(a) Vol.116,No.1, K61-K64

U. DILLNER and W. SCHNELLE
Electrical and Galvanomagnetic Properties of Undoped and Doped Polycrystalline Bismuth Films. III. Analysis in an Anisotropic ...
73.60; 72.10; S4
(a) Vol.116,No.2, 715-724

T. KANADANI and A. SAKAKIBARA
Effect of an Addition of Ge on the Formation of GP Zones in Al-4mass%Zn Alloy
61.70; S4
(a) Vol.116,No.2, K153-K156

L. JAHN, J.S. PASTUSCHENKOV, and V. CHRISTOPH
On the Angular Dependence of the Coercivity of NdFeB Hard Magnets
75.70; 75.60; S4
(a) Vol.116,No.2, K179-K184

S5. Nonmetallic elements

V.A. KULBACHINSKII
Galvanomagnetic Properties of Graphite Fibers
72.60 and 74.90; S5
(b) Vol.151,No.1, 185-192

I. MUEHLING and P. BLAUDECK
Multiple Scattering Approach to the Density of States and Electron Energy Loss Spectra of Amorphous Carbon
61.40 and 71.20; S5
(b) Vol.151,No.2, 545-554

S.N. SAHU and A. BOURDILON
First Non-Aqueous Electrocodeposition in InP, In, and P Films
68.55; 78.65; S4; S5; S7.11
(a) Vol.111,No.2, K179-K184

I.M. GOLBAN
Determination of Deformation Potential Constants of Arsenic
72.20; S5
(b) Vol.152,No.1, 225-230

W. SZYSZKO and J. SIELANKO
Monte Carlo Calculation of Depth Distribution of Implanted Ions and Sputtering Yield. Comparison for Some of the ...
61.80 and 79.20; S5; S5.11
(a) Vol.112,No.1, 315-318

S. ERKOÇ
A New Empirical Many-Body Potential Energy Function: Application to Microclusters
63.10; 61.55; S1.2; S1.3; S2; S5
(b) Vol.152,No.2, 447-454

C. MARTI, C. DE BEAUVAIS, T. CEVA, B. CROSET, and M. GOLDMANN
Incommensurate-Commensurate Transition Driven by Concentration in Rare Gas Mixtures Adsorbed on Graphite
64.60; 68.45; S5
(b) Vol.152,No.2, 463-474

G.B. SMITH, D.R. MCKENZIE, and P.J. MARTIN
An XPS Study of Chemical Order in Hydrogenated Amorphous Silicon-Carbon Alloy Films
61.40 and 68.55; 78.70; S5; S5.11
(b) Vol.152,No.2, 475-480

W. SCHARFF, K. HAMMER, B. EIBISCH, O. STENZEL, S. ROTH, and T. FRAUENHEIM
Deposition of Hyrdogenated Amorphous Carbon Films by a ECR Microwave Plasma
61.80 and 68.55; 73.60; 78.65; S5
(a) Vol.112,No.2, 667-678

M.M. KAZANIN and V.V. KUTASOV
Fe Impurity in alpha- and beta-Boron and in Some Boron Compounds
71.55; 72.20; 76.70; S5
(a) Vol.113,No.1, 143-150

O.A. GOLIKOVA
Amorphous Boron and Amorphous-Like Borides
61.40 and 72.20; S5
(a) Vol.113,No.1, K1-K4

S. SATPATHY and A.P. PATHAK
Planar Channeling of Electrons and Positrons in Crystals. Local-Density Calculation of Planar Averaged Potential and ..
61.80 and 78.70; S5; S5.11; S5.12
(b) Vol.153,No.2, 455-464

H.-J. AUGUST and J. WERNISCH
Analytical Expressions for the Electron Backscattering Coefficient
79.20; S1; S5; S5.11
(a) Vol.114,No.2, 629-634

S. ERKOÇ
A New Empirical Many-Body Potential Energy Function. Application to Microclusters: Elements in B.C.C., F.C.C., and H.C.P. Structures
63.10; 61.55; S1.2; S2; S5
(b) Vol.155,No.2, 461-466

A.F. BURENKOV, V.S. VARYCHENKO, A.M. ZAITSEV, F.F. KOMAROV, G.G. KONOPLYANIK, A.A. MELNIKOV, V.F. STELMAKH, and V.V. TKACHEV
Depth Distribution of Defects and Impurities in Diamond and Cubic Boron Nitride after High-Energy Ion Implantation
61.80; 61.70; 78.60; S5; S7.1
(a) Vol.115,No.2, 427-436

O. STENZEL, B. RAU, and G. SCHAARSCHMIDT
Optical Properties of Amorphous Carbon Deposited by an Ion-Plating Technique
78.65; S5
(a) Vol.115,No.2, K247-K250

F. BECHSTEDT and D. REICHARDT
Calculation of Surface-Induced Core-Level Shifts for Covalent Semiconductors C, Si, Ge, and alpha-Sn. I.(111)2x1 Surfaces
68.20 and 73.20; S5; S5.11; S5.12
(b) Vol.156,No.2, 471-486

S5.11. Silicon

V. HOLÝ and J. KUBĚNA
On the Integrated Intensity of X-Ray Diffraction in Crystals with Randomly Distributed Defects
61.10 and 61.70; S5.11
(b) Vol.151,No.1, 23-28

J.-P. ZOELLNER and G. PAASCH
Conditions for the Occurence of the Electric Quantum Limit in Inversion Layers
71.25 and 73.40; S5.11; S5.12; S7.1
(b) Vol.151,No.1, 145-158

A.A. BUGAI, I.N. KRAVTSOVA, V.M. MAKSIMENKO, B.D. SHANINA, and V.I. SHAKHOVTSOV
Properties of the Si-S1 Center in $Si_{1-x}Ge_x$ Solid Solutions
71.55; 76.30; S5.11; S5.12
(b) Vol.151,No.1, 159-166

S.V. KOVESHNIKOV, E.B. YAKIMOV, N.A. YARYKIN, and V.A. YUNKIN
Plasma Stimulated Impurity Redistribution in Silicon
61.80 and 71.55; 73.40; S5.11
(a) Vol.111,No.1, 81-88

A.A. KARANOVICH and A.V. DVURECHENSKII
The EPR Study of Inhomogeneous Deformations in Neutron-Irradiated Silicon
61.70; 76.30; S5.11
(a) Vol.111,No.1, 257-264

G.A. ADEGBOYEGA
Schottky Barrier Height Modification on n-Type Silicon by Wet Chemical Etching
73.30; S5.11
(a) Vol.111,No.1, K31-K36

MIN GONG and ZHI-PU YOU
The Deep Levels in Nickel-Doped Silicon.
71.55; S5.11
(a) Vol.111,No.1, K49-K52

M. KLENNER, C. FALTER, and W. LUDWIG
A Double Bond Charge Model for Si
63.20; S5.11
(b) Vol.151,No.2, 503-512

C. ENGLER
Potential Surfaces for Chemisorption on Semiconductor Surfaces from Pair Potential Considerations
68.20; 71.10; 73.20; S5.11
(b) Vol.151,No.2, 555-564

N.J. CASTELLANI and W.R.L. LAMBRECHT
Core Hole Relaxation Energy in Semiconductors. Using a Pseudopotential Description of the Core Hole Perturbation
71.10; S5.11; S7.1; S8.1
(b) Vol.151,No.2, 565-570

N. AZZIZ and P.C. MURLEY
Effective Charge of Moving Ions in Matter
61.80; S5.11
(b) Vol.151,No.2, K105-K108

V.M. AROLE, M.G. TAKWALE, V.G. BHIDE, A.D. SHALIGRAM, and S.K. DAVID
Hydrogen-Induced Structural Changes in Polycrystalline Silicon as Revealed by Positron Lifetime Spectroscopy
61.70; 72.20; S5.11
(a) Vol.111,No.2, 457-462

R.M. DE LA CRUZ and R. PAREJA
Porous Silicon Investigated by Positron Annihilation
61.70; 78.70; S5.11
(a) Vol.111,No.2, 463-468

P. MORFOULI and G. PANANAKAKIS
Electrical Study of Thin SiO_2 Layers Obtained by Anodic Oxidation of p-Type Silicon in Pure Water Using I-U, C-U,and ...
73.60; 73.40; S5.11; S10.1
(a) Vol.111,No.2, 529-540

R. NIES and F.R. KESSLER
Coherent and Incoherent Magnetooptical Reflection and Transmission Effects of Multilayer Structures
78.20; 78.50; S5.11
(a) Vol.111,No.2, 639-654

M.V.S. LAKSHMI and K. RAMKUMAR
I-U Characteristics of MOS Structures on Polycrystalline Silicon
73.40 and 85; S5.11
(a) Vol.111,No.2, 667-674

V.V. PETROV and V.V. CHORNYI
Recharging and Negative Annealing of Divacancies in Neutron-Irradiated Silicon
61.70 and 78.50; S5.11
(a) Vol.111,No.2, K141-K144

V.P. MARKEVICH and L.I. MURIN
Thermal Donor Formation in Pre-Heat-Treated n-Si:O Crystals
61.70; S5.11
(a) Vol.111,No.2, K149-K154

K.T. AVETIAN, S.H. ANCHARAKIAN, and M.M. ARAKELIAN
Stress Distribution at Biaxial Stressed Conditions
62.20; S5.11
(a) Vol.111,No.2, K161-K164

K. SCHMALZ, F.-G. KIRSCHT, and J. REICHEL
On Quenching Induced Deep Levels in Czochralski p-Silicon
71.55; S5.11
(a) Vol.111,No.2, K195-K200

K. NATARAJAN, K. RAMKUMAR, and M. SATYAM
Capacitance of p-n Junctions under Electrical Breakdown
73.40 and 85; S5.11
(a) Vol.111,No.2, K269-K272

V.V. KAPLIN, V.I. GRIDNEV, V.G. KHLABUTIN, E.I. ROZUM, and S.A. VOROBIEV
Volume Capture of 1.5 to 5.7 MeV Electrons in Quantum Channeling States
61.80; S5.11
(b) Vol.152,No.1, 61-66

GONG YU, GUANGHUA CHEN, and FANGQING ZHANG
Theoretical Study of Raman Spectrum in a-$Si_{1-x}N_x$:H Films
61.40 and 68.30; 78.30; S5.11
(b) Vol.152,No.1, 73-78

P.A. APANASEVICH, V.D. ASIMOVA, V.A. ZAPOROZHCHENKO, R.G. ZAPOROZHCHENKO, and I.V. PILIPOVICH
Energetic and Temporal Characteristics of Picosecond Pulse Self-Diffraction in Silicon
78.70; 71.25; S5.11
(b) Vol.152,No.1, 347-356

V. GEIST and D. STEPHAN
Investigation of Lattice Strain in Proton-Irradiated Semiconductor Materials
61.80; 61.10; S5.11; S5.12; S7.1
(a) Vol.112,No.1, 79-88

J. GYULAI
Some Recently Emerged Problems of Ion Implantation: A Review
61.80; 68.55; S5.11
(a) Vol.112,No.1, 237-244

F.F. KOMAROV and A.P. NOVIKOV
Self-Annealing in Ion-Implanted Silicon
61.80; S5.11
(a) Vol.112,No.1, 273-288

K. PAPROCKI, I. BRYŁOWSKA, W. ANDRAE, and J. KRYNICKI
Recoil Implantation from Sb Thin Films under Ion Bombardment
61.80 and 79.20; S5.11
(a) Vol.112,No.1, 301-304

G. BARDOS
An Approximation for Solving the Transport Equation of the Angular Flux of Ions in Matter
61.80; S1.2; S5.11
(a) Vol.112,No.1, 311-314

W. SZYSZKO and J. SIELANKO
Monte Carlo Calculation of Depth Distribution of Implanted Ions and Sputtering Yield. Comparison for Some of the ...
61.80 and 79.20; S5; S5.11
(a) Vol.112,No.1, 315-318

J. SIELANKO
Influence of the Primary Ion Beam on the Depth Distribution of Implanted Atoms Measured by SIMS
61.80; 79.20; S5.11
(a) Vol.112,No.1, 319-322

F.F. KOMAROV, A.P. NOVIKOV, E.V. KOTOV, and E.A. PODLIPKO
The Behaviour of Arsenic in Crystals of Silicon during Self-Annealing Ion Implantation
61.80; S5.11
(a) Vol.112,No.1, 323-326

J. HLAVKA and J. GRNO
Annealing Stages of Implanted Silicon
61.70; 72.40; S5.11
(a) Vol.112,No.1, 327-330

L. JONIKAS, K. JARASIUNAS, and J. VAITKUS
Study of Carrier Dynamics and Radiation Defects in Ion-Implanted Silicon by Transient Grating Technique
71.55 and 78.20; S5.11
(a) Vol.112,No.1, 375-380

V.F. STELMAKH, YU.R. SUPRUN-BELEVICH, and A.R. CHELYADINSKII
Effect of Radiation Defects and Elastic Incompatibility Stresses on the Electrical Activation and Diffusion of Boron ...
66.30; 61.80; 62.20; S5.11
(a) Vol.112,No.1, 381-384

D. JAWORSKA, E. TARNOWSKA, A.P. KOBZEV, and R.A. ILKHAMOV
Depth Distribution of the Gettered Au Atoms Measured by Means of RBS and Radioactive Methods
61.80; 66.30; S5.11
(a) Vol.112,No.1, 385-390

C. ASCHERON, A. SCHINDLER, R. FLAGMEYER, and G. OTTO
Characterization of Irradiation-Induced Lattice Disorder by Dilatometric Measurements
61.50 and 61.80; S5.11; S5.12; S7.11; S7.12
(a) Vol.112,No.1, 391-394

J. KRYNICKI and H. RZEWUSKI
Ionization Assisted Transformations of Defects in Carbon Implanted Silicon
61.70; S5.11
(a) Vol.112,No.1, 405-410

A.V. KONDRACHUK
Model of Subsurface Redistribution of Impurity Atoms and Radiation Defects upon Light Ion Irradiation at Elevated Temperature
61.80; 68.20; S5.11
(a) Vol.112,No.1, 411-414

K. MOJEJKO-KOTLINSKA, J. KOTLINSKI, and M. SUBOTOWICZ
Deep Traps in Doubly Implanted p-Type Silicon. DLTS and C-U Measurements
61.80; 71.55; S5.11
(a) Vol.112,No.1, 415-418

D. HOEHL and H. BARTSCH
Results of In-Situ Annealing of Ion-Implanted Silicon in a High Voltage Electron Microsope
61.70 and 61.80; 61.14; 68.48; S5.11
(a) Vol.112,No.1, 419-424

R. SELLMANN and J. MIMKES
Diffusion of Thallium in Silicon
66.30; S5.11
(a) Vol.112,No.1, K5-K8

V.S. LYSENKO, A.N. NAZAROV, G.A. NAUMOVETS, V.B. POPOV, and A.S. TKACHENKO
Manifestation of Hydrogen in Al-SiO_2-Si Structures Subjected to a RF Plasma Annealing
61.70; 71.55; 73.40; S5.11
(a) Vol.112,No.1, K9-K12

S.P. ZHVAVYI and O.L. SADOVSKAYA
Nanosecond Laser Annealing of Implanted Silicon. Simulation of Dynamics
64.70; 66.70; S5.11
(a) Vol.112,No.1, K19-K22

P. ALTANZOG, D. BATSUUR, S. BOLD, B. BURMAA, SH. CHADRAABAL, and A.G. KASANSKII
Temperature Dependence of Dark Conductivity and Steady-State Photoconductivity of a-Si:H Samples
72.40 and 73.60; S5.11
(a) Vol.112,No.1, K23-K28

G.B. SMITH, D.R. MCKENZIE, and P.J. MARTIN
An XPS Study of Chemical Order in Hydrogenated Amorphous Silicon-Carbon Alloy Films
61.40 and 68.55; 78.70; S5; S5.11
(b) Vol.152,No.2, 475-480

E.L. MATHE and J.C. DESOYER
Implantation of Aluminum in Silicon: Simultaneous Showing of Self-Si Defects and of Al Precipitates
61.70; S5.11
(a) Vol.112,No.2, 511-518

V.V. PETROV and V.S. PROSOLOVICH
Thermal Defects in Silicon Doped with Rare-Earth Elements
61.70 and 71.55; 72.20; 78.30; S5.11
(a) Vol.112,No.2, 561-568

W. SKORUPA, R. GROETZSCHEL, and H. BARTSCH
Recent Advances of Ion Beam Synthesis for Silicon-on-Insulator Structures
61.70 and 61.80; S5.11
(a) Vol.112,No.2, 661-666

J. AULEYTNER
Structural Changes in Solids Induced by Heavy Ions at Energies of MeV/n. Results of X-Ray and Electron Microscopic Studies
61.80; 61.10; 61.16; S5.11; S11.2
(a) Vol.112,No.2, 679-688

P.W. ŻUKOWSKI, S.B. KANTOROW, D. MĄCZKA, and W.F. STELMAKH
Procecces of Radiation Defect Interaction and Amorphisation of Silicon at Large Implantation Doses
61.70; 79.20; S5.11
(a) Vol.112,No.2, 695-698

YA.V. FATTACHOV, I.B. KHAIBULLIN, R.M. BAYAZITOV, and E.M. MISYURYOV
Anisotropic Local Melting of Semiconductors by Incoherent Light Pulses
79.20; 64.70; S5.11; S5.12
(a) Vol.112,No.2, 699-702

P.I. GAIDUK, F.F. KOMAROV, and V.S. SOLOVEV
Structure Changes in Ion Implanted (111)-Si Layers during Pulsed Nanosecond Annealing and Following Thermal Treatment
61.80 and 64.75; S5.11
(a) Vol.112,No.2, 703-706

H. BARTSCH, D. HOEHL, and P. WERNER
TEM Cross-Sectional Investigations of the Amorphous/Crystalline Phase Transition of Ion-Implanted Silicon
61.80 and 68.55; 68.48; S5.11
(a) Vol.112,No.2, 715-720

R. BANISCH, B. TILLACK, H.H. RICHTER, B. HUNGER, P. BARNA, V. SCHILLER, A. ADAM, and J. GYULAI
On the Characteristics of CMOS-Transistors in Thick SOI Films
73.40; 68.55; S5.11
(a) Vol.112,No.2, 721-726

J. CZĘRBNIAK, M. MONETA, and B. PAWŁOWSKI
The Equilibrium Charge of Planar Channeled 50 to 100 keV Protons
61.80; S5.11
(a) Vol.112,No.2, 773-776

R.A. ILKHAMOV, LI JEN HO, S.S. KHUSMURODOV, D. MACHAJDIK, G.M. OSETINSKIJ, M. PAJEK, R. SANDRIK, and D.M. SHIROKOV
Hydrogen Depth Profiling by Elastic Recoil Detection
61.70 and 61.80; S5.11
(a) Vol.112,No.2, 781-784

V.I. GAVRILENKO, N.I. KLYUI, V.G. LITOVCHENKO, and B.N. ROMANYUK
Investigation of the Ion Implanted Silicon by Electroreflectance Spectroscopy
61.70; 78.20; S5.11
(a) Vol.112,No.2, 805-810

K.D. GLINCHUK, N.M. LITOVCHENKO, and V.YU. PTITSIN
Influence of Germanium on the Formation of Oxygen-Induced Recombination Centres in Heat-Treated Silicon
61.70; 71.55; S5.11
(a) Vol.112,No.2, K65-K70

H. GDANITZ and K. SCHMALZ
On the Shape of Aus Depth Profile in Heat Treated Si Wafers
66.30; S5.11
(a) Vol.112,No.2, K75-K80

J. BELAN and V. MIKOLAJ
Voltage Noise of the Breakdown Region of Subsurface Zener Diodes
72.70; 85; S5.11
(a) Vol.112,No.2, K119-K120

J. TATARKIEWICZ
Optical Effects of High Energy Implantations in Semiconductors
61.80 and 78.20; 78.30; 78.50; 78.55; S5.11; S7.12
(b) Vol.153,No.1, 11-48

J.M. CHEN, L.M. XIE, G.R. BAI, and Y.G. ZHAO
Temperature Effects of the Several Intense IR Absorption Peaks in NTD Si:H
63.20 and 78.30; S5.11
(b) Vol.153,No.1, 107-114

R.V. KONAKOVA, YU.S. MELNIKOVA, E.V. MOZDOR, and V.I. FINEBERG
Breakdown of Silicon p^+-n-n^+ Diodes
73.40 and 85; S5.11
(a) Vol.113,No.1, 215-222

K. SCHMALZ and K. TITTELBACH-HELMRICH
On the Hydrogen Diffusion in Silicon at about 100 °C
66.30; S5.11
(a) Vol.113,No.1, K9-K14

Y. SHI, F.M. WU, D.X. SHEN, M.K. DENG, K.J. CHENG, and C.H. WANG
Influence of Previous Defects on the Formation of Irradiated Defects in NTD Si
61.80; S5.11
(a) Vol.113,No.1, K15-K18

A. POGGI and E. SUSI
Effect of Passivating Oxides on the Surface Recombination Velocity in Silicon
73.25; S5.11
(a) Vol.113,No.1, K61-K66

N.M. FICZA and A.L. TOTH
The Effect of Lead Impurity on Amorphous Si Thin Film Transistors
61.70 and 85; 68.55; S5.11; S10.1
(a) Vol.113,No.1, K147-K148

S. SATPATHY and A.P. PATHAK
Planar Channeling of Electrons and Positrons in Crystals. Local-Density Calculation of Planar Averaged Potential and ..
61.80 and 78.70; S5; S5.11; S5.12
(b) Vol.153,No.2, 455-464

H. RUECKER, R. ENDERLEIN, and F. BECHSTEDT
Strain Effects on the Band Structure of $Si/Si_{1-x}Ge_x(001)$ Superlattices
71.25 and 73.60; S5.11; S5.12
(b) Vol.153,No.2, 595-610

G.L. GUTSEV, G.S. MYAKENKAYA, V.V. FROLOV, and V.B. GLAZMAN
Nature of Hydrogen Bonding in Si-A Center
71.55; 61.70; S5.11
(b) Vol.153,No.2, 659-666

S.D. KOUIMTZI
Localized States in Crystalline Semiconductors
61.80 and 71.55; S5.11; S5.12; S7.1
(b) Vol.153,No.2, 675-682

YANG-FANG CHEN
Comment on the Determination of the Midgap Density of States in Amorphous Semiconductors Using the Space-Charge-Limited-Current ...
71.20 and 72.20; S5.11
(b) Vol.153,No.2, 695-702

T. DRUESEDAU, M. ECKLER, R. BINDEMANN, and H. FIEDLER
Enhancement of Photoconductivity in Magnetron-Sputtered a-Si:H Doped with Nitrogen
68.55 and 73.60; 72.40; 78.65; S5.11
(b) Vol.153,No.2, K119-K124

A.V. ZASTAVNOY and V.M. KOROL
Radiation Defects in n-Silicon Doped with Lithium and Sodium
61.70; 71.55; S5.11
(a) Vol.113,No.2, 277-284

S. KOPTA, R. HAJDUK, J. LEKKI, B. RAJCHEL, and A.Z. HRYNKIEWICZ
Experimental Determination of the Energy Loss of Protons Channeled along the (111) Axis through a Silicon Single Crystal
61.80; S5.11
(a) Vol.113,No.2, 295-306

L.A. KAZAKEVICH, P.F. LUGAKOV, and I.M. FILIPPOV
The Role of Dislocations during Charge-Carrier Recombination at Radiation Defects in p-Si
61.70 and 72.20; 61.80; S5.11
(a) Vol.113,No.2, 307-314

V.V. BOLOTOV, V.M. EMEKSUZYAN, V.N. SPIRIDONOV, K. SCHMALZ, and M. TRAPP
Radiation-Induced Redistribution of Gold in the SiO_2-Si Structures
61.80 and 73.20; 68.48; S5.11
(a) Vol.113,No.2, 315-320

R. KOEGLER, E. WIESER, J. ALBRECHT, and P. KNOTHE
Investigation of As-O Clustering in Si
61.70; 72.20; S5.11
(a) Vol.113,No.2, 321-330

A.V. LATYSHEV, A.L. ASEEV, A.B. KRASILNIKOV, and S.I. STENIN
Initial Stages of Silicon Homoepitaxy Studied by in situ Reflection Electron Microscopy
68.55; 61.16; S5.11
(a) Vol.113,No.2, 421-430

XIN-FAN HUANG, XI-MAO BAO, and JIAN CHEN
Hydrogen Passivation and CO_2 Laser Annealing of Interface between Ar^+ Laser Recrystallized Poly-Si and SiO_2 Substrate
68.48 and 73.40; S5.11
(a) Vol.113,No.2, 453-458

C.A. LONDOS
Defect States in Electron Bombarded n-Type Silicon
71.55; S5.11
(a) Vol.113,No.2, 503-510

A.V. DVURECHENSKII, V.A. DRAVIN, and A.I. YAKIMOV
Electrical Properties of Ion-Doped Amorphous Silicon
72.20; 72.60; S5.11
(a) Vol.113,No.2, 519-528

V.S. LYSENKO, A.N. NAZAROV, S.A. VALIEV, I.M. ZARITSKII, T.E. RUDENKO, and A.S. TKACHENKO
EPR and TSCR Investigations of Implanted Al-SiO_2-Si Systems Treated with RF Plasma Discharge
73.40 and 85; 76.30; S5.11; S10.1
(a) Vol.113,No.2, 653-666

G.A. KACHURIN, I.E. TYSHCHENKO, V.P. POPOV, and S.A. TIJS
Interaction of Nitrogen Implanted into Silicon with Defects in the Buried Layer
61.70; S5.11
(a) Vol.113,No.2, K165-K170

L. BISCHOFF and R. KOEGLER
Rapid Thermal Annealing of Neutron Transmutation Doped Silicon
61.70; 72.80; S5.11
(a) Vol.113,No.2, K185-K188

WANLU WANG and KEJUN LIAO
Persistent Photoconductivity in a-Si:H/a-SiN_x:H Multilayer Films at Low Temperature
72.40 and 73.60; S5.11; S6
(a) Vol.113,No.2, K223-K226

S.L. PINJARE, N. BALASUBRAMANYAM, and VIKRAM KUMAR
Photoluminescence at 0.944 eV from Heat Treated n-Type Silicon
78.55; 61.70; 71.55; S5.11
(a) Vol.113,No.2, K261-K264

E. SOBESLAVSKY and W. SKORUPA
Modeling of Nitrogen High Dose Implantation into Silicon in the Energy Range of 150 to 330 keV
61.80; S5.11
(a) Vol.114,No.1, 135-144

N.T. BAGRAEV, I.S. POLOVTSEV, and K. SCHMALZ
An ONP Investigation of Deep Level Defects Produced by Heat Treatment at 600 and 800 °C in Cz-Si
61.70 and 71.55; 76.60; S5.11
(a) Vol.114,No.1, 233-246

G.A. ADEGBOYEGA, A. POGGI, E. SUSI, and M. REICHE
The Effects of High Temperature Anneal on the Electrical Activity of Iron in B-Doped Cz Sìlicon
71.55; 72.20; 78.30; S5.11
(a) Vol.114,No.1, 247-252

H. WIECZOREK and W. FUHS
Deep Trapping of Carriers in a-Si:H Solar Cells Studied by Transient Photocurrents
72.40 and 73.40; 71.55; S5.11
(a) Vol.114,No.1, 413-418

J. BUDINAVICIUS, L. PRANEVICIUS, and S. TAMULEVIČIUS
Peculiarities of the Ag-Si Interface Formed under Simultaneous Deposition and Ion Beam Irradiation
61.80 and 68.48; S5.11
(a) Vol.114,No.1, K25-K28

M.YA. BAKIROV, B.A. NAYAFOV, V.S. MAMEDOV, and R.S. MADATOV
Electrical and Optical Properties of Amorphous $Ge_{1-x}Si_x$ Semiconductors
72.80 and 78.40; S5.11; S5.12
(a) Vol.114,No.1, K45-K48

B. SUKUMAR and K. NAVANEETHAKRISHNAN
Shallow Donor Binding Energies in Semiconductors in a Magnetic Field
71.55; 71.38; S5.11; S7.11
(b) Vol.154,No.2, 643-648

YONG-LIANG WANG and H. TEICHLER
LCAO Recursion Approach for the Bound Electron States at the 90°-Partial Dislocation in Silicon
71.55; S5.11
(b) Vol.154,No.2, 649-660

G.S. KOUSIK, C.M. VAN VLIET, G. BOSMAN, and HORNG-JYE LUO
Quantum 1/f Noise Associated with Intervalley Scattering in Nondegenerate Semiconductors. I. Analytical Calculations
72.10 and 72.70; S5.11
(b) Vol.154,No.2, 713-726

J. TATARKIEWICZ and W. WITTHUHN
Hydrogen Vibrations in Oxygen-Doped Silicon
61.70 and 63.20; S5.11
(b) Vol.154,No.2, K99-K102

A. BECKMANN, K. DETTMER, and F.R. KESSLER
Thermoabsorption Studies of the Indirect Exciton Bandgap of Si and GaP
78.30 and 78.40; 71.35; S5.11; S7.11
(b) Vol.154,No.2, K191-K196

M. TOULEMONDE, J. DURAL, G. NOUET, P. MARY, J.F. HAMET, M.F. BEAUFORT, J.C. DESOYER, C. BLANCHARD, and J. AULEYTNER
High Energy Heavy Ion Irradiation of Silicon
61.80; S5.11
(a) Vol.114,No.2, 467-474

L. TESAŘ and R. LENHARD
Evaluation of Silicon Wafers with Different Concentrations of Interstitial Oxygen
61.70 and 71.55; 73.40; S5.11
(a) Vol.114,No.2, 475-480

G. AMATO, G. BENEDETTO, R. SPAGNOLO, and M. TURNATURI
Photoacoustic Measurements of Doped Silicon Wafers
65.70 and 66.70; 72.80; S5.11
(a) Vol.114,No.2, 519-524

K. HEILIG, E. KOLBIG, and J. REINEKE
Electrical Properties of Silicon/Lead Boro-Alumino-Silicate Glass Interfaces: Insulator Charge and Distribution of Interface States
73.20 and 73.40; S5.11; S10
(a) Vol.114,No.2, 579-586

V. HEERA
A Critical Discussion of the Vacancy Diffusion Model of Ion Beam Induced Epitaxial Crystallization
68.48 and 68.55; S5.11
(a) Vol.114,No.2, 599-608

H.-J. AUGUST and J. WERNISCH
Analytical Expressions for the Electron Backscattering Coefficient
79.20; S1; S5; S5.11
(a) Vol.114,No.2, 629-634

J. PROKEŠ and O. STEFAN,JR.
Determination of the Capture Cross Section from DLTS Measurements
71.55; S5.11
(a) Vol.114,No.2, 643-650

H. BUDINOV, V. STAVROV, and R. BURKOVA
Flash Lamp Annealing of Phosphorus-Implanted Silicon
61.70; 79.20; S5.11
(a) Vol.114,No.2, K131-K134

B.M.S. LEWERENTZ and B.G. SVENSSON
Capacitance-Voltage Characteristics of Au-Si Schottky Diodes: Influence of Electron Irradiation
61.80 and 73.30; S5.11
(a) Vol.114,No.2, K147-K152

A.G. BUDREVICH, A.S. LYUTOVICH, and B.L. OKSENGENDLER
Radiation Sieve Effect during Epitaxy from Low Energy Ion-Molecular Flow
68.55; 61.80; S5.11
(a) Vol.114,No.2, K163-K166

H.V. ANANDA and M. SATYAM
Effect of Field Plate on the Electrical Breakdown Characteristics of Bipolar Transistors
73.40 and 85; S5.11
(a) Vol.114,No.2, K259-K262

T.D. DZHAFAROV
Radiation-Stimulated Diffusion in Semiconductors (Review Article)
61.80 and 66.30; S5.11
(b) Vol.155,No.1, 11-52

H. AOURAG, A. BELAIDI, T. KOBAYASI, R.N. WEST, and B. KHELIFA
Positron Annihilation in Si and Ge
71.10 and 78.70; S5.11; S5.12
(b) Vol.155,No.1, 191-200

V.S. LYSENKO, A.N. NAZAROV, I.M. ZARITSKII, G. SERFÓZO, G. BATTISTIG, J. GYULAI, and L. DÓZSA
RF Plasma Modification of Heavy Destroyed Ion Implanted Subsurface Silicon Layers
61.80; 73.40; S5.11
(a) Vol.115,No.1, 75-80

J.D. STEPHENSON
Simulated Quenching of Silicon WSR Monochromators Using WSR Section Topography
61.70; 65.70; S5.11; S5.12
(a) Vol.115,No.1, 97-106

S. BAUNACK and A. ZEHE
A Study of UV/Ozone Cleaning Procedure for Silicon Surfaces
68.20; 68.55; 79.20; S5.11
(a) Vol.115,No.1, 223-228

L.N. ALEKSANDROV, V.YU. BALANDIN, A.V. DVURECHENSKII, and O.A. KULYASOVA
A Two-Dimensional Model of SOI Structure Crystallization by Pulse Nanosecond Heating
68.55 and 79.20; S5.11; S10.1
(a) Vol.115,No.1, K23-K26

I.G. MARCHENKO
The Effect of Radiation Defects on the Differential Resistance of p-n Junctions during Avalanche Breakdown
61.80 and 73.40; S5.11
(a) Vol.115,No.1, K39-K42

B.C. CHAKRAVARTY, S.K. SHARMA, S.N. SINGH, B.K. DAS, RAVI KUMAR, and B.R. CHAKRABORTY
Effect of the Growth of Thin SiO Interlayers on the Performance of Si Solar Cell
68.55 and 85; 73.40; S5.11
(a) Vol.115,No.1, K125-K128

I. ZOLOMY
Theory of the Negative Resistance Avalanche MIS Tunnel Diode
73.40 and 85; S5.11
(a) Vol.115,No.1, K129-K132

V. HOLY and J. KUBENA
Characterization of Microdefects in Silicon by Means of X-Ray Reflection Curves
61.10 and 61.70; S5.11
(b) Vol.155,No.2, 339-348

O.A. GOLIKOVA and A.M. GREKHOV
Structural Network and Some Photoelectrical and Optical Parameters of Amorphous Hydrogenated Silicon
71.20; 73.60; 78.65; S5.11
(b) Vol.155,No.2, 475-480

V.I. GAVRILENKO, J. HUMLIČEK, N.I. KLYUI, and V.G. LITOVCHENKO
Electron Energy Structure and Optical Properties of Microcrystalline Silicon
78.20; 71.25; S5.11
(b) Vol.155,No.2, 723-732

F.S. NASREDINOV, M.M. MEZDROGINA, P.P. SEREGIN, U.ZH. ABDUMANAPOV, and U.S. TURSUNOV
Nature of the Impurity States Arising from Transition Metals in Hydrogenated Amorphous Silicon
71.55; S5.11
(b) Vol.155,No.2, K113-K118

J. HAERTWIG and S. GROSSWIG
Measurement of X-Ray Diffraction Angles of Perfect Monocrystals with High Accuracy Using a Single Crystal Diffractometer
61.10; S5.11
(a) Vol.115,No.2, 369-382

R. FABBRI, M. SERVIDORI, A. ZANI, and F. CEMBALI
Two-Stage Recovery of Lattice Damage Induced in Silicon by Ion Implantation below Amorphization Threshold. Analysis by...
61.80; 61.10; S5.11
(a) Vol.115,No.2, 437-444

H. BERGNER and T. DAMM
Binary States in Integrated Circuits Investigated by Laser Scanning Microscopy with Optical Beam Induced Current
72.40 and 85; S5.11
(a) Vol.115,No.2, 607-616

R.M. IMAMOV, A.A. LOMOV, and D.V. NOVIKOV
Grazing Incidence Diffraction X-Ray Topography
61.10; 68.20; S5.11
(a) Vol.115,No.2, K133-K134

K. NATARAJAN, K. RAMKUMAR, and M. SATYAM
Reverse Characteristics of Phosphorous Passivated Polysilicon p-n Junction Diodes
73.40 and 85; 61.70; S5.11
(a) Vol.115,No.2, K265-K268

M. TOMÁŠEK, Š. PICK, G.V. GADIYAK, and YU.N. MOROKOV
Simple Analysis of Atomic Forces Acting on Silicon Surfaces
68.25 and 68.40; 73.20; S5.11
(b) Vol.156,No.1, 243-252

S. NOOR MOHAMMAD and M. ABDUS SOBHAN
Doping Dependent Bandgap Narrowing in Silicon and Germanium
71.25; 71.55; S5.11; S5.12
(b) Vol.156,No.1, 287-302

G.L. GUTSEV and G.S. MYAKENKAYA
The Acceptor States of Hydrogen in Silicon
71.55; S5.11
(b) Vol.156,No.1, 319-324

M. HOEHNE
EPR of Lithium-Induced Silicon-$5d^n$ Pairs in Silicon
71.55 and 76.30; S5.11
(b) Vol.156,No.1, 325-332

V.V. TOKAREV, V.E. BORISENKO, A.I. DEMCHENKO, and T.M. PYATKOVA
Growth of Epitaxial Nickel Disilicide during Rapid Thermal Processing of Argon-Implanted Nickel Films on Silicon
68.55; 61.70; S1.61; S5.11
(a) Vol.116,No.1, 331-336

N. MAZUMDER, J.P. BANERJEE, and S.K. ROY
Variation of High Frequency Negative Resistance of Silicon n^+pp^+ and GaAs p^+nn^+ IMPATT Diodes with Enhancement of Reverse Saturation ...
73.40 and 85; S5.11; S7.12
(a) Vol.116,No.1, 415-424

I.V. ANTONOVA, A.V. VASILEV, V.I. PANOV, and S.S. SHAIMEEV
Formation of Clusters in Gold Doped Silicon
61.70; S5.11
(a) Vol.116,No.1, K33-K36

K. SCHMALZ, K. TITTELBACH, V.V. EMTSEV, and YU.N. DALUDA
Effects of Oxygen Clustering at 600 °C on the Annealing of A-Centers in Cz Silicon
61.70; S5.11
(a) Vol.116,No.1, K37-K42

L.N. ALEKSANDROV, S.L. BABENKOVA, and V.S. ZINOVEV
Thermoelastic Stresses in Multilayered Structures of Silicon-on-Insulator Type (Computational Experiment)
65.70 and 68.25; 64.30; S5.11; S10.1
(a) Vol.116,No.1, K57-K60

P. KOSTIAL, I. JAMNICKY, and J. DURCEK
Contribution to the Experimental Methods of Deep Level Acoustic Transient Spectroscopy
68.25 and 71.55; 72.50; S5.11; S10
(a) Vol.116,No.1, K77-K80

K. VAVŘINA and J. KODES
Simulation of Transient Phenomena in n-Type Si
72.20; S5.11
(a) Vol.116,No.1, K85-K90

G. SCHRAMM
Quantitative Analysis of Sb Doped Silicon by Photoluminescence
78.55; S5.11
(a) Vol.116,No.1, K97-K100

F. BECHSTEDT and D. REICHARDT
Calculation of Surface-Induced Core-Level Shifts for Covalent Semiconductors C, Si, Ge, and alpha-Sn. I.(111)2x1 Surfaces
68.20 and 73.20; S5; S5.11; S5.12
(b) Vol.156,No.2, 471-486

B. RADJENOVIC and D. TJAPKIN
The Influence of Hybridization of the Impurity and the Conduction Bands on the Density of States of Doped Semiconductors
71.20; 71.30; S5.11
(b) Vol.156,No.2, 487-496

H. AOURAG, A. BELAIDI, T. KOBAYASI, R.N. WEST, and B. KHELIFA
Electron and Positron Distribution for the Plane ($1\bar{1}0$) in Si and GaAs
71.20 and 71.60; S5.11; S7.12
(b) Vol.156,No.2, 497-506

I.YA. DUBOVSKAYA and G.K. SAVCHUK
The Parametric X-Ray Radiation in Case of Three-Wave Asymmetric Diffraction
78.70; S5.11
(b) Vol.156,No.2, 687-696

A.E. KIV, V.V. KOVALCHUK, and V.A. YANCHUK
H Adsorption by Disordered Si Clusters
68.45; S5.11
(b) Vol.156,No.2, K101-K104

Y.S. TANG
A Lineshape Analysis of the near 1.54 μm Emission Band of Er^{3+} in Silicon
78.55; S5.11
(b) Vol.156,No.2, K153-K156

D.K. AN, K. MADL, A. BARNA, G. BATTISTIG, and J. GYULAI
The Simultaneous Diffusion of Gold and Boron into Silicon. Push Effect of Gold to Boron
66.30; 61.70; S5.11
(a) Vol.116,No.2, 561-570

H.U. JAEGER, T. FEUDEL, and S. ULBRICHT
Modeling of Defect-Phosphorus Pair Diffusion in Phosphorus-Implanted Silicon
61.70; 66.30; S5.11
(a) Vol.116,No.2, 571-582

D.A. PAVLOV, A.F. KHOKHLOV, R.V. KUDRYAVTSEVA, and A.V. ERSHOV
Structure and Properties of Amorphous Silicon Doped with Isovalent Impurities
68.55 and 73.60; S5.11
(a) Vol.116,No.2, 697-702

V.A. GURTOV and P.A. RAIKERUS
Radiation Charge in Silicon Dioxide Considering Diffusion Currents
72.20; 71.55; 85; S5.11
(a) Vol.116,No.2, 769-776

V.I. KHRUPA, V.V. NIKOLAEV, and M.YA. SKOROKHOD
On the Transition between Dynamical and Kinematical X-Ray Diffraction in Thin Crystals with Randomly Distributed Dislocations
61.10; 61.70; S5.11
(a) Vol.116,No.2, K141-K146

G. SCHRAMM
Determination of Free Exciton Capture Cross Section of Si:Sb by Photoluminescence (PL)
78.55; S5.11
(a) Vol.116,No.2, K209-K212

I. ZOLOMY
Erratum to Theory of the Negative Resistance Avalanche MIS Tunnel Diode (phys. stat. sol. (a) 115, K129(1989))
73.40 and 85; S5.11
(a) Vol.116,No.2, K223-K224

S5.12. Germanium

J.-P. ZOELLNER and G. PAASCH
Conditions for the Occurence of the Electric Quantum Limit in Inversion Layers
71.25 and 73.40; S5.11; S5.12; S7.1
(b) Vol.151,No.1, 145-158

A.A. BUGAI, I.N. KRAVTSOVA, V.M. MAKSIMENKO, B.D. SHANINA, and V.I. SHAKHOVTSOV
Properties of the Si-S1 Center in $Si_{1-x}Ge_x$ Solid Solutions
71.55; 76.30; S5.11; S5.12
(b) Vol.151,No.1, 159-166

G. WEBER and L.E. OLIVEIRA
On the Binding Energies of D^- States in Doped Germanium. Effects of Valley-Orbit Corrections and Compressive Uniaxial Stress
71.55; S5.12
(b) Vol.151,No.2, 605-614

V. GEIST and D. STEPHAN
Investigation of Lattice Strain in Proton-Irradiated Semiconductor Materials
61.80; 61.10; S5.11; S5.12; S7.1
(a) Vol.112,No.1, 79-88

C. ASCHERON, A. SCHINDLER, R. FLAGMEYER, and G. OTTO
Characterization of Irradiation-Induced Lattice Disorder by Dilatometric Measurements
61.50 and 61.80; S5.11; S5.12; S7.11; S7.12
(a) Vol.112,No.1, 391-394

WEN-CAI WANG and YU CHEN
Study on the Short-Range Order Structures of Fe-Ge Amorphous Thin Films by EXAFS
61.40 and 68.55; S1.1; S5.12
(a) Vol.112,No.2, 541-550

YA.V. FATTACHOV, I.B. KHAIBULLIN, R.M. BAYAZITOV, and E.M. MISYURYOV
Anisotropic Local Melting of Semiconductors by Incoherent Light Pulses
79.20; 64.70; S5.11; S5.12
(a) Vol.112,No.2, 699-702

J. UTZIG
Moessbauer Spectroscopy of Cobalt in Germanium
76.80; S1.2; S5.12
(b) Vol.153,No.1, 377-382

S. SATPATHY and A.P. PATHAK
Planar Channeling of Electrons and Positrons in Crystals. Local-Density Calculation of Planar Averaged Potential and ..
61.80 and 78.70; S5; S5.11; S5.12
(b) Vol.153,No.2, 455-464

H. RUECKER, R. ENDERLEIN, and F. BECHSTEDT
Strain Effects on the Band Structure of $Si/Si_{1-x}Ge_x(001)$ Superlattices
71.25 and 73.60; S5.11; S5.12
(b) Vol.153,No.2, 595-610

S.D. KOUIMTZI
Localized States in Crystalline Semiconductors
61.80 and 71.55; S5.11; S5.12; S7.1
(b) Vol.153,No.2, 675-682

FANG-QING ZHANG and GUANG-HUA CHEN
Study of the Hydrogen Content and Profiles in a-Ge:H Films by Nuclear Resonant Reactor Analysis
61.70 and 68.60; S5.12
(b) Vol.154,No.1, K39-K42

D.V. NOVIKOV, E.A. KONDRASHKINA, and S.A. STEPANOV
Study of X-Ray Surface Backscattering
61.10; 68.20; S5.12
(a) Vol.114,No.1, K7-K10

M.YA. BAKIROV, B.A. NAYAFOV, V.S. MAMEDOV, and R.S. MADATOV
Electrical and Optical Properties of Amorphous $Ge_{1-x}Si_x$ Semiconductors
72.80 and 78.40; S5.11; S5.12
(a) Vol.114,No.1, K45-K48

I.P. KOGUTYUK, P.N. KEIVAN, and I.M. RARENKO
A Study of Interstitial Impurity Diffusion by the Linear Response Theory
66.30; S5.12
(b) Vol.154,No.2, 513-518

G.A. MATVEEV, I.M. TSIDILKOVSKII, and A.T. LONCHAKOV
Piezomagnetoresistance Anisotropy of n-Ge in Strong Magnetic Fields near the Metal-Insulator Transition
71.30 and 72.20; S5.12
(b) Vol.154,No.2, 703-712

H. AOURAG, A. BELAIDI, T. KOBAYASI, R.N. WEST, and B. KHELIFA
Positron Annihilation in Si and Ge
71.10 and 78.70; S5.11; S5.12
(b) Vol.155,No.1, 191-200

J.D. STEPHENSON
Simulated Quenching of Silicon WSR Monochromators Using WSR Section Topography
61.70; 65.70; S5.11; S5.12
(a) Vol.115,No.1, 97-106

V.V. LITVINOV, G.V. PALCHIK, and V.I. URENEV
On the Radiation-Enhanced Mechanism of the Thermal Donor Formation Process in Germanium
61.70 and 71.55; S5.12
(a) Vol.115,No.1, K9-K12

D.A. ARONOV, B.R. MAMATKULOV, and R. MAMATKULOV
The Influence of Diffusion on the Carrier Exclusion Effect in High-Purity (Intrinsic) Semiconductors with Non-Injecting Contacts
73.40; S5.12
(a) Vol.115,No.2, K167-K170

S. NOOR MOHAMMAD and M. ABDUS SOBHAN
Doping Dependent Bandgap Narrowing in Silicon and Germanium
71.25; 71.55; S5.11; S5.12
(b) Vol.156,No.1, 287-302

F.R. KESSLER and J. NIEDER
Impact Ionization of Indirect Excitons in Crossed Electric and Magnetic Fields in Germanium
71.35; 72.20; 72.40; S5.12
(b) Vol.156,No.1, K45-K50

B. POHORYLES
Helium Pressure Effects on Photoconductivity and Position of Dislocation Deep Electron States in Plastically Deformed Ge
61.70 and 71.55; 72.40; S5.12
(a) Vol.116,No.1, 349-356

F. BECHSTEDT and D. REICHARDT
Calculation of Surface-Induced Core-Level Shifts for Covalent Semiconductors C, Si, Ge, and alpha-Sn. I.(111)2x1 Surfaces
68.20 and 73.20; S5; S5.11; S5.12
(b) Vol.156,No.2, 471-486

S5.2. Solidified Gases

I.YA. FUGOL, A.M. RATNER, YU.I. RYBALKO, and YU.L. STESHENKO
On the Mechanism of Vibrational Relaxation of Two-Centre Excitations in Solid Neon
63.90; 71.45; 78.55; S5.2
(b) Vol.153,No.2, 539-546

S.K. MISHRA and T.N. SINGH
Crystal Dynamics of F.C.C. Inert Gas Solids
63.20; S5.2
(b) Vol.154,No.1, 127-134

O.S. BINBREK, J.F. HIGGS, and A. ANDERSON
Lattice Dynamics of Nitrogen and Carbon Dioxide
63.20; S5.2
(b) Vol.155,No.2, 427-436

N.S. SULLIVAN and Y. LIN
Low-Energy Excitations in the Orientational Glass State of Solid Hydrogen
71.20; 76.60; S5.2
(b) Vol.155,No.2, K109-K112

V.B. KOKSHENEV and N.N. ZHOLONKO
Effects of Correlated Rotation of Linear Molecules in the Low-Temperature Dynamics and Thermodynamics of Weak Molecular Solution
63.10 and 65.40; 65.70; S5.2
(b) Vol.156,No.1, 119-130

S6. Group IV compounds
(for transition metal compounds, see S1.61)

H. PAGNIA and N. SOTNIK
Metal Influence on Switching MIM Diodes
73.40 and 85; S1.3; S6; S10
(a) Vol.111,No.1, 387-394

WANLU WANG and KEJUN LIAO
Persistent Photoconductivity in a-Si:H/a-SiN_x:H Multilayer Films at Low Temperature
72.40 and 73.60; S5.11; S6
(a) Vol.113,No.2, K223-K226

GUANGHUA CHEN, FANGQING ZHANG, WEI JIA, and WEI CHEN
Study on Defects in Reactively Sputtered a-GeN_x:H Films
68.60 and 76.30; S6
(a) Vol.114,No.2, K171-K174

R. ARORA and A. KUMAR
Dielectric Relaxation in Glassy Se and $Se_{100-x}Te_x$ Alloys
77.40; S6
(a) Vol.115,No.1, 307-314

V.V. LOPATIN, A.V. KABYSHEV, and L.S. BUSHNEV
Modification of Polycrystalline BN, AlN, and Si_3N_4 Surfaces by Ion Beams
61.70 and 68.20; S6; S7
(a) Vol.116,No.1, K69-K72

V.A. NADOLINNYI, V.V. VASILEV, and I.P. MIKHAILOVSKII
Photo-Induced Metastable States in Silicon Nitride Amorphous Layers
76.30 and 78.55; S6
(a) Vol.116,No.1, K105-K110

A. MITTAS, N. GEORGOULAS, D. GIRGINOUDI, and A. THANAILAKIS
Properties of Amorphous Non-Hydrogenated Silicon-Tin Alloys
73.60 and 78.65; S6
(a) Vol.116,No.2, 725-728

L.I. BEREZHINSKII, S.I. VLASKINA, V.E. RODIONOV, and H.A. SHAMURATOV
Initial Stages of the Epitaxial Silicon Carbide Film Growth on Silicon Substrates
68.55; S6
(a) Vol.116,No.2, K169-K172

S7. Group V compounds
(for transition metal compounds, see S1.62)

L.E. SOSHNIKOV and A.U. SHELEG
Elastic Properties and Phase Transitions in beta-CdP_2
62.20 and 64.60; S7
(a) Vol.111,No.2, 485-490

K. BARTKOWSKI, G. POMPE, and E. HEGENBARTH
Specific Heat of Single-Crystalline Cd_3As_2, Cd_3P_2, and Zn_3P_2 at Low Temperatures
65.40; S7
(a) Vol.111,No.2, K165-K170

A. HUPFER, D. HIRSCH, and S. SCHULZE
Photoemission on $A^{II}_3B^{V}_2$ Semiconductor Material
71.20 and 79.60; S7
(b) Vol.152,No.2, 505-518

E. HENSEL, H. SOMMER, P. KNOTHE, and E. RICHTER
Silicon Nitride Layers on Tool Steel Produced by Ion Beam Mixing and Ion Beam Assisted Deposition
68.48 and 68.55; S1.1; S7
(a) Vol.112,No.2, 533-540

K. JEZIERSKI, J. MISIEWICZ, and F. KROLICKI
Influence of the Surface Quality on the Reflectance for Zn_3P_2
78.40; S7
(a) Vol.112,No.2, K135-K138

V.V. SOBOLEV, A.I. KOZLOV, YU.I. POLYGALOV, V.E. TUPITSYN, and A.S. POPLAVNOI
Reflectivity Spectra and Band Structure of the Zinc and Cadmium Diphosphides
71.25 and 78.40; 71.35; S7
(b) Vol.154,No.1, 377-388

K.P. GHATAK
On the Carrier Contribution to Elastic Constants of $A^{II}_3B^{V}$ Semiconductors
62.20; S7
(b) Vol.154,No.1, K29-K34

I.S. GORBAN, N.S. KORETS, TS.A. KRYSKOV, and M.V. CHUKICHEV
Cathodoluminescence in Doped CdP_2 and $CdSiP_2$ Crystals
78.60; 71.55; S7
(a) Vol.115,No.2, 555-560

N.S. KORETS, N.V. STUCHINSKAYA, and M.V. CHUKICHEV
Cathodoluminescence of Cadmium Diphosphide Crystals under Uniaxial Deformation
78.60; S7
(a) Vol.115,No.2, 561-568

V.A. RUBTSOV, E.M. SMOLARENKO, V.M. TRUKHAN, V.N. YAKIMOVICH, and L.K. ORLIK
Phase Diagram of the CdP_2-$CdAs_2$ System
64.70; S7
(a) Vol.115,No.2, K155-K158

V.V. LOPATIN, A.V. KABYSHEV, and L.S. BUSHNEV
Modification of Polycrystalline BN, AlN, and Si_3N_4 Surfaces by Ion Beams
61.70 and 68.20; S6; S7
(a) Vol.116,No.1, K69-K72

S7.1. III-V compounds

J.-P. ZOELLNER and G. PAASCH
Conditions for the Occurence of the Electric Quantum Limit in Inversion Layers
71.25 and 73.40; S5.11; S5.12; S7.1
(b) Vol.151,No.1, 145-158

N.J. CASTELLANI and W.R.L. LAMBRECHT
Core Hole Relaxation Energy in Semiconductors. Using a Pseudopotential Description of the Core Hole Perturbation
71.10; S5.11; S7.1; S8.1
(b) Vol.151,No.2, 565-570

J. MISIEWICZ, Z. GUMIENNY, R. CYWINSKI, and E. MUGENSKI
The Luminescence of Zn_3P_2 in the 0.84 to 1.1 μm Wavelength Range
78.55; 71.55; S7.1
(a) Vol.111,No.2, K249-K252

V. GEIST and D. STEPHAN
Investigation of Lattice Strain in Proton-Irradiated Semiconductor Materials
61.80; 61.10; S5.11; S5.12; S7.1
(a) Vol.112,No.1, 79-88

H.-MATSUO KAGAYA, H. KOTOKU, and T. SOMA
Thermal Pressure and Pressure-Volume Relations of III-V Covalent Crystals
62.20; 63.20; S7.1
(b) Vol.153,No.1, K17-K20

S.D. KOUIMTZI
Localized States in Crystalline Semiconductors
61.80 and 71.55; S5.11; S5.12; S7.1
(b) Vol.153,No.2, 675-682

C. PRASAD and M. SAHAY
Electronic Structure and Properties of Boron Phosphide and Boron Arsenide
71.25; S7.1
(b) Vol.154,No.1, 201-208

P. ENDERS and U. SCHADE
Addendum to: Atomic Diffusion in Tetrahedral Semiconductors. Improved Approximate Diffusion Profiles
66.30; S7.1
(b) Vol.154,No.1, K5-K8

R.K. RAM, S.S. KUSHWAHA, and A. SHUKLA
Phonon Assignments in II-VI and III-V Semiconductor Compounds Having Zincblende-Type Structure
63.20; 78.30; S7.1; S8.1
(b) Vol.154,No.2, 553-564

H. BOUDRIOT and H.A. SCHNEIDER
Influence of Dipole Scattering on the Free Carrier Absorption
78.30; 72.20; S7.1
(b) Vol.154,No.2, K175-K178

A.F. BURENKOV, V.S. VARYCHENKO, A.M. ZAITSEV, F.F. KOMAROV, G.G. KONOPLYANIK, A.A. MELNIKOV, V.F. STELMAKH, and V.V. TKACHEV
Depth Distribution of Defects and Impurities in Diamond and Cubic Boron Nitride after High-Energy Ion Implantation
61.80; 61.70; 78.60; S5; S7.1
(a) Vol.115,No.2, 427-436

E.M. SHISHONOK, V.B. SHIPILO, A.I. LUKOMSKII, and T.V. RAPINCHUK
The Nature of Some Luminescence Centers in Cubic Boron Nitride
78.60; S7.1
(a) Vol.115,No.2, K237-K242

S7.11. Phosphides

H.-J. WUENSCHE and H. WENZEL
Influence of Surface Recombination on the Rate of Electron Leakage in Ridge-Waveguide Injection Lasers
73.40 and 85; S7.11; S7.15
(a) Vol.111,No.1, 377-386

K. SIERANSKI and J. SZATKOWSKI
Electrical Properties of Zn_3P_2 at Room Temperature
72.80; S7.11
(a) Vol.111,No.1, K57-K60

T.O. BUDKO, E.V. GUSHCHINSKAYA, YU.S. EMELYANENKO, and S.A. MALYSHEV
Diffusion of Zinc into Unpassivated Surface of Indium Phosphide
66.30; S7.11
(a) Vol.111,No.2, 451-456

S.N. SAHU and A. BOURDILON
First Non-Aqueous Electrocodeposition in InP, In, and P Films
68.55; 78.65; S4; S5; S7.11
(a) Vol.111,No.2, K179-K184

E. WENDLER, W. WESCH, and G. GOETZ
Defects in Weakly Damaged Ion Implanted GaAs and Other III-V Semiconductors
61.80; 78.50; S7.11; S7.12
(a) Vol.112,No.1, 289-300

C. ASCHERON, A. SCHINDLER, R. FLAGMEYER, and G. OTTO
Characterization of Irradiation-Induced Lattice Disorder by Dilatometric Measurements
61.50 and 61.80; S5.11; S5.12; S7.11; S7.12
(a) Vol.112,No.1, 391-394

G. WAGNER, V. GOTTSCHALCH, H. RHAN, and P. PAUFLER
Pseudomorphic Growth and Nucleation of Misfit Dislocations in the Epitaxial System (001) $InP/In_{1-x}Ga_xAs$. I. Pseudomorphic Growth ...
61.70 and 68.55; S7.11; S7.12; S7.15
(a) Vol.112,No.2, 519-532

A. KOZANECKI
Backscattering and Optical Studies of Ytterbium Implanted into $A^{III}-B^{V}$ Compounds
61.80; 78.55; S7.11; S7.15
(a) Vol.112,No.2, 777-780

J.S. NKOMA
Radiation Induced Luminescence from Dipoles Immersed in a Solid with a Rough Surface
68.20 and 78.55; S7.11
(b) Vol.153,No.1, 383-390

T. KAWAKUBO and M. OKADA
Optical and Electron Paramagnetic Resonance Studies on Neutron-Irradiated GaP Crystals
71.55; 76.30; 78.50; S7.11
(b) Vol.153,No.1, K93-K96

S.K. NIKOGOSYAN and V.A. SAAKYAN
The Effect of Electron Radiation on Thermal and Electrophysical Properties of n-InP
61.80; 66.70; 72.20; S7.11
(a) Vol.113,No.1, 43-50

G. WAGNER, V. GOTTSCHALCH, H. RHAN, and P. PAUFLER
Pseudomorphic Growth and Nucleation of Misfit Dislocations in the Epitaxial System (001) $InP/In_{1-x}Ga_xAs$. II. Critical Thickness and ...
61.70 and 68.55; S7.11; S7.15
(a) Vol.113,No.1, 71-82

L.F. ZAKHARENKOV, V.F. KOBELEV, A.YA. NIKOLAICH, and V.A. OPARIN
Analysis of Precipitate Phases in Fe-Doped Indium Phosphide Crystals
64.75; S7.11
(a) Vol.113,No.1, K41-K44

L. HERNANDEZ and C. PELOSI
InGaAs/InP Schottky Barrier Diode
73.30 and 73.40; S7.11; S7.15
(a) Vol.113,No.2, 677-684

L.L. KULYUK, N.F. PERELMAN, S.I. RADAUTSAN, A.V. SIMINEL, and E.E. STRUMBAN
About the Possibility of Double Centre Formation in InP:Fe:O
71.55 and 72.40; 76.30; 78.55; S7.11
(a) Vol.113,No.2, K227-K230

E. GOLDYS
Photoionisation and Inter (Conduction) Band Absorption in S and Te Doped Gallium Phosphide
78.50; S7.11
(b) Vol.154,No.1, 397-404

G. OELGART, V. GOTTSCHALCH, H. HAEFNER, R. HEILMANN, W. KLEIN, and B. RHEINLAENDER
Characterization of GaInAsP/InP Double Heterostructure Laser-Wafers
73.40 and 85; 78.60; S7.11; S7.15
(a) Vol.114,No.1, 419-430

B. SUKUMAR and K. NAVANEETHAKRISHNAN
Shallow Donor Binding Energies in Semiconductors in a Magnetic Field
71.55; 71.38; S5.11; S7.11
(b) Vol.154,No.2, 643-648

A. BECKMANN, K. DETTMER, and F.R. KESSLER
Thermoabsorption Studies of the Indirect Exciton Bandgap of Si and GaP
78.30 and 78.40; 71.35; S5.11; S7.11
(b) Vol.154,No.2, K191-K196

V.G. MALESH, YU.M. AZHNYUK, D.B. GOYER, and A.V. GOMONNAI
Disorder-Activated First-Order Raman Scattering by Acoustic Phonons in Electron-Irradiated GaP Crystals
61.80 and 78.30; S7.11
(b) Vol.154,No.2, K197-K200

V.N. BRUDNYI, V.V. PESHEV, and S.V. SMORODINOV
New Metastable W-Center in Electron-Irradiated n-Type InP
61.80; S7.11
(a) Vol.114,No.2, K139-K142

G.J. PAPAIOANNOU, E. ANAGNOSTAKIS, P.C. EUTHYMIOU, C.D. KOURKOUTAS, and G. ZARDAS
On the Impurity Photoconductivity of Semi-Insulating GaP:Cr
72.40; 78.50; S7.11
(a) Vol.114,No.2, K215-K218

J. KREISSL, W. ULRICI, and W. GEHLHOFF
Electron Paramagnetic Resonance Study of the Dynamical Jahn-Teller Effect on the $3d^1$ Impurities Ti^{3+} and V^{4+} in GaP
71.55; 76.30; S7.11
(b) Vol.155,No.2, 597-608

S.I. RADAUTSAN, I.M. TIGINYANU, V.V. URSAKII, and N.B. PYSHNAYA
Investigation of Traps in InP:Zn Single Crystals Implanted by Phosphorus Ions
71.55; S7.11
(a) Vol.115,No.2, K191-K194

M. PESSA, H. ASONEN, J. VARRIO, and A. SALOKATVE
A Comparative Study of Molecular Beam Epitaxy Methods
(Review Article)
68.55; S7.11; S7.12
(a) Vol.116,No.2, 443-456

J. KANIEWSKI, M. KANIEWSKA, and M. JEŻEWSKI
Interpretation of DLTS Emission and Capture Data of Co Related Deep Level in GaP
71.55; S7.11
(a) Vol.116,No.2, 745-754

V. RIEDE, H. SOBOTTA, H. NEUMANN, C. ASCHERON, C. NEELMEIJER, and A. SCHINDLER
Depth Dependence of Localized Mode Absorptions in Proton-Implanted InP
61.80 and 78.30; 63.20; S7.11
(a) Vol.116,No.2, K147-K152

L. PRANEVICIUS, V. SARGUNAS, and G. ZUBAUSKAS
Component Autodiffusion into Protective Coatings in Ion-Implanted InP
66.30 and 68.48; S7.11
(a) Vol.116,No.2, K161-K164

S7.12. Arsenides

J. SABIN DEL VALLE, J. LOPEZ GONDAR, and M. DE DIOS LEYVA
Landau Levels and Cyclotron Masses in GaAs-AlGaAs Quantum Wells
71.25; S7.12; S7.15
(b) Vol.151,No.1, 127-132

J. HEROLD, H. BARTSCH, W. WESCH, and G. GOETZ
Investigation of Low-Temperature Epitaxial Regrowth of Ion-Implanted Amorphous GaAs
68.48; 61.16; 61.80; S7.12
(a) Vol.111,No.1, 59-70

J. PASTRNAK, J. OSWALD, I. GREGORA, V. VORLICEK, and M. BABINSKY
Lateral Distribution of Si Related Defects and Carrier Density in GaAs Crystals
78.55; 71.55; 78.30; S7.12
(a) Vol.111,No.1, 345-360

A. SAKALAS and S. ZHUKAUSKAS
Unified Inverted Defect Model for GaAs Schottky Barrier Formation
73.30; S7.12
(a) Vol.111,No.1, K43-K48

J. GONG and S.H. YANG
Annealing Temperature Dependence of the 1/f Noise in Si-Implanted GaAs
61.70 and 72.70; S7.12
(a) Vol.111,No.1, K53-K56

J.D. COLLINS, G.A. GLEDHILL, R. MURRAY, P.S. NANDHRA, and R.C. NEWMAN
The Selective Trapping of Arsenic Interstitial Atoms by Impurities in Gallium Arsenide
61.70; 78.50; S7.12
(b) Vol.151,No.2, 469-478

L.B. GUOBA, A.A. VITKAUSKAS, J.V. KAMENECKAS, V.R. SARGUNAS, and A.P. SALALAS
The Influence of Ion-Irradiation on Interdiffusion and Annealing Behavior of Ohmic Contacts to GaAs
68.48; 79.20; S7.12
(a) Vol.111,No.2, 507-514

R. FLAGMEYER, G. OELGART, K. KREHER, R. SCHWABE, F. BUGGE, L. LEHMANN, and K. JACOBS
Characterization of MOCVD Grown (Al,Ga)As/GaAs Single Quantum Well Structures by Rutherford Backscattering and Photoluminescence
73.40 and 85; 78.55; S7.12
(a) Vol.111,No.2, 541-550

I. LUKES, J. HUMLICEK, V. VORLICEK, and M. ZAVETOVA
Infrared Reflectance of the GaAlAs/GaAs System
78.30; 73.40; S7.12
(a) Vol.111,No.2, 655-666

L. ZAKHARENKO, V. KOZLOVSKII, and YA. PILKEVICH
Role of Irradiation Temperature in the Conductivity Compensation of n-GaAs by Radiation Defects
61.80 and 72.80; S7.12
(a) Vol.111,No.2, K215-K218

B.C. DE COOMAN and C.B. CARTER
The Formation of Faulted Dipoles in Plastically Deformed GaAs
61.70; 61.16; 62.20; S7.12
(a) Vol.112,No.1, 41-54

W.I. KHAN
A New Model for the Negative Resistances in Semiconductor Devices
73.40 and 85; S7.12
(a) Vol.112,No.1, 221-228

E. WENDLER, W. WESCH, and G. GOETZ
Defects in Weakly Damaged Ion Implanted GaAs and Other III-V Semiconductors
61.80; 78.50; S7.11; S7.12
(a) Vol.112,No.1, 289-300

V.F. KOROTOV, V.I. KHITKO, V.A. YURCHENKO, and V.S. PIVOVAR
On the Specific Features of Gallium Arsenide Doping with Donor by Ion Implantation
61.80 and 73.40; S7.12
(a) Vol.112,No.1, 331-334

C. ASCHERON, A. SCHINDLER, R. FLAGMEYER, and G. OTTO
Characterization of Irradiation-Induced Lattice Disorder by Dilatometric Measurements
61.50 and 61.80; S5.11; S5.12; S7.11; S7.12
(a) Vol.112,No.1, 391-394

S.R. BLIGHT and H. THOMAS
Observation of Bulk Traps in Ion-Implanted GaAs MESFETs
71.55; 85; S7.12
(a) Vol.112,No.1, K29-K34

W. SIEGEL, H. WITTE, and G. KUEHNEL
Inhomogeneity of Semi-Insulating GaAs Wafers with Nearly Uniform Resistivity Distribution
72.80; S7.12
(a) Vol.112,No.1, K35-K38

B.C. DE COOMAN and C.B. CARTER
High-Resolution Microscopy of Dissociated Screw Dislocations in GaAs
61.70; 61.16; S7.12
(a) Vol.112,No.2, 473-492

G. WAGNER, V. GOTTSCHALCH, H. RHAN, and P. PAUFLER
Pseudomorphic Growth and Nucleation of Misfit Dislocations in the Epitaxial System (001) InP/$In_{1-x}Ga_xAs$. I. Pseudomorphic Growth ...
61.70 and 68.55; S7.11; S7.12; S7.15
(a) Vol.112,No.2, 519-532

J.J. LIOU
Theoretical Study of Forward-Voltage Heterojunction Space-Charge-Region Capacitance Including Interface-State Effects
73.40 and 85; S7.12
(a) Vol.112,No.2, 651-660

M. KULIK and K. ZUKOWSKA
The Multilayer Model of Implanted GaAs. Ellipsometric Investigation
61.80; 78.20; S7.12
(a) Vol.112,No.2, 785-788

M. KULIK and J. ZUK
Ellipsometric Investigation of the Kinetics of Native Oxide Growth on Aluminium and Argon Implanted GaAs
61.80 and 68.55; 78.20; S7.12
(a) Vol.112,No.2, 789-792

M. KULIK, D. JAWORSKA, and J. TRZCIANKA
Determination of the Damage Profiles in Implanted GaAs from the Change of (n,k) Optical Parameters
61.80; 78.20; S7.12
(a) Vol.112,No.2, 793-798

T. SKRABKA, M.J. SŁABY, and B. MIZERA
Ohmic Contacts to n-Type GaAs Using RTA
73.40; 85; S7.12
(a) Vol.112,No.2, K105-K110

J. TATARKIEWICZ
Optical Effects of High Energy Implantations in Semiconductors
61.80 and 78.20; 78.30; 78.50; 78.55; S5.11; S7.12
(b) Vol.153,No.1, 11-48

T. FUKUMORI and K. FUTAGAMI
Nonuniformity on GaAs Wafers Revealed by an X-Ray Double-Crystal Method
61.10 and 61.70; S7.12
(a) Vol.113,No.1, K19-K22

J. KLEJST and J. BRACINIK
Photoacoustic Study of Semi-Insulating GaAs:Cr
62.65 and 66.70; S7.12
(a) Vol.113,No.1, K37-K40

K. BHATTACHARYYA
Alloy Scattering Limited Mobility of Two-Dimensional Electron Gas in $SD-Al_{0.3}Ga_{0.7}As/Ga_{0.87}In_{0.13}As/GaAs$ Structures
72.20 and 73.40; S7.12; S7.15
(b) Vol.153,No.2, K135-K140

H. WEINERT, A. ZUKAUSKAS, G. TAMULAITIS, and V. STEPANKEVIČIUS
Some Hints on the Influence of the Barrier on Excitonic Optical Properties of GaAs-GaAlAs Superlattices Obtained by ...
78.55; 71.35; 78.30; S7.12; S7.15
(b) Vol.153,No.2, K193-K198

D. BAITHER and D. PANKNIN
Residual Defects in Implanted GaAs after Rapid Thermal Annealing with Incoherent Light
61.70; S7.12
(a) Vol.113,No.2, 331-336

Z.G. IVANOV
Superconducting Transport Properties of Ohmic Contacts
73.40 and 74.50; S7.12
(a) Vol.113,No.2, 439-446

T. HARA, A. SUGA, and R. ICHIKAWA
Properties of CVD WSi_x Films and WSi_x/GaAs Schottky Barriers
73.30 and 73.60; 73.20; S1.61; S7.12
(a) Vol.113,No.2, 459-466

K.D. GLINCHUK, N.S. ZAYATS, and A.V. PROKHOROVICH
Effect of Proton Bombardment on GaAs Luminescence
71.55 and 78.55; 61.80; S7.12
(a) Vol.113,No.2, 497-502

K. GERMANOVA, V. DONCHEV, V. VALCHEV, and CH. HARDALOV
Characterization of Medium-Doped n-Type GaAs by Hall Measurements
72.80; S7.12
(a) Vol.113,No.2, K231-K234

M.O. MANARESH, D.W. FISCHER, and W.C. MITCHEL
The EL2 Defect in GaAs: Some Recent Developments (Review Article)
61.70 and 71.55; 72.40; 78.30; 78.50; 78.55; S7.12
(b) Vol.154,No.1, 11-42

D. CHATTOPADHYAY and A. KABASI
Warm Electron Transport in GaAs Quantum Wells
72.20 and 73.40; S7.12
(b) Vol.154,No.1, 281-286

N.D. THEODORE, B.C. DE COOMAN, and C.B. CARTER
Extrinsic Dissociation of a Dislocation in GaAs in the Presence of Point Defects
61.70; S7.12
(a) Vol.114,No.1, 105-112

I.I. KHODOS, M.SH. SHIKHSAIDOV, I.I. SNIGHIREVA, A.P. USHAKOVA, and V.I. NIKOLAICHIK
The Dissociation of Dislocations in Pure and Doped GaAs Crystals
61.70; S7.12
(a) Vol.114,No.1, 113-126

G. E. ZARDAS, Z. HATZOPOULOS, and P.H. YANNAKOPOULOS
Effect of Low Dose Electron Irradiation on AlGaAs/GaAs at Low Temperatures
61.80 and 72.40; S7.12; S7.15
(a) Vol.114,No.1, K41-K44

F. BECHSTEDT and H. GERECKE
Long-Wavelength Optical Phonons in GaAs-AlAs Superlattices
63.20 and 68.30; 73.40; S7.12
(b) Vol.154,No.2, 565-582

R. GOLDHAHN and H. ROEPPISCHER
Electroreflectance Lineshape of Step-Like Doped GaAs Structures
78.20; S7.12
(b) Vol.154,No.2, 769-776

V.N. BRUDNYI, A.D. POGREBNYAK, YU.P. SUROV, and A.S. RUDNEV
Positron Trapping at Point Defects in Electron-Irradiated GaAs
61.70; 61.80; 78.70; S7.12
(a) Vol.114,No.2, 481-490

K. SOMOGYI, D.N. KORBUTYAK, L.N. LASHKEVICH, and I. POZSGAI
Cu Diffusion into GaAs from Water Based Solution
66.30 and 73.40; 78.65; S7.12
(a) Vol.114,No.2, 635-642

Y.N. MOHAPATRA and V. KUMAR
Temperature Dependence of Photocurrent in Undoped Semi-Insulating Gallium Arsenide
72.40; 71.55; S7.12
(a) Vol.114,No.2, 659-664

A. BROZI
A New Computer Model of the MOCVD Process
68.55; S7.12
(a) Vol.114,No.2, K167-K170

H. WEINERT, K. JARAŠIŪNAS, and B. SCHUMANN
Carrier Diffusion and Recombination in Type II GaAs/$Ca_xSr_{1-x}F_2$ Quantum Well Heterostructures Studied by the Transient Gating Technique
73.40; 68.55; S7.12; S9.15
(a) Vol.114,No.2, K183-K188

K.S. SRIVASTAVA and A. TANDON
Dispersion Relation for Coupled Surface Waves in Thin Polar Semiconducting Foils
68.30; 71.45; S7.12
(b) Vol.155,No.1, 127-136

M. AIDARALIEV, N.V. ZOTOVA, S.A. KARANDASHOV, B.A. MATVEEV, N.M. STUS, and G.N. TALALAKIN
Spontaneous and Stimulated Emission from InAsSbP/InAs Heterostructures
78.45 and 78.55; S7.12; S7.15
(a) Vol.115,No.1, K117-K120

EN-GE WANG, LI-YUAN ZHANG, and HUAI-YU WANG
Electronic Properties of Vacancy-Induced Structures in AlAs/GaAs Superlattices
71.55; 71.20; 73.20; S7.12
(b) Vol.155,No.2, 587-596

T. KITANO, T. ISHIKAWA, and J. MATSUI
Contrast Formation Mechanism around the Cell Walls in Equi-Lattice-Spacing Mapping X-Ray Topographs for an Undoped...
61.10; S7.12
(a) Vol.115,No.2, 383-388

K. LENKEIT, A.K. GUTAKOVSKII, YU.O. KANTER, R. FLAGMEYER, S.M. PINTUS, S.V. RUBANOV, A.A. FEDOROV, and V.P. POPOV
Study of InGaAs-GaAs Strained-Layer Superlattices by TEM and RBS Techniques
68.48; 61.80; S7.12; S7.15
(a) Vol.115,No.2, 413-426

E. NEBAUER and U. LOHSE
Lamp Furnace-Annealed Au/Ni/AuGe Ohmic Contacts for GaAs MESFETs
73.40; 85; S7.12
(a) Vol.115,No.2, K171-K174

F. BECHSTEDT and H. GERECKE
Relationship of Microscopic and Macroscopic Theories for Long-Wavelength Optical Phonons in GaAs-AlAs Superlattices
63.20; 68.30; S7.12; S7.15
(b) Vol.156,No.1, 151-170

I.YU. GOLINEY and V.N. RUDKO
Surface Electromagnetic Waves in Quasiperiodic Superlattices
78.65; S7.12; S7.15
(b) Vol.156,No.1, 211-224

B.G. MULIMANI, C.H. RAMAKRISHNA, B.S. KRISHNAMURTHY, and C.K. SARKAR
Transient Response of Hot Electrons in Semiconductor Quantum Well Wires
72.10 and 72.40; S7.12
(b) Vol.156,No.1, 253-258

A. ZEHE and A. JUAREZ
Double-Peak Formation as a Doping-Superlattice Effect in Donor-Acceptor Pair-Luminescence Spectra
71.55 and 78.55; 73.40; S7.12
(b) Vol.156,No.1, K89-K92

A.M. AFANASEV, R.M. IMAMOV, E.KH. MUKHAMEDZHANOV, and A.A. BZHEUMIKHOV
Double-Channel X-Ray Standing Wave Technique
61.10; 78.70; S7.12
(a) Vol.116,No.1, 197-202

N. MAZUMDER, J.P. BANERJEE, and S.K. ROY
Variation of High Frequency Negative Resistance of Silicon n^+pp^+ and GaAs p^+nn^+ IMPATT Diodes with Enhancement of Reverse Saturation ...
73.40 and 85; S5.11; S7.12
(a) Vol.116,No.1, 415-424

L. HRIVNAK
Compositional Dependences of Band-Gap Energy and Electron and Light Hole Effective Masses in $In_xGa_{1-x}As$
71.25; S7.12; S7.15
(a) Vol.116,No.1, K73-K76

W. HOERIG, V.F. ORLENKO, H. NEUMANN, and C. ASCHERON
Near-Edge Optical Absorption Spectra of Proton-Implanted GaAs
61.70 and 78.50; S7.12
(a) Vol.116,No.1, K119-K124

P.C. EUTHYMIOU, P.C. BANBURY, G.J. PAPAIOANNOU, and G.E. ZARDAS
Dependence of I-V Characteristics of GaAs MESFET on Temperature and Eletron Irradiation
73.40 and 85; 61.80; S7.12
(a) Vol.116,No.1, K133-K136

M. ASCHE, L. EAVES, K. FRIEDLAND, H. KOSTIAL, and O.G. SARBEY
Transverse Transport in $n^+GaAs/Al_xGa_{1-x}As/n$-GaAs Structures
73.40; S7.12; S7.15
(b) Vol.156,No.2, 461-470

H. AOURAG, A. BELAIDI, T. KOBAYASI, R.N. WEST, and B. KHELIFA
Electron and Positron Distribution for the Plane (1$\bar{1}$0) in Si and GaAs
71.20 and 71.60; S5.11; S7.12
(b) Vol.156,No.2, 497-506

D. VIGNAUD and J.L. FARVACQUE
Free-Carrier Optical Absorption Induced by Dislocation Scattering Mechanisms in Semiconductors: Application to GaAs
78.20; 61.70; S7.12
(b) Vol.156,No.2, 717-724

S. ERKOÇ, T. HALICIOĞLU, and W.A. TILLER
Adsorption of Gold Adatom on the GaAs(110) Surface: A Molecular-Dynamics Simulation
68.45; S7.12
(b) Vol.156,No.2, K105-K108

M. PESSA, H. ASONEN, J. VARRIO, and A. SALOKATVE
A Comparative Study of Molecular Beam Epitaxy Methods (Review Article)
68.55; S7.11; S7.12
(a) Vol.116,No.2, 443-456

S7.13. Antimonides

P. AGHAMKAR and P.K. SEN
Stimulated Brillouin Scattering in Doped Centrosymmetric Semiconductors
78.35; S7.13
(b) Vol.151,No.2, 739-746

V.I. ALEKSEENKO, V.I. BARBASHOV, L.A. ZILBERMAN, V.M. MOSTOVOI, and M.YA. SKOROKHOD
Effects of Dislocation Loop Interaction in Indium Antimonide. Dynamics and Geometric Aspects
61.70; S7.13
(a) Vol.111,No.2, K145-K148

R. MICHALEC, B. CHALUPA, and J. VAVRA
Neutron Acoustic Resonance in Perfect InSb Single Crystals
61.12 and 62.65; S7.13
(a) Vol.112,No.1, 29-34

W. KRAAK, S. TROTT, and R. HERRMANN
Weak Localization Effects in n-Type Inversion Layers of InSb Grain Boundaries under High Hydrostatic Pressure
73.40; S7.13
(b) Vol.152,No.2, 481-488

M. DEY(BOSE) and S. GHOSH
Amplificaion of Acoustic Waves in Magnetised Piezoelectric Semiconductors. Effect of Nonlinearity in Electron Effective Mass ...
72.50; 71.25; S7.13
(b) Vol.152,No.2, 625-632

A. NEOGI and S. GHOSH
Parametric Amplification in Magnetised Piezoelectric Semiconducting Plasmas
78.20; 77.60; S7.13
(b) Vol.152,No.2, 691-700

V.YA. KOSTYUCHENKO, E.M. SKOK, and S.A. STUDENIKIN
Magneto-Resonant Absorption from the Local Center in Degenerate n-InSb
78.20; 71.55; S7.13
(b) Vol.152,No.2, K59-K62

J. GALIBERT, P. PERRIER, and S. ASKENAZY
Résonance magnétophonon avec retournement de spin dans l'antimoniure d'indium
63.20 and 71.70; 72.20; S7.13
(b) Vol.153,No.2, 653-658

E. MAJKOVÁ, J. ČERVENÁK, J. KREMPASKY, and P. DUHAJ
Temperature Dependence of the Seebeck Coefficient in InSb Prepared by Rapid Quenching
72.20; S7.13
(b) Vol.153,No.2, K147-K150

DOAN NHAT QUANG
On the Theory of Auger Recombination in Disordered Semiconductors. A Green's Function Approach
71.25 and 72.20; S7.13
(b) Vol.154,No.1, 237-248

J. TATARKIEWICZ, M. CARDONA, A. BREITSCHWERDT, C.-T. LIN, R. GROETZSCHEL, and W. WITTHUHN
In Search of Hydrogen Local Vibrational Modes in CuCl, CuBr, CuI, ZnS, and AlSb
63.20; 78.30; S7.13; S8.11; S9
(b) Vol.155,No.1, 77-88

M. MEBARKI
GaSb(n)-$Ga_{0.83}Al_{0.17}Sb$(p) Heterojunction by Capacitance-Voltage Measurements
73.40; 73.20; S7.13; S7.15
(a) Vol.115,No.1, 173-178

V.I. BARBASHOV and T.V. RODZINA
Polarity of the Screw-Dislocation Motion in InSb Crystals
61.70; S7.13
(a) Vol.115,No.2, K143-K146

A. NEOGI and S. GHOSH
Stimulated Brillouin Scattering in a Transversely Magnetised Doped Centrosymmetric Semiconductor
78.35; S7.13
(b) Vol.156,No.2, 705-716

S7.15. Mixed crystals

J. SABÍN DEL VALLE, J. LÓPEZ GONDAR, and M. DE DIOS LEYVA
Landau Levels and Cyclotron Masses in GaAs-AlGaAs Quantum Wells
71.25; S7.12; S7.15
(b) Vol.151,No.1, 127-132

J. ŠIKULA, P. VAŠINA, and I. HUETTEL
The Noise of the GaAlAs Super-LED Radiation
72.70 and 73.40; 78.60; S7.15
(a) Vol.111,No.1, 367-370

H.-J. WUENSCHE and H. WENZEL
Influence of Surface Recombination on the Rate of Electron Leakage in Ridge-Waveguide Injection Lasers
73.40 and 85; S7.11; S7.15
(a) Vol.111,No.1, 377-386

C. GHEZZI, D. MARTIN, A. PARISINI, J.L. STAEHLI, and L. TARRICONE
Investigation of (AlGa)As/GaAs Quantum Wells by Optical and Phototransport Measurements
73.40; 71.35; 72.40; 78.55; S7.15
(b) Vol.152,No.1, 101-110

W. THIELEMANN, B. RHEINLAENDER, P. NEVERMANN, and G. OELGART
Determination of the Internal Quantum Yield and of the Injection Effiency of Red-Light Emitting (Al,Ga)As-DHS Diodes
73.40 and 78.60; 68.55; 78.55; S7.15
(a) Vol.112,No.1, 229-236

G. WAGNER, V. GOTTSCHALCH, H. RHAN, and P. PAUFLER
Pseudomorphic Growth and Nucleation of Misfit Dislocations in the Epitaxial System (001) InP/$In_{1-x}Ga_xAs$. I. Pseudomorphic Growth ...
61.70 and 68.55; S7.11; S7.12; S7.15
(a) Vol.112,No.2, 519-532

K. ŽĎAŇSKÝ
Threshold Current Optimization of InGaAsP/InP Lasers with Waveguiding Layers
73.40 and 85; S7.15
(a) Vol.112,No.2, 639-650

A. KOZANECKI
Backscattering and Optical Studies of Ytterbium Implanted into A^{III}-B^{V} Compounds
61.80; 78.55; S7.11; S7.15
(a) Vol.112,No.2, 777-780

V. RIEDE, H. NEUMANN, N. SHARIF, F.S. HASOON, and H. SOBOTTA
Structural and Infrared Optical Properties of $(CuInSe_2)_{1-x}(2InAs)_x$ Solid Solutions
61.60 and 78.30; S7.15; S8.15
(a) Vol.112,No.2, K139-K144

N.A. GUSHCHINA, V.K. NIKULIN, and YU.N. TZAREV
On the Band Structure of Ultrathin Layer Semiconductor Superlattices
73.40; 71.25; S7.15
(b) Vol.153,No.1, 161-166

G. WAGNER, V. GOTTSCHALCH, H. RHAN, and P. PAUFLER
Pseudomorphic Growth and Nucleation of Misfit Dislocations in the Epitaxial System (001) $InP/In_{1-x}Ga_xAs$. II. Critical Thickness and ...
61.70 and 68.55; S7.11; S7.15
(a) Vol.113,No.1, 71-82

K. BHATTACHARYYA
Alloy Scattering Limited Mobility of Two-Dimensional Electron Gas in SD-$Al_{0.3}Ga_{0.7}As/Ga_{0.87}In_{0.13}As/GaAs$ Structures
72.20 and 73.40; S7.12; S7.15
(b) Vol.153,No.2, K135-K140

H. WEINERT, A. ŽUKAUSKAS, G. TAMULAITIS, and V. STEPANKEVIČIUS
Some Hints on the Influence of the Barrier on Excitonic Optical Properties of GaAs-GaAlAs Superlattices Obtained by ...
78.55; 71.35; 78.30; S7.12; S7.15
(b) Vol.153,No.2, K193-K198

N.N. FALEEV, L.I. FLAKS, and S.G. KONNIKOV
Double Crystal X-Ray Diffraction Characterization of Multilayer Heteroepitaxial Structures with Submicron Layers
68.55; 61.10; S7.15
(a) Vol.113,No.2, 431-438

L. HERNANDEZ and C. PELOSI
InGaAs/InP Schottky Barrier Diode
73.30 and 73.40; S7.11; S7.15
(a) Vol.113,No.2, 677-684

U. PIETSCH and J. WOLF
The Thermal Expansion Coefficient of Lattice Matched $Ga_{1-x}Al_xAs$ Layers Grown on a [001] GaAs Substrate
65.70; S7.15
(a) Vol.113,No.2, K203-K206

R. FLAGMEYER, U. PIETSCH, H. RHAN, M. HOERICKE, and B. JENICHEN
Strain Relaxation in $In_xGa_{1-x}As/GaAs$ Heterostructures
68.55; S7.15
(a) Vol.113,No.2, K211-K216

C.P. Chang, K.F. PAI, C.S. FANG, and W.S. TSE
Dispersion Curves of Phonon-Plasmon Coupled Polaritons in n-Type $Al_xGa_{1-x}As$
63.20; 71.36; 78.30; S7.15
(b) Vol.154,No.1, 135-142

P.S. GHOSH, D.K. GHOSH, and L.K. SAMANTA
Studies of Optical Properties and Applications of Some Mixed Ternary Semiconductors
78.20; 85; S7.15
(a) Vol.114,No.1, 381-386

G. OELGART, V. GOTTSCHALCH, H. HAEFNER, R. HEILMANN, W. KLEIN, and B. RHEINLAENDER
Characterization of GaInAsP/InP Double Heterostructure Laser-Wafers
73.40 and 85; 78.60; S7.11; S7.15
(a) Vol.114,No.1, 419-430

G. E. ZARDAS, Z. HATZOPOULOS, and P.H. YANNAKOPOULOS
Effect of Low Dose Electron Irradiation on AlGaAs/GaAs at Low Temperatures
61.80 and 72.40; S7.12; S7.15
(a) Vol.114,No.1, K41-K44

K.P. GHATAK, B. MITRA, and A. GHOSHAL
On the Effective Electron Mass in Superlattices of III-V Semiconductors in the Presence of a Quantizing Magnetic Field
73.40; S7.15
(b) Vol.154,No.2, K121-K126

O. ZIEMANN, B. RHEINLAENDER, G. SCHWARZ, and R. SCHULZ
Influence of Time-Dependent Lateral Temperature Profile on Power-Current Characteristics of (Al,Ga)As Oxide Stripe Lasers
78.45; S7.15
(a) Vol.114,No.2, K237-K240

M. MEBARKI
GaSb(n)-$Ga_{0.83}Al_{0.17}Sb$(p) Heterojunction by Capacitance-Voltage Measurements
73.40; 73.20; S7.13; S7.15
(a) Vol.115,No.1, 173-178

G. OELGART, B. LIPPOLD, R. HEILMANN, H. NEUMANN, and B. JACOBS
Determination of the Germanium Acceptor Ionisation Energy of $Al_xGa_{1-x}As$ ($0<x<0.40$) by Hall Effect and Luminescence
71.55; 72.80; 78.55; S7.15
(a) Vol.115,No.1, 257-266

M. AIDARALIEV, N.V. ZOTOVA, S.A. KARANDASHOV, B.A. MATVEEV, N.M. STUS, and G.N. TALALAKIN
Spontaneous and Stimulated Emission from InAsSbP/InAs Heterostructures
78.45 and 78.55; S7.12; S7.15
(a) Vol.115,No.1, K117-K120

Ľ. MALACKÝ, J. KUZMÍK, and V. CAMBEL
Optical Properties of Metal/GaInAs Contacts for Schottky Photodiodes
78.65 and 85; S7.15
(a) Vol.115,No.1, K121-K124

K. LENKEIT, A.K. GUTAKOVSKII, YU.O. KANTER, R. FLAGMEYER, S.M. PINTUS, S.V. RUBANOV, A.A. FEDOROV, and V.P. POPOV
Study of InGaAs-GaAs Strained-Layer Superlattices by TEM and RBS Techniques
68.48; 61.80; S7.12; S7.15
(a) Vol.115,No.2, 413-426

F. BECHSTEDT and H. GERECKE
Relationship of Microscopic and Macroscopic Theories for Long-Wavelength Optical Phonons in GaAs-AlAs Superlattices
63.20; 68.30; S7.12; S7.15
(b) Vol.156,No.1, 151-170

I.YU. GOLINEY and V.N. RUDKO
Surface Electromagnetic Waves in Quasiperiodic Superlattices
78.65; S7.12; S7.15
(b) Vol.156,No.1, 211-224

K. ŽĎÁNSKÝ and THAN BINH LE
Kinetics of DX Centers with Illumination. The Justification of the Band Structure Model
71.55; S7.15
(a) Vol.116,No.1, 357-366

Ľ. HRIVNÁK
Compositional Dependences of Band-Gap Energy and Electron and Light Hole Effective Masses in $In_xGa_{1-x}As$
71.25; S7.12; S7.15
(a) Vol.116,No.1, K73-K76

M. ASCHE, L. EAVES, K. FRIEDLAND, H. KOSTIAL, and O.G. SARBEY
Transverse Transport in $n^+GaAs/Al_xGa_{1-x}As/n$-GaAs Structures
73.40; S7.12; S7.15
(b) Vol.156,No.2, 461-470

C.K. SARKAR and P. BANERJI
Estimation of the Alloy Scattering Potential in $Ga_{0.47}In_{0.53}As$ Using Cyclotron Resonance Measurements
76.30; 72.20; S7.15
(b) Vol.156,No.2, K145-K148

S7.16. Higher compounds

M. LEVALOIS and G. ALLAIS
Etude par diffraction de R-X de la densité de charge de valence dans les deux semi-conducteurs tétraédriques $ZnSiAs_2$ et $ZnGeAs_2$
71.20; S7.16
(a) Vol.111,No.1, 211-222

V.YU. RUD, YU.V. RUD, M. SERGINOV, and M.A. TAIROV
Production of and Physical Processes in n-p-$CdSiAs_2$ Structures
72.40 and 73.40; 85; S7.16
(a) Vol.113,No.1, 207-214

G. ALLAIS, A. HAIRIE, M. LEVALOIS, R. MADELON, and E. PAUMIER
La densité de charge de valence dans $ZnSiAs_2$ et $ZnGeAs_2$. Comparison des resultats de pseudo-potentieal avec l'experience
71.25; S7.16
(a) Vol.113,No.2, 489-496

G.A. MEDVEDKIN, YU.V. RUD, and M.A. TAIROV
Photoelectric Anisotropy of II-IV-V_2 Ternary Semiconductors (Review Article)
72.40; 78.20; S7.16
(a) Vol.115,No.1, 11-50

D. MEIMARIS, F. VIRAS, and P. BEKRIS
A Study of the Structural Changes in Cd-Ge-Sb-As Glass during Transition to the Crystalline State Induced by Thermal Treatment
64.70 and 72.80; 61.40; S7.16
(a) Vol.115,No.2, 483-490

G. BABONAS and S. MARCINKEVIČIUS
Photovoltaic Effects in $CdGeP_2$
72.40; S7.16
(a) Vol.115,No.2, K211-K214

S8. Chalcogenides
(for transition metal compounds, see S1.63)

S. BISWAS and R. BHATTACHARYA
Two Valence Subbands in Single Crystals of Bismuth Telluride Doped with Lead and Its Electrical Properties
72.20; 71.25; S8
(b) Vol.151,No.1, 193-202

C.D. LOKHANDE and S.H. PAWAR
Electrodeposition of Thin Film Semiconductors (Review Article)
68.55; 73.40; 85; S8
(a) Vol.111,No.1, 17-40

J.C. BERNÈDE, A. MALLOUKY, and J. POUZET
Etude de la résistivité de couches minces polycristallines de $MoSe_2$
73.60; S8
(a) Vol.111,No.1, 181-192

F.M. KAMARLY, N.G. GUSEINOV, and A.YU. YANGIROV
Magnetic Susceptibility and Electrophysical Properties of SiTe Single Crystals
61.70; 72.80; 75.50; S8
(a) Vol.111,No.1, K17-K20

V.I. GARASIM, D.M. ZAYACHUK, R.D. IVANCHUK, V.B. LOTOTSKII, P.M. STARIK, and V.A. SHENDEROVSKII
Peculiarities of Physical Properties of Lead and Tin Selenide Solid Solutions
72.80 and 75.50; S8
(a) Vol.111,No.1, K99-K104

M. MOCKER, F. LEMKE, and P. SELBMANN
Direction Dependence of Auger Recombination in Low-Dimensional PbSnTe Structures
72.10; S8
(b) Vol.151,No.2, K151-K154

A.M. SANDAUER
Anomalous Hall Effect in PbSnMnTe
72.20; S8
(a) Vol.111,No.2, K219-K224

J. CALAS, S. ABDEL-MAKSOUD, and M. AVEROUS
Temperature Dependence of Carrier Trapping in Bulk Vitreous $GeSe_3$ from Transit Time Measurements
71.55 and 72.40; S8
(b) Vol.152,No.1, 209-216

A.M. ELKORASHY
Photoconductivity in Germanium Selenide Single Crystals
72.40; S8
(b) Vol.152,No.1, 249-260

F.M. GASHIMZADE and E.V. TAGIROV
Free-Carrier Absorption for Intervalley Phonon Scattering
78.20; S8
(b) Vol.152,No.1, 329-334

I.D. KOZMIK, I.I. GRIGORCHAK, Z.D. KOVALYUK, and I.Z. MARCHUK
Kinetic and Optical Properties of H_xInSe and H_xGaSe Proton Intercalates and Their Thermodynamic Model
61.70; 72.80; 78.50; S8
(a) Vol.112,No.2, K61-K64

R.A. YAKSHIBAEV, M.KH. BALAPANOV, N.N. MUKHAMADEVA, and G.R. AKMANOVA
Partial Conductivity of Cations of Different Kinds in the Alloys of Cu_2X-Ag_2X (X=Se,Te) Mixed Conductors
66.30; S8
(a) Vol.112,No.2, K97-K100

E.E. MATYAS, T.E. BORISENKO, N.N. KOREN, and V.A. IVANOV
Spectral Distribution of Photoresponse in PbTe-SnS Films
72.40; S8
(a) Vol.112,No.2, K111-K112

N.R. LOKHANDE, V.B. SAPRE, and C. MANDE
Energy Gap and Chemical Shift of X-Ray K Absorption Discontinuity of Arsenic in Its Compounds
71.25 and 78.70; S8; S10.1
(b) Vol.153,No.1, K35-K38

E. PAVLIDOU and O. VALASSISDES
Multicarrier Investigation of $Pb_{0.95}Ge_{0.05}Te$ Crystals by the Magnetoconductivity Tensor Component $\sigma_{xy}(B)$ Analysis
72.20; S8
(b) Vol.153,No.1, K51-K56

M. GUYMONT, A. TOMAS, M.-P. PARDO, and M. GUITTARD
Electron Microscopy Study of gamma-Ga_2S_3
61.16 and 61.60; S8
(a) Vol.113,No.1, K5-K8

K. SCHMIDT, L. BAUCH, G. KLUGE, and P. SUEPTITZ
Preparation and Properties of Spin Coated Amorphous $GeSe_2$ Layers
68.55; 78.65; S8
(a) Vol.113,No.1, K55-K60

S.A. BELOKON, S.O. KLIMONSKII, and V.D. KUZNETSOV
The Influence of $Pb_{0.99}Mn_{0.01}Te$ Solid Solution Ageing and Annealing on Its Paramagnetic Curie Temperature
75.50; S8
(a) Vol.113,No.1, K123-K126

P.E. MOZOL, N.A. SKUBENKO, P.A. SKUBENKO, YU.P. GNATENKO, E.A. SALKOV, and Z.D. KOVALYUK
Effect of Laser Irradiation on Low-Temperature Photoconductivity and Photoluminescence Spectra of Gallium Selenide
71.35 and 71.55; 72.40; 78.55; S8
(b) Vol.153,No.2, 667-674

G. BABONAS, R. DAGYS, and G. PUKINSKAS
Electronic Structure and Magnetooptical Effects in gamma-Ln_2S_3
78.20; S8
(b) Vol.153,No.2, 741-750

M. ZAŁUZNY and W. OKULSKI
On Electron States of Quantum Wells with Anisotropic and Tilted Valleys in Magnetic Field
71.25 and 73.60; S8
(b) Vol.153,No.2, K115-K118

I. YU, H.L. PARK, H.K. KIM, S.K. CHANG, and C.H. CHUNG
Sodium Related Luminescence Centers in $CaS:Bi^{3+}$ Phosphor
78.55; S8
(b) Vol.153,No.2, K183-K186

G. KLUGE, A. THOMAS, R. KLABES, and P. SUEPTITZ
Applicability of Silver Implanted Amorphous $GeSe_2$ as Inorganic Negative Resist
61.80; 66.30; 79.20; S8
(a) Vol.113,No.2, K171-K176

J. NOTHOLT and J. LIESE
The Influence of the Preparation Conditions on the Structure of Evaporated As_2Se_3 Layers
68.55; S8
(a) Vol.114,No.1, 207-214

G. MICOCCI, A. RIZZO, P. SICILIANO, and A. TEPORE
Deep Level Transient Spectroscopy Measurements in InSe Single Crystals
71.55; S8
(a) Vol.114,No.1, 253-258

S. BISWAS and R. BHATTACHARYA
Electrical Properties of Bismuth Telluride Doped with Bismuth Iodide
72.80; S8
(a) Vol.114,No.1, 277-284

V.A. KUTASOV and L.N. LUKYANOVA
The Conduction Band Parameters and Scattering Mechanisms in Solid Solutions Based on Bi_2Te_3
72.10; S8
(b) Vol.154,No.2, 669-678

P.I. SAVITSKII, A.V. KONSTANTINOVICH, and Z.D. KOVALYUK
The Te Intercalation Effect on the Electrical Properties of Indium Monoselenide
72.80; S8
(b) Vol.154,No.2, K151-K156

K.N. MADHUSOODANAN, K. NANDAKUMAR, J. PHILIP, S.S.K. TITUS, S. ASOKAN, and E.S.R. GOPAL
Photoacoustic Investigation of Glass Transition in As_xTe_{1-x} Glasses
64.70; 61.40; 65.70; 72.80; S8
(a) Vol.114,No.2, 525-530

E. HAJTO, P.J.S. EWEN, P.G. HILL, and A.E. OWEN
Spectroscopic Investigation of Spin-Deposited As-S Amorphous Films
78.65; 68.55; S8
(a) Vol.114,No.2, 587-598

J.W. TOMM, J. GRIESCHE, H. SCHMIDT, D. GENZOW, and J. SIEMON
Strain in Lead Salt Layers. A Luminescence Study
68.55; 78.55; S8
(a) Vol.114,No.2, 621-628

T.A. CHRISTAKUDI, G.CH. CHRISTAKUDIS, and L.D. BORISOVA
Thermoelectric Power of Bismuth Doped Lead Telluride
72.20; S8
(a) Vol.114,No.2, K201-K204

S.A. HUSSEIN and A.T. NAGAT
Characteristic Features of Electrical Properties of In_2Te_3 Single Crystals
72.80; S8
(a) Vol.114,No.2, K205-K210

T.D. IBRAGIMOV, G.G. KURBANOVA, and V.S. GORELIK
Raman Scattering in Gallium Selenide and Sulfide Films
63.20 and 68.30; 78.30; S8
(b) Vol.155,No.1, 113-116

V.S. STEPANYUK, A. SZÁSZ, O.V. FARBEROVICH, A.A. GRIGORENKO, A.V. KOZLOV, and V.V. MIKHAILIN
An Electronic Band Structure Calculation and the Optical Properties of Alkaline-Earth Sulphides
71.25; 78.20; S8
(b) Vol.155,No.1, 215-220

P. LOŠŤÁK, Z. STARÝ, J. HORÁK, and J. PANCIŘ
Substitutional Defects in Sb_2Te_3 Crystals
61.70; 72.20; S8
(a) Vol.115,No.1, 87-96

O. BUDKE and P. ŠVEC
Application of Thermal Conductivity Measurement by the Relaxation Method to Crystallization Kinetics of Glassy As_2Se_3+1mol%In
64.70; 66.70; S8
(a) Vol.115,No.1, 143-148

S.I. DRAPAK, V.N. KATERINCHUK, Z.D. KOVALYUK, and V.A. MANASSON
The Injection Peculiarities of Minority Charge Carriers in the Anisotype SIS Structure
73.40 and 78.60; S8; S10.1
(a) Vol.115,No.1, K35-K38

S. SHIGETOMI, T. IKARI, H. NAKASHIMA, and H. OHKUBO
Deep Levels of Cd-Doped InSe
71.55; S8
(a) Vol.115,No.1, K43-K46

PARAMDEEP KHURANA and K.K. SRIVASTAVA
Pressure Dependence of Capacitance and Conductance for a-$Ge_{17}Te_{83}$ at Moderate Temperatures
72.80; S8
(a) Vol.115,No.1, K57-K62

A.A. ZAHAB, M. ABD-LEFDIL, and M. CADENE
Effect of Annealing on Structure and Electrical Conductivity of Evaporated Indium Telluride
68.55 and 73.60; S8
(a) Vol.115,No.2, 491-496

A.V. KOLOBOV, V.M. LYUBIN, and J. TROELTZSCH
On the Mechanism of Anomalous Surface Diffusion of Silver in an Amorphous Chalcogenide/Silver Structure
66.30; S8
(a) Vol.115,No.2, K139-K142

S.D. CHATURVEDI, S.B. SHARMA, P. PALIWAL, and M. KUMAR
Analysis of Crystal Binding and Structural Phase Transition in Alkaline-Earth and Alkali Chalcogenides
62.20 and 64.60; S8
(b) Vol.156,No.1, 171-180

P. KONSIN and T. OERD
Influence of Hydrostatic Pressure on the Ferroelectric Phase Transition Temperature in (Pb,Sn)Te
62.50 and 63.75; 63.70; 77.80; S8
(b) Vol.156,No.1, K37-K40

R. MAMY and X. ZAOUI
Growth of Au on MoS_2 Studied by Photoemission
68.55 and 79.60; S8
(b) Vol.156,No.2, K109-K112

A.SH. ABDINOV, A.M. GUSEINOV, YU.G. NURULLAEV, and O.M. SADYKHOV
The Influence of Gadolinium Doping on the Switching Effect in Indium Selenide Single Crystals
72.40; S8
(a) Vol.116,No.2, K173-K178

S8.1. II-VI compounds

SANG-BO ZHAO, HUI-SU WANG, and JUN-KAI XIE
Coordination Symmetry and Energy Spectrum Analysis of Fe^{2+} in ZnS:Fe^{2+} Crystals
71.70; S8.1
(b) Vol.151,No.1, 167-174

N.J. CASTELLANI and W.R.L. LAMBRECHT
Core Hole Relaxation Energy in Semiconductors. Using a Pseudopotential Description of the Core Hole Perturbation
71.10; S5.11; S7.1; S8.1
(b) Vol.151,No.2, 565-570

YUAN LI and P.J. LIN-CHUNG
Semiempirical Construction of LCAO Parameters for II-VI Compound Semiconductors
71.10; 71.25; S8.1
(b) Vol.153,No.1, 215-224

R.K. RAM, S.S. KUSHWAHA, and A. SHUKLA
Phonon Assignments in II-VI and III-V Semiconductor Compounds Having Zincblende-Type Structure
63.20; 78.30; S7.1; S8.1
(b) Vol.154,No.2, 553-564

R.A. VARDANYAN and S.G. VESELKO
Dislocation Recharging in the Plastically Deformed Crystals of the A_2B_6-Type
61.70 and 71.55; 72.90; S8.1
(b) Vol.156,No.2, 527-536

S8.11. Sulfides

C.P. HILTON, J.J. DAVIES, J.E. NICHOLLS, and O. GOEDE
Optically-Detected Magnetic Resonance of Excitons Bound at Te_2 Isoelectronic Centres in $CdS_{1-x}Te_x$
71.35 and 71.55; 76.90; 78.55; S8.11; S8.15
(b) Vol.151,No.1, 175-184

Y. MERA, K. MAEDA, and S. HYODO
Microscopic Origin of Electric Conduction Anisotropy in Plastically Deformed CdS Single Crystals
72.20; 61.70; S8.11
(a) Vol.111,No.1, 223-232

A.N. GEORGOBIANI, A.N. GRUZINTSEV, A.V. SPITSYN, and I.M. TIGINYANU
Change of the Optical Properties of CdS Single Crystals upon High Dose Indium and Gallium Implantation
78.20 and 78.55; 78.60; S8.11
(a) Vol.111,No.1, 301-308

K. LOEBE, R. BOYN, and R. REETZ
Stability of Sm Centers in Electroluminescent ZnS:SmF$_3$ MISIM Structures during Ageing
78.60; S8.11
(a) Vol.111,No.1, K129-K132

L. MAGAFAS, A.N. ANAGNOSTOPOULOS, and J.G. ANTONOPOULOS
On the Conductivity of Amorphous CdS Films
73.60; S8.11
(a) Vol.111,No.2, K175-K178

C. LEVADE, J.J. COUDERC, and A. KARA
Plastic Deformation of Natural Sphalerite Single Crystals between 300 and 620 K
62.20; 61.70; S8.11
(a) Vol.112,No.1, 89-98

V.YA. REZNICHENKO, M.I. STRASHNIKOVA, and V.V. CHERNY
Dispersion Properties of Submicron-Thin CdS Crystals
78.20; S8.11
(b) Vol.152,No.2, 675-684

M.YA. GAMARNIK
On a Possibility of Sphalerite-Wurtzite Phase Transition Caused by Change in the Crystal Shape
64.70; S8.11
(a) Vol.113,No.1, K45-K48

U. BRAATZ, E. KRAUSE, and J. KUNSCH
Optoelectronic Switching in ZnSe and ZnS
72.40; S8.11; S8.12
(a) Vol.113,No.1, K79-K84

J. SCHLOESSER and A. STAHL
A Symmetry-Adapted RPA: Dynamics of the Electronic Density Matrix in a Semiconductor
71.25 and 78.20; S8.11
(b) Vol.153,No.2, 773-786

M. JAYACHANDRAN, M.J. CHOCKALINGAM, and V.K. VENKATESAN
Characteristics of Electrodeposited SnO_2-CdS/Au Schottky Devices
73.30; 85; S8.11
(a) Vol.113,No.2, K217-K222

B. MITRA and K.P. GHATAK
Electronic Effects in Elastic Constants of II-VI Semiconductors
62.20; S8.11
(b) Vol.154,No.1, K35-K38

K. SWIATEK, M. GODLEWSKI, D. HOMMEL, and H. HARTMANN
On the Incorporation of Rare Earth Ions into II-VI Compounds: ZnS:Eu
61.70; 76.30; S8.11
(a) Vol.114,No.1, 127-134

M.K. JAYARAJ and C.P.G. VALLABHAN
Low Voltage Driven White Light Emitting Thin Film EL Device
78.60; S8.11
(a) Vol.114,No.2, K233-K236

J. TATARKIEWICZ, M. CARDONA, A. BREITSCHWERDT, C.-T. LIN, R. GROETZSCHEL, and W. WITTHUHN
In Search of Hydrogen Local Vibrational Modes in CuCl, CuBr, CuI, ZnS, and AlSb
63.20; 78.30; S7.13; S8.11; S9
(b) Vol.155,No.1, 77-88

A. FELIMBAN and M. ABDEL NABY
Effect of Annealing and Doping on the Performance of CdS/ZnTe Heterojunctions
73.40; S8.11; S8.13
(a) Vol.115,No.1, K27-K30

J. DEXPERT-GHYS, Y. CHARREIRE, P. ESTEBAN-PUGES, L. ALBERT, and H. DEXPERT
Optical and EXAFS Investigations of ZnS:Tm^{3+} Powders
71.55; 61.10; 78.55; S8.11
(b) Vol.155,No.2, 571-580

E. CHIMCZAK and M. BERTRANDT-ŻYTKOWIAK
Blocking Effect in Electroluminescence of ZnS:Mn Driven by Short Voltage Pulses
78.60; 78.65; S8.11
(a) Vol.115,No.2, 581-586

H. DONKER, N. YAMASHITA, W.M.A. SMIT, and G. BLASSE
Luminescence Decay Times of the Sb^{3+}, Pb^{2+}, and Bi^{3+} Ions in Alkaline-Earth Sulfides
71.70; 78.55; S8.11
(b) Vol.156,No.2, 537-544

U. BRAATZ, J. KUNSCH, E. KRAUSE, and E. JAEGER
Sweep Drive Circuit for Streak Cameras Based on ZnS and ZnSe Optoelectronic Switches
72.40 and 85; S8.11; S8.12
(a) Vol.116,No.2, K219-K222

S8.12. Selenides

S.L. VINOGRADOV, O.K. MALINOVSKAYA, and V.A. TOLOKONNIKOV
Relaxation of Trapping Sites in ZnSe Films of Au-ZnSe-SiO_2-Si and Au-ZnSe-Ge Structures
71.55 and 73.30; S8.12
(a) Vol.111,No.1, 201-210

F. FIRSZT and H.L. OCZKOWSKI
Distribution of Traps in ZnSe:Cu Crystals Obtained by Fractional Glow Technique
71.55 and 78.60; S8.12
(a) Vol.111,No.1, K113-K118

R. BALTRAMIEJŪNAS, V. GAVRYUSHIN, G. RAČIUKAITIS, V. KUBERTAVIČIUS, and A. KAZLAUSKAS
Two-Photon Absorption via Deep Local Levels as Virtual Intermediate States in Semiconductors
71.55 and 78.50; S8.12; S10.1
(b) Vol.151,No.2, 721-732

U. PAL, S. SAHA, B.K. SAMANTARAY, H.D. BANERJEE, and A.K. CHOUDHURY
X-Ray Line Broadening and Electron Microscopic Studies on ZnTe Thin Films
68.55; S8.12
(a) Vol.111,No.2, 515-522

U. BRAATZ, E. KRAUSE, and J. KUNSCH
Optoelectronic Switching in ZnSe and ZnS
72.40; S8.11; S8.12
(a) Vol.113,No.1, K79-K84

V.A. KASIYAN, D.D. NEDEOGLO, A.V. SIMASHKEVICH, and I.N. TIMCHENKO
Conductivity and the Metal-Dielectric Transition in Compensated n-ZnSe
72.20; 71.30; 71.55; S8.12
(b) Vol.154,No.1, 287-296

V.A. KASIYAN, D.D. NEDEOGLO, A.V. SIMASHKEVICH, and I.N. TIMCHENKO
Critical Behaviour of n-ZnSe Parameters in the Vicinity of the Metal-Insulator Transition
72.20; 71.30; S8.12
(b) Vol.154,No.2, 691-702

A.A. KLYUKANOV, E.A. SENOKOSOV, and L.E. CHIBOTARU
The Shape of Multi-Plasmon Radiation Band in Direct Gap Semiconductors
71.45 and 78.55; S8.12
(b) Vol.155,No.1, 295-308

V.K. LUKYANYUK, S.P. VORONYUK, and Z.D. KOVALYUK
Alkali-Metal-Intercalated Indium and Gallium Selenides:
Non-Monotonous Shift of Exciton Lines
78.20; S8.12
(b) Vol.155,No.2, 717-722

V.P. GRIBKOVSKII, L.G. ZIMIN, S.V. GAPONENKO, I.E. MALINOVSKII,
P.I. KUZNETSOV, and G.G. YAKUSHCHEVA
Nonlinear Laser Spectroscopy of Exciton Absorption in MOCVD Grown
ZnSe Monocrystalline Films
71.35 and 78.65; S8.12
(b) Vol.156,No.2, 449-454

U. BRAATZ, J. KUNSCH, E. KRAUSE, and E. JAEGER
Sweep Drive Circuit for Streak Cameras Based on ZnS and ZnSe
Optoelectronic Switches
72.40 and 85; S8.11; S8.12
(a) Vol.116,No.2, K219-K222

S8.13. Tellurides

A.V. SIMASHKEVICH, L.V. GORCHAK, E.V. GILAN, and K.D. SUSHKEVICH
Study of p-n Junctions Fabricated by Phosphorus-Ion Implantation into
CdTe
61.80 and 73.40; S8.13
(a) Vol.112,No.1, 305-310

U. BECKER, H. ZIMMERMANN, P. RUDOLPH, and R. BOYN
Optical Study of Impurity Distribution in Vertical-Bridgman Grown
CdTe Crystals
71.55; 64.75; 78.50; 78.55; S8.13
(a) Vol.112,No.2, 569-578

Y. MADHAVAN, K. RAMACHANDRAN, and T.M. HARIDASAN
On the Mean Square Displacements of Hg and Te in HgTe
63.20; 61.70; S8.13
(b) Vol.153,No.2, 579-582

Y. MADHAVAN, K. RAMACHANDRAN, and T.M. HARIDASAN
Diffusion Studies in CdTe and HgTe
66.30; S8.13
(b) Vol.154,No.1, 55-60

S. SAHA, U. PAL, A.K. CHAUDHURI, V.V. RAO, and H.D. BANERJEE
Optical Properties of CdTe Thin Films
78.65; S8.13
(a) Vol.114,No.2, 721-730

B.B. ISMAIL and R.D. GOULD
Structural and Electronic Properties of Evaporated Thin Films of Cadmium Telluride
68.55 and 73.60; S8.13
(a) Vol.115,No.1, 237-246

A. FELIMBAN and M. ABDEL NABY
Effect of Annealing and Doping on the Performance of CdS/ZnTe Heterojunctions
73.40; S8.11; S8.13
(a) Vol.115,No.1, K27-K30

K. SADAIYANDI and K. RAMACHANDRAN
Diffusion Studies in ZnTe for an Active Role of Vacancies
66.30; S8.13
(b) Vol.155,No.2, K85-K88

S8.15. Mixed crystals

C.P. HILTON, J.J. DAVIES, J.E. NICHOLLS, and O. GOEDE
Optically-Detected Magnetic Resonance of Excitons Bound at Te_2 Isoelectronic Centres in $CdS_{1-x}Te_x$
71.35 and 71.55; 76.90; 78.55; S8.11; S8.15
(b) Vol.151,No.1, 175-184

O. GOEDE, D. BACKS, W. HEIMBRODT, and M. KANIS
EPR Study of the Antiferromagnetic Phase Transition in (Cd,Mn)S
76.30; 75.30; S8.15
(b) Vol.151,No.1, 311-318

K. KUMAZAKI
Resonant Raman Scattering near $E_1+\Delta_1$ in $Cd_xHg_{1-x}Se$
78.30; S8.15
(b) Vol.151,No.1, 353-362

T. KIYOSAWA, S. TAKAHASHI, and N. KOGUCHI
Solid Solubility Range and Lattice Parameter of Semiconducting $Pb_{1-x}Cd_xS_{1-y}Se_y$ for Mid-Infra-Red Lasers
64.70 and 64.75; S8.15
(a) Vol.111,No.1, K21-K26

A.I. DMITRIEV and G.V. LASHKAREV
Nonthermal Spread of Quantum Levels in Solids Due to Defects
72.10; S8.15
(b) Vol.151,No.2, 633-644

K.P. GHATAK and A. GHOSAL
On the Effective Electron Mass in n-Channel Inversion Layers on Ternary Semiconductors in the Presence of a Parallel ...
71.25 and 73.40; S8.15
(b) Vol.151,No.2, K135-K140

C. RAZZETTI, T. BESAGNI, S. BINI, and P.P. LOTTICI
Defective Pseudoternary Alloys. The $CdGa_2Se_4$-$CdIn_2Se_4$ System
61.60; 64.60; S8.15
(a) Vol.111,No.2, 411-416

R.A. YAKSHIBAEV, V.N. ZABOLOTSKII, and R.F. ALMUKHAMATOV
Structural Features and Ionic Transport in the Ag_xNbSe_2 Two-Dimensional Superionic Conductor
61.80 and 66.90; S8.15
(a) Vol.111,No.2, 431-442

A. ROGALSKI and K. JÓŹWIKOWSKI
The Intrinsic Carrier Concentration in $Pb_{1-x}Sn_xTe$, $Pb_{1-x}Sn_xSe$, and $PbS_{1-x}Se_x$
72.20; S8.15
(a) Vol.111,No.2, 559-566

QIGUANG LI, SHAN JIANG, WEI LU, XIAOMING FANG, S.C. CHEN, and W. GIRIRAT
Optical Absorption Experiments of Cycling through the Structural Transition under Hydrostatic Pressure in $Cd_{1-x}Mn_xSe$
62.50 and 78.40; S8.15
(b) Vol.152,No.1, K33-K36

V. RIEDE, H. NEUMANN, N. SHARIF, F.S. HASOON, and H. SOBOTTA
Structural and Infrared Optical Properties of $(CuInSe_2)_{1-x}(2InAs)_x$ Solid Solutions
61.60 and 78.30; S7.15; S8.15
(a) Vol.112,No.2, K139-K144

V.K. DUGAEV and P.P. PETROV
Spinodal Decomposition in Semiconductor Alloys
64.60; 61.60; S8.15
(b) Vol.153,No.1, 115-122

B.S. ACHARYA, S.K. SINGH, and B.B. NAYAK
X-Ray Emission Spectroscopy of the Electronic States in $Pb_{1-x}Hg_xS$ Thin Films
71.20 and 73.60; 78.70; S8.15
(b) Vol.153,No.1, 153-160

S.Z. KORCZAK, W. KORCZAK, M. SUBOTOWICZ, and H. WASIEWICZ
EPR Linewidth in $Pb_{1-x}Mn_xTe$
76.30; S8.15
(b) Vol.153,No.1, 361-366

C. BENECKE, W. BUSSE, H.-E. GUMLICH, H. HOFFMANN, S. HOFFMANN, A. KROST, and H. WALDMANN
On the Excitation and Emission of Infrared Luminescence of Wide Band Gap Mn II-VI Semimagnetic Semiconductors
78.55; 71.70; S8.15
(b) Vol.153,No.1, 391-402

F.F. SIZOV and S.D. DARCHUK
Electrical Characteristics of Narrow-Gap Multilayer Semiconductor Structures
73.40; S8.15
(a) Vol.113,No.1, 107-112

C. ALBERS and A. JACOBI
Temperature Correction of x-Value Determination in $Hg_{1-x}Cd_xTe$ by Optical Transmission Measurements
78.20; S8.15
(a) Vol.113,No.1, K135-K138

C. GOURDON and P. LAVALLARD
Exiton Transfer Between Localized States in $CdS_{1-x}Se_x$ Alloys
71.35 and 78.55; S8.15
(b) Vol.153,No.2, 641-652

K. KUMAZAKI
Resonant Raman Scattering in $Zn_xHg_{1-x}Se$
78.30; S8.15
(b) Vol.153,No.2, 751-756

YU.V. LILENKO, K.V. SHASTOV, A.S. PETROV, A.V. VOITSEKHOVSKII, V.S. KULIKAUSKAS, V.N. KUZNETSOV, and A.P. MAMONTOV
Radation-Induced Defects in Implanted $Hg_{1-x}Cd_xTe$ Crystals
61.80; 72.20; 79.20; S8.15
(a) Vol.113,No.2, 285-294

M.J. MALACHOWSKI, J. PIOTROWSKI, and A. ROGALSKI
Influence of Dislocaions on the Performance of 3 to 5 μm $Hg_{1-x}Cd_xTe$ Graded Gap Photoresistors
72.40 and 73.60; S8.15
(a) Vol.113,No.2, 467-476

K. PONNURAJU, P.R. VAYA, and J. SOBHANADRI
The Effect of Substrate Temperature on the Optical Properties of $Pb_{1-x}Sn_xTe$ Films Formed by Molecular Beam Epitaxy
78.65; 68.55; S8.15
(a) Vol.113,No.2, 477-482

P. LOŠŤÁK, R. NOVOTNÝ, E. URBANOVÁ, and J. HORÁK
Optical Properties of $Bi_{2-x}In_xSe_3$ Single Crystals
78.30; S8.15
(a) Vol.113,No.2, 615-622

G. BABONAS, R.R. GAŁĄZKA, G. PUKINSKAS, and A. ŠILEIKA
Crystal Structure and Magnetooptical Properties of $Cd_{1-x}Mn_xTe$ with High Mn Concentrations
78.20; 61.60; S8.15
(b) Vol.154,No.1, 389-396

W. HEIMBRODT, C. BENECKE, O. GOEDE, and H.E. GUMLICH
Optical Effects Due to Antiferromagnetic Spin Ordering in (Cd,Mn)S at Highest Mn Concentrations
78.20; 75.30; S8.15
(b) Vol.154,No.1, 405-416

O.I. VASIN, A.E. KLIMOV, I.G. NEIZVESTNYI, and V.N. SHUMSKII
The Three-Layer Structure of $Pb_{0.8}Sn_{0.2}Te$ Epitaxial Films Prepared by MBE Technique on BaF_2
73.60; 68.55; S8.15
(a) Vol.114,No.1, 161-166

P. BROGOWSKI, H. MUCHA, and J. PIOTROWSKI
Modification of Mercury Cadmium Telluride, Mercury Manganese Telluride, and Mercury Zinc Telluride by Ion Etching
61.80 and 72.80; 85; S8.15
(a) Vol.114,No.1, K37-K40

G.D. KHATTAK, V. KEITH, and PH. MARTIN
Magnetization of $Cd_{1-x}Mn_xSe$
75.60; S8.15
(a) Vol.114,No.1, K71-K74

A.G. ARESHKIN, L.G. SUSLINA, and D.L. FEDOROV
On Concentration Dependences of the Band Parameters in $A^{II}B^{VI}$ Solid Solutions with Structural Phase Transitions
64.70 and 71.25; 71.35; 78.40; S8.15
(b) Vol.154,No.2, 609-622

A. SCHENK, M. STAHL, and H.-J. WUENSCHE
Calculation of Interband Tunneling in Inhomogeneous Fields
72.10 and 73.40; 77.50; S8.15
(b) Vol.154,No.2, 815-826

K. SADAIYANDI and K. RAMACHANDRAN
Variation of the Mean Square Amplitude with Concentration in $Hg_{1-x}Cd_xTe$
63.20; S8.15
(b) Vol.154,No.2, K113-K116

K.H. HERRMANN, U. JANK, and K.-P. MOELLMANN
Bipolar Generation within the Exponential Tail of the Interband Absorption Edge in HgCdTe
72.40 and 78.30; S8.15
(a) Vol.114,No.2, K249-K254

E. FLACH, H.-E. GUMLICH, CH. JUNG, and M. KRAUSE
Reflectance Spectroscopy on $Zn_{1-x}Mn_xS$ in the Energy Range from 11 to 20 eV by Use of Synchrotron Radiation
78.40; 71.25; 71.35; S8.15
(b) Vol.155,No.1, 317-328

R.K. SHARMA, S.K. MEHTA, V.K. SINGH, G. PRASAD, and B.B. SHARMA
Stoichiometric Annealing and Electrical Properties of $Hg_{0.8}Cd_{0.2}Te$ Grown by Solid State Recrystallization
64.75; 72.20; S8.15
(a) Vol.115,No.1, 135-142

M. QUINTERO, R. TOVAR, M. DHESI, and J.C. WOOLLEY
$(CuIn)_x(AgIn)_yMn_{2z}Te_2$ Alloys: T(z) Phase Diagram and Optical Energy Gap Values
64.70; S8.15
(a) Vol.115,No.1, 157-164

G.I. ZHOVNIR, S.V. KLETSKII, and N.V. SOCHINSKII
Computer Simulation of the Inverse Stefan Problem in Conformity with the Liquid Phase Epitaxy of $Hg_{1-x}Mn_xTe$
68.55; S8.15
(a) Vol.115,No.1, K31-K34

L.A. MANAKOVA and S.P. IONOV
Correlation Mechanism to Change Lattice Parameters and Phase Transitions in Compounds with Intermediate Valency
64.60; 71.25; S8.15
(b) Vol.155,No.2, 397-406

A. KLIMAKOW and M. SCHENK
Grain-Boundary Diffusion of Mercury in p-Type (Hg,Cd)Te
61.70 and 66.30; S8.15
(a) Vol.115,No.2, K135-K138

I. ILIEV, T. DIMOV, D. RIBAROV, and H. LANGE
Composition Determination of CdS_xSe_{1-x} Mixed Crystals by Optical Dispersion
78.20; S8.15
(a) Vol.115,No.2, K243-K246

Y. TOKUMITSU, H. KITAYAMA, A. KAWABUCHI, T. IMURA, and Y. OSAKA
Raman Spectra of ZnS_xTe_{1-x}/GaAs(100)
78.30; S8.15
(a) Vol.115,No.2, K261-K264

V.V. KARYAGIN and I.I. LYAPILIN
Thermo-Galvanomagnetic Phenomena in Spin Quantum Wells
71.10 and 73.40; 72.20; S8.15
(b) Vol.156,No.1, 235-242

K. KUMAZAKI, L. VIÑA, C. UMBACH, and M. CARDONA
Interband Critical Point Parameters Determined by Ellipsometry in $Cd_xHg_{1-x}Se$
71.25 and 78.20; 77.20; S8.15
(b) Vol.156,No.1, 371-376

A.P. ZDEBSKII, G. GARAYAGDEV, K. ORAZOV, and G.N. POLISSKII
Ultrasound-Stimulated Increase of Photosensitivity and Luminescence Intensity of II-VI Compound Solid Solutions
62.65 and 72.40; S8.15
(a) Vol.116,No.1, K91-K96

U. STUTENBAEUMER, H.-E. GUMLICH, and H. ZUBER
The Energy Transfer between Free Excitons, Bound Excitons, and the 3d-Shell of Mn^{2+} in $Zn_{1-x}Mn_xSe$ Mixed Crystals and s-d Type Exchange
71.35 and 71.70; 78.55; S8.15
(b) Vol.156,No.2, 561-568

V.M. MIKHEEV
Impurity Scattering in Weakly Compensated Gapless Semiconductors at Low Temperatures
72.10; 72.20; S8.15
(b) Vol.156,No.2, 577-582

S8.16. Higher compounds

V.M. BURLAKOV and M.R. YAKHEEV
Raman Scattering in Ferroelectric $TlGaSe_2$. Soft Mode and Order Parameter
78.30; 77.80; S8.16
(b) Vol.151,No.1, 337-346

G.A. MEDVEDKIN, YU.V. RUD, and M.A. TAIROV
Fundamental Optical Absorption Edge in $MnIn_2Te_4$ Single Crystals
78.20; S1.63; S8.16
(a) Vol.111,No.1, 289-300

D.D. ASKEROV, T.D. DZHAFAROV, A.M. AGAEV, S.G. ABDINOVA, and M.S. SADYGOV
Photoelectric Properties of $CuInSe_2$-GaAs Heterojunctions
72.40 and 73.40; S8.16
(a) Vol.111,No.1, K73-K76

I.A. AKSENOV, A.I. LUKOMSKII, and L.A. MAKOVETSKAYA
Edge Absorption and Cathodoluminescence Study of $CuAl_xIn_{1-x}S_2$ Solid Solutions
78.40 and 78.60; S8.16
(a) Vol.111,No.1, K109-K112

H. NEGISHI, S. OHARA, and M. INOUE
Effect of Pressure on Lattice Parameters of Intercalation Compounds M_xTiS_2 (M=Fe and Ni)
61.60; S8.16
(b) Vol.151,No.2, 441-444

G.A. MEDVEDKIN, V.D. PROCHUKHAN, YU.V. RUD, and M.A. TAIROV
Anisotropy of the Optical Band Edge Absorption in $CuInTe_2$ Single Crystals
78.20; S8.16
(b) Vol.151,No.2, 711-720

A.D. PRINS, K.R. ALLAKHVERDIEV, S.S. BABAEV, S.G. GUSEINOV, E. MEKHTIEV, M.M. SHIRINOV, M.M. TAGYEV, and D.J. DUNSTAN
Effect of High Pressure on the Optical Transmission Spectra of $A^{III}B^{III}C_2^{V}$ Crystals
78.20; S8.16
(b) Vol.151,No.2, 759-764

R. TOVAR, M. QUINTERO, P. GRIMA, and J.C. WOOLLEY
Phase Diagram, Lattice Parameter, and Optical Energy Gap Values for the $Zn_{2x}(AgIn)_{1-x}Te_2$ Alloys
61.60; 64.75; S8.16
(a) Vol.111,No.2, 405-410

G.CH. CHRISTAKUDIS, S.K. PLACHKOVA, and L.E. SHELIMOVA
Influence of Defects on the $(GeTe)_{1-x}[(Ag_2Te)_{1-y}(Sb_2Te_3)_y]_x$ Phonon Thermal Resistivity
61.70 and 66.70; S8.16
(a) Vol.111,No.2, 469-476

S.A. AL-KUHAIMI
Optical and Compositional Properties of $CuInSe_2$ Thin Films at High Temperature
68.55 and 78.65; S8.16
(a) Vol.111,No.2, 523-528

T.K. KASUMOV and F.I. MAMEDOV
Photocurrent Oscillations Due to Optical Phonons in $AgGaSe_2$ Single Crystals
72.40; 63.20; S8.16
(a) Vol.111,No.2, K205-K208

G. ANTONIOLI, P.P. LOTTICI, and C. RAZZETTI
The Structure of the Defect Chalcopyrite $ZnGa_2Se_4$ Studied by EXAFS
61.60 and 78.70; S8.16
(b) Vol.152,No.1, 39-50

K.R. ALLAKHVERDIEV, S.S. BABAEV, M.M. TAGIEV, and M.M. SHIRINOV
Low-Temperature IR and Raman Scattering Spectra of TlInS Layered Crystal
64.60 and 78.30; S8.16
(b) Vol.152,No.1, 317-328

H. SOMMER, D.W. BULLETT, H. NEUMANN, and G. KUEHN
Valence Band Density of States in $LiInSe_2$
71.20; 79.60; S8.16
(b) Vol.152,No.2, K47-K50

O. UEMURA, T. AKAI, Y. KAMEDA, and T. SATOW
The Molten Structure of $CuTlSe_2$ and $CuTlTe_2$
61.40; 61.12; S8.16
(a) Vol.112,No.2, 467-472

M. JURCZYK, G.S. BURCHANOV, and O.D. CHISTYAKOV
Magnetic and Structural Properties of $Y_2Fe_{14-x}Nb_xB$ Alloys
75.50; S8.16
(a) Vol.112,No.2, K121-K126

N.M. GASANLY, B.A. NATIG, A.E. BAKHYSHOV, and K.E. SHIRINOV
Long-Wave Optical Phonons in $InSbSe_3$ Layer Single Crystals
63.20 and 78.30; S8.16
(b) Vol.153,No.1, K89-K92

YU.I. DURNEV, B.S. KULBUZHEV, A.U. MALSAGOV, L.M. RABKIN, V.I. TORGASHEV, and YU.I. YUZYK
Vibrational Spectra and Phase Transitions in Layered Semiconducting Ferroelectrics with $TlGaSe_2$ Structure. I. Temperatur Dependence ...
64.60 and 78.30; S8.16
(b) Vol.153,No.2, 517-528

YU.G. GONCHAROV, G.V. KOZLOV, B.S. KULBUZHEV, V.B. SHIROKOV, V.I. TORGASHEV, A.A. VOLKOV, and YU.I. YUZYK
Vibration Spectra and Phase Transitions in Layered Semiconducting Ferroelectrics with $TlGaSe_2$ Structure. II. Thermodynamic ...
64.60 and 78.30; S8.16
(b) Vol.153,No.2, 529-538

V.M. BURLAKOV, A.P. RYABOV, M.R. YAKHEEV, E.A. VINOGRADOV, N.N. MELNIK, and N.M. GASANLY
Raman Spectroscopy of Soft and Rigid Modes in Ferroelectric $TlInS_2$
78.30; 77.80; S8.16
(b) Vol.153,No.2, 727-740

A.A. ABDURRAGIMOV, M.A. ALYANOV, N.G. GUSEINOV, and A.G. GUSEINOV
Heat Capacity of $CoCr_2S_4$
65.40; S8.16
(a) Vol.113,No.2, K207-K210

T.K. KASUMOV and F.I. MAMEDOV
Combined Photoconductivity in $AgGaSe_2$ Single Crystals
72.40; S8.16
(a) Vol.113,No.2, K241-K244

YU.O. DERID, E.I. KOZHACHENKO, S.I. RADAUTSAN, and I.M. TIGINYANU
Radiative Recombination in $ZnGa_2S_4$ Single Crystals
78.55; S8.16
(a) Vol.113,No.2, K265-K266

S.I. RADAUTSAN, I.M. TIGINYANU, V.N. FULGA, and YU.O. DERID
Quasi-Continuously Distributed Traps and Photoluminescence in $ZnGa_2Se_4$ Single Crystals
71.55; 78.55; S8.16
(a) Vol.114,No.1, 259-264

V.K. LUKYANYUK, M.V. TOVARNITSKII, and Z.D. KOVALYUK
Hall-Coefficient Sign Inversion, Electrical Conductivity, and Its Anisotropy in Highly Intercalated Na_xInSe Single Crystals
72.80; 71.55; S8.16
(a) Vol.114,No.1, 293-298

P. KASHYAP, M. JAIN, O.P. AGNIHOTRI, and H.K. SEHGAL
Effect of Composition on the Optical and Electrical Characteristics of Flash Evaporated $Cu_{2-x}In_xSe_2$ Thin Films
73.60 and 78.65; S8.16
(a) Vol.114,No.1, K29-K32

A.K. ABIEV, E.Z. ALIEV, B.A. NATIG, and R.M. SARDARLY
Electron Irradiation Effect on Dielectric Properties of $TlInS_2$ Single Crystals
77.80; 61.80; S8.16
(a) Vol.114,No.1, K119-K122

G. ANTONIOLI, P.P. LOTTICI, and C. RAZZETTI
A Study of the Second and Third Neighbour Structure in Chalcopyrite $CuGaSe_2$ by EXAFS
61.60; 78.70; S8.16
(b) Vol.154,No.2, 461-468

P.P. LOTTICI
Breathing-Like A-Symmetry Vibrations in $ZnGa_2(S_xSe_{1-x})_4$ and $CdGa_2(S_xSe_{1-x})_4$
63.20; 78.30; S8.16
(b) Vol.154,No.2, K117-K120

A. GALDIKAS, S. GREBINSKII, T. AMINOV, and L. OCHERTIANOVA
Interrelation of Piezo- and Magnetoresistance Anisotropy in Ferromagnetic Semiconductors p-$HgCr_2Se_4$ and p-$CdCr_2Se_4$
72.20 and 72.80; S8.16
(a) Vol.114,No.2, K211-K214

D.A. GUSEINOV, A.M. ABDULLAEV, A.B. ABDULLAEV, and S.M. YAVADOVA
Thermodynamic Properties and Heat Capacity of $CoFe_2Se_4$
65.40 and 65.50; S8.16
(a) Vol.115,No.1, K5-K8

I.A. AKSENOV, I.R. GULAKOV, V.I. LIPNITSKII, A.I. LUKOMSKII, and L.A. MAKOVETSKAYA
Cathodoluminescence and Electrical Properties of $CuAlS_2$
72.80 and 78.60; S8.16
(a) Vol.115,No.1, K113-K116

G.CH. CHRISTAKUDIS
Phonon Thermal Resistivity in $(GeTe)_{1-x}[(Ag_2Te)_{1-y}(Sb_2Te_3)_y]_x$ Solid Solutions
66.70; S8.16
(b) Vol.155,No.2, K89-K92

E. GRILLI, M. GUZZI, and M. PASTORI
Time Resolved Photoluminescence of $CdIn_2S_4$ Single Crystals
78.55; 71.55; S8.16
(a) Vol.115,No.2, 599-606

Y.SATYANARAYANA MURTHY, B.SRINIVASULU NAIDU, and P.JAYARAMA REDDY
Electrical Conductivity of Single Phase $AgGaSe_2$ Thin Films
73.60; S8.16
(a) Vol.115,No.2, K175-K180

PREETI KASHYAP, MUKESH JAIN, O.P. AGNIHOTRI, and H.K. SEHGAL
Effect of Composition on the Optical and Electrical Parameters of Flash Evaporated $Cu_{2-y}In_yTe_2$ Thin Films
73.60 and 78.65; S8.16
(a) Vol.115,No.2, K181-K184

S.K. PLACHKOVA, O. GEORGIEVA, K. PETROV, and D. KOVACHEVA
X-Ray Study of Defects in $GeTe-Ag_2Te-Bi_2Te_3$ System
61.60 and 61.70; S8.16
(a) Vol.116,No.1, 213-220

K.G. NIKIFOROV, A.G. GUREVICH, L.YA. PASENKO, S.I. RADAUTSAN, and L.M. EMIRYAN
Ferromagnetic Resonance and Cr^{4+} Ions in the Magnetic Semiconductor $HgCr_2Se_4$
76.50; S8.16
(a) Vol.116,No.2, K185-K190

S9. Halides

G. MIZUTANI and N. NAGASAWA
Exciton-Grating Induced Light Mixing at the Giant Two-Photon Resonances of Excitonic Molecules in CuBr
71.35 and 78.20; S9
(b) Vol.151,No.2, 591-598

V.A. BIBIK and N.A. DAVYDOVA
Transformation of Excitonic Spectra of PbI_2 Layer Crystals under Order-Disorder-Order Structural Transitions
71.35 and 78.20; S9
(b) Vol.151,No.2, 733-738

A. MANAR, M.J.M. GOMES, B. HOENERLAGE, and R. LEVY
Time Evolution of the Refractive Index of CuCl under Picosecond Pulsed Excitation
71.35 and 78.40; S9
(b) Vol.151,No.2, 747-758

C. TRALLERO-GINER and R. RIERA
Multiple LO-Phonon Raman Scattering and Hot Excitons
71.35 and 78.30; S9
(b) Vol.152,No.1, 357-368

J. MAIER
Thin Film Conductance: Evaluation and Correlation with Defect Formation Parameters
73.60; S9
(a) Vol.112,No.1, 115-122

G. MIZUTANI and N. NAGASAWA
Exciton-Grating Induced Metamorphoses of the Degenerate Four-Wave Mixing Spectra in the Giant Two-Photon Resonances of the Ecxitonic ...
71.35 and 78.20; S9
(b) Vol.152,No.2, 489-498

R.S. EACHUS, M.T. OLM, R. JANES, and M.C.R. SYMONS
Observation by ESR of Electrons Localised at Intrinsic Shallow Traps in Silver Halide Systems
71.50 and 76.30; 71.38; 76.40; S9
(b) Vol.152,No.2, 583-592

A. LIPOWSKI and Z. ONYSZKIEWICZ
Elementary Excitations in Two-Sublattice Spin Model with Three-Ion Interaction
75.30; S9
(b) Vol.153,No.1, 315-322

P. GILLIOT, J.C. MERLE, R. LEVY, M. ROBINO, and S. HOENERLAGE
Laser Induced Absorption of CuCl Microcrystallites in a Glass Matrix
78.20; S9
(b) Vol.153,No.1, 403-410

ZHENGJU YANG and F. LUTY
Theoretical Investigation of Ferroelastic Phase Transition of Pure and Mixed Alkali Cyanide Systems by Elastic Dipole Model (II)
64.60; S9
(b) Vol.154,No.1, 181-194

SHENG-LI GUO and KE-QIN LU
EPR Analysis of Fe^{3+} in CsCl and NH_4Cl
76.30; S9; S9.12
(b) Vol.154,No.1, K67-K72

A.D. STEVENS, M.C.R. SYMONS, and R.S. EACHUS
ESR Study of Silver Clusters in AgCl and AgBr. An Attempt to Detect Photographic Latent Images
61.70; 76.30; S9
(b) Vol.154,No.2, 483-496

J. WĄSICKI, W. NAWROCIK, Z. PAJĄK, I. NATKANIEC, and A.V. BELUSHKIN
Neutron Scattering Study of Some Pyridinium Salts
64.70; 61.12; S9
(a) Vol.114,No.2, 497-504

J. TATARKIEWICZ, M. CARDONA, A. BREITSCHWERDT, C.-T. LIN, R. GROETZSCHEL, and W. WITTHUHN
In Search of Hydrogen Local Vibrational Modes in CuCl, CuBr, CuI, ZnS, and AlSb
63.20; 78.30; S7.13; S8.11; S9
(b) Vol.155,No.1, 77-88

V. KUMAR and G.M.PRASAD
Electronic Properties of A_mB_n Complex Crystals
71.45; S9; S10
(b) Vol.155,No.1, 209-214

S.M.J. AKHTAR, D. RISTAU, J. EBERT, and H. WELLING
Characterization of Dielectric Films and Damage Threshold at 1.064 µm
78.65; 77.55; S9; S10
(a) Vol.115,No.1, 191-198

C.-M. LIEGENER
Calculation of the Auger Spectra of Clusters Modelling Solid Hydrogen Fluoride
79.20; S9
(b) Vol.156,No.2, 441-448

T. SHIOZAWA and T. KOBAYASHI
High Resolution Electron Microscopy of the Orientational Growth of Silver Precipitates to the Silver Gold Sulfide Specks Formed on ...
61.50 and 61.60; S9
(a) Vol.116,No.2, 513-520

S9.11. Alkali halides

A. STOLOVICH and G. LIIDJA
V_K Centres in Alkali Halides: Direct Spin-Lattice Relaxation and Electron-Phonon Interaction
61.70; S9.11
(b) Vol.151,No.1, 29-38

R. CYWIŃSKI, E. MUGEŃSKI, W. NOWY-WIECHUŁA, and J. WIECHUŁA
Suzuki Phase Precipitates in Doubly Doped $KCl:Pb^{2+},Mn^{2+}$ Crystals
61.70; 76.30; 78.50; S9.11
(b) Vol.151,No.1, 47-52

E. KOENIGSBERGER and H. SCHRUNNER
On Lattice Parameters and Enthalpies of Mixing of Alkali Halide Solid Solutions
61.50 and 64.75; 65.50; S9.11
(b) Vol.151,No.1, 101-110

J. OTTO
Electric Field Induced Splitting of the Stretching Vibration Lines SH^-, SeH^-, and TeH^- Centres in Alkali Halides
78.30 and 78.50; S9.11
(b) Vol.151,No.1, 363-374

N. NISHIMAKI, S. SHIMANUKI, and Y. NASU
Copper Centers Induced by Optical Bleaching of Cu^- Center in KCl and KBr Crystals
78.60; S9.11
(b) Vol.151,No.1, K81-K86

W.P. AMBROSE and A.J. SIEVERS
Determination of the Orientation of NO_2^- in KI by Persistent IR Spectral Hole Burning
63.20 and 78.50; S9.11
(b) Vol.151,No.1, K97-K100

T. OHGAKU and N. TAKEUCHI
Relation between Plastic Deformation and the Blaha Effect for Alkali Halide Crystals
62.20; 62.40; S9.11
(a) Vol.111,No.1, 165-172

R. SUBRAHMANYAM and Y.V.G.S. MURTI
Laser Induced F-F^- Conversions in KCl Crystals
61.70; 71.55; 78.50; S9.11
(b) Vol.151,No.2, 445-452

F. DE MATTEIS, A. SCACCO, F. SOMMA, G. BALDACCHINI, E. GIOVENALE, and U.M. GRASSANO
Complex Centres in Additively Coloured RbCl:Li^+
61.70; 78.50; 78.55; S9.11
(b) Vol.151,No.2, 453-462

I.I. GEGUSIN, V.N. DATSYUK, A.D. YUMATOFF, and R.V. VEDRINSKII
Atomic and Solid State Effects in X-Ray Absorption Spectra of Ionic Crystals
71.25 and 78.70; S9.11
(b) Vol.151,No.2, 581-590

R. FOLLATH and H. JEX
Temperature Dependent Frequency Shifts in the Resonance Absorption and Resonance Fluorescence Scattering of Moessbauer gamma-Quanta
76.80; S9.11
(b) Vol.151,No.2, 703-710

V.K. JAIN, N.V. VUGMAN, and M.L.N. GRILLO
Electron Paramagnetic Resonance Studies of Irradiated Single Crystals of Cesium Chloride Doped with Nickel Cyanide Complexes
76.30; S9.11
(a) Vol.111,No.2, 625-628

T. KUBOZOE and Y. WATANABE
Thermally Stimulated Exoelectron Emission from LiF Single Crystals Excited with Electrons and X-Rays and Subsequently with UV Light
71.55 and 79.75; S9.11
(a) Vol.111,No.2, K201-K204

V. TOPA, A. IOAN, B. ILIESCU, and G. MITROAICA
The Conversion of Electronic Centres into U and OH^- Centres
61.70; S9.11
(b) Vol.152,No.1, 51-60

L. BOSI and M. NIMIS
On the Decay Properties of the M′-Center in LiF Crystals
61.70; 78.50; S9.11
(b) Vol.152,No.1, 67-72

V.D. LAKHNO and G.N. CHUEV
Local Phonons in Crystals with Electron Centres
63.20; 71.38; S9.11
(b) Vol.152,No.1, 79-88

R. CYWIŃSKI and M. MANFREDI
Aggregation Processes in the NaCl:Eu^{2+} System Induced by the High-Temperature Annealing
78.20; 71.55; S9.11
(a) Vol.112,No.1, 215-220

P. CZUBA, A. PORADZISZ, Z. POSTAWA, T. DOHNALIK, and M. SZYMOŃSKI
Electronic and Cascade Processes in Ion-Induced Sputtering of Alkali Halides
79.20; S9.11
(a) Vol.112,No.1, 431-436

S. DZHUMANOV and P.K. KHABIBULLAEV
The Coexistence of Unstable,Metastable, and Separated Frenkel Pair Defects in Solids. III. Theory of the Non-Impact Mechanism for ...
61.70; 71.35; 71.38; S9.11
(b) Vol.152,No.2, 395-414

G. LIIDJA, V. NAGIRNII, T. SOOVIK, and S. ZAZUBOVICH
New Off-Centre System: Jahn-Teller Mercury-Like Impurities in Alkali Halides
71.55 and 78.55; S9.11
(b) Vol.152,No.2, 563-576

R.V. GOPALA RAO and A. SATPATHY
Structural Studies of Charged Hard-Sphere Mixtures through Blums Generalized Gamma Parameter. Application to ...
61.25 and 61.50; S9.11
(b) Vol.153,No.2, 421-434

S. EMURA and M. ISHIGURO
Effects of Hydrostatic Pressure on the Cu^{+} Center in Potassium Halides
71.55; 78.50; 78.55; S9.11
(b) Vol.153,No.2, 683-694

S.S. CAMARGO,JR. and J.P. VON DER WEID
Triplet Molecular Excitations in Potassium Halide Cyanide Crystals. I. The CN- Impurity Centre
71.35 and 78.60; S9.11
(b) Vol.153,No.2, 757-764

J. E. MUÑOZ SANTIUSTE and J. GARCÍA-SOLÉ
Lifetime of Eu^{2+} Luminescence in Alkali Halides. Dependence on the Host Matrix
78.55; S9.11
(b) Vol.153,No.2, K187-K192

M.V. GALUSTASHVILI, D.G. DRIYAEV, I.A. POLITOV, and Z.K. SARALIDZE
Changes in the Electrical Properties of LiF Crystals in the Process of Post-Radiation Annealing
61.70; S9.11
(a) Vol.114,No.1, 99-104

B.P. CHANDRA and D.P. BISEN
Theoretical Approach to the Mechanoluminescence of Thermoluminescent Crystals
78.60; S9.11
(a) Vol.114,No.1, K123-K126

H.-G. FRITSCHE
Particle Size Effects of Ion Crystallites
68.40; 62.20; S9.11
(b) Vol.154,No.2, 603-608

A.A. O'CONNEL-BRONIN and A.B. TRESHCHALOV
Dynamics of Exciton Reflection Spectra on the Change of the Incidence Angle in NaI Crystals
78.40; 71.35; S9.11
(b) Vol.154,No.2, 763-768

E.R. LÓPEZ-TÉLLEZ, D. CÁRDENAS-GARCÍA, and C. RUÍZ-MEJÍA
F_A-Center Optical Absorption in CsF
71.55 and 78.50; S9.11
(b) Vol.154,No.2, K171-K174

S. LAKSHMI RAGHAVAN, N. RAMESH, and K. SRINIVASAN
Dispersion of the Quadratic Electro-Optical (Kerr) Constants in $KCl:Cd^{2+}$ and $KCl:Sr^{2+}$
78.20; 61.70; S9.11
(a) Vol.114,No.2, 749-756

G.A. BIELIG
The Peaking Effect in Nominally Pure KCl Single Crystals
62.40; S9.11
(a) Vol.114,No.2, K153-K158

V. BICHEVIN, G. ALSEITOV, M. GAPANOV, and H. KAAMBRE
Excimer-Laser-Induced Exoemission and Thermoluminescence of Some Alkali Halides
78.60 and 79.75; 61.80; S9.11
(a) Vol.114,No.2, K241-K244

A.K. PACHAURI, A.K. SHARMA, and H.C. SAXENA
Volume Dependence of Short Range Force Constant and Dielectric Properties of Alkali Halides
63.90; 77.20; S9.11
(b) Vol.155,No.1, 73-76

F. MAES, F. CALLENS, P. MATTHYS, and E. BOESMAN
Electron Paramagnetic Resonance of an O_3^- Centre in KBr
76.30; 61.70; S9.11
(b) Vol.155,No.1, K55-K58

V. BICHEVIN and H. KÄÄMBRE
Low-Temperature Thermo- and Photostimulated Exoemission (TSEE and PSEE) of KBr
79.75; S9.11
(a) Vol.115,No.1, K109-K112

A.A. MIKHALENKO, V.M. KHULUGUROV, E.I. DMITRIEVA, and V.N. SALOMATOV
F^{2+} Centers in LiF-Mg Crystals
61.70 and 71.55; S9.11
(b) Vol.155,No.2, 359-364

A.V. EGRANOV, A.I. NEPOMNYACHIKH, V.V. OTROSHOK, and B.P. CHERNYAGO
H_2^--Centre in LiF-H^-
61.70; 76.30; 78.40; S9.11
(b) Vol.155,No.2, 365-372

V.K. ADAMCHUK, A.E. KUCHMA, and S.L. MOLODTSOV
Effects of Low-Energy Electron Transport in Insulators
71.25; 79.60; S9.11
(b) Vol.155,No.2, 525-534

S.N. MAITI, S. SENGUPTA, and D. ROY
Theoretical Interpretation of EXAFS Results for the Relaxation in Mixed Crystals
64.75; 61.10; S9.11
(b) Vol.156,No.1, 83-92

F.A. KHWAJA, M.S.K. RAZMI, and S.H. NAQVI
A Comparative Study of the Phenomenological Potentials for Ionic Crystals of Alkali Halides (II)
64.10; 64.60; S9.11
(b) Vol.156,No.1, 111-118

B.K. RATH and P.N. RAM
Gap Modes and Local Density of States of Impurities in RbCl and KBr
63.20 and 71.55; 63.50; S9.11
(b) Vol.156,No.1, 137-144

L. BOSI and M. NIMIS
On the Criterions for F_A(II) Center Formation in Alkali Halide Crystals
61.70; S9.11
(b) Vol.156,No.1, K5-K10

YU.M. ANNENKOV, V.F. STOLARENKO, and T.S. FRANGULIAN
The Effect of Temperature on the Formation and Accumulation of X_3^- Centers in KBr Crystals under Powerful Pulsed Electron Irradiation
61.70; S9.11
(b) Vol.156,No.1, K23-K26

B. SRINIVASA RAO and S.P. SANYAL
High Pressure Elastic Constants of NaCl
62.20; S9.11
(b) Vol.156,No.1, K27-K32

A.A. ALYBAKOV, V.A. GUBANOVA, M.ZH. ZHANIBEKOV, and N. TOICHEV
EPR Spectra of X-Irradiated KBr:CrO_4^{2-}- Ba^{2+} Single Crystals
76.30; 61.80; S9.11
(b) Vol.156,No.1, K77-K80

K.-E. PEIPONEN, P. SILFSTEN, P. KETOLAINEN, and T. TSUBOI
A Remark on the Thermo-Optical Properties of the F_A(II) Centre in KCl:Li Crystals
61.70 and 78.50; S9.11
(b) Vol.156,No.1, K81-K84

R. KERN and M. KROHN
Interaction of Metal Clusters with Surface Steps
68.20; 61.70; 62.20; S9.11
(a) Vol.116,No.1, 23-38

H. HAEFKE and H. STENZEL
Cross-Section Decoration. A Method for Determination of Diffusion Coefficients
66.30 and 68.48; S9.11
(a) Vol.116,No.1, 39-46

P.B. BARNA, M. EL HITI, L. MALICSKÓ, and F.M. REICHA
On the Orientation Relationships of Decorating Gold Particles on Alkali Halides
68.48 and 68.55; S1.3; S9.11
(a) Vol.116,No.1, 55-66

F. APPEL
On the Intersection of Oblique Screw Dislocations in NaCl Single Crystals
61.70; 62.20; S9.11
(a) Vol.116,No.1, 153-164

A. GHAZI, J. BASHIR, M.M. BEG, N.M. BUTT, and Q.H. KHAN
Temperature Factors of RbCl by the Powder Elastic Neutron Diffraction Method
63.70; S9.11
(a) Vol.116,No.1, K47-K50

V. TABAKOVA
Emission Spectra of Tallium Centres in Barium Chloride Single Crystals
78.55; S9.11
(a) Vol.116,No.1, K101-K104

GH. STOICESCU, S.V. NISTOR, and C.D. MATEESCU
Aggregation of Bismuth in NaCl Crystals
61.70; 66.30; 78.50; S9.11
(b) Vol.156,No.2, 411-418

S9.12. Alkaline earth halides

M. KIKNADZE, R. MIRIANASHVILI, and T. SANADZE
RFDS Studies of Thermally Non-Stable Trigonal Centers of Tm^{2+} in SrF_2 and BaF_2 Single Crystals
76.70; S9.12
(b) Vol.151,No.1, K49-K52

T. TSUBOI
Vibronic Structure in the Absorption Spectra of $CaF_2:Sm^{2+}$
63.20 and 78.40; S9.12
(b) Vol.151,No.1, K91-K96

WEN-CHEN ZHENG
A Study of the Spin-Hamiltonian Parameters of Isoelectronic Impurity Ions in the Same Host Crystal
71.70; 76.30; S9.12; S10; S10.1
(b) Vol.151,No.2, K145-K150

G.A. KOUROUKLIS and E. ANASTASSAKIS
Pressure-Induced Phase Transitions and Anharmonicity Study of Alkaline-Earth Fluorides
64.60; 63.20; 78.30; S9.12
(b) Vol.152,No.1, 89-100

M. KIKNADZE, R. MIRIANASHVILI, and T. SANADZE
Ligand Hyperfine Interaction and Lattice Distortion for Cubic Er^{3+} Centers in BaF_2
76.70; S9.12
(b) Vol.152,No.1, K13-K16

R. SINGH and J. SHANKER
Three-Body Force and Shell Model Contributions to Third-Order-Elastic Constants for Fluorite Structure Crystals
62.20; S9.12
(b) Vol.153,No.1, 123-130

T.M. HARIDASAN and P. SEKAR
Thermodynamic Parameters of Anti-Frenkel Formation in Alkaline Earth Fluorides
65.50; 63.90; S9.12
(b) Vol.153,No.1, 131-138

N. GUSKOS
Ground State of the Gd^{3+} Ion in MF_2 Single Crystals under the Influence of External Static Electric Field
71.70; 76.30; S9.12
(b) Vol.153,No.1, 243-248

SHENG-LI GUO and KE-QIN LU
EPR Analysis of Fe^{3+} in CsCl and NH_4Cl
76.30; S9; S9.12
(b) Vol.154,No.1, K67-K72

V.A. MEZRIN
Phase Transitions in High-Temperature Fluorite-Structure Superionic Conductors
64.60; 61.70; S9.12; S10.1
(a) Vol.114,No.1, 145-156

K.P. OBOTH, F.J. LOHMEIER, and F. FISCHER
VUV and UV Spectroscopy of Pb^{2+} and Bi^{2+} Centres in Alkaline-Earth Fluorides
78.50; 78.55; S9.12
(b) Vol.154,No.2, 789-804

C.N. MISHRA, R.P. GOYAL, and B.R.K. GUPTA
Photoelastic Behaviour of Alkaline Earth Fluoride Crystals
62.20 and 78.20; S9.12
(b) Vol.155,No.1, K81-K84

F. FERMI, A. NAGORNYI, and C. PARACCHINI
Defect Annealing in Pure CdF_2 Crystals
61.70 and 78.60; S9.12
(a) Vol.115,No.1, 325-334

R. DUBOURG, M. SCHVOERER, R. BERGER, and B. DUMERCQ
Mécanismes de thermoluminescence dans des fluorines CaF_2 naturelles et de synthèse
78.60; 61.70; S9.12
(a) Vol.115,No.1, 335-346

R.V. GOPALA RAO and U. BANDYOPADHYAY
Evaluation of Partial Structures and Radial Distribution Functions of CaF_2 and $BaCl_2$
61.25; S9.12
(a) Vol.115,No.2, 357-368

A. THOMAS, A. ZEHE, B. SCHRECKENBACH, J.-W. ERBEN, and R. GROETZSCHEL
Ion Beam Lithography of Polycrystalline CaF_2 Films Deposited on Silicon
61.80 and 68.55; 77.55; S9.12
(a) Vol.116,No.2, 735-744

S9.15. Mixed crystals

S.S. CAMARGO,JR. and J.P. VON DER WEID
Triplet Molecular Excitations in Potassium Halide Cyanide Crystals. II. The $K(CN)_xCl_{1-x}$ Mixed System
71.35 and 78.60; S9.15
(b) Vol.153,No.2, 765-772

WEN-CHEN ZHENG
Investigation of the Pressure and the Stress Dependence of Low-Temperature EPR Spectra for $ZnSiF_6 \cdot 6H_2O:V^{2+}$ Crystals
76.30; 71.70; S9.15
(b) Vol.154,No.1, K73-K78

H. WEINERT, K. JARAŠIŪNAS, and B. SCHUMANN
Carrier Diffusion and Recombination in Type II $GaAs/Ca_xSr_{1-x}F_2$ Quantum Well Heterostructures Studied by the Transient Gating Technique
73.40; 68.55; S7.12; S9.15
(a) Vol.114,No.2, K183-K188

R.S. DE BIASI
Comments on Concentration Peculiarities of a Set of Impurity Centres in Fluorite Crystals Activated by Gadolinium
61.70; 76.30; 78.60; S9.15
(b) Vol.155,No.1, K3-K4

J. SHANKER, R. SINGH, and M. KUMAR
Analysis of Combination Rules for Elastic Constants of Mixed Ionic Crystals
62.20; S9.15
(b) Vol.156,No.1, 131-136

S9.16. Higher compounds

A. MARTÍN, F.J. LÓPEZ, and H. AREND
Luminescence Properties of $\{(CH_3)_4N\}_2ZnCl_4$ Single Crystals
78.55 and 78.60; S9.16
(b) Vol.151,No.1, 331-336

L.K. AMINOV, A.V. VINOKUROV, I.N. KURKIN, YU.YU. KOSTETSKII, S.L. KORABLEVA, I.KH. SALIKHOV, S.I. SAKHAEVA, and M.A. TEPLOV
Distortion of a Crystal Field in the van Vleck Paramagnet $LiTmF_4$ under Activation by Isovalent Impurities
71.70; 76.30; 78.50; S9.16
(b) Vol.152,No.1, 191-202

KE-QIN LU, SHENG-LI GUO, and PING-FENG WU
An Interpretation of the Pressure Dependence of EPR Zero-Field Splitting of Mn^{2+} in $Na_2ZnCl_4 \cdot 3\ H_2O$
71.70 and 76.30; S9.16
(b) Vol.152,No.2, 577-582

J. CZAPLICKI, N. WEIDEN, and A. WEISS
Proton Reorientations and Tunneling in $Cd(NH_3)_6Cl_2$ from NMR
76.60; S9.16
(a) Vol.113,No.1, 163-168

T. TSUBOI, M. CHIBA, and R. LAIHO
Magnetic Susceptibility of One-Dimensional Ferromagnetic $CsFeCl_3$ Crystals
75.30 and 75.50; S9.16
(b) Vol.153,No.2, K159-K162

A. MARTÍN, F.J. LÓPEZ, and H. AREND
X-Ray Induced Optical Absorption in Cs_2CdBr_4
78.50; S9.16
(b) Vol.153,No.2, K199-K202

A.A. KAMINSKII, M.I. DEMCHUK, N.V. ZHAVORONKOV, and V.P. MICHAILOV
Ultrashort Pulses of Stimulated Emission by Nd^{3+}-Doped ($^4F_{3/2}$-$^4I_{11/2}$ Channel) Anisotropic Fluorides, Acentric Dispersed Silicates, ...
78.45; S9.16; S11.2
(a) Vol.113,No.2, K257-K260

A. LÓPEZ-ECHARRI, I. RUIZ-LARREA, and M.J. TELLO
Thermodynamics of the Phase Transition Sequence in the Incommensurate Compound $[N(CH_3)_4]_2CuBr_4$
64.60; 65.40; S9.16
(b) Vol.154,No.1, 143-152

L.E. MISIAK, S.K. MISRA, and U. ORHUN
Study of Temperature Variation of EPR Linewidth of Gd^{3+}-Doped $LiYb_xY_{1-x}F_4$ Single Crystals. Estimation of Yb^{3+} Spin-Lattice
71.70 and 76.30; S9.16
(b) Vol.154,No.1, 249-258

A.S. KROCHUK, O.R. ONUFRIV, and Z.P. CHORNIY
Characteristic Properties of the Radiation Colouring Mechanism in MeFX Compounds (Me = Sr,Ba, X = Cl,Br)
61.70; 78.50; S9.16
(b) Vol.154,No.1, K9-K12

M. PAWŁOWSKA, W. STRĘK, and I. TRABJERG
Temperature Dependence of Luminescence of $RbMnCl_3:Sm^{3+}$ Crystal
78.55; S9.16
(b) Vol.154,No.1, K89-K92

V.I. TORGASHEV, YU.I. YUZYUK, F. SMUTNÝ, P. VANĚK, and B. BŘEZINA
Raman Spectra of the Incommensurate and Ferroelectric Phases of Rb_2CoCl_4 Single Crystals
78.30; 77.80; S9.16
(b) Vol.154,No.2, 777-788

A.B. VASSILIKOU-DOVA, H. MATTHIES, K. RECKER, F. WALLRAFEN, and G. LEHMANN
EPR of Mn^{2+} in Ba_2ZnF_6 Single Crystals
76.30; 71.55; S9.16
(b) Vol.155,No.1, K59-K62

M. HIDAKA, Z.Y. ZHOU, and B.M. WANKLYN
Structural Phase Transitions of Layer Compounds $KFeF_4$, $KTiF_4$, and KVF_4
64.70; S9.16
(a) Vol.115,No.1, 149-156

S.N. BALYAKIN, V.N. EROFEEV, D.N. MOGILYANSKII, and V.SH. SHEKHTMAN
Investigation of the Structure and Properties of KCl-NaCl Crystals at Elevated Temperatures
64.70; 62.20; S9.16
(a) Vol.115,No.1, 199-206

D. TALAPATRA and R.K. MUKHERJEE
Optical Studies of Electronic Transitions of Mn^{2+} Ion in $Na_2Mn_3Cl_8$ Single Crystal
71.70; 78.20; S9.16
(b) Vol.155,No.2, 541-548

C.K. JAYASANKAR, M.F. REID, and F.S. RICHARDSON
Comparative Crystal-Field Analyses of $4f^N$ Energy Levels in $LiYF_4:Ln^{3+}$ Systems
71.70; S9.16
(b) Vol.155,No.2, 559-570

WEN-CHEN ZHENG
Theoretical Considerations on the Axial Crystal-Field Splitting and Its Dependence on the Bonding Angle Distortion in Tetragonal ...
71.70 and 76.30; S9.16
(b) Vol.155,No.2, K149-K152

S.K. GUPTA, V.A. KIRIKOV, V.V. GLADKII, and B. BREZINA
Effect of Mechanical Stresses on Anomalies of Dielectric Properties of the Incommensurate Phase in Rb_2CoCl_4 Crystals
77.80; S9.16
(a) Vol.115,No.2, 569-574

Y. FUKUDA, A. TOMITA, and N. TAKEUCHI
Thermoluminescence and Thermally Stimulated Exoelectron Emission of $Ca_2B_5O_9Cl$ Activated with Divalent Europium or Lead
78.60 and 79.75; S9.16
(a) Vol.115,No.2, K251-K254

YU.N. MOSKVICH, B.I. CHERKASOV, A.M. POLYAKOV, A.A. SUKHOVSKII, and R.L. DAVIDOVICH
NMR Study of Internal Motions in a New Family of Ionic Conductors M_2SbF_5
76.60; 66.30; 72.20; S9.16
(b) Vol.156,No.2, 615-632

M. KRUPSKI
Rigid Sphere Model of the Phase Transitions in Complex Antifluorite and Elpasolite Compounds
64.60; S9.16
(a) Vol.116,No.2, 657-668

S10. Oxides

J.J. BARA and B.F. BOGACZ
Investigation of ^{57}Fe-Doped theta-$Al_xV_2O_5$ Bronzes
61.60; 76.80; S10
(a) Vol.111,No.1, 53-58

E. SCHULZ, G. DRAEGER, W. CZOLBE, and O. BRUEMMER
X-Ray Spectroscopic Investigations of Titanium Oxides and Basic Materials for Ferroelectric Ceramics
78.70; 71.25; S10
(a) Vol.111,No.1, 361-366

H. PAGNIA and N. SOTNIK
Metal Influence on Switching MIM Diodes
73.40 and 85; S1.3; S6; S10
(a) Vol.111,No.1, 387-394

WEN-CHEN ZHENG
A Study of the Spin-Hamiltonian Parameters of Isoelectronic Impurity Ions in the Same Host Crystal
71.70; 76.30; S9.12; S10; S10.1
(b) Vol.151,No.2, K145-K150

J. SUÑÉ, I. PLACENCIA, N. BARNIOL, E. FARRÉS, and X. AYMERICH
Degradation and Breakdown of Gate Oxides in VLSI Devices
73.40 and 77.50; 73.60; 85; S10
(a) Vol.111,No.2, 675-686

A.A. KAMINSKII, A.V. BUTASHIN, A.A. DEMIDOVICH, V.G. KOPTEV, B.V. MILL, and A.P. SHKADAREVICH
Broad-Band Tunable Stimulated Emission from Octahedral Cr^{3+} Ions in New Acentric Crystals with Ca-Gallogermanate Structure
78.20 and 78.45; 78.55; S10
(a) Vol.112,No.1, 197-206

A.A. KAMINSKII, A.V. BUTASHIN, I.A. MASLYANITSIN, and V.D. SHIGORIN
Nonlinear Optical Properties of Acentric Crystals with Ca-Gallogermanate Structure
78.20; S10
(a) Vol.112,No.1, K49-K52

W. DZIURDA, L. GABLA, and M. TULETA
Bombardment-Induced Photon Emission from Modified Surface of Glass
61.80 and 78.55; S10
(a) Vol.112,No.2, 811-814

A.R. URMANOV, V.G. GRIGORIEV, and A.A. PUZANOV
A Kinetic Theory for the Motion of Accelerated Particles in Solids. II. Dechanneling in Complex Crystals
61.80 and 79.20; S10
(b) Vol.153,No.1, 65-78

Z.C. KANG, D. MORIN, L. PIERRE, O. MONNEREAU, and C. BOULESTEIX
AEM Study of the High Temperature Superconductor $Bi_2Sr_2CaCu_2O_8$
61.60 and 74.70; 61.16; S10
(a) Vol.113,No.1, 23-34

M.E. ABU-ZEID, Y.A. YOUSSEF, A.E. RAKHSHANI, and J. VARGHESE
Investigation of Cu_2O Thin Film Using Thermal Interferometry
68.55; 78.65; S10
(a) Vol.113,No.1, 123-130

J. PIECHOTA, Z. KOZIOL, A. PAJACZKOWSKA, and H. SZYMCZAK
AC Susceptibility in Ceramic and Powdered $YBa_2Cu_3O_{7-\delta}$
74.70; S10
(a) Vol.113,No.1, 151-158

V.I. KUZNETSOV, E.N. YURCHENKO, M.T. PROTASOVA, E.A. TARABAN, O.P. KRIVORUCHKO, and R.A. BUYANOV
Moessbauer Investigation of the Influence of Al(III) Ions on the Properties of Mixed Fe(III)-Al(III) Hydroxides and Oxides
64.75; 76.80; S10
(a) Vol.113,No.2, 359-364

N.C. HALDER and L.C. NAGARAJAN
AC Conduction in Metal-Oxide Cermets. II. Sn-Oxide Resistors
73.60; S10
(a) Vol.113,No.2, 511-518

R.A. RUPP, A.E. KRUMINS, and K. KERPELIN
Dielectric, Electrooptic, and Holographic Properties Close to the Diffuse Phase Transition of PLZT 10/65/35
78.20; 64.60; 77.80; S10
(a) Vol.113,No.2, 637-652

I.O. TROYANCHUK, V.N. DERKACHENKO, and E.F. SHAPOVALOVA
Magnetic Properties of $A_2Mn_2O_7$ Pyrochlores
75.60; S10
(a) Vol.113,No.2, K249-K252

M. GASGNIER
Rare Earth Compounds (Oxides, Sulfides ...)as Thin Films and Thin Crystals. Physico-Chemical Properties and Applications(Review Article)
68.55 and 73.40; 75.70; 78.65; S1.4; S1.63; S10
(a) Vol.114,No.1, 11-72

Z.C. KANG, C. BOULESTEIX, G. NIHOUL, O. MONNEREAU, F. REMY, and A. CASALOT
High Resolution Image of a New Mixed Valence Compound: $Al_2Ti_7O_{15}$
61.60; 61.16; S10
(a) Vol.114,No.1, 73-82

C. ZALDO, L. CONTRERAS, L. ARIZMENDI, and E. DIEGUEZ
Optical Properties of $Bi_2Ge_3O_9$ Single Crystals. X-Ray Irradiation Effects
78.50; 78.55; S10
(a) Vol.114,No.1, 397-406

A. MORAWIEC
Calculation of Polycrystal Elastic Constants from Single-Crystal Data
62.20; S1; S10
(b) Vol.154,No.2, 535-542

Z.C. KANG, F. REMY, O. MONNEREAU, A. CASALOT, and C. BOULESTEIX
Plane Defects in a New Mixed Valence Compound $Al_2Ti_7O_{15}$
61.60; 61.16; 61.70; S10
(a) Vol.114,No.2, 441-450

K. HEILIG, E. KOLBIG, and J. REINEKE
Electrical Properties of Silicon/Lead Boro-Alumino-Silicate Glass Interfaces: Insulator Charge and Distribution of Interface States
73.20 and 73.40; S5.11; S10
(a) Vol.114,No.2, 579-586

A.A. KAMINSKII, A. KHOLOV, P.V. KLEVTSOV, and S.KH. KHAFIZOV
Spectroscopy and Stimulated Emission of Nd^{3+}-Doped Tetragonal $NaBi(MoO_4)_2$ and $NaBi(WO_4)_2$ Disordered Crystals
78.45 and 78.55; 71.55; 78.50; S10
(a) Vol.114,No.2, 713-720

JIPING YANG, S. RENTSCH, H. WABNITZ, and R. ERIKASHVILI
Transient Behaviour of Ce^{3+}- and Ce^{3+}/Nd^{3+}-Doped Silica Glasses after Excitation with Picosecond UV Pulses
78.20; S10
(a) Vol.114,No.2, 757-764

Y. FUKUDA, A. TOMITA, and N. TAKEUCHI
Thermoluminescence and Thermally Stimulated Exoelectron Emission of Sintered MgB_4O_7 Doped with $DyCl_3$ and Tm_2O_3
78.60 and 79.75; S10
(a) Vol.114,No.2, K245-K248

J.F.J. JORDAN, J. KALUS, U. SCHMELZER, and G. ECKOLD
Anharmonic Phonon Frequency Shifts and Widths in Solid Deuterated Anthracene-d_{10}
63.20; S10
(b) Vol.155,No.1, 89-106

V. KUMAR and G.M.PRASAD
Electronic Properties of A_mB_n Complex Crystals
71.45; S9; S10
(b) Vol.155,No.1, 209-214

R. MAŃKA and A. MOLAK
The Soliton Generalization of Bipolaronic Model Proposed for the High-Tc Superconductivity in Lanthanum and Yttrium Cupric Oxides
74.20; 71.38; S10
(b) Vol.155,No.1, 231-240

S.L. KAKANI and R.K. PALIWAL
Theoretical Study of Density of States and Specific Heat of Itinerant Ferromagnetic Superconductors
74.10 and 75.40; S10
(b) Vol.155,No.1, 241-248

S. FLACH, N.M. PLAKIDA, and V.L. AKSENOV
A Theoretical Study of Soft Mode Degeneracy in La_2CuO_4
63.20 and 63.75; S10
(b) Vol.155,No.1, K9-K12

P. WESOŁOWSKI, W. JAKUBOWSKI, and J.L. NOWIŃSKI
Electrical Properties of Superionic Silver-Borate Glasses Doped with AgI
66.30; S10
(a) Vol.115,No.1, 81-86

S.M.J. AKHTAR, D. RISTAU, J. EBERT, and H. WELLING
Characterization of Dielectric Films and Damage Threshold at 1.064 μm
78.65; 77.55; S9; S10
(a) Vol.115,No.1, 191-198

B.I. BAYRACHNYI, L.V. LIASHOK, V.P. GOMOZOV, and L.I. SKATKOV
Changes in Structure and Properties of Nb_2O_5 Anodic Films Caused by Generating Anion Defects on Their Surface
68.55; 73.60; S10
(a) Vol.115,No.1, 207-212

J.L. ZHANG and B.R. LI
Anomalous Electrical Conduction of $Cd_8Sb_2O_{10}$ Ceramics
64.70 and 72.80; S10
(a) Vol.115,No.1, K67-K70

A.V. CHZHAN, YU.M. FEDOROV, and T.I. ISAEVA
Features of the Temperature Dependence of the Initial Magnetic Susceptibility in $FeBO_3$
75.60; S10
(a) Vol.115,No.1, K101-K104

A. CONAN, A. BONNET, and M. MORSLI
Electrical and Structural Properties of the Tungsten Bronze $Bi_4W_{20}O_{62}$
72.20; 61.60; 71.20; S10
(b) Vol.155,No.2, 609-614

B.M. ANGHELOV
Energy Levels and Intensity Parameters of $LiTbP_4O_{12}$
71.30 and 78.20; S10
(b) Vol.155,No.2, K165-K168

V.G. TSIRELSON, V.A. STRELTSOV, R.P. OZEROV, and K. YVON
Electron Density Distribution in 3d-Metal Sesquioxides
71.20; 61.55; S10
(a) Vol.115,No.2, 515-522

K. SUN, M.X. LIN, T. HE, and H.L. LUO
The Moessbauer Study of alpha-Fe_2O_3 Fine Particles with and without Adsorbed Cobalt
75.60; 76.80; S10
(a) Vol.115,No.2, 539-546

A.I. GUSEV
Short-Range Order in Nonstoichiometric Transition Metal Carbides Nitrides and Oxides (Review Article)
61.60 and 75.30; 76.60; S1.61; S1.62; S10
(b) Vol.156,No.1, 11-40

E. PIPPEL and J. WOLTERSDORF
Electron Microscopy of TiC Containing Alumina Ceramics
61.70; 61.16; S1.61; S10
(a) Vol.116,No.1, 165-178

JUN CHEN and DUAN FENG
Crystallographic Shear Planes and Stacking Disorders in Perovskites $KNbO_3$ and $NaNbO_3$
61.70; 77.80; S10
(a) Vol.116,No.1, 231-236

SH.M. EFENDIEV, A.A. AGASIEV, V.E. BAGIEV, and YA.YU. GUSEINOV
Nonlinear Current-Voltage Characteristics in Me-Bi_2O_3-Me Structure
73.40; 72.20; S10
(a) Vol.116,No.1, 305-312

R.B. BECK and B. MAJKUSIAK
The Initial Growth Rate of Thermal Silicon Oxide
68.55; S10
(a) Vol.116,No.1, 313-330

P. KOŠTIAL, I. JAMNICKY, and J. ĎURČEK
Contribution to the Experimental Methods of Deep Level Acoustic Transient Spectroscopy
68.25 and 71.55; 72.50; S5.11; S10
(a) Vol.116,No.1, K77-K80

N. MLIKI, M. BEN SALEM, and B. YANGUI
Parois de domaines ferroelastiques dans B-Ln_2O_3:Déscription en terme de densité de dislocations
64.70; S10
(a) Vol.116,No.2, 669-680

S10.1. Simple oxides

WEN-CHEN ZHENG
A Further Study of Tetragonal Distortion of the Cr^{3+}-V_c Centre in MgO from Optical and EPR Spectra
61.70 and 71.70; 76.30; S10.1
(b) Vol.151,No.1, K17-K22

M. VALLADE and M. PETIT
Stress-Induced Defects in the Incommensurate Phase of Quartz
64.60; S10.1
(a) Vol.111,No.1, 155-164

I.D. MAKUTA, S.K. POZNYAK, A.I. KULAK, and E.A. STRELTSOV
Photocurrent Generation and Optical Transitions on Degenerate Cadmium Oxide Photoanodes
68.45 and 72.40; S10.1
(a) Vol.111,No.1, 193-200

I. GODMANIS and W. HOHENAU
On the Nature of Afterglow of the X-Ray Induced Luminescence in Crystalline and Glassy SiO_2
78.60; S10.1
(a) Vol.111,No.1, 335-344

P. STEGGER and G. LEHMANN
Dynamic Effects in a New Substitutional Center of Trivalent Iron in Quartz
61.70; 76.30; S10.1
(b) Vol.151,No.2, 463-468

R. BALTRAMIEJŪNAS, V. GAVRYUSHIN, G. RAČIUKAITIS, V. KUBERTAVIČIUS, and A. KAZLAUSKAS
Two-Photon Absorption via Deep Local Levels as Virtual Intermediate States in Semiconductors
71.55 and 78.50; S8.12; S10.1
(b) Vol.151,No.2, 721-732

WEN-CHEN ZHENG
A Study of the Spin-Hamiltonian Parameters of Isoelectronic Impurity Ions in the Same Host Crystal
71.70; 76.30; S9.12; S10; S10.1
(b) Vol.151,No.2, K145-K150

M. MATLAK
Curie Temperature of EuO under Pressure
75.50; S10.1
(b) Vol.151,No.2, K187-K190

P. MORFOULI and G. PANANAKAKIS
Electrical Study of Thin SiO_2 Layers Obtained by Anodic Oxidation of p-Type Silicon in Pure Water Using I-U, C-U,and ...
73.60; 73.40; S5.11; S10.1
(a) Vol.111,No.2, 529-540

H. DINTNER and F. THRUM
The Homogeneous Distribution of Non-Metallic Bonds in As-Deposited NiCr-O Resistive Films
72.15 and 73.60; S1.2; S10.1
(a) Vol.111,No.2, 551-558

SANG-BO ZHAO and KANG-WEI ZHOU
Strong-Field Scheme for $C_{3v}(d^3)$ and the Spectra of Ruby and Cr_2O_3
71.70; 78.50; S10.1
(b) Vol.152,No.1, 181-190

D.F. MARIANI, F. GONZÁLES, J. VIGNOLO, M. JIMÉNEZ DE CASTRO, and A. IBARRA
Activation Energies for the Thermoluminescence Processes of Al_2O_3 Gamma-Irradiated at Room Temperature
78.60; S10.1
(a) Vol.112,No.1, K53-K56

J. PIPREK
Numerical Analysis of TVS Ionic Current Applied to the Na^+-Motion in Implanted MOS Oxide Layers
66.30 and 73.40; S10.1
(a) Vol.112,No.2, 551-560

N.C. HALDER
AC Conduction in Metal-Oxide Cermets. I. Ta-Oxide Resistors
73.60; S10.1
(a) Vol.112,No.2, 589-600

N.R. LOKHANDE, V.B. SAPRE, and C. MANDE
Energy Gap and Chemical Shift of X-Ray K Absorption Discontinuity of Arsenic in Its Compounds
71.25 and 78.70; S8; S10.1
(b) Vol.153,No.1, K35-K38

E. FARRÉS, J. SUÑÉ, I. PLACENCIA, N. BARNIOL, and X. AYMERICH
Si-SiO_2 Interfacial Atomic Scale Roughness Caused by Inhomogeneous Thermal Oxidation
68.55 and 73.40; 68.48; S10.1
(a) Vol.113,No.1, 83-96

G. RAATZ and J. WOLTERSDORF
Structure of Metal Deposits on Ceramic Materials Studied in the Ni/MgO System
68.55; 68.48; S1.2; S10.1
(a) Vol.113,No.1, 131-142

H. DUŃSKI and H. SUGIER
The Study of Temperature Programmed Desorption of Oxygen from Magnesium Oxide
68.45; S10.1
(a) Vol.113,No.1, K49-K54

XIAO HUA CHEN, YE NING WANG, and HUI MING SHEN
Elastic Behaviour Related to the Superconductivity in $YBa_2Cu_3O_x$
62.20 and 74.70; S10.1
(a) Vol.113,No.1, K85-K88

N.M. FICZA and A.L. TOTH
The Effect of Lead Impurity on Amorphous Si Thin Film Transistors
61.70 and 85; 68.55; S5.11; S10.1
(a) Vol.113,No.1, K147-K148

SHENG-LI GUO, KE-QIN LIU, LIN-RUI PING, and GAO-CHAO LIU
Analysis of the EPR of Cr^{3+}-Vacancy Center MgO
71.70 and 76.30; S10.1
(b) Vol.153,No.2, K167-K170

J. CASTRO, J. RIVAS, H.J. BLYTHE, J. IÑIGUEZ, and C. DE FRANCISCO
Orientation and Diffusion Magnetic After-Effects in Vacancy-Doped Polycrystalline Magnetite in the Temperature Range 300 to 700 K
75.60; S10.1
(a) Vol.113,No.2, 541-550

V.S. LYSENKO, A.N. NAZAROV, S.A. VALIEV, I.M. ZARITSKII, T.E. RUDENKO, and A.S. TKACHENKO
EPR and TSCR Investigations of Implanted Al-SiO_2-Si Systems Treated with RF Plasma Discharge
73.40 and 85; 76.30; S5.11; S10.1
(a) Vol.113,No.2, 653-666

M.I. SHMIGLYUK, P.I. BARDETSKII, and S.D. TIRON
Nonlinear Optical Nutation on Interexciton Transition in Cu_2O. An Analogue of Equidistant Three-Level System
71.35; 78.20; S10.1
(b) Vol.154,No.1, 209-218

V.A. MEZRIN
Phase Transitions in High-Temperature Fluorite-Structure Superionic Conductors
64.60; 61.70; S9.12; S10.1
(a) Vol.114,No.1, 145-156

E. FARRES, J. SUNE, I. PLACENCIA, N. BARNIOL, and X. AYMERICH
The Thermal Growth of Very Thin SiO_2 Films. A Diffusion-Controlled Process
68.55; S10.1
(a) Vol.114,No.1, 167-176

E. WOLSKA and U. SCHWERTMANN
Selective X-Ray Line Broadening in the Goethite-Derived Hematite Phase
61.10 and 64.70; 61.60; S10.1
(a) Vol.114,No.1, K11-K16

V. PARKHUTIK, V. SOKOL, V. SHERSHULSKII, E. UTKINA, and A. VOROBEVA
Hopping Electrical Conductivity in the Al-Anodic Al_2O_3-Al System
73.40; S4; S10.1
(a) Vol.114,No.1, K33-K36

R. LACROIX
Theoretical Considerations on the Splitting of the Ruby R_1 Line under Stress
71.70; 62.20; S10.1
(b) Vol.154,No.2, K143-K146

L.N. SKUJA
Photoluminescence of Intrinsic Defects in Glassy GeO_2. Twofold Coordinated Ge and Nonbridging Oxygen
78.55; S10.1
(a) Vol.114,No.2, 731-738

Y. KIRSH, J.E. TOWNSEND, and P.D. TOWNSEND
Kinetic Analysis of the New Sensitization Effect in the TL of Silica Optical Fibres
78.60; S10.1
(a) Vol.114,No.2, 739-748

V.S. STEPANYUK, A. SZASZ, B.L. GRIGORENKO, O.V. FARBEROVICH, and A.A. KATSNELSON
Electronic Structure and Optical Properties of MgO: Band Structure Calculation and Cluster Model
71.25; 78.20; S10.1
(b) Vol.155,No.1, 179-184

A.V. ZALESSKII, V.G. KRIVENKO, V.S. LUTOVINOV, V.A. MURASHOV, and T.A. KHIMICH
NMR Study of the Supertransferred Hyperfine Fields at ^{57}Fe Nuclei and Morin Transition in Al-Substituted Hematite Crystals
76.60; S10.1
(b) Vol.155,No.1, K63-K66

M.A. MOHAMMED and D.V. MORGAN
Reliability of Tantalum Oxide Film Capacitors
73.60; S10.1
(a) Vol.115,No.1, 213-222

A. POEPPL and G. VOELKEL
ESR Investigation of the Oxygen Vacancy in Pure and Bi_2O_3-Doped ZnO Ceramics
71.55; 76.30; S10.1
(a) Vol.115,No.1, 247-256

L.N. ALEKSANDROV, V.YU. BALANDIN, A.V. DVURECHENSKII, and O.A. KULYASOVA
A Two-Dimensional Model of SOI Structure Crystallization by Pulse Nanosecond Heating
68.55 and 79.20; S5.11; S10.1
(a) Vol.115,No.1, K23-K26

S.I. DRAPAK, V.N. KATERINCHUK, Z.D. KOVALYUK, and V.A. MANASSON
The Injection Peculiarities of Minority Charge Carriers in the Anisotype SIS Structure
73.40 and 78.60; S8; S10.1
(a) Vol.115,No.1, K35-K38

YU.A. VIDADI, V.E. BAGIEV, and YA.YU. GUSEINOV
Frequency-Compensating Effect in the Electric Conductivity of Bi_2O_3 Films
72.80; S10.1
(a) Vol.115,No.1, K53-K56

M.K. JAYARAJ and C.P.G. VALLABHAN
AC Conductivity in Europium Oxide Thin Films
72.80; S10.1
(a) Vol.115,No.1, K63-K66

E.M. IBRAGIMOVA and V.N. STEPANOV
Surface Defect States at Silica-Polymer Interfaces in Fiber-Optic Waveguides
61.80 and 78.60; S10.1; S12
(a) Vol.115,No.2, K255-K260

V.I. LYGIN, A.D. SERAZETDINOV, E.A. RYABENKO, and O.I. CHERTIKHINA
Quantum-Chemical Calculations of Models for Surface Defects in Dehydroxylated Silica
68.20; S10.1
(b) Vol.156,No.1, 205-210

L.N. ALEKSANDROV, S.L. BABENKOVA, and V.S. ZINOVEV
Thermoelastic Stresses in Multilayered Structures of Silicon-on-Insulator Type (Computational Experiment)
65.70 and 68.25; 64.30; S5.11; S10.1
(a) Vol.116,No.1, K57-K60

M. LIBRA, L. BARDOŠ, and H. BARANKOVA
Influence of Post-Deposition Annealing on the Electrical Conductivity of Indium-Tin-Oxide Films
73.60; S10.1
(a) Vol.116,No.1, K65-K68

A.K. ABASS and E.M. JABOORI
Optical Properties of Cr_2O_3 Coatings in the Fundamental Absorption Region
78.65; S10.1
(a) Vol.116,No.1, K111-K114

N.C. BACALIS and A.B. KUNZ
Ab Inition Calculations on the Adsorption of H, O, and CO atop Cu^{2+} on MgO(001)
68.45; S10.1
(b) Vol.156,No.2, 545-560

P.V. CHERNOV, E.M. DIANOV, V.N. KARPECHEV, L.S. KORNIENKO, I.O. MOROZOVA, A.O. RYBALTOVSKII, V.O. SOKOLOV, and V.B. SULIMOV
Spectroscopic Manifestations of Self-Trapped Holes in Silica: Theory and Experiment
78.55; 71.55; S10.1
(b) Vol.156,No.2, 663-676

M. KUMEDA, M. KIWAKI, and T. SHIMIZU
Erratum to Temperature Dependence of the E′ Center Creation in Silica Glasses (phys. stat. sol. (b) 147, K1 (1988))
61.70 and 76.30; S10.1
(b) Vol.156,No.2, K163-K164

F.G. REQUEJO, A.G. BIBILONI, C.P. MASSOLO, J. DESIMONI, and M. RENTERIA
TDPAC Study of Stoichiometric and Reduced MoO_3. Localization of Indium Trapped Oxygen Single Vacancies
61.60; 61.70; 76.80; S10.1
(a) Vol.116,No.2, 503-512

S.A. LITVINENKO and V.I. SOKOLOV
Formation of Pores in Thermal SiO_2 Due to Internal Mechanical Stress Relaxation
61.70 and 64.75; S10.1
(a) Vol.116,No.2, 615-620

H. DUNSKI and H. SUGIER
Temperature-Programmed Desorption of Oxygen from gamma-Irradiated Polycrystalline Magnesium Oxide
61.80 and 68.45; S10.1
(a) Vol.116,No.2, K165-K168

S10.15. Mixed crystals

V. DVORAK
Landau Theory of the Structural Phase Transition in $YBa_2Cu_3O_{7-x}$
64.70; 74.70; S10.15
(b) Vol.151,No.1, K23-K28

A. KRELL and W. POMPE
Grain Boundary Microdamage Due to Visco-Elastic Relaxation of Residual Stresses in Alumina Ceramics
61.70 and 62.20; S10.15
(a) Vol.111,No.1, 109-118

O. SMRCKOVÁ, D. SYKOROVA, L. SMRCKA, and P. VASEK
Preparation of Pb-Doped Bi-Sr-Ca-Cu-O Superconductors
74.70; S10.15
(a) Vol.111,No.1, K77-K82

B.B. BOIKO, A.I. AKIMOV, V.I. GATALSKAYA, S.E. DEMYANENKOV, L.A. KUROCHKIN, M.L. PETROVSKII, E.K. STRIBUK, and V.M. FINSKAYA
Anomalous Electrical Resistance of Y-Ba-Cu-O System Ceramics
74.70; S10.15
(a) Vol.111,No.1, K89-K92

K. BENTE, E. SCHWARZMANN, and D. SIEBERT
EPR Studies of Cr^{2+} from Impurity Phases in the High Temperature Superconducting $YBa_2Cu_3O_{7-x}$
74.70 and 76.30; S10.15
(a) Vol.111,No.1, K93-K98

K. TABATA and H. KIDO
Preparation and Magnetic and Electrical Properties of $La_{1-x}Ce_xCoO_3$ Compounds
72.80 and 75.50; S10.15
(a) Vol.111,No.1, K105-K108

F. LAGNEL, B. POUMELLEC, and C. PICARD
Electron Transport in Amorphous $(Ti,Nb)O_2$ Solid Solutions
73.60; S10.15
(b) Vol.151,No.2, 531-544

A.A. BLINKIN, V.N. GOLOVIN, V.V. DEREVYANKO, G.F. TIKHINSKII, and V.A. FINKEL
Substitution Dependence of the Crystal Structure and Superconductivity in High-T_c Superconductors
74.70; S10.15
(b) Vol.151,No.2, K161-K166

W. ANDRAE, H. DANAN, and R. HERGT
Anisotropy of Superconducting Properties of a $YBa_2Cu_3O_{7-x}$ Crystal Investigated by Torque Measurements
74.60 and 74.70; S10.15
(a) Vol.111,No.2, 583-590

B.B. BOIKO, A.I. AKIMOV, V.I. GATALSKAYA, S.E. DEMYANOV, A.L. KARPEI, L.A. KUROCHKIN, YU.N. LEONOVICH, A.K. LETKO, M.N. MURAYA et al.
High T_c Superconductivity in Tl-Ba-Ca-Cu-O Ceramics
74.70; S10.15
(a) Vol.111,No.2, K209-K214

M. PĘKAŁA, K. PĘKAŁA, and A. PEJĄCZKOWSKA
Normal State Thermoelectric Power of Bi-Sr-Ca-Cu-O Superconductors
72.20 and 74.70; S10.15
(b) Vol.152,No.1, K1-K4

R. HERRMANN, U. PREPPERNAU, CH. RELLIN, A. KRAPF, and H. DWELK
Normal State Thermoelectric Power of Bi-Sr-Ca-Cu-O High-T_c Superconductor
72.20 and 74.70; S10.15
(b) Vol.152,No.1, K5-K8

N. GUSKOS, TH. LEVENTOURI, CH. TRIKALINOS, M. CALAMIOTOU, S.M. PARASKEVAS, J. KURIATA, and L. SADLOWSKI
On the Possibility of the Excistence of Cu^{3+} Ions in High-T_c Superconducting Samples of $YBa_2Cu_3O_x$
74.70 and 76.30; S10.15
(b) Vol.152,No.1, K9-K12

I.O. TROYANCHUK, A.S. CHERNYI, and E.F. SHAPOVALOVA
Magnetic Transitions in Perovskites $Cu(Mn_{3-x}Cu_x)Mn_4O_{12}$
75.30; S10.15
(a) Vol.112,No.1, 155-160

J.TH. ZETTLER, J. BOLLMANN, K.H. HECKNER, and H.A. KLOSE
Radiation-Induced Defects in Silicon Due to Low-Energetic Ion-Assisted Deposition of Indium Tin Oxide
61.80 and 68.55; 73.40; S10.15
(a) Vol.112,No.1, 395-404

J. BASZYŃSKI
Superconducting $YBa_2Cu_3O_{7-\delta}$ Textured Films on MgO Substrates
68.55 and 74.70; S10.15
(a) Vol.112,No.1, K39-K42

A.G. SAIF and J. CHELA-FLORES
Nucleation of Superconductivity in Ceramic Oxides
74.20; 74.70; S10.15
(b) Vol.152,No.2, 617-624

H. GRUBER, E. KRAUTZ, H.P. FRITZER, K. GATTERER, and A. POPITSCH
Electrical Conductivity Magnetic Susceptibility and Infrared Spectra in the Superconducting Systems Bi-Sr-Ca-Cu-O and Tl-Ba-Cu-O
74.70; 75.30; 78.30; S10.15
(b) Vol.153,No.1, 307-314

N. PRUSS, W. KRAAK, N. KUBICKI, H. DWELK, and R. HERRMANN
The Influence of Boron on the Electrical Properties of the Ceramic Superconductor $YBa_2Cu_3O_{7-\delta}$
74.70; S10.15
(b) Vol.153,No.1, K63-K68

C. LANDRON, J.P. COUTURES, R. ERRE, and P. LEHUEDE
Etude par ESCA de l'oxygene dans les verres silicates alcalins
61.40; 64.75; S10.15
(a) Vol.113,No.1, 35-42

H. DWELK, D. FREUDE, R. HERRMANN, H. PFEIFFER, and N. PRUSS
Substitution of Yttrium for Boron in the Structure of $YBa_2Cu_3O_{7-\delta}$
72.20 and 74.70; 76.60; S10.15
(a) Vol.113,No.1, 159-162

S.E. SARKISOV, K.K. PUKHOV, A.A. KAMINSKII, A.G. PETROSYAN, and T.I. BUTAEVA
Manifestation of Electron-Phonon Interaction in Insulating Crystals Doped with Pr^{3+} Ions
63.20 and 78.55; 71.55; S10.15
(a) Vol.113,No.1, 193-198

S. NOWAK, T. KACZOR, and J. ŻMIJA
Photochromic Phenomenon Kinetics
61.70 and 78.50; 71.55; S10.15
(a) Vol.113,No.1, 199-206

K. FUKUSHIMA and H. SATO
Possible Effect of Two-Dimensionality on the Superconducting Pairing Interaction
74.20; S10.15
(b) Vol.153,No.2, K141-K146

M. ANWAR, C.A. HOGARTH, and K.A.K. LOTT
ESR Studies of MoO_3/In_2O_3 Amorphous Thin Films Deposited by Co-Evaporation
68.55 and 76.30; S10.15
(a) Vol.113,No.2, 483-488

E.M. JABOORI and A.K. ABASS
Optical Properties of Chemically Deposited Fe-Cr Mixed Oxide Coatings
78.40; S10.15
(a) Vol.113,No.2, 579-584

B.B. BOIKO, F.P. KORSHUNOV, G.V. GATALSKII, A.I. AKIMOV, V.I. GATALSKAYA, and E.K. STRIBUK
Effect of Irradiation by Fast Electrons and ^{60}Co-gamma Rays on Superconducting and Electrical Properties of Thallium ...
61.80 and 74.70; S10.15
(a) Vol.113,No.2, K235-K240

C. GIOVANELLA
The Superconductor Glasses (SuG). A Fractal Description
74.50; 64.70; S10.15
(b) Vol.154,No.1, 273-280

JIANSHENG XIA, LIEZHAO CAO, SHIFANG SUN, QIRUI ZHANG, JIAN CHEN, ZUYAO CHEN, XIANHUI CHEN, and YITAI QIAN
Superconducting Behavior of $Bi_{1.5}Pb_{0.5}Ca_2Sr_2Cu_3O_y$ at Low Magnetic Field
74.70; S10.15
(b) Vol.154,No.1, K51-K54

C.D. MUKHERJEE, R. RANGANATHAN, A.K. RAYCHAUDHURI, and N. CHATTERJEE
Thermoelectric Power of the Ni and Cd Substituted YBCO System
72.15 and 74.70; S10.15
(b) Vol.154,No.1, K55-K58

M. ANWAR and C.A. HOGARTH
A Study of Optical Absorption in MoO_3/SiO Amorphous Thin Films
78.65; 68.55; S10.15
(a) Vol.114,No.1, 215-224

M. ANWAR, C.A. HOGARTH, and K.A.K. LOTT
Electron Spin Resonance Study of Thermally Evaporated Amorphous Thin Films of MoO_3/SiO
68.55 and 76.30; S10.15
(a) Vol.114,No.1, 225-232

S.J. WANG, X.H. LI, Y.L. CHEN, S.Q. LI, G.H. FONG, Z. WANG, A. CHEN, and B.R. LI
Structural Changes between 100 and 300 K of the High-T_c Superconductor Y-Ba-Cu-O Studied by Positron Annihilation
74.70 and 78.70; S10.15
(a) Vol.114,No.1, 273-276

M.N. KHAN
The Electronic and Spectroscopic Studies of Semiconducting Germanium-Tellurite Glasses Containing Iron Oxide
72.20 and 78.30; S10.15
(a) Vol.114,No.1, 285-292

A. BALS, S.I. GORBACHUK, J. KLIAVA, and N.G. KAKAZEI
EPR Study of the Structure of Disordered Polycrystals. Mn^{2+} in Lithographic Stone
76.30; S10.15
(a) Vol.114,No.1, 305-314

T. GOLDMANN, E. GERDES, and J. SIMANOWSKI
Dielectric Properties of Passivation Glasses in the System $PbO-B_2O_3-SiO_2-Al_2O_3-CeO_2$ Depending on Glass Composition
77.20 and 77.40; S10.15
(a) Vol.114,No.1, 387-396

E.M. GOLOLOBOV, N.A. PRYTKOVA, ZH.M. TOMILO, D.M. TURTSEVICH, and N.M. SHIMANSKAYA
High T_c Superconductivity of Bi-Sr-Ca-Cu Based Metal Oxides Doped by Magnesium
74.70; S10.15
(a) Vol.114,No.1, K57-K62

S.P. MUNTYAN, G.A. KIOSSE, and F.M. MUNTYANU
Superconductivity in the Ba-Y-Cu-Pb-O System
74.70; S10.15
(a) Vol.114,No.1, K63-K66

T. KOPTE and E. HEGENBARTH
Behaviour of the Thermal Conductivity of High Temperature Superconductors around the Critical Temperature
66.70 and 74.70; S10.15
(a) Vol.114,No.1, K67-K70

I.N. ZAKHARCHENKO, E.S. NIKITIN, V.M. MUKHORTOV, YU.I. GOLOVKO, M.G. RADCHENKO, and V.P. DUDKEVICH
Peculiarities of Ferroelectric Phase Transition in $(Ba,Sr)TiO_3$ Heteroepitaxial Films Obtained by HF Cathode Sputtering
64.70 and 68.55; 77.80; S10.15
(a) Vol.114,No.2, 559-566

Š. CHROMIK, H. TOLLET, C. CREMERS, D. MACHAJDIK, V. SMATKO, Y. BRUYNSERAEDE, and S. BEŇAČKA
The Composition of High T_c Superconducting Y-Ba-Cu-O Thin Films Prepared by Vacuum Coevaporation
68.55 and 74.70; S10.15
(a) Vol.114,No.2, 567-572

JINGKUI LIANG, XIAOLONG CHEN, YULING ZHANG, SISHEN XIE, JINQI HUANG, and YONGMIN NI
The Crystal Structure and Superconductivity of the Superconducting Phase $TlSr_2(Ca,La)Cu_2O_7$
61.60 and 74.70; S10.15
(a) Vol.114,No.2, 651-658

R. RICHTER, T. BREMER, P. HERTEL, and E. KRAETZIG
Refractive Index and Concentration Profiles of Proton-Exchanged $LiNbO_3$ Waveguides
78.20; 61.70; S10.15
(a) Vol.114,No.2, 765-774

JIRONG JIN, WUMING CHEN, XIN JIN, YITONG ZHANG, MU LU, YUBAO SHENG, JIABEN JI, GUOSHENG HAN, YI TANG, and XIXIAN YAO
Effect of Neutron Irradiation on the Superconductivity of Y-Ba-Cu-O
74.70; S10.15
(a) Vol.114,No.2, K189-K192

H.-J. DIETZE and S. BECKER
Laser-Induced Plasma Deposition of Superconducting Bi-Sr-Ca-Cu-O Thin Films on Silicon
68.55 and 74.70; S10.15
(a) Vol.114,No.2, K193-K196

A.S. SIDORENKO, L.A. KONOPKO, A.E. KOSHELEV, and A.YU. SIMONOV
Resistive Transition Broadening of Metal-Oxide Superconductors in Magnetic Fields
74.70; S10.15
(b) Vol.155,No.1, K45-K50

C.A. KLEINT, U. STOEPEL, and A. ROST
X-Ray Diffraction and Conductivity Investigations of Lanthanum-Doped Barium Titanate Ceramics
64.70 and 77.80; 64.80; S10.15
(a) Vol.115,No.1, 165-172

G.V. BAZUEV, K.R. KRYLOV, A.I. PONOMAREV, V.I. TSIDILKOVSKII, I.M. TSIDILKOVSKII, and T.B. CHARIKOVA
Peculiarities of Galvanomagnetic and Thermoelectric Properties of $YBa_{2-x}La_xCu_3O_{7-\delta}$ Solid Solutions
74.70; 72.20; S10.15
(a) Vol.115,No.1, 267-272

M. OZVOLDOVA
Optical Properties of Uranium in Potassium-Alumino-Phosphate Glasses
78.40 and 78.60; S10.15
(a) Vol.115,No.1, 315-324

S.X. WANG, Z.H. HE, X.H. CHEN, M.R. JI, Y.T. QIAN, Z.Y. CHEN, and Q.R. ZHANG
The Effect of Substitution of Cd for Ca in the Bi-Sr-Ca-Cd-Cu-O System Studied by X-Ray Photoemission Spectroscopy
74.70 and 79.60; S10.15
(a) Vol.115,No.1, K71-K74

FANCHUN ZHENG, JIANYI JIANG, YUPIN SUN, HUAQING YIN, and JIAJU DU
Effects of Cooling Rate on the Properties and the Oxygen Content of the Superconductor $Bi_2Sr_2CaCu_2O_y$
74.70; 62.40; S10.15
(a) Vol.115,No.1, K75-K78

E.M. GOLOLOBOV, N.A. PRYTKOVA, ZH.M. TOMILO, D.M. TURTSEVICH, M.S. TSELUEVSKII, and N.M. SHIMANSKAYA
Influence of Aluminium and Indium Addition on the Properties of Superconducting $Bi_2Sr_2CaCu_2O_y$ Metal Oxide
61.70 and 74.70; S10.15
(a) Vol.115,No.1, K79-K82

E.R. KORUTCHEVA
Mean-Field Theory of an Extended Hubbard Model
74.20; S10.15
(b) Vol.155,No.2, K119-K122

M. PĘKAŁA, R. RATAJ, A. PAJĄCZKOWSKA, and B. GEGENHEIMER
Normal State Transport Study of the Ba-K-Bi-O Superconductor
72.80 and 74.70; 72.20; S10.15
(b) Vol.155,No.2, K123-K126

V.E. BORISENKO, S.YU. NIKITIN, A.B. FILONOV, and A.L. PUSHKARCHUK
Electronic Property Simulation of the $YBa_2Cu_3O_7$ Films
71.25 and 74.70; S10.15
(b) Vol.155,No.2, K127-K130

E.V. KHOLOPOV
Hole Polarisation Nature of Boson Modes Forming High-T_c Superconductivity
74.20; S10.15
(b) Vol.155,No.2, K131-K136

M. TAIBI, J. ARIDE, E. ANTIC-FIDANCEV, M. LEMAITRE-BLAISE, and P. PORCHER
Crystal Field Parameters of Eu^{3+} and Nd^{3+} in Ln_2BaZnO_5 Phases (Ln=La,Nd,Y)
71.70; 78.50; 78.55; S10.15
(a) Vol.115,No.2, 523-532

M. CHYBICKI, J. LIWO, and K. TRZEBIATOWSKI
Electrical Surface Conductivity Modifications of Lead-Silicate Glasses
73.25; S10.15
(a) Vol.115,No.2, K185-K190

N.YU. ARUTYUNOV and V.YU. TRASHCHAKOV
1D-Angular Distributions of Annihilation Photons in RE-Ba-Cu-O Oxides (RE=Y,Nd,Lu)
74.70 and 78.70; S10.15
(a) Vol.115,No.2, K201-K204

A.I. AKIMOV, B.B. BOIKO, S.I. BOROVITSKII, V.I. GATALSKAYA, V.D. GELIKONOVA, S.E. DEMYANOV, A.M. KLUSHIN, L.A. KUROCHKIN et al.
Observation on the Josephson Effect in TlBaCaCuO-TlBaCaCuO Junctions at Temperatures above 100 K
74.70; S10.15
(a) Vol.115,No.2, K205-K206

G. GŁADYSZEWSKI, A. SMAL, Z. MITURA, and M. SUBOTOWICZ
Formation of Superconducting Bi-Sr-Ca-Cu-O Thin Films by Flash Evaporation
68.55 and 74.70; S10.15
(a) Vol.115,No.2, K207-K210

I.O. TROYANCHUK and S.N. PASTUSHONOK
Antiferromagnetic-Ferromagnetic Transition in Orthomanganites
75.50; S10.15
(a) Vol.115,No.2, K225-K228

R. KIEMEL, W. WISCHERT, and S. KEMMLER-SACK
Dependence of T_c on the Hole Concentration in the High-T_c Superconductors $Bi_{1.5}Pb_{0.5}Sr_2Ca_{1-x}Gd_xCu_2O_{8+z}$
74.70; 71.30; 72.20; S10.15
(b) Vol.156,No.1, 339-344

I.S. SMIRNOVA
Twins and Dislocations in $RBa_2Cu_3O_7$ Crystals
61.70; 74.70; S10.15
(a) Vol.116,No.1, 237-254

WENHUI SU, CHANGQING JIN, LANYING HE, YIFENG WANG, HONGJIAN LIU, and XIANG LONG
Study on Oxygen Hole and Superconductivity in $Y_1Ba_2Cu_3O_{7-\delta}$ Material
64.70 and 74.70; 61.10; S10.15
(a) Vol.116,No.1, 267-274

Z. DAMM, J. KLAMUT, T. MORAWSKA-KOWAL, J. OLEJNICZAK, J. STEPIEŃ-DAMM, CZ. SUŁKOWSKI, and E. TROJNAR
Some Properties of Superconducting $(RE)Ba_2Cu_3O_7$ Compounds with Different Bulk Densities
74.70; S10.15
(a) Vol.116,No.1, 367-370

CHUN-SHENG JIA and YAN-JUN YANG
Research on Heat Treatment of Tl-Ba-Ca-Cu-O Superconducting Ceramics
74.70; S10.15
(a) Vol.116,No.1, 371-376

HONG ZHANG, XIAO-GANG WANG, LE LUO, and SHU-HUI HU
A Study on the Positron Annihilation of Glassy $Bi_1Sr_1Ca_1Cu_2O_x$ Quenched from the Molten Liquid
61.40 and 78.70; 74.70; S10.15
(a) Vol.116,No.1, K7-K12

E.N. IZAKOVICH, N.A. ZHELTOVA, YU.M. KOROLEV, ZH.D. SOKOLOVSKAYA, V.N. SPECTOR, and M.L. KHIDEKEL
Study of Isostructural Compounds Y_2BaCuO_5 and $(Y_{1.4}Cu^{3+}_{0.6})BaCuO_5$
61.50; 74.70; S10.15
(a) Vol.116,No.1, K13-K18

OM PARKASH, CH.DURGA PRASAD, and DEVENDRA KUMAR
Preparation and Characterisation of the System $Sr_{1-x}La_xTi_{1-x}Co_xC_3$
61.50 and 72.20; S10.15
(a) Vol.116,No.1, K81-K84

T. KOPTE and E. HEGENBARTH
Thermal Conductivity of $BaCuO_{2+x}$ at Low Temperatures and High Magnetic Fields
66.70; S10.15
(b) Vol.156,No.2, K97-K100

H. BOERNER, M. WURLITZER, H.-CH. SEMMELHACK, B. LIPPOLD, R.-M. BOETTCHER, and M. KROETZSCH
A Method for the Preparation of Single-Phase $YBa_2Cu_3O_x$ Ceramics
61.50; 74.70; S10.15
(a) Vol.116,No.2, 537-544

A. MEHDAOUI, D. BOLMONT, and B. LOEGEL
Dynamical Critical Fields Aging Effects and Restoring in Some $YBa_2Cu_3O_x$ Compounds Studied by Complex Susceptibility Measurements
74.70; 75.30; S10.15
(a) Vol.116,No.2, 777-786

Y. UMAKOSHI, W. TAKAHARA, T. YAMANE, and K.A. TANAKA
Effect of Laser Irradiation on the Superconductive Properties of $(Y_{0.95}Sm_{0.05})Ba_2Cu_3O_x$
74.70; 61.80; S10.15
(a) Vol.116,No.2, 787-792

N. PRUSS, W. KRAAK, H.-U. MUELLER, A. JACOBI, H. DWELK, and R. HERRMANN
Influence of the Sintering Temperature on the Physical Properties of $YBa_2Cu_3O_{7-\delta}$ Ceramic Samples
72.20 and 74.20; S10.15
(a) Vol.116,No.2, 793-802

M.V. RAMANA and G.S. SASTRY
Thermoluminescence in Oxy-Fluoro-Borate Glasses
78.60; S10.15
(a) Vol.116,No.2, K205-K208

S11. Other higher compounds

F. POVOLO and E.B. HERMIDA
Analysis of Stress Relaxation and Creep Curves for a Log-Normal Distribution of Relaxation Times
62.20; S4; S11
(b) Vol.151,No.1, 71-84

R.A. EVARESTOV, V.P. SMIRNOV, and S.A. EGOROV
Band Corepresentations of Magnetic Space Groups
75.10; 71.25; 71.50; S11
(b) Vol.151,No.1, 275-282

M. CASALBONI, R. FRANCINI, F.J. LOHMEIER, and U.M. GRASSANO
Two-Photon Absorption of Bismuth Germanate
78.20; S11
(b) Vol.151,No.1, 347-352

S. WAPLAK and T. KRAJEWSKI
EPR Studies of $LiNaSO_4$ Chromium Cr^{6+} Doped Monocrystals after X-Ray Irradiation
76.30; 61.80; 66.70; S11
(a) Vol.111,No.1, 271-278

J.L. RAO, M.R. REDDY, and S.V.J. LAKSHMAN
Electronic Absorption Spectrum in Ni^{2+} Ions Doped in Zinc Thallium Sulphate Hexahydrate
71.55 and 78.50; S11
(b) Vol.151,No.2, 599-604

R.M. KRISHNA, J.L. RAO, PREM CHAND, and S.V.J. LAKSHMAN
Electron Paramagnetic Resonance Study of Copper(II) Ions in Single Crystals of Thallium Magnesium Sullphate Hexahydrate
71.10 and 76.30; S11
(b) Vol.151,No.2, 615-622

K.P. AREFIEV, O.V. BOEV, and YU.P. SUROV
Positron Annihilation Processes in Mordenite Samples
78.70; S11
(a) Vol.111,No.2, K243-K248

D. GOURIER, T. GBEHI, R. VISOCEKAS, J. THERY, and D. VIVIEN
Color Centres in Barium Hexaaluminate (Phase I). Study by EPR, Optical Absorption, and Thermally Stimulated Luminescence
61.70; 76.30; 78.50; 78.60; S11
(b) Vol.152,No.2, 415-430

A.E. BOCQUET, J.F. DOBSON, P.C. HEALY, S. MYHRA, and J.G. THOMPSON
X-Ray Photoelectron Spectra of High T_c Bismuth Compounds: General Trends for Layer Structures
71.45 and 79.60; 74.10; S11
(b) Vol.152,No.2, 519-532

R.A. EVARESTOV and V.P. SMIRNOV
Tensor Fields in Crystals and Band Representations of Space Groups
75.10; 71.25; 71.50; S11
(b) Vol.152,No.2, 633-638

P.K. BAJPAI
Ferroelectric Phase Transition in $(NH_4)_2BeF_4$. Possibility of Molecular Distortion Type of Transition
64.60; 77.80; S11
(b) Vol.153,No.1, 101-106

J.L. RAO, G.L. NARENDRA, B. SREEDHAR, and S.V.J. LAKSHMAN
Electron Spin Resonance and Optical Absorption Spectra of Mn^{2+} Ions in K_2SO_4-$ZnSO_4$ Glasses
71.70; 76.30; 78.50; S11
(b) Vol.153,No.1, 257-262

R. POPRAWSKI and S.A. TARASKIN
Calorimetric Studies of the Phase Transition in Nd_4DSeO_4 Crystals
64.70; 65.40; S11
(a) Vol.113,No.1, K31-K36

CZ. PAWLACZYK and A. PAWLOWSKI
Electric Capacitance Anomaly near the Superionic Phase Transition in $(NH_4)_4H_2(SeO_4)_3$ Single Crystals
73.40 and 77.40; S11
(a) Vol.113,No.2, 447-452

Y. KIRSH, S. SHOVAL, S. YARIV, and N. KRISTIANPOLLER
Natural and Beta Induced Thermoluminescence in Microcline
78.60; S11
(a) Vol.113,No.2, 631-636

D. BAHADUR, RAM BILAS, and B.S.R. MURTY
Electronic Properties of the Defect Perovskite System $La_{1-x}Li_xMO_3$
61.60 and 72.20; 75.50; S11
(b) Vol.154,No.1, 297-304

S. MISHRA, A.V. KRISHNA RAO, and K.V. RAO
Dielectric Properties of Tourmaline under Different Conditions
77.20 and 77.40; 77.70; S11
(a) Vol.114,No.1, K115-K118

M. POLOMSKA and F. SMUTNÝ
Rb$_4$LiH$_3$(SeO$_4$)$_4$: A New Tetragonal Crystal with Low-Temperature Ferroelastic Phase
64.70; 77.20; S11
(b) Vol.154,No.2, K103-K106

E.L. SAW, M.H. KUOK, C.T. YAP, and J.B. WARD
Temperature Dependence of the Moessbauer Spectra of $CsSnCl_3$
76.80; S11
(a) Vol.114,No.2, K223-K228

H. ZANGAR, J.L. MIANE, C. COURSEILLE, N.B. CHANH, M. COUZI, and Y. MLIK
Structural Phase Transition in the Perovskite-Type Layer Compound $(C_3H_7NH_3)_2PbCl_4$
64.70; 61.14; 77.20; S11
(a) Vol.115,No.1, 107-118

X. GERBAUX, A. HADNI, and A. KITADE
Far IR Spectra of Ferroelectric Rochelle Salt and Sodium Ammonium Tartrate
78.30; 77.80; S11
(a) Vol.115,No.2, 587-598

S11.1. Salts of oxy-acids

WEN-CHEN ZHENG
A Study of Distortion of the Immediate Environment of the Chromium Ion in Alums
61.70; 62.20; 76.30; S11.1
(b) Vol.151,No.1, K11-K16

V.A. IVANSHIN, I.N. KURKIN, W. BRUNNER, and G. VOELKEL
Electron Spin-Lattice Relaxation of the $(SeO_2)^-$ Radical in X-Irradiated $KH_3(SeO_3)_2$
76.30 and 77.80; S11.1
(b) Vol.151,No.1, K53-K58

SH.M. EFENDIEV, A.M. MAMEDOV, and N.G. DARVISHOV
VUV Reflection Spectra of $PbWO_4$ Single Crystals
78.40; 71.25; S11.1
(b) Vol.151,No.1, K87-K90

LING CHEN, MINGKANG TENG, GUANGHOU WANG, XIAOYUN LI, and TIANCHANG LU
Positron Annihilation Lifetime Study of Positive Temperature Coefficient $BaTiO_3$
78.70; S11.1
(a) Vol.111,No.1, K123-K128

N. KRISTOFFEL and M. KLOPOV
Influence of Soft Modes on the Local Dynamics of Defects ($SrTiO_3$)
63.20; S11.1
(b) Vol.151,No.2, K119-K122

A.A. SUKHOVSKII, I.P. ALEKSANDROVA, and YU. N. MOSKVICH
NMR Confirmation of the Existence of Superstructure in the Ferroelectric Phase of NH_4HSeO_4
64.70; 76.60; S11.1
(a) Vol.111,No.2, 477-484

P.S.S. PRASAD and S. RADHAKRISHNA
Chromium Centres in 3 $CdSO_4 \cdot 8\ H_2O$ Single Crystals
61.80 and 71.55; 78.50; S11.1
(b) Vol.152,No.1, 171-180

G.G. BOGACHEV, M.N. ILIEV, and V. PETROV
Raman Scattering of $ScBO_3$
78.30; S11.1
(b) Vol.152,No.1, K29-K32

K. ISHII, N. MORITA, H. NAKAYAMA, S. TSUNEKAWA, and T. FUKUDA
Raman Spectra of $LaNbO_4$ in the Ferroelastic Phase and the Relaxation after the State Shift
78.30; S11.1
(a) Vol.112,No.1, 207-214

P.E. TOMASZEWSKI
On the Effect of Doping on the Mechanism of Phase Transition of Some Inorganic Compounds (Comment)
64.70 and 77.80; 77.70; S11.1
(a) Vol.112,No.1, K13-K14

A.K. PRADHAN and R.N.P. CHOUDHARY
Raman and Infrared Effect of Sm^{3+} in $YAsO_4$ Crystals
78.30; 63.20; S11.1
(b) Vol.152,No.2, K69-K72

M. VERWERFT, D. VAN DYCK, V.A.M. BRABERS, J. VAN LANDUYT, and S. AMELINCKX
Electron Microscopic Study of the Phase Transformations in $AgNbO_3$
61.16 and 64.60; S11.1
(a) Vol.112,No.2, 451-466

P. CORDIER, B. BOULOGNE, N. DOUKHAN, and J.C. DOUKHAN
Lattice Defects in Lithium Tantalate
61.70; 77.60; S11.1
(a) Vol.112,No.2, 493-510

R. NAVIL KUMAR and C.P.G. VALLABHAN
Effect of Deuteration on the Electrical Conductivity, Dielectric Constant, and Phase Transition in $(NH_4)_3H(SO_4)_2$
72.80 and 77.20; 64.70; S11.1
(a) Vol.112,No.2, K113-K118

SHENG-LI GUO and KE-QIN LU
Analysis of the Spectra of Mn^{2+} in Rhodonite
71.55 and 78.50; S11.1
(b) Vol.153,No.1, K45-K50

L. KOVÁCS, R. SOMMERFELDT, YE MING, and E. KRAETZIG
Light-Induced Charge Transport in $LiNbO_3$:Mg,Cr Crystals
61.70 and 72.40; 78.50; S11.1
(a) Vol.113,No.1, K75-K78

L. HOLTMANN
A Model for the Nonlinear Photoconductivity of $BaTiO_3$
72.40; S11.1
(a) Vol.113,No.1, K89-K94

I.O. TROYANCHUK, A.A. SHEMYAKOV, and V.K. PROKOPENKO
Magnetic Structure of $NiMnO_3$ and $CoMnO_3$ Ilmenites
75.30 and 76.60; S11.1
(a) Vol.113,No.1, K107-K110

WEN-CHEN ZHENG
Local Thermal Expansion around the Impurity in Chromium-doped alpha-$LiIO_3$ Crystals
65.70; 71.70; 76.30; S11.1
(b) Vol.153,No.2, K103-K108

R. MIZERIS, J. GRIGAS, R.N.P. CHOUDHARY, and T.V. NARASAIAH
Microwave Dielectric Investigations of TlH_2PO_4.
77.20; 77.80; S11.1
(a) Vol.113,No.2, 597-602

XIAODING QI, ZUNDU LUO, QINGZHEN HUANG, JINGKUI LIANG, JIMING CHENG, YICHUAN HUANG, MINWANG QIU, and HONG ZHANG
A New Laser Host Crystal. Lathanum Tungstate $La_2(WO_4)_3$
78.45; 78.30; 78.55; S11.1
(a) Vol.114,No.1, K127-K130

O. HUDÁK
Thermodynamics of Localized Magnetic Singlets and 1/2-Spins in the Plane: Paramagnetic Phase of La_2CuO_4
75.40; S11.1
(b) Vol.154,No.2, 743-746

WEN-CHEN ZHENG
Theoretical Investigation of the Spin-Lattice Coupling Coefficients G_{11} and G_{44} for Cr^{3+} in Cubic $SrTiO_3$ Crystals
71.70 and 76.30; S11.1
(b) Vol.154,No.2, K167-K170

V.A. GANSHIN, YU.N KORKISHKO, T.V. MOROZOVA, and V.V. SARAIKIN
The Study of Proton Exchange in Lithium Tantalate Crystals
61.70; 78.20; S11.1
(a) Vol.114,No.2, 457-466

P. PISKUNOWICZ, T. BRĘCZEWSKI, and T. WOŁEJKO
Thermal Expansion of $Rb_4LiH_3(SO_4)_4$ and $K_4LiH_3(SO_4)_4$ Crystals in the Temperature Range from 80 to 300 K
64.70 and 65.70; S11.1
(a) Vol.114,No.2, 505-510

I.P. ALEKSANDROVA, PH. COLOMBAN, F. DÉNOYER, N. LE CALVÉ, A. NOVAK, B. PASQUIER, and A. ROZYCKI
Phase Transitions in Ammonium Hydrogen Selenates NH_4HSeO_4 and ND_4DSeO_4
64.70; 78.30; S11.1
(a) Vol.114,No.2, 531-544

C.K. JAYASANKAR and F.S. RICHARDSON
Spin-Correlated Crystal-Field Analyses of the $4f^2(Pr^{3+})$ Energy Levels in $LiYF_4:Pr^{3+}$ and $LiBiF_4:Pr^{3+}$
71.70; S11.1
(b) Vol.155,No.1, 221-230

O. HUDÁK
Reentrant Paramagnetic-AFM-Paramagnetic Phase Transitions in La_2CuO_4
75.30 and 75.50; S11.1
(b) Vol.155,No.1, 257-262

B.K. CHOUDHARY, R.N.P. CHOUDHARY, and K.V. RAO
Dielectric Properties of $TbAsO_4$ Single Crystals
77.20 and 77.40; S11.1
(a) Vol.115,No.1, 301-306

V.S. KISHAN KUMAR, S.B.S. SASTRY, and B S V.S.R. ACHARYULU
Optical Studies on Europium Doped K_2SO_4
61.70 and 78.50; 78.55; S11.1
(b) Vol.155,No.2, 679-684

W. KUCHARCZYK and P. GÓRSKI
Bond Polarizability Calculation of the Quadratic Electrooptic Effect in KDP
78.20; S11.1
(b) Vol.155,No.2, K157-K160

B. MRÓZ and R. LAIHO
Elastic Properties of $K_4LiH_3(SO_4)_4$ Crystals
77.60; 62.20; S11.1
(a) Vol.115,No.2, 575-580

S. AHMED, A.M. SHAMAH, A. IBRAHIM, and F. HANNA
X-Ray Studies of the High Temperature Phase Transition of Ammonium Sulphate Crystals
64.70; 77.80; S11.1
(a) Vol.115,No.2, K149-K154

D. BRACH, P. GLAREMIN, and H.-J. WEBER
Optical Investigations of the Phase Transition in Pure and Cr^{3+}-Doped Potassium Alum
64.60; 78.20; 78.55; S11.1
(b) Vol.156,No.1, 195-204

H. DONKER, W.M.A. SMIT, and G. BLASSE
The Luminescence of $CaCO_3:Sn^{2+}$
78.55; 78.50; S11.1
(b) Vol.156,No.1, K93-K96

S.K. ARORA and T. MATHEW
Dielectric Studies of $CuWO_4$ Crystals
77.20 and 77.40; S11.1
(a) Vol.116,No.1, 405-414

SH. EFENDIEV, V. GAVRYUSHIN, G. RACIUKAITIS, G. PUZONAS, A. KAZLAUSKAS, N. DARVISHOV, and S. SHAKHDGAN
Two-Photon Spectroscopy of $PbMoO_4$ Single Crystals. Indirect Interband Transitions
78.40; 71.25; S11.1
(b) Vol.156,No.2, 697-704

W. ZAPART, M.B. ZAPART, and A.I. OTKO
EPR Detection of the Ferroelastic Phase Transition in $KSc(WO_4)_2$
76.30; 64.70; S11.1
(a) Vol.116,No.2, K191-K194

F. BALLER, B. GATHER, B. HELLERMANN, H. HESSE, and E. KRAETZIG
The Influence of Stress on the Dielectric Properties of KTN Crystals
77.20; 77.80; S11.1
(a) Vol.116,No.2, K195-K198

S11.2. Spinels, garnets, ferrites

V.N. BERZHANSKII and V.I. IVANOV
Critical Relaxation of Uniform Precession of Magnetization in Cubic Ferrimagnets
75.30 and 75.50; 76.50; S11.2
(b) Vol.151,No.1, 259-268

O.A. BAYUKOV, V.P. IKONNIKOV, and A.F. SAVITSKII
Magnetic and Moessbauer Studies of Anion-Substituted Nickel Ferrites
75.30; 76.80; S11.2
(b) Vol.151,No.1, 269-274

D. VLADIKOVA, L. ILKOV, and S. KARBANOV
Influence of the Microstructure on Some Microwave Properties of Substituted Nickel Ferrites
64.80; 72.20; 75.50; S11.2
(a) Vol.111,No.1, 145-154

J. WIEHL, R. HIRRLE, W. WISCHERT, and S. KEMMLER-SACK
Cathodo- and Photoluminescence in the Mn^{2+} Activated Garnets $Cd_{3-x}Ca_xAl_2Ge_3O_{12}$
78.55 and 78.60; 71.55; S11.2
(a) Vol.111,No.1, 315-324

M.A. AHMED
Electrical Properties of Co-Zn Ferrites
72.80; S11.2
(a) Vol.111,No.2, 567-572

R. HIRRLE, J. WIEHL, W. WISCHERT, and S. KEMMLER-SACK
Cathodo- and Photoluminescence in the Tb^{3+} Activated Garnets of Type $Cd_{3-x}Ca_xAl_2Ge_3O_{12}$
78.55 and 78.60; 71.55; S11.2
(a) Vol.111,No.2, 629-638

G.V. BICH, V.M. KAMYSHOV, A.N. MEN, and YU.A. PETROV
Structural Transformation during Spinel Phase Oxidation in the Ge-Fe-O System
64.70; S11.2
(a) Vol.111,No.2, K171-K174

E.Z. KATSNELSON, A.G. KAROZA, and M.M. CHERVINSKII
Photomagnetization of Ferrite Films and Its Effect on Optical Spectra
75.70 and 78.20; S11.2
(a) Vol.111,No.2, K233-K236

S. PHANJOUBAM, D. KOTHARI, and J.S. BAIJAL
Electrical Conductivity of Tetravalent Titanium Substituted Lithium Zinc Ferrite
72.20; S11.2
(a) Vol.112,No.1, 131-136

YU.G. CHUKALKIN, V.R. SHTIRZ, and B.N. GOSHCHITSKII
The Structure and Magnetism of Amorphous $Y_3Fe_5O_{12}$
61.40 and 75.50; S11.2
(a) Vol.112,No.1, 161-174

C. DE FRANCISCO, J.M. MUÑOZ, J. AYALA, and J.I. IÑIGUEZ
An Investigation of the Magnetic Disaccommodation in Zinc Ferrites Sintered under Very Low Oxidation Pressures
75.60; S11.2
(a) Vol.112,No.1, K43-K48

A.D. BALAEV, O.A. BAYUKOV, and A.F. SAVITSKII
Magnetic and Moessbauer Studies of Magnesium-Chromium Ferrites
75.50 and 76.80; S11.2
(b) Vol.152,No.2, 639-644

E.Z. KATSNELSON, A.G. KAROZA, L.A. MELESHCHENKO, and L.A. BASHKIROV
IR-Reflection Spectra of Manganese-Zinc Ferrites
78.30; S11.2
(b) Vol.152,No.2, 657-666

O.A. BAYUKOV and A.F. SAVITSKII
Application of the Superexchange Model to the Evaluation Method of Exchange and 3d-Electron Transfer Parameters in Magnetodielectrics
75.30; S11.2
(b) Vol.152,No.2, K51-K54

C. DJEGA-MARIADASSOU, T. MERCERON, J.L. DORMANN, M. PORTE, and P. RENAUDIN
Verwey Transition in Two Series of Substituted Magnetites. $Li_xFe_{3-x}O_4$ and $Mg_xFe_{3-x}O_4$
75.30 and 75.50; S11.2
(a) Vol.112,No.2, 601-616

J. AULEYTNER
Structural Changes in Solids Induced by Heavy Ions at Energies of MeV/n. Results of X-Ray and Electron Microscopic Studies
61.80; 61.10; 61.16; S5.11; S11.2
(a) Vol.112,No.2, 679-688

V.N. SHAMBALEV, L.A. BASHKIROV, V.V. PANKOV, A.G. DUTOV, and A.N. IGUMENTSEV
Self-Diffusion of Cations in the Solid Solution System. Ti-Substituted Mn-Zn Ferrites
66.30; S11.2
(a) Vol.112,No.2, K71-K74

I. TOMÁŠ, D. TOMÁŠ, and T. DOLEŽAL
A Quick Estimate of Saturation Magnetization Value from the Bitter Powder Picture of Open-Flux Domain Structures
75.60; S11.2
(a) Vol.113,No.1, K95-K98

J. HORN and H.-J. ZACHARIAS
The Lines of Constant Vacancy Content in the Phase Diagram of the Al-Fe Ferrite System
61.70 and 75.50; S11.2
(a) Vol.113,No.1, K103-K106

A.I. MITSEK and V.N. PUSHKAR
The Topology of Magnetic Inhomogeneities in Thin Ferromagnetic Layers. The Model of the Magnetic Vortex
75.60 and 75.70; S1.1; S1.2; S11.2
(a) Vol.113,No.2, 529-540

M. JIRSA, G. VÉRTESY, and J. PAČES
Temperature Dependence of the FMR in Garnet Films with Bubble-Domain Lattice
75.70 and 76.50; 75.60; S11.2
(a) Vol.113,No.2, 573-578

S.E. SARKISOV, A.A. KAMINSKII, and A.V. BUTASHIN
Optical Phonons and Lattice-Activator Interactions of Li^{3+} Ions in the Disordered Oxide Single Crystals with Ca-Gallogermanate ...
78.30; 71.70; S11.2
(a) Vol.113,No.2, 585-596

K. HISATAKE, I. MATSUBARA, K. MAEDA, T. FUJIHARA, S. SUZUKI, H. WAKAO, T. HIGASHI, and K. UEMATSU
Gamma-Ray Induced Change of the Initial Permeability in Yttrium Iron Garnet at 77 K
75.50; S11.2
(a) Vol.113,No.2, K245-K248

A.A. KAMINSKII, M.I. DEMCHUK, N.V. ZHAVORONKOV, and V.P. MICHAILOV
Ultrashort Pulses of Stimulated Emission by Nd^{3+}-Doped ($^4F_{3/2}-^4I_{11/2}$ Channel) Anisotropic Fluorides, Acentric Dispersed Silicates, ...
78.45; S9.16; S11.2
(a) Vol.113,No.2, K257-K260

TH. KLUPSCH, W. SCHUEPPEL, and M. ROESLER
Magnetization Curve and Magnetic Characterization of Planar Hexaferrite Powders Prepared by Glass Crystallization
75.60; S11.2
(a) Vol.114,No.1, 327-340

M.A. AHMED, M.K. EL-NIMR, A. TAWFIK, and A.M. ABOELATA
Dielectric Behaviour in Co-Zn Ferrites
77.20 and 77.40; S11.2
(a) Vol.114,No.1, 377-380

R. HOEHNE
On the Instability of Relaxation Processes of Induced Magnetic Anisotropy in Gallium-Iron Ferrites
75.30 and 75.50; S11.2
(a) Vol.114,No.1, K93-K98

O.A. BAYUKOV and A.F. SAVITSKII
The Prognostication Possibility of Some Magnetic Properties for Dielectrics on the Basis of Covalency Parameters of Ligand-Cation ...
75.30; S11.2
(b) Vol.155,No.1, 249-256

Z. ŠIMŠA and P. GOERNERT
Preparation and Faraday Rotation Spectra of YIG:Pb,Pt Garnet Films
68.55 and 78.20; S11.2
(a) Vol.115,No.1, 229-236

V.M. TALANOV
Anion Ordering in Spinels
61.60; S11.2
(a) Vol.115,No.1, K1-K4

N.P. KOLMAKOVA, R.Z. LEVITIN, and V.N. ORLOV
Anomalies of Crystal Lattice Parameters in Paramagnetic Rare-Earth Garnets at Low Temperatures
65.70 and 75.80; S11.2
(a) Vol.115,No.1, K87-K90

SHOUYE ZHANG, GUANGYAO FONG, and MING MAO
Spin-Canting Effect and Magnetic Moment of Li-Zn Ferrite
75.50; S11.2
(b) Vol.155,No.2, 647-652

M.M. SAVOSTA. V.D. DOROSHEV, P. NOVÁK, and I. VELTRUSKY
NMR of ^{167}Er and Magnetic Structure of $Er_3Fe_5O_{12}$ Garnet
76.60; 75.25; S11.2
(b) Vol.155,No.2, 669-678

N.A. MIRONOVA, V.N. SKVORTSOVA, A.S. SMIRNOV, D.V. RIEKSTINYA, G.A. GRINVALDS, and L.A. LITVINOV
Anomalous Behaviour of $MgO \cdot 2.5Al_2O_3$ under Radiation
61.80 and 78.40; 78.55; S11.2
(a) Vol.116,No.1, K125-K128

A.YA. FISHMAN, N.M. KOVTUN, V.YA. MITROFANOV, V.K. PROKOPENKO, and A.A. SHEMYAKOV
Supertransferred Hyperfine Fields at Nuclei of Diamagnatic Copper Ions in Chalcogenide Spinels $Fe_{1-x}Cu_xCr_2S_4$
75.30; 76.60; S11.2
(b) Vol.156,No.2, 597-604

S12. Organic compounds

T. OVSENIK, R. BLINC, M. ČOPIČ, G. LAHAJNAR, I. ZUPANČIČ, P. GUENTER, H. AREND, and CH. BOSSHARD
Self-Diffusion in an Undercooled Organic Melt near the Liquid-Glass Transition
66.30; 64.70; S12
(b) Vol.151,No.1, K1-K6

A.A. ROMANYUKHA, YU.N. SHVACHKO, A.V. SKRIPOV, and V.V. USTINOV
Conduction Electron Spin Resonance in the New Organic Superconductor $(ET)_2Cu(SCN)_2$
76.30; S12
(b) Vol.151,No.1, K59-K64

YU.N. SHVACHKO, A.A. ROMANYUKHA, A.V. SKRIPOV, and V.V. USTINOV
Effects of Electron Irradiation on ESR Properties of β-$(ET)_2I_3$
76.30; S12
(b) Vol.151,No.1, K65-K70

S.S. BAWA, K. SAXENA, and S. CHANDRA
Polarization Reversal Current Behaviour for Parallel and Antiparallel Rubbed Surface Stabilized Ferroelectric Liquid Crystals
77.80; S12
(a) Vol.111,No.1, 309-314

R. JAKUBAS, P.E. TOMASZEWSKI, and L. SOBCZYK
Phase Transition in $(CH_3NH_3)_3Bi_2Cl_9$
64.70; 77.20; S12
(a) Vol.111,No.1, K27-K30

B. MALLIK
Electrical Conduction in Ferrocene with Adsorbed Vapours
72.80; 73.25; S12
(a) Vol.111,No.1, K61-K66

P.L. MUTHAL, S.V. MOHARIL, and B.T. DESHMUKH
Radiation-Induced Sensitization in Polyethylene
78.60; S12
(a) Vol.111,No.1, K119-K122

M. GAAFAR, M.A. LASHEEN, A.A. ALY, and A.M. AHMED
Optical Properties of $C_6H_{6-y}X_y$ (y=2,3 and X=Cl,Br)
78.20; S12
(a) Vol.111,No.2, K259-K264

H. KRESSE, H. STETTIN, F. GOUDA, and G. ANDERSSON
Dielectric Relaxation of a Mixture of a Nematic Liquid Crystal and a Non-Liquid Crystalline Compound
61.25 and 77.20; S12
(a) Vol.111,No.2, K265-K268

C. DERVOS, P.D. BOURKAS, and E.A. KAYAFAS
Potential Profile of Dielectrics during High-Voltage Pulse Application
72.20 and 77.50; S12
(a) Vol.112,No.1, 123-130

S.A. RÓŻAŃSKI, W. KUCZYŃSKI, and J. MAŁECKI
Biaxiality of the Ferroelectric Liquid Crystal DOBAMBC in Electric Field
77.80 and 78.20; 61.30; S12
(a) Vol.112,No.2, 617-624

Z. ZIKMUND and J. FOUSEK
Stabilization of Polarization Vector in Uniaxial Ferroelectrics by Means of Chiral Dopants. Macroscopic Symmetry Conditions
77.30 and 77.80; S12
(a) Vol.112,No.2, 625-634

M. ABD-LEFDIL and M. CADÈNE
Optical Studies of Oriented Undoped Polyacetylene
78.20; S12
(a) Vol.112,No.2, 635-638

M. KISZA and E. OLESZKIEWICZ
Modified PMMA-Polymers for Masks in Ion Beam Lithography
61.40 and 61.80; S12
(a) Vol.112,No.2, 761-764

W. RUDOLPH, D. GRAMBOLE, C. HEISER, F. HERRMANN, and C. NEELMEIJER
Compositional Changes of PMMA Layers during ^{15}N Bombardment
61.40 and 61.80; S12
(a) Vol.112,No.2, 765-768

R. MUEHLE and T. HARTUNG
Ion Beam Exposure Characteristics of Nitrocellulose
61.70 and 61.80; S12
(a) Vol.112,No.2, 769-772

R. RIESENBERG, A. UNKROTH, and D. DIETRICH
Stress Controlled Adhesion and Magnetic Properties of Ion Implanted Co Films
62.20; 61.80; 75.60; S12
(a) Vol.112,No.2, K85-K90

V.A. OSIPOV, G. BOCHNACKA, V.K. FEDYANIN, and J. MALEK
Polarons in Trans-Polyacetylene and Polyyne Chains
71.38; S12
(b) Vol.153,No.1, 207-214

M. HAASE and M. TAUT
Plasmon Dispersion in Trans-Polyacetylene
71.45; S12
(b) Vol.153,No.1, 225-234

R. SIGNERSKI, J. KALINOWSKI, P. DI MARCO, and G. GIRO
Photoconductivity of Polycrystalline Thionaphtenindol
72.40 and 73.60; S12
(b) Vol.153,No.1, 281-294

T. BLUM and H. BAESSLER
Light-Induced Inter-Chain Charge Transfer in Crystalline Polydiacetylene (PTS)
72.40; S12
(b) Vol.153,No.1, K57-K62

K. INOUE, K. IKEZAKI, and M. KATO
Thermally Stimulated Currents in Plasma Treated FEP Films
73.60; S12
(a) Vol.113,No.1, K67-K70

R. CACH and Z. CZAPLA
Dielectric Nonlinearity of Diglycine Nitrate (DGN) Crystal in the Paraelectric Phase
77.20; 77.80; S12
(a) Vol.113,No.1, K139-K142

B.D. SHRIVASTAVA, S.K. JOSHI, and K.B. PANDEYA
X-Ray K-Absorption Studies of Cobalt(II) Complexes in Solid State
78.70; S12
(a) Vol.113,No.2, 603-608

M. OLSZOWY
Electret State in the Polyethylene Terephthalate Films Implanted by Oxygen and Bismuth Ions
77.30; 61.80; S12
(a) Vol.113,No.2, 609-614

M. PALPACUER, M. ROLLAND, and C. BENOIT
Polarized Spectra of Oriented Polyacetylene Doped with Iodine
78.30; 78.65; S12
(a) Vol.113,No.2, 623-630

M.M. KHALIFA, M.N. AVASTHI, and S.M. EL-MASHRI
Heat of Transport of an Organo-Metallic Superionic Material
66.70; S12
(a) Vol.113,No.2, K177-K180

J. SCHLEIFER, J. KALUS, U. SCHMELZER, and G. ECKOLD
Phonon Dispersion in an alpha-Perylen d_{12}-Crystal at 10 K
63.20; S12
(b) Vol.154,No.1, 153-166

S.K. MISRA and CHUNZHENG WANG
EPR and Optical-Absorption Studies of Cu^{2+}-Doped $Mg(CH_3COO)_2 \cdot 4H_2O$ Single Crystal. Core Polarization and Molecular-Orbital Coefficients
71.70; 76.30; 78.50; S12
(b) Vol.154,No.1, 259-272

H. ISOTALO, J. LAAKSO, P. KUIVALAINEN, H. STUBB, J.-E. OESTERHOLM and P. YLI-LAHTI
Charge Transport in $FeCl_3$-Doped Polyethylenevinylacetate-Polyoctylthiophene Blends
72.20; 72.60; 76.30; S12
(b) Vol.154,No.1, 305-314

M. TAYA, T. ASAJI, and D. NAKAMURA
A Low-Temperature Phase Transition in Ferroelectric $(CH_3NH_3)HgCl_3$ Definitely Located by Pyroelectric and Dielectric Measurements
64.70 and 77.80; S12
(a) Vol.114,No.1, 157-160

T.P. BEALES, S. QUERESHI, and M.R. WILLIS
The Effect of Pressure on the Electrical Conductivity of $[(C_6H_5)_3PCH_3]^+(TCNQ)_2$
72.80; S12
(a) Vol.114,No.1, 299-304

S.S. BAWA, K. SAXENA, and S. CHANDRA
Surface and Bulk Switching Dynamics in Surface Stabilized Ferroelectric Liquid Crystals
77.30 and 77.80; S12
(a) Vol.114,No.1, 369-376

YUAN-CHAO CHU, T. TSANG, and L. YIN
Structures of Cyano-Biphenyl Liquid Crystals
61.30; S12
(a) Vol.114,No.1, K1-K6

S.N. FEDOSOV
Voltage Recovery in Open-Circuit Polar Electret
77.30; S12
(a) Vol.114,No.1, K103-K106

I. GŁOWACKI and J. ULAŃSKI
Precipitation of NIPC Dopant in Polycarbonate Matrix as Detected by Thermoluminescence
78.60; S12
(a) Vol.114,No.1, K107-K110

B. KOSTUREK and R. JAKUBAS
Optical Birefringence of $(CH_3NH_3)_5Bi_2Br_{11}$
78.20; S12
(a) Vol.114,No.1, K111-K114

A. HEILMANN, C. HAMANN, G. KAMPFRATH, and DO NGOC UAN
TEM Investigations on Plasma Polymer Silver Composite Films
68.55; S1.3; S12
(a) Vol.114,No.2, 551-558

TH. ERGE, D. MICHEL, J. PETERSSON, and W. WINDSCH
^{35}Cl NMR Study in the Paraelectric Phase of Tris-Sarcosine Calcium Chloride
76.60; 77.80; S12
(a) Vol.114,No.2, 705-712

T. TANABE, R. IKEDA, and D. NAKAMURA
The Highest-Temperature Crystalline Phase of Tetramethylammonium Thiocyanate with the Cations Performing Three-Dimensional ...
66.30 and 76.60; S12
(a) Vol.114,No.2, K143-K146

W. UNGER and W. POSSART
An XPS Investigation of the Interface between Thin Polymethylmethacrylate (PMMA) Films and Natively Oxidized Al
68.48 and 79.60; S4; S12
(a) Vol.114,No.2, K175-K178

V.G. PODOPRIGORA, I.A. REMIZOV, V.F. SHABANOV, and L.V. SHMELEVA
Dynamic and Electrooptic Parameters of Monolayers in Ethane Adsorbed on Graphite
68.30; S12
(b) Vol.155,No.1, 117-126

S.N. FEDOSOV
Deep Trapping of Injected Carriers in Ferroelectric Polymer
77.80; 77.30; S12
(a) Vol.115,No.1, 293-300

R. JAKUBAS and H. PYKACZ
Dilatometric Study of the Ferroelectric Phase Transition in $(CH_3NH_3)_5Bi_2Br_{11}$
64.70 and 65.70; S12
(a) Vol.115,No.1, K17-K18

A.V. VANNIKOV, A.YU. KRYUKOV, A.G. TYURIN, and T.S. ZHURAVLEVA
Influence of the Medium Polarity on Electron Transport in Polymer Systems
72.80; S12
(a) Vol.115,No.1, K47-K52

S. ICHINOSE
Two-Component Solitary Excitations in alpha-Helical Protein Molecules
64.10 and 71.45; S12
(b) Vol.155,No.2, 467-474

J. SERNA, J.CH. ABBÉ, and G. DUPLÂTRE
Size of Voids in Polyethylene
61.40 and 61.70; 78.70; S12
(a) Vol.115,No.2, 389-398

L. LEJČEK, A. FOUSKOVÁ, and M. PAVEL
Note on Light Scattering by Dislocations in TGS
61.70 and 78.20; S12
(a) Vol.115,No.2, 445-450

N.N. DINH, V.T. BICH, L.Q. MINH, and N.T. TOAN
Preparation of Lu-Diphthalocyanine Thin Films by Electron Beam Deposition and Electrochromic Effect
68.55 and 78.65; S12
(a) Vol.115,No.2, K163-K166

E.M. IBRAGIMOVA and V.N. STEPANOV
Surface Defect States at Silica-Polymer Interfaces in Fiber-Optic Waveguides
61.80 and 78.60; S10.1; S12
(a) Vol.115,No.2, K255-K260

A. SCHAPER, D. ZENKE, E. SCHULZ, R. HIRTE, and M. TAEGE
Structure-Property Relationships of High-Performance Polyethylene Fibres
61.40; 61.16; 62.20; S12
(a) Vol.116,No.1, 179-196

P.S.R. PRASAD and H.D. BIST
Vibrational Spectroscopic Study of Order-Disorder Transition in Ethylammonium Tetrachloro Cadmate
64.70 and 78.30; S12
(a) Vol.116,No.1, 275-286

O.G. VLOKH, A.V. KITYK, O.M. MOKRY, and V.G. GRYBYK
Pressure-Temperature Phase Diagram of $\{N(CH_3)_4\}_2CuCl_4$ Crystals. Birefringent and Elastic Properties
64.60; 62.20; 78.20; S12
(a) Vol.116,No.1, 287-294

H. OTOP
Some Experimental Observations of Exoemission from Ammonium Hydrogen Selenates
77.80 and 79.75; S12
(a) Vol.116,No.1, K115-K118

B.P. GORSHUNOV, M.G. KAPLUNOV, N.D. KUSHCH, and O.I. SYROTYNSKII
Temperature Dependence of the Optical Properties of the Organic Superconductor $(ET)_2Cu(NCS)_2$ in the Submillimeter Range
78.20; 74.70; S12
(a) Vol.116,No.1, K129-K132

G. HAUCK, E. KUKLEVA, and H.D. KOSWIG
Electrically Induced Deformations in Cholesteric Systems with Pretilt
61.30; 64.70; S12
(a) Vol.116,No.2, 545-554

XIANG-FENG JIN, BING MAO, SHU-QIN ZHOU, REN-YUAN QIAN, J.S. PARK, and P. EHRLICH
Carrier Transport Properties of Trans-Polyphenylacetylene
73.60; 71.55; S12
(a) Vol.116,No.2, 709-714

C. HAMANN, R. LAIHO, and A. MRWA
Surface Analysis on Lead Phthalocyanine with STM
68.20 and 68.55; S12
(a) Vol.116,No.2, 729-734

R. CACH, S. DACKO, and Z. CZAPLA
Dielectric Properties and State Equation Parameters of $(CH_3)_2NH_2Al(SO_4)_2 \cdot 6H_2O$ Ferroelectric Crystals
77.80; 77.20; S12
(a) Vol.116,No.2, 827-832

ZHONGMING LIU, XIAOHUA CHEN, HUIMIN SHEN, YENING WANG, HUIFANG YOUNG, and P.C.W. FUNG
Ferroelectric Domain Wall Motion and Related Internal Friction in TGS Crystal
62.20 and 77.80; S12
(a) Vol.116,No.2, K199-K204

J.J. DONAGHY and K.E. VAN NESS
A Trapping Model Analysis of the Positron Lifetime Spectra of Polymers
78.70; S12
(a) Vol.116,No.2, K213-K218

S12.1. Quasi-one-dimensional systems

RENYUAN QIAN, SONGQING SHI, TONGMING LIU, and DAOBEN ZHU
Carrier Transport Properties of Poly(1,6-Bis-9-Carbazolyl-2,4-Hexadinyne)
72.80; 72.40; 78.40; S12.1
(a) Vol.111,No.1, K67-K72

M.G. OSTAPENKO and E.G. PETROV
The Donor-Acceptor Electron Transfer with Participation of Short Polymer Chains
72.80; 71.55; S12.1
(b) Vol.152,No.1, 239-248

J. MERTSCHING
Double Periodic Peierls Phases
61.65 and 71.25; S12.1
(b) Vol.152,No.2, 551-562

A.N. CHUVYROV, L.A. KOSYKH, G.V. LEPLYANIN, V.N. SALIMGAREEVA, and N.S. SANNIKOVA
Magnetically Ordered Phase in Quasi-One-Dimensional Systems Based on Polyacetylene
75.50; S12.1
(a) Vol.113,No.1, K111-K116

M.A. YURISHCHEV
Ising Model Consisting of Central and Three Side Chains
75.10; S12.1
(b) Vol.153,No.2, 703-710

JIE LIU, DE-GANG ZHANG, and JIN YANG
Effects of Alternating Potential on the Dimerisation of Diatomic Peierls Systems
71.45; 71.38; S12.1
(b) Vol.153,No.2, K131-K134

L. FIRLEJ, A. GRAJA, J. WOLAK, R.N. LYUBOVSKAYA, and L.M. GOLDENBERG
ESR Studies of Single Crystals of Bis(ethylenedithio)tetrahiafulvalene Based Conductor with Octaiodinemercurate Anion $(BEDT\text{-}TTF)_4Hg_3I_8$
64.70 and 76.30; 72.20; S12.1
(b) Vol.154,No.1, 333-340

A.B. FAZAKAS and R. PITIS
Non-Zero Momentum Pairing and Peierls Order in One Dimensional Systems
71.30; S12.1
(b) Vol.154,No.2, 637-642

A.M. VAINRUB, G.S. ZAVT, and A. GRAJA
Absence of the Gap in the Optical Excitation Spectrum of TCNQ Salts with Methyl Derivatives of Pyridinium
71.45; 63.20; S12.1
(b) Vol.155,No.1, 147-154

A.P. POPOV and A.A. LAVRENTEV
The Rebuilding of the Electron Spectra in One-Dimensional Structures by Increasing the Electron-Phonon Binding
71.38; S12.1
(b) Vol.155,No.1, 185-190

V.A. OSIPOV, G. BOCHNACKA, and V.K. FEDYANIN
Polarons in the Finite-Band Continuum Model of Diatomic Polymers
71.38; S12.1
(b) Vol.155,No.1, K29-K32

P. ENDERS
Comment on: Effects of Alternating Potential on the Dimerisation of Diatomic Peierls Systems by Liu et al.(phys.stat.sol.(b)153,K131(1989)
71.45; 71.38; S12.1
(b) Vol.156,No.2, K123-K124

Author Index

A.K. Abass	(a) 113, 579	78.40; S10.15
	(a) 116,K111	78.65; S10.1
J.Ch. Abbé	(a) 115, 389	61.40 and 61.70; 78.70; S12
M. Abd-Lefdil	(a) 112, 635	78.20; S12
	(a) 115, 491	68.55 and 73.60; S8
M. Abdel Naby	(a) 115, K27	73.40; S8.11; S8.13
S. Abdel-Maksoud	(b) 152, 209	71.55 and 72.40; S8
A.Sh. Abdinov	(a) 116,K173	72.40; S8
S.G. Abdinova	(a) 111, K73	72.40 and 73.40; S8.16
A.B. Abdullaev	(a) 115, K5	65.40 and 65.50; S8.16
A.M. Abdullaev	(a) 115, K5	65.40 and 65.50; S8.16
U.Zh. Abdumanapov	(b) 155,K113	71.55; S5.11
A.A. Abdurragimov	(a) 113,K207	65.40; S8.16
T.Sh. Abesadze	(b) 156, K67	76.60
A.K. Abiev	(a) 114,K119	77.80; 61.80; S8.16
S. Abis	(a) 114, 511	64.75; 72.15; S1.2; S4
A.M. Aboelata	(a) 114, 377	77.20 and 77.40; S11.2
M. Abraham	(b) 151, 383	78.65
G.E. Abrosimova	(a) 111,K237	75.60; S1.1
M.E. Abu-Zeid	(a) 113, 123	68.55; 78.65; S10
B.N. Narahari Achar	(b) 154, K59	72.20 and 73.60
B.S. Acharya	(b) 153, 153	71.20 and 73.60; 78.70; S8.15
B.S.V.S.R. Acharyulu	(b) 155, 679	61.70 and 78.50; 78.55; S11.1
A. Adam	(a) 112, 721	73.40; 68.55; S5.11
V.K. Adamchuk	(b) 155, 525	71.25; 79.60; S9.11
B. Adamczyk	(a) 112, 829	79.40
V.E. Adamian	(b) 156, 633	64.75 and 75.30; S1.4
G.A. Adegboyega	(a) 111, K31	73.30; S5.11
	(a) 114, 247	71.55; 72.20; 78.30; S5.11
A.M. Afanasev	(a) 113,K153	61.10
	(a) 116, 197	61.10; 78.70; S7.12
A.M. Agaev	(a) 111, K73	72.40 and 73.40; S8.16
R.P. Agarwala	(a) 114,K159	62.20; S1.1
A.A. Agasiev	(a) 116, 305	73.40; 72.20; S10
P. Aghamkar	(b) 151, 739	78.35; S7.13
O.P. Agnihotri	(a) 114, K29	73.60 and 78.65; S8.16
	(a) 115,K181	73.60 and 78.65; S8.16
R. Agrawal	(a) 116, 837	73.40 and 85
P.K. Ahluwalia	(b) 151, 225	72.10 and 72.15
M. Ahmad	(b) 154,K163	75.30; S2
S. Ahmad	(b) 155, 445	64.60; S1.5
A.M. Ahmed	(a) 111,K259	78.20; S12
M.A. Ahmed	(a) 111, 567	72.80; S11.2
	(a) 114, 185	73.60
	(a) 114, 377	77.20 and 77.40; S11.2
S. Ahmed	(a) 115,K149	64.70; 77.80; S11.1
B.L. Ahuja	(b) 152, 145	71.25 and 78.70; S1
M. Aidaraliev	(a) 115,K117	78.45 and 78.55; S7.12; S7.15
T. Akai	(a) 112, 467	61.40; 61.12; S8.16
M.T. Akchulakov	(a) 115, 179	61.80 and 68.20; S1.61; S1.62
S.M.J. Akhtar	(a) 115, 191	78.65; 77.55; S9; S10
A.I. Akimov	(a) 111, K89	74.70; S10.15

A.I. Akimov	(a) 111,K209	74.70; S10.15
	(a) 113,K235	61.80 and 74.70; S10.15
	(a) 115,K205	74.70; S10.15
O. Akinlade	(b) 156, 569	72.15; S2
G.R. Akmanova	(a) 112, K97	66.30; S3
I.A. Aksenov	(a) 111,K109	78.40 and 78.60; S8.16
	(a) 115,K113	72.80 and 78.60; S8.16
V.L. Aksenov	(b) 155, K9	63.20 and 63.75; S10
S.A. Al-Kuhaimi	(a) 111, 523	68.55 and 78.65; S8.16
C. Albers	(a) 113,K135	78.20; S8.15
L. Albert	(b) 155, 571	71.55; 61.10; 78.55; S8.11
J. Albrecht	(a) 113, 321	61.70; 72.20; S5.11
E.L. Albuquerque	(b) 151,K109	63.20 and 76.80
	(b) 154,K179	75.80 and 78.20
L.N. Aleksandrov	(a) 115, K23	68.55 and 79.20; S5.11; S10.1
	(a) 116, K57	65.70 and 68.25; 64.30; S5.11; S10.1
Yu.V. Aleksandrov	(b) 155, 201	71.45; 78.70
I.P. Aleksandrova	(a) 111, 477	64.70; 76.60; S11.1
	(a) 114, 531	64.70; 78.30; S11.1
V.I. Alekseenko	(a) 111,K145	61.70; S7.13
S.V. Alekseev	(b) 153, K11	61.70; 62.20; S4
E.Z. Aliev	(a) 114,K119	77.80; 61.80; S8.16
G. Allais	(a) 111, 211	71.20; S7.16
	(a) 113, 489	71.25; S7.16
K.R. Allakhverdiev	(b) 151, 759	78.20; S8.16
	(b) 152, 317	64.60 and 78.30; S8.16
R. Allub	(b) 154, 347	75.20; 75.30
R.F. Almukhamatov	(a) 111, 431	61.80 and 66.90; S8.15
G. Alseitov	(a) 114,K241	78.60 and 79.75; 61.80; S9.11
P. Altanzog	(a) 112, K23	72.40 and 73.60; S5.11
A.A. Aly	(a) 111,K259	78.20; S12
M.A. Alyanov	(a) 113,K207	65.40; S8.16
A.A. Alybakov	(b) 156, K77	76.30; 61.80; S9.11
L. Amaral	(a) 111, 173	61.80 and 68.20; S1.1; S4
G. Amato	(a) 114, 519	65.70 and 66.70; 72.80; S5.11
J.P. Ambroise	(a) 116, 529	61.55; 61.12; 76.80; S1.1
W.P. Ambrose	(b) 151, K97	63.20 and 78.50; S9.11
S. Amelinckx	(a) 112, 451	61.16 and 64.60; S11.1
L.K. Aminov	(b) 152, 191	71.70; 76.30; 78.50; S9.16
T. Aminov	(a) 114,K211	72.20 and 72.80; S8.16
D.K. An	(a) 116, 561	66.30; 61.70; S5.11
E. Anagnostakis	(a) 114,K215	72.40; 78.50; S7.11
A.N. Anagnostopoulos	(a) 111,K175	73.60; S8.11
M.S. Anand	(a) 114,K159	62.20; S1.1
H.V. Ananda	(a) 114,K259	73.40 and 85; S5.11
C. Anastasiadis	(a) 114, 491	61.70; 77.30
E. Anastassakis	(b) 152, 89	64.60; 63.20; 78.30; S9.12
S.H. Ancharakian	(a) 111,K161	62.20; S5.11
A. Anderson	(b) 155, 427	63.20; S5.2
G. Andersson	(a) 111,K265	61.25 and 77.20; S12
W. Andrae	(a) 111, 583	74.60 and 74.70; S10.15
	(a) 112, 301	61.80 and 79.20; S5.11
G. André	(a) 116, K29	61.12; S1.61
A.V. Andreev	(a) 111, 285	75.50; S1.2; S1.4
	(b) 153, K69	71.70 and 75.60; S1.2
G.A. Andreev	(a) 116, 457	62.20
B.M. Anghelov	(b) 155,K165	71.30 and 78.20; S10
Vo Hong Anh	(b) 154, 661	72.50; 72.55

Yu.M. Annenkov	(b) 156, K23	61.70; S9.11
E. Antic-Fidancev	(a) 115, 523	71.70; 78.50; 78.55; S10.15
G. Antonioli	(b) 152, 39	61.60 and 78.70; S8.16
	(b) 154, 461	61.60; 78.70; S8.16
C. Antonione	(a) 115, 459	61.40 and 64.60; S1.1
J.G. Antonopoulos	(a) 111,K175	73.60; S8.11
I.V. Antonova	(a) 116, K33	61.70; S5.11
M. Anwar	(a) 113, 483	68.55 and 76.30; S10.15
	(a) 114, 215	78.65; 68.55; S10.15
	(a) 114, 225	68.55 and 76.30; S10.15
S. Anzai	(a) 115,K215	75.50; S1.63
H. Aourag	(b) 155, 191	71.10 and 78.70; S5.11; S5.12
	(b) 156, 497	71.20 and 71.60; S5.11; S7.12
P.A. Apanasevich	(b) 152, 347	78.70; 71.25; S5.11
L. Apekis	(a) 111, 325	77.30
S.S. Aplesnin	(b) 153, K79	75.10
A.V. Apostolov	(a) 116, 627	62.20; 61.40; S1.1
F. Appel	(a) 116, 153	61.70; 62.20; S9.11
K. Aradhana	(b) 156, 77	63.20; 62.20; S3
S. Arajs	(a) 113, K71	72.15; S1.2
M.M. Arakelian	(a) 111,K161	62.20; S5.11
V.H. Arakelian	(b) 156, 275	71.36 and 78.70
K.P. Arefiev	(a) 111,K243	78.70; S11
H. Arend	(b) 151, 331	78.55. and 78.60; S9.16
	(b) 151, K1	66.30; 64.70; S12
	(b) 153,K199	78.50; S9.16
A.G. Areshkin	(b) 154, 609	64.70 and 71.25; 71.35; 78.40; S8.15
J. Aride	(a) 115, 523	71.70; 78.50; 78.55; S10.15
L. Arizmendi	(a) 114, 397	78.50; 78.55; S10
G. Arlt	(b) 155, 309	77.40
V.M. Arole	(a) 111, 457	61.70; 72.20; S5.11
D.A. Aronov	(a) 115,K167	73.40; S5.12
R. Arora	(a) 115, 307	77.40; S6
S.K. Arora	(a) 116, 405	77.20 and 77.40; S11.1
A.A. Artsruni	(b) 156, 633	64.75 and 75.30; S1.4
N.Yu. Arutyunov	(a) 115,K201	74.70 and 78.70; S10.15
T. Asaji	(a) 114, 157	64.70 and 77.80; S12
M. Asche	(b) 156, 461	73.40; S7.12; S7.15
C. Ascheron	(a) 112, 391	61.50 and 61.80; S5.11; S5.12; S7.11; S7.12
	(a) 116,K119	61.70 and 78.50; S7.12
	(a) 116,K147	61.80 and 78.30; 63.20; S7.11
A.L. Aseev	(a) 113, 421	68.55; 61.16; S5.11
V.D. Asimova	(b) 152, 347	78.70; 71.25; S5.11
S. Askenazy	(b) 153, 653	63.20 and 71.70; 72.20; S7.13
B.M. Askerov	(b) 151,K157	72.20
D.D. Askerov	(a) 111, K73	72.40 and 73.40; S8.16
S. Asokan	(a) 114, 525	64.70; 61.40; 65.70; 72.80; S8
H. Asonen	(a) 116, 443	68.55; S7.11; S7.12
D. Ast	(a) 116, 399	75.40; 65.40; 71.20; S1.2
H.-J. August	(a) 114, 629	79.20; S1; S5; S5.11
J. Auleytner	(a) 112, 679	61.80; 61.10; 61.16; S5.11; S11.2
	(a) 114, 467	61.80; S5.11
M.N. Avasthi	(a) 113,K177	66.70; S12
M. Averous	(b) 152, 209	71.55 and 72.40; S8
K.T. Avetian	(a) 111,K161	62.20; S5.11

A.H. Awad	(b) 151, 495	66.70
J. Ayala	(a) 112, K43	75.60; S11.2
X. Aymerich	(a) 111, 675	73.40 and 77.50; 73.60; 85; S10
	(a) 113, 83	68.55 and 73.40; 68.48; S10.1
	(a) 114, 167	68.55; S10.1
Yu.M. Azhnyuk	(b) 154,K197	61.80 and 78.30; S7.11
N. Azziz	(b) 151,K105	61.80; S5.11
M.M. Babaev	(b) 151,K157	72.20
S.S. Babaev	(b) 151, 759	78.20; S8.16
	(b) 152, 317	64.60 and 78.30; S8.16
Z.M. Babaev	(b) 156, 419	64.60 and 65.40; S1.1; S1.2
S.L. Babenkova	(a) 116, K57	65.70 and 68.25; 64.30; S5.11; S10.1
M. Babiker	(b) 153, 145	68.40; S1.2; S3; S4
M. Babinský	(a) 111, 345	78.55; 71.55; 78.30; S7.12
G. Babonas	(b) 153, 741	78.20; S8
	(b) 154, 389	78.20; 61.60; S8.15
	(a) 115,K211	72.40; S7.16
N.C. Bacalis	(b) 156, 545	68.45; S10.1
D. Backs	(b) 151, 311	76.30; 75.30; S8.15
K. Baerner	(a) 113, 403	64.70; 75.30; S1.2
H. Baessler	(b) 153, K57	72.40; S12
L.A. Bagdassarian	(b) 156, 275	71.36 and 78.70
V.E. Bagiev	(a) 115, K53	72.80; S10.1
	(a) 116, 305	73.40; 72.20; S10
N.T. Bagraev	(a) 114, 233	61.70 and 71.55; 76.60; S5.11
D. Bahadur	(b) 154, 297	61.60 and 72.20; 75.50; S11
G.R. Bai	(b) 153, 107	63.20 and 78.30; S5.11
E.Ya. Baiguzin	(a) 116, 595	62.20
J.S. Baijal	(a) 112, 131	72.20; S11.2
D. Baither	(a) 113, 331	61.70; S7.12
P.K. Bajpai	(b) 153, 101	64.60; 77.80; S11
R.K. Bakanas	(a) 112, 579	72.20; 61.70
A.E. Bakhyshov	(b) 153, K89	63.20 and 78.30; S8.16
M.Ya. Bakirov	(a) 114, K45	72.80 and 78.40; S5.11; S5.12
H. Bakker	(b) 155, 107	65.50; 64.60
A.D. Balaev	(b) 152, 639	75.50 and 76.80; S11.2
	(b) 155, 639	75.25; 75.30
V.Yu. Balandin	(a) 115, K23	68.55 and 79.20; S5.11; S10.1
M.Kh. Balapanov	(a) 112, K97	66.30; S3
L.L. Balashova	(b) 155, 289	71.45 and 78.70; 79.20
G. Balasubramanian	(a) 111,K229	75.80; 76.50; S1.1
N. Balasubramanyam	(a) 113,K261	78.55; 61.70; 71.55; S5.11
T. Balcerzak	(b) 154, 583	73.60
	(b) 156, 355	75.10 and 75.30
G. Baldacchini	(b) 151, 453	61.70; 78.50; 78.55; S9.11
S. Balle	(b) 152, 543	71.20 and 71.25; 71.50; S1.4
F. Baller	(a) 116,K195	77.20; 77.80; S11.1
A. Bals	(a) 114, 305	76.30; S10.15
R. Baltramiejūnas	(b) 151, 721	71.55 and 78.50; S8.12; S10.1
D. Baltrūnas	(b) 154, 341	76.80; 61.55; S1.63
S. Balúch	(a) 113,K143	76.80 and 78.70; S1.1
A.V. Baluev	(a) 115,K195	74.70; 78.30; 79.20; 79.60
S.N. Balyakin	(a) 115, 199	64.70; 62.20; S9.16
P.C. Banbury	(a) 116,K133	73.40 and 85; 61.80; S7.12

U. Bandyopadhyay	(a) 115, 357	61.25; S9.12
H.D. Banerjee	(a) 111, 515	68.55; S8.12
	(a) 114, 721	78.65; S8.13
J.P. Banerjee	(a) 116, 415	73.40 and 85; S5.11; S7.12
P. Banerji	(b) 156,K145	76.30; 72.20; S7.15
N.S. Banger	(b) 154, K19	63.20; S2
R. Banisch	(a) 112, 721	73.40; 68.55; S5.11
J. Bánkuti	(a) 111,K185	73.60
Xi-Mao Bao	(a) 113, 453	68.48 and 73.40; S5.11
J.J. Bará	(a) 111, 53	61.60; 76.80; S10
J.M. Barandiarán	(a) 111, 279	75.80; S1.1; S1.2
	(a) 113, 187	75.50; S1.1; S1.2
H. Baránková	(a) 116, K65	73.60; S10.1
P.I. Baranskii	(a) 112, 11	72.20
V.I. Barbashov	(a) 111,K145	61.70; S7.13
	(a) 115,K143	61.70; S7.13
T. Barbaszewski	(a) 112, 347	61.80 and 68.20; S1.1
A.F. Barbosa	(b) 151, K45	72.15; S1.4
P.I. Bardetskii	(b) 154, 209	71.35; 78.20; S10.1
G. Bárdoš	(a) 112, 311	61.80; S1.2; S5.11
L. Bárdoš	(a) 116, K65	73.60; S10.1
O.I. Barkalov	(a) 115, K19	64.70; S4
A. Barna	(a) 116, 561	66.30; 61.70; S5.11
P. Barna	(a) 112, 721	73.40; 68.55; S5.11
P.B. Barna	(a) 116, 55	68.48 and 68.55; S1.3; S9.11
J. Barnas	(b) 151, K71	75.70; 76.50
	(b) 155, 581	71.45 and 71.70
N. Barniol	(a) 111, 675	73.40 and 77.50; 73.60; 85; S10
	(a) 113, 83	68.55 and 73.40; 68.48; S10.1
	(a) 114, 167	68.55; S10.1
V. Bârsan	(b) 153,K151	72.15
K. Bartkowski	(a) 111,K165	65.40; S7
H. Bartsch	(a) 111, 59	68.48; 61.16; 61.80; S7.12
	(a) 112, 419	61.70 and 61.80; 61.14; 68.48; S5.11
	(a) 112, 661	61.70 and 61.80; S5.11
	(a) 112, 715	61.80 and 68.55; 68.48; S5.11
A.Yu. Basai	(a) 112,K131	79.20
J. Bashir	(a) 116, K47	63.70; S9.11
L.A. Bashkirov	(b) 152, 657	78.30; S11.2
	(a) 112, K71	66.30; S11.2
F.G. Bass	(a) 112, 579	72.20; 61.70
F. Bassani	(b) 153, 611	71.35 and 71.38
P. Bastie	(a) 116, 529	61.55; 61.12; 76.80; S1.1
A.N. Basu	(b) 154,K107	62.65 and 65.70; 63.20
B.K. Basu	(b) 151,K115	65.40; S4
P.P. Basu	(b) 151,K129	72.20 and 73.40
J. Baszyński	(a) 112, K39	68.55 and 74.70; S10.15
I.L. Bataronov	(b) 151, 121	68.20 and 72.15
D. Batsuur	(a) 112, K23	72.40 and 73.60; S5.11
L. Battezzati	(a) 115, 459	61.40 and 64.60; S1.1
G. Battistig	(a) 115, 75	61.80; 73.40; S5.11
	(a) 116, 561	66.30; 61.70; S5.11
L. Bauch	(a) 113, K55	68.55; 78.65; S8
I.J.R. Baumvol	(a) 112, 105	68.48 and 68.55; S1.1; S4
S. Baunack	(a) 115, 223	68.20; 68.55; 79.20; S5.11
S.S. Bawa	(a) 111, 309	77.80; S12
	(a) 114, 369	77.30 and 77.80; S12

E.M. Bayadilov	(a) 115, 179	61.80 and 68.20; S1.61; S1.62
R.M. Bayazitov	(a) 112, 699	79.20; 64.70; S5.11; S5.12
B.I. Bayrachnyi	(a) 115, 207	68.55; 73.60; S10
O.A. Bayukov	(b) 151, 269	75.30; 76.80; S11.2
	(b) 152, 639	75.50 and 76.80; S11.2
	(b) 152, K51	75.30; S11.2
	(b) 155, 249	75.30; S11.2
A.I. Bazhin	(a) 112, 437	79.20; 75.50; S1.2
I.V. Bazhin	(b) 156, 309	71.20; S1.61
G.V. Bazuev	(a) 115, 267	74.70; 72.20; S10.15
T.P. Beales	(a) 114, 299	72.80; S12
M.F. Beaufort	(a) 114, 467	61.80; S5.11
C. de Beauvais	(b) 152, 463	64.60; 68.45; S5
F. Bechstedt	(b) 153, 595	71.25 and 73.60; S5.11; S5.12
	(b) 154, 565	63.20 and 68.30; 73.40; S7.12
	(b) 156, 151	63.20; 68.30; S7.12; S7.15
	(b) 156, 471	68.20 and 73.20; S5; S5.11; S5.12
R.B. Beck	(a) 116, 313	68.55; S10
S. Becker	(a) 114,K193	68.55 and 74.70; S10.15
U. Becker	(a) 112, 569	71.55; 64.75; 78.50; 78.55; S8.13
A. Beckmann	(b) 154,K191	78.30 and 78.40; 71.35; S5.11; S7.11
S. Bednarski	(b) 154, 543	63.20; 62.20; S1.3
M.M. Beg	(a) 116, K47	63.70; S9.11
M. Behar	(a) 111, 173	61.80 and 68.20; S1.1; S4
D.L. Beke	(b) 155, 385	61.70
P. Bekris	(a) 115, 483	64.70 and 72.80; 61.40; S7.16
A. Belaidi	(b) 155, 191	71.10 and 78.70; S5.11; S5.12
	(b) 156, 497	71.20 and 71.60; S5.11; S7.12
J. Belan	(a) 112,K119	72.70; 85; S5.11
I.T. Belash	(a) 115, K19	64.70; S4
O. Belhachemi	(b) 156, 569	72.15; S2
B.I. Belitskii	(b) 155,K161	78.30
S.A. Belokon	(a) 113,K123	75.50; S8
A.V. Belousov	(b) 155, 535	71.55; 63.20
A.V. Belushkin	(a) 114, 497	64.70; 61.12; S9
V.I. Belyavskii	(a) 116, 255	62.40; S1.2
Š. Beňačka	(a) 114, 567	68.55 and 74.70; S10.15
A. Benaissa	(b) 156, 633	64.75 and 75.30; S1.4
C. Benecke	(b) 153, 391	78.55; 71.70; S8.15
	(b) 154, 405	78.20; 75.30; S8.15
G. Benedetto	(a) 114, 519	65.70 and 66.70; 72.80; S5.11
C. Benoit	(a) 113, 623	78.30; 78.65; S12
K. Bente	(a) 111, K93	74.70 and 76.30; S10.15
L.I. Berezhinskii	(a) 116,K169	68.55; S6
R. Berger	(a) 115, 335	78.60; 61.70; S9.12
H. Bergner	(a) 115, 607	72.40 and 85; S5.11
S.I. Beril	(b) 156, 225	71.38 and 73.20
J.C. Bernède	(a) 111, 181	73.60; S8
M. Bertrandt-Żytkowiak	(a) 115, 581	78.60; 78.65; S8.11
V.N. Berzhanskii	(b) 151, 259	75.30 and 75.50; 76.50; S11.2
T. Besagni	(a) 111, 411	61.60; 64.60; S8.15
V.G. Beshenkov	(a) 114, 191	64.80 and 68.55; 79.20; S1.61
A. van den Beukel	(a) 112, 35	61.40 and 64.70; 62.20; 72.15; S1.1; S1.2
A.P. Bezirganyan	(a) 116, 469	61.10
P.A. Bezirganyan	(a) 116, 469	61.10

P.H. Bezirganyan	(a) 111, 41	61.10
	(a) 116, 483	61.10 and 78.70
	(a) 116, 493	61.10 and 78.70
S.E. Bezirganyan	(a) 116, 469	61.10
G.V. Bezrukov	(a) 116, 603	64.10
R. Bhattacharya	(b) 151, 193	72.20; 71.25; S8
	(a) 114, 277	72.80; S8
A. Bhattacharyya	(b) 151,K129	72.20 and 73.40
K. Bhattacharyya	(b) 153,K135	72.20 and 73.40; S7.12; S7.15
V.G. Bhide	(a) 111, 457	61.70; 72.20; S5.11
R.S. de Biasi	(b) 155, K3	61.70; 76.30; 78.60; S9.15
V.A. Bibik	(b) 151, 733	71.35 and 78.20; S9
A.G. Bibiloni	(a) 116, 503	61.60; 61.70; 76.80; S10.1
G.V. Bich	(a) 111,K171	64.70; S11.2
V.T. Bich	(a) 115,K163	68.55 and 78.65; S12
V. Bichevin	(a) 114,K241	78.60 and 79.75; 61.80; S9.11
	(a) 115,K109	79.75; S9.11
L. Biegała	(b) 152, K17	75.30
G.A. Bielig	(a) 114,K153	62.40; S9.11
Ram Bilas	(b) 154, 297	61.60 and 72.20; 75.50; S11
O.S. Binbrek	(b) 155, 427	63.20; S5.2
R. Bindemann	(b) 153,K119	68.55 and 73.60; 72.40; 78.65; S5.11
S. Bini	(a) 111, 411	61.60; 64.60; S8.15
J. Bischof	(a) 114,K229	75.40; S1.4
L. Bischoff	(a) 113,K185	61.70; 72.80; S5.11
D.P. Bisen	(a) 114,K123	78.60; S9.11
I.K. Bishay	(a) 111, 137	62.20; S4
H.D. Bist	(a) 116, 275	64.70 and 78.30; S12
S. Biswas	(b) 151, 193	72.20; 71.25; S8
	(a) 114, 277	72.80; S8
J. Blaesing	(b) 152, 311	75.30 and 75.50; 75.40
C. Blanchard	(a) 114, 467	61.80; S5.11
A.G. Blashuk	(a) 115, K95	75.50; S1.1; S1.2
G. Blasse	(b) 156, K93	78.55; 78.50; S11.1
	(b) 156, 537	71.70; 78.55; S8.11
P. Blaudeck	(b) 151, 545	61.40 and 71.20; S5
S.R. Blight	(a) 112, K29	71.55; 85; S7.12
R. Blinc	(b) 151, K1	66.30; 64.70; S12
A.A. Blinkin	(b) 151,K161	74.70; S10.15
Yu.F. Blinov	(a) 116, 555	61.80
M. Blomberg	(a) 112, K81	63.20; S1.62
T. Blum	(b) 153, K57	72.40; S12
H.J. Blythe	(a) 113, 541	75.60; S10.1
A. Bobák	(b) 155,K143	75.10
G. Bochnacka	(b) 153, 207	71.38; S12
	(b) 155, K29	71.38; S12.1
H. Bock	(a) 114,K255	78.30; S1
A.E. Bocquet	(b) 152, 519	71.45 and 79.60; 74.10; S11
J. Bodzenta	(b) 153, 573	62.65 and 78.70; 62.60
J.L. de Boer	(a) 114, 91	61.60; S1.63
H. Boerner	(a) 116, 537	61.50; 74.70; S10.15
E. Boesman	(b) 155, K55	76.30; 61.70; S9.11
A. Boettcher	(a) 113, 551	75.60; S1.1
R.-M. Boettcher	(a) 116, 537	61.50; 74.70; S10.15
H. Boettger	(b) 151, 521	63.50
	(b) 156, 431	63.50
O.V. Boev	(a) 111,K243	78.70; S11
G.G. Bogachev	(b) 152, K29	78.30; S11.1

B.F. Bogacz	(a) 111, 53	61.60; 76.80; S10
D.B. Bogomolov	(a) 112, 357	61.80; 68.20; S1.1
A. Bohdziewicz	(a) 114, K49	74.70; S1.4
B.B. Boiko	(a) 111, K89	74.70; S10.15
	(a) 111,K209	74.70; S10.15
	(a) 113,K235	61.80 and 74.70; S10.15
	(a) 115,K205	74.70; S10.15
G.G. Bokuchava	(b) 155,K153	76.60
S. Bold	(a) 112, K23	72.40 and 73.60; S5.11
J. Bollmann	(a) 112, 395	61.80 and 68.55; 73.40; S10.15
D. Bolmont	(a) 116, 777	74.70; 75.30; S10.15
V.V. Bolotov	(a) 113, 315	61.80 and 73.20; 68.48; S5.11
A.V. Boltushkin	(a) 111,K225	75.70; S1.2
A. Bonnet	(b) 155, 609	72.20; 61.60; 71.20; S10
N.I. Borgardt	(a) 113, K23	61.16 and 61.70
	(a) 113, 251	61.16 and 61.70
J.A.T. Borges da Costa	(a) 112, 105	68.48 and 68.55; S1.1; S4
T.E. Borisenko	(a) 112,K111	72.40; S8
V.E. Borisenko	(a) 111, 499	61.70 and 68.55; S1.61
	(b) 155,K127	71.25 and 74.70; S10.15
	(a) 116, 331	68.55; 61.70; S1.61; S5.11
L.D. Borisova	(a) 114,K201	72.20; S8
Th. Born	(b) 151, K41	72.15; S4
V.A. Borodin	(b) 154, 497	61.70
S.I. Borovitskii	(a) 115,K205	74.70; S10.15
L. Bosi	(b) 152, 67	61.70; 78.50; S9.11
	(b) 156, K5	61.70; S9.11
G. Bosman	(b) 154, 713	72.10 and 72.70; S5.11
Ch. Bosshard	(b) 151, K1	66.30; 64.70; S12
K.Z. Botros	(a) 114, 83	61.16 and 61.70; S1.2
H. Boudriot	(b) 154,K175	78.30; 72.20; S7.1
P. Boukal	(a) 112, K81	63.20; S1.62
C. Boulesteix	(a) 113, 23	61.60 and 74.70; 61.16; S10
	(a) 114, 73	61.60; 61.16; S10
	(a) 114, 441	61.60; 61.16; 61.70; S10
B. Boulogne	(a) 112, 493	61.70; 77.60; S11.1
K.K. Bourdelle	(a) 112, 707	61.80; 64.70; S1.61
A. Bourdilon	(a) 111,K179	68.55; 78.65; S4; S5; S7.11
P.D. Bourkas	(a) 112, 123	72.20 and 77.50; S12
M. Bourrous	(a) 112, 181	75.60; 75.50; S1.2
	(a) 113, 169	75.60; 75.50; S1.2
	(a) 113, 383	61.40 and 64.60; 75.60; S1.2
R. Boyn	(a) 111,K129	78.60; S8.11
	(a) 112, 569	71.55; 64.75; 78.50; 78.55; S8.13
U. Braatz	(a) 113, K79	72.40; S8.11; S8.12
	(a) 116,K219	72.40 and 85; S8.11; S8.12
V.A.M. Brabers	(a) 112, 451	61.16 and 64.60; S11.1
D. Brach	(b) 156, 195	64.60; 78.20; 78.55; S11.1
J. Braciník	(a) 113, K37	62.65 and 66.70; S7.12
I. Bradl	(b) 151, 521	63.50
N.B. Brandt	(a) 114, 265	61.60 and 72.80; 75.50; S1.4
W. Brauer	(b) 156, K85	79.20; S4
T. Breczewski	(a) 114, 505	64.70 and 65.70; S11.1
A. Breitschwerdt	(b) 155, 77	63.20; 78.30; S7.13; S8.11; S9
T. Bremer	(a) 114, 765	78.20; 61.70; S10.15
B. Březina	(b) 154, 777	78.30; 77.80; S9.16

B. Březina	(a) 115, 569	77.80; S9.16
P. Brogowski	(a) 114, K37	61.80 and 72.80; 85; S8.15
A. Brozi	(a) 114,K167	68.55; S7.12
V.N. Brudnyi	(a) 114, 481	61.70; 61.80; 78.70; S7.12
	(a) 114,K139	61.80; S7.11
O. Bruemmer	(a) 111, 361	78.70; 71.25; S10
W. Brunner	(b) 151, K53	76.30 and 77.80; S11.1
Y. Bruynseraede	(a) 114, 567	68.55 and 74.70; S10.15
I. Bryłowska	(a) 112, 301	61.80 and 79.20; S5.11
	(a) 112, 353	61.80; 68.20; S1.1
	(a) 112, 743	61.80 and 68.48; S1.61
M. Bucki	(a) 112, 335	61.80 and 68.20; S1
I.S. Buda	(a) 112, 11	72.20
J. Budinavičius	(a) 114, K25	61.80 and 68.48; S5.11
H. Budinov	(a) 114,K131	61.70; 79.20; S5.11
O. Budke	(a) 115, 143	64.70; 66.70; S8
T.O. Budko	(a) 111, 451	66.30; S7.11
A.G. Budrevich	(a) 114,K163	68.55; 61.80; S5.11
A.A. Bugai	(b) 151, 159	71.55; 76.30; S5.11; S5.12
F. Bugge	(a) 111, 541	73.40 and 85; 78.55; S7.12
L.L. Buishvili	(b) 155,K153	76.60
D.W. Bullett	(b) 152, K47	71.20; 79.60; S8.16
G.S. Burchanov	(a) 112, 335	61.80 and 68.20; S1
	(a) 112,K121	75.50; S8.16
A.F. Burenkov	(a) 115, 427	61.80; 61.70; 78.60; S5; S7.1
J.P. Burger	(b) 154, 321	72.15; 75.50; S1.4
R. Burkova	(a) 114,K131	61.70; 79.20; S5.11
V.M. Burlakov	(b) 151, 337	78.30; 77.80; S8.16
	(b) 153, 727	78.30; 77.80; S8.16
B. Burmaa	(a) 112, K23	72.40 and 73.60; S5.11
E. Burzo	(a) 113,K253	61.55 and 75.50; S1.1
L.S. Bushnev	(a) 116, K69	61.70 and 68.20; S6; S7
G. Busiello	(b) 153, 561	63.75 and 64.60; 75.50
W. Busse	(b) 153, 391	78.55; 71.70; S8.15
T.I. Butaeva	(a) 113, 193	63.20 and 78.55; 71.55; S10.15
A.V. Butashin	(a) 112, 197	78.20 and 78.45; 78.55; S10
	(a) 112, K49	78.20; S10
	(a) 113, 585	78.30; 71.70; S11.2
P.N. Butcher	(b) 154, 229	71.38; 73.40
M.Z. Butt	(a) 113,K189	62.20; S1.2
N.M. Butt	(a) 116, K47	63.70; S9.11
D.J. Buttle	(a) 112, 55	61.80; 75.50; S1.1
P. Butvin	(a) 115, 533	75.60; S1.1; S1.2
B. Butvinová	(a) 115, 533	75.60; S1.1; S1.2
R.A. Buyanov	(a) 113, 359	64.75; 76.80; S10
S.E. Bykhover	(a) 114, K99	75.30; S1.4
A.A. Bzheumikhov	(a) 116, 197	61.10; 78.70; S7.12
B. Čabrić	(b) 153,K125	71.35
R. Cach	(a) 113,K139	77.20; 77.80; S12
	(a) 116, 827	77.80; 77.20; S12
M. Cadène	(a) 112, 635	78.20; S12
	(a) 115, 491	68.55 and 73.60; S8
J.L. Cadorin	(b) 153, K73	75.10
M. Calamiotou	(b) 152, K9	74.70 and 76.30; S10.15
J. Calas	(b) 152, 209	71.55 and 72.40; S8

F. Callens	(b) 155, K55	76.30; 61.70; S9.11
N. le Calvé	(a) 114, 531	64.70; 78.30; S11.1
S.S. Camargo,Jr.	(b) 153, 757	71.35 and 78.60; S9.11
	(b) 153, 765	71.35 and 78.60; S9.15
V. Cambel	(a) 115,K121	78.65 and 85; S7.15
L.S. Campana	(b) 151,K177	75.10
Liezhao Cao	(b) 154, K51	74.70; S10.15
P. Čapková	(a) 112, K81	63.20; S1.62
D. Cardenas-Garcia	(b) 154,K171	71.55 and 78.50; S9.11
M. Cardona	(b) 155, 77	63.20; 78.30; S7.13; S8.11; S9
	(b) 156, 371	71.25 and 78.20; 77.20; S8.15
C.B. Carter	(a) 112, 41	61.70; 61.16; 62.20; S7.12
	(a) 112, 473	61.70; 61.16; S7.12
	(a) 114, 105	61.70; S7.12
M. Casalboni	(b) 151, 347	78.20; S11
A. Casalot	(a) 114, 73	61.60; 61.16; S10
	(a) 114, 441	61.60; 61.16; 61.70; S10
N.J. Castellani	(b) 151, 565	71.10; S5.11; S7.1; S8.1
J. Castro	(a) 113, 541	75.60; S10.1
M. Celasco	(a) 116, 813	75.60
F. Cembali	(a) 115, 437	61.80; 61.10; S5.11
B. Çendlewska	(a) 114, K49	74.70; S1.4
F. Černý	(a) 112, 335	61.80 and 68.20; S1
J. Červenák	(b) 153,K147	72.20; S7.13
L. de Cesare	(b) 153, 561	63.75 and 64.60; 75.50
T. Ceva	(b) 152, 463	64.60; 68.45; S5
N.F. Chaban	(a) 114, 265	61.60 and 72.80; 75.50; S1.4
Z.B. Chachkhiani	(a) 114, K87	76.60; S1.2
G.S. Chaddha	(b) 151, 283	75.10 and 75.40
Sh. Chadraabal	(a) 112, K23	72.40 and 73.60; S5.11
I.A. Chaikovskii	(b) 152, 593	72.20
B.R. Chakraborty	(a) 115,K125	68.55 and 85; 73.40; S5.11
B.C. Chakravarty	(a) 115,K125	68.55 and 85; 73.40; S5.11
B. Chalupa	(a) 111,K133	61.12
	(a) 112, 29	61.12 and 62.65; S7.13
Prem Chand	(b) 151, 615	71.10 and 76.30; S11
B.P. Chandra	(a) 114,K123	78.60; S9.11
S. Chandra	(a) 111, 309	77.80; S12
	(a) 114, 369	77.30 and 77.80; S12
C.P. Chang	(b) 154, 135	63.20; 71.36; 78.30; S7.15
S. K. Chang	(b) 153,K183	78.55; S8
N.B. Chanh	(a) 115, 107	64.70; 61.14; 77.20; S11
I.M. Chapnik	(b) 155, K39	74.70
M.S.R. Chari	(a) 116, 377	72.15; S1.1; S1.2
T.B. Charikova	(a) 115, 267	74.70; 72.20; S10.15
Y. Charreire	(b) 155, 571	71.55; 61.10; 78.55; S8.11
N. Chatterjee	(b) 154, K55	72.15 and 74.70; S10.15
D. Chattopadhyay	(b) 151,K129	72.20 and 73.40
	(b) 154, 281	72.20 and 73.40; S7.12
S.D. Chaturvedi	(b) 156, 171	62.20 and 64.60; S8
N.H. Chau	(a) 114, 361	75.30 and 75.80; S1.1; S1.2; S1.4
A.K. Chaudhuri	(a) 114, 721	78.65; S8.13
V.N. Chebotar	(b) 155, 535	71.55; 63.20
L.A. Chebotkevich	(a) 114, K75	75.70; S1.2
E.S. Chebukov	(a) 115, 179	61.80 and 68.20; S1.61; S1.62
M.A. Chechelnitskii	(b) 156,K157	64.70 and 78.50
N.G. Chechenin	(a) 112, 707	61.80; 64.70; S1.61

V.I. Chechernikov	(a) 114, K87	76.60; S1.2
G.M. Chechin	(b) 152, 431	61.50 and 64.60
J. Chela-Flores	(b) 152, 617	74.20; 74.70; S10.15
A.R. Chelyadinskii	(a) 112, 381	66.30; 61.80; 62.20; S5.11
A. Chen	(a) 114, 273	74.70 and 78.70; S10.15
Chuan-Yu Chen	(b) 154, K137	71.38
Gang Chen	(b) 156, 145	63.20 and 78.30; 71.36
Guanghua Chen	(b) 152, 73	61.40 and 68.30; 78.30; S5.11
	(b) 154, K39	61.70 and 68.60; S5.12
	(a) 114, K171	68.60 and 76.30; S6
Hong Chen	(b) 152, 499	71.25; 73.40
J.M. Chen	(b) 153, 107	63.20 and 78.30; S5.11
Jian Chen	(a) 113, 453	68.48 and 73.40; S5.11
	(b) 154, K51	74.70; S10.15
Jun Chen	(a) 116, 231	61.70; 77.80; S10
Ling Chen	(a) 111, K123	78.70; S11.1
Q.M. Chen	(a) 116, K51	64.70; S1.1
S.C. Chen	(b) 152, K33	62.50 and 78.40; S8.15
Wei Chen	(a) 114, K171	68.60 and 76.30; S6
Wuming Chen	(a) 114, K189	74.70; S10.15
X. Chen	(a) 113, K161	61.70 and 78.70; S1.2
X.H. Chen	(a) 115, K71	74.70 and 79.60; S10.15
Xianhui Chen	(b) 154, K51	74.70; S10.15
Xiaohua Chen	(a) 113, K85	62.20 and 74.70; S10.1
	(a) 116, K199	62.20 and 77.80; S12
Xiaolong Chen	(a) 114, 651	61.60 and 74.70; S10.15
Y.L. Chen	(a) 114, 273	74.70 and 78.70; S10.15
	(a) 115, K91	76.80; S1.1
Yang-Fang Chen	(b) 153, 695	71.20 and 72.20; S5.11
Yu Chen	(a) 112, 541	61.40 and 68.55; S1.1; S5.12
Z.Y. Chen	(a) 115, K71	74.70 and 79.60; S10.15
Zuyao Chen	(b) 154, K51	74.70; S10.15
B.L. Cheng	(a) 115, 119	62.40; S4
Jiming Cheng	(a) 114, K127	78.45; 78.30; 78.55; S11.1
K.J. Cheng	(a) 113, K15	61.80; S5.11
Yifan Cheng	(b) 152, 33	61.14; 61.70; S1.2; S4
B.I. Cherkasov	(b) 156, 615	76.60; 66.30; 72.20; S9.16
P.V. Chernov	(b) 156, 663	78.55; 71.55; S10.1
V.V. Cherny	(b) 152, 675	78.20; S8.11
B.P. Chernyago	(b) 155, 365	61.70; 76.30; 78.40; S9.11
A.S. Chernyi	(a) 112, 155	75.50; S10.15
O.I. Chertikhina	(b) 156, 205	68.20; S10.1
M.M. Chervinskii	(a) 111, K233	75.70 and 78.20; S11.2
M. Chiba	(b) 153, K159	75.30 and 75.50; S9.16
L.E. Chibotaru	(b) 155, 295	71.45 and 78.55; S8.12
E. Chimczak	(a) 115, 581	78.60; 78.65; S8.11
O.D. Chistyakov	(a) 112, K121	75.50; S8.16
	(a) 114, K219	75.50; S1.1
	(a) 115, K229	75.50; S1.2
M.J. Chockalingam	(a) 113, K217	73.30; 85; S8.11
Z.P. Chorniy	(b) 154, K9	61.70; 78.50; S9.16
V.V. Chornyi	(a) 111, K141	61.70 and 78.50; S5.11
B.K. Choudhary	(a) 115, 301	77.20 and 77.40; S11.1
R.N.P. Choudhary	(b) 152, K69	78.30; 63.20; S11.1
	(a) 113, 597	77.20; 77.80; S11.1
	(a) 115, 301	77.20 and 77.40; S11.1
A.K. Choudhury	(a) 111, 515	68.55; S8.12
T.A. Christakudi	(a) 114, K201	72.20; S8
G.Ch. Christakudis	(a) 111, 469	61.70 and 66.70; S8.16

G.Ch. Christakudis	(a) 114,K201	72.20; S8
	(b) 155, K89	66.70; S8.16
C. Christodoulides	(a) 111, 325	77.30
V. Christoph	(b) 151, 687	75.60
	(b) 156,K149	75.60 and 75.90
	(a) 116,K179	75.70; 75.60; S4
S. Chromik	(a) 114, 567	68.55 and 74.70; S10.15
Yuan-Chao Chu	(a) 114, K1	61.30; S12
Y.C. Chuang	(a) 112, 175	75.50; S1.1; S1.2
	(a) 115,K219	75.50; 75.30; S1.1
G.N. Chuev	(b) 152, 79	63.20; 71.38; S9.11
Yu.G. Chukalkin	(a) 112, 161	61.40 and 75.50; S11.2
M.V. Chukichev	(a) 115, 555	78.60; 71.55; S7
	(a) 115, 561	78.60; S7
O.V. Chumanova	(b) 155, 289	71.45 and 78.70; 79.20
C. H. Chung	(b) 153,K183	78.55; S8
I.E. Chupis	(b) 156,K135	75.80
Yu.I. Churakov	(b) 154, 507	61.80
A.N. Chuvyrov	(a) 113,K111	75.50; S12.1
M. Chybicki	(a) 115,K185	73.25; S10.15
S.I. Chykhrii	(a) 114, 265	61.60 and 72.80; 75.50; S1.4
A.V. Chzhan	(a) 115,K101	75.60; S10
G. Cocco	(a) 115, 459	61.40 and 64.60; S1.1
J.D. Collins	(b) 151, 469	61.70; 78.50; S7.12
Ph. Colomban	(a) 114, 531	64.70; 78.30; S11.1
A. Conan	(b) 155, 609	72.20; 61.60; 71.20; S10
A.A. O'Connel-Bronin	(b) 154, 763	78.40; 71.35; S9.11
R.F. O'Connell	(b) 153, 343	71.25 and 75.10
L. Contreras	(a) 114, 397	78.50; 78.55; S10
B.C. de Cooman	(a) 112, 41	61.70; 61.16; 62.20; S7.12
	(a) 112, 473	61.70; 61.16; S7.12
	(a) 114, 105	61.70; S7.12
M. Čopić	(b) 151, K1	66.30; 64.70; S12
P. Cordier	(a) 112, 493	61.70; 77.60; S11.1
B.V. Costa	(b) 155, 663	75.10; 75.50
J. Costa-Quintana	(b) 152, 543	71.20 and 71.25; 71.50; S1.4
M.G. Cottam	(b) 153, 145	68.40; S1.2; S3; S4
J.J. Couderc	(a) 112, 89	62.20; 61.70; S8.11
C. Courseille	(a) 115, 107	64.70; 61.14; 77.20; S11
J.P. Coutures	(a) 113, 35	61.40; 64.75; S10.15
M. Couzi	(a) 115, 107	64.70; 61.14; 77.20; S11
B.I. Craig	(b) 154,K127	71.10; 62.20; S1.61
I. Creanga	(a) 113,K253	61.55 and 75.50; S1.1
C. Cremers	(a) 114, 567	68.55 and 74.70; S10.15
B. Croset	(b) 152, 463	64.60; 68.45; S5
H.R. da Cruz	(b) 151, 675	75.30 and 75.50
R.M. de la Cruz	(a) 111, 463	61.70; 78.70; S5.11
A.E. Curzon	(a) 114, 199	68.60; 68.55; S1.2
R. Cywiński	(b) 151, 47	61.70; 76.30; 78.50; S9.11
	(a) 111,K249	78.55; 71.55; S7.1
	(a) 112, 215	78.20; 71.55; S9.11
Z. Czapla	(a) 113,K139	77.20; 77.80; S12
	(a) 116, 827	77.80; 77.20; S12
J. Czaplicki	(a) 113, 163	76.60; S9.16
D. Czekaj	(a) 114, K21	79.20; S1.2
J. Częrbniak	(a) 112, 773	61.80; S5.11
	(b) 154, K13	61.80
R. Černý	(a) 113,K181	66.30
W. Czolbe	(a) 111, 361	78.70; 71.25; S10

P. Czuba	(a) 112, 431	79.20; S9.11
P. Czurratis	(a) 111, 417	61.55; 62.20; 64.70; S4
A.Caramico D'Auria	(b) 151,K177	75.10
J. Dąbek	(a) 112, 829	79.40
M. Dąbrowski	(a) 112, 347	61.80 and 68.20; S1.1
S. Dacko	(a) 116, 827	77.80; 77.20; S12
R. Dagys	(b) 153, 741	78.20; S8
Yu.N. Daluda	(a) 116, K37	61.70; S5.11
T. Damm	(a) 115, 607	72.40 and 85; S5.11
Z. Damm	(a) 116, 367	74.70; S10.15
H. Danan	(a) 111, 583	74.60 and 74.70; S10.15
J. Danilovich	(a) 112, 733	68.55; 61.80; S1.61
J.N. Daou	(b) 154, 321	72.15; 75.50; S1.4
D. Daoukaki-Diamanti	(a) 111, 325	77.30
S.D. Darchuk	(a) 113, 107	73.40; S8.15
A. Dargys	(b) 155, 615	72.10; 71.25
C.N.W. Darlington	(a) 113, 63	64.70; 61.10; 77.80
N. Darvishov	(b) 156, 697	78.40; 71.25; S11.1
N.G. Darvishov	(b) 151, K87	78.40; 71.25; S11.1
B.K. Das	(a) 115,K125	68.55 and 85; 73.40; S5.11
V.N. Datsyuk	(b) 151, 581	71.25 and 78.70; S9.11
S.K. David	(a) 111, 457	61.70; 72.20; S5.11
R.L. Davidovich	(b) 156, 615	76.60; 66.30; 72.20; S9.16
J.J. Davies	(b) 151, 175	71.35 and 71.55; 76.90; 78.55; S8.11; S8.15
N.A. Davydova	(b) 151, 733	71.35 and 78.20; S9
E. Dębowska	(b) 156, 303	71.25; 78.70; S1.2; S1.3
J. Degro	(a) 111, 603	75.50; S1.1
A.I. Demchenko	(a) 116, 331	68.55; 61.70; S1.61; S5.11
M.I. Demchuk	(a) 113,K257	78.45; S9.16; S11.2
A.A. Demidovich	(a) 112, 197	78.20 and 78.45; 78.55; S10
S.E. Demyanenkov	(a) 111, K89	74.70; S10.15
S.E. Demyanov	(a) 111,K209	74.70; S10.15
	(a) 115,K205	74.70; S10.15
Hui-Fang Deng	(b) 156, K11	66.30
M.K. Deng	(a) 113, K15	61.80; S5.11
F. Dénoyer	(a) 114, 531	64.70; 78.30; S11.1
V.V. Derevyanko	(b) 151,K161	74.70; S10.15
Yu.O. Derid	(a) 113,K265	78.55; S8.16
	(a) 114, 259	71.55; 78.55; S8.16
V.N. Derkachenko	(a) 113,K249	75.60; S10
C. Dervos	(a) 112, 123	72.20 and 77.50; S12
B.T. Deshmukh	(a) 111,K119	78.60; S12
J. Desimoni	(a) 116, 503	61.60; 61.70; 76.80; S10.1
J.C. Desoyer	(a) 112, 511	61.70; S5.11
	(a) 114, 467	61.80; S5.11
V. Detcheva	(b) 152, 163	71.25
I. Detemple	(a) 116, 681	62.20 and 68.25; S1.2
K. Dettmer	(b) 154,K191	78.30 and 78.40; 71.35; S5.11; S7.11
H. Dexpert	(b) 155, 571	71.55; 61.10; 78.55; S8.11
J. Dexpert-Ghys	(b) 155, 571	71.55; 61.10; 78.55; S8.11
M. Dey(Bose)	(b) 152, 625	72.50; 71.25; S7.13
M. Dhesi	(a) 115, 157	64.70; S8.15
E.M. Dianov	(b) 156, 663	78.55; 71.55; S10.1
E. Dieguez	(a) 114, 397	78.50; 78.55; S10

D. Dietrich	(a) 112, K85	62.20; 61.80; 75.60; S12
H.-J. Dietze	(a) 114,K193	68.55 and 74.70; S10.15
U. Dillner	(a) 115, 505	73.60; S4
	(a) 116, 337	73.60; 72.10; S4
	(a) 116, 715	73.60; 72.10; S4
V.I. Dimitrov	(a) 116, 627	62.20; 61.40; S1.1
T. Dimov	(a) 115,K243	78.20; S8.15
Dihua Ding	(a) 116, 203	61.40; 75.60; 76.80; S1.1
N.N. Dinh	(a) 115,K163	68.55 and 78.65; S12
H. Dintner	(a) 111, 551	72.15 and 73.60; S1.2; S10.1
M. de Dios Leyva	(b) 151, 127	71.25; S7.12; S7.15
M. Diviš	(b) 153, K69	71.70 and 75.60; S1.2
	(a) 114,K229	75.40; S1.4
R.N. Dixit	(b) 151,K115	65.40; S4
C. Djega-Mariadassou	(a) 112, 601	75.30 and 75.50; S11.2
A.I. Dmitriev	(b) 151, 633	72.10; S8.15
E.I. Dmitrieva	(b) 155, 359	61.70 and 71.55; S9.11
J.F. Dobson	(b) 152, 519	71.45 and 79.60; 74.10; S11
T. Dohnalik	(a) 112, 431	79.20; S9.11
T. Doležal	(a) 113, K95	75.60; S11.2
V.A. Dolina	(b) 155, 639	75.25; 75.30
J.J. Donaghy	(a) 116,K213	78.70; S12
V. Donchev	(a) 113,K231	72.80; S7.12
H. Donker	(b) 156, K93	78.55; 78.50; S11.1
	(b) 156, 537	71.70; 78.55; S8.11
P.E. Donovan	(a) 115, 61	61.16; S1.2; S4
J.L. Dormann	(a) 112, 601	75.30 and 75.50; S11.2
A. Doroba	(b) 152, 275	75.10 and 75.30
V.D. Doroshev	(b) 155, 669	76.60; 75.25; S11.2
Y. Dossmann	(a) 116, 399	75.40; 65.40; 71.20; S1.2
J.C. Doukhan	(a) 112, 493	61.70; 77.60; S11.1
N. Doukhan	(a) 112, 493	61.70; 77.60; S11.1
N.A. Doumanis	(b) 153, K1	61.80; S4
L. Dózsa	(a) 115, 75	61.80; 73.40; S5.11
G. Draeger	(a) 111, 361	78.70; 71.25; S10
H. Draganovská	(a) 115, 533	75.60; S1.1; S1.2
T. Dragon	(a) 116, 803	75.60; S1.1
V.M. Drako	(a) 111, K13	61.80 and 68.48; S1.2
S.I. Drapak	(a) 115, K35	73.40 and 78.60; S8; S10.1
V.A. Dravin	(a) 113, 519	72.20; 72.60; S5.11
D.G. Driyaev	(a) 114, 99	61.70; S9.11
T. Drusedau	(b) 153,K119	68.55 and 73.60; 72.40; 78.65; S5.11
M. Drwiega	(a) 112, 347	61.80 and 68.20; S1.1
Jiaju Du	(a) 115, K75	74.70; 62.40; S10.15
Mao-Lu Du	(b) 153, 249	71.70; 76.30; S1.2
R. Dubourg	(a) 115, 335	78.60; 61.70; S9.12
I.Ya. Dubovskaya	(b) 155, 685	78.70
	(b) 156, 687	78.70; S5.11
N.H. Duc	(a) 114, 361	75.30 and 75.80; S1.1; S1.2; S1.4
V.P. Dudkevich	(a) 114, 559	64.70 and 68.55; 77.80; S10.15
V.K. Dugaev	(b) 153, 115	64.60; 61.60; S8.15
P. Duhaj	(a) 111, 597	75.80; S1.1; S1.2
	(b) 153,K147	72.20; S7.13
B. Dumercq	(a) 115, 335	78.60; 61.70; S9.12
L.J. Dunne	(b) 154, 727	72.10; 72.15
H. Duński	(a) 113, K49	68.45; S10.1

H. Dunski	(a) 116,K165	61.80 and 68.45; S10.1
D.J. Dunstan	(b) 151, 759	78.20; S8.16
G. Duplâtre	(a) 115, 389	61.40 and 61.70; 78.70; S12
J. Dural	(a) 114, 467	61.80; S5.11
J. Durček	(a) 116, K77	68.25 and 71.55; 72.50; S5.11; S10
K. Durczewski	(b) 153, 331	75.10 and 75.30
Yu.I. Durnev	(b) 153, 517	64.60 and 78.30; S8.16
A.G. Dutov	(a) 112, K71	66.30; S11.2
V. Dvořák	(b) 151, K23	64.70; 74.70; S10.15
A.V. Dvurechenskii	(a) 111, 257	61.70; 76.30; S5.11
	(a) 113, 519	72.20; 72.60; S5.11
	(a) 115, K23	68.55 and 79.20; S5.11; S10.1
H. Dwelk	(b) 152, K5	72.20 and 74.70; S10.15
	(b) 153, K63	74.70; S10.15
	(a) 113, 159	72.20 and 74.70; 76.60; S10.15
	(a) 116, 793	72.20 and 74.20; S10.15
R. Dwivedi	(a) 116, 837	73.40 and 85
D. van Dyck	(a) 112, 451	61.16 and 64.60; S11.1
A. Dygo	(a) 112, 343	61.80; 61.70; S1.1
M.I. Dzhafarov	(b) 151,K157	72.20
T.D. Dzhafarov	(a) 111, K73	72.40 and 73.40; S8.16
	(b) 155, 11	61.80 and 66.30; S5.11
F.S. Dzheparov	(b) 151, 663	76.30
S. Dzhumanov	(b) 152, 395	61.70; 71.35; 71.38; S9.11
W. Dziurda	(a) 112, 811	61.80 and 78.55; S10
I.P. Dzyub	(b) 151, 61	63.10 and 63.50
R.S. Eachus	(b) 152, 583	71.50 and 76.30; 71.38; 76.40; S9
	(b) 154, 483	61.70; 76.30; S9
L. Eaves	(b) 156, 461	73.40; S7.12; S7.15
Y. Ebata	(b) 152, 533	61.50 and 71.45
J. Ebert	(a) 115, 191	78.65; 77.55; S9; S10
M. Eckler	(b) 153,K119	68.55 and 73.60; 72.40; 78.65; S5.11
G. Eckold	(b) 154, 153	63.20; S12
	(b) 155, 89	63.20; S10
Sh. Efendiev	(b) 156, 697	78.40; 71.25; S11.1
Sh.M. Efendiev	(b) 151, K87	78.40; 71.25; S11.1
	(a) 116, 305	73.40; 72.20; S10
N.N. Efimova	(b) 154, 353	75.30 and 75.50
K. Eftaxias	(b) 156, 393	66.30; S1.2
S.A. Egorov	(b) 151, 275	75.10; 71.25; 71.50; S11
V.E. Egorushkin	(b) 154, K43	71.20 and 71.25; S1.2
A.V. Egranov	(b) 155, 365	61.70; 76.30; 78.40; S9.11
J. Ehlert	(a) 113, K99	75.60
	(a) 116, 389	75.60
P. Ehrlich	(a) 116, 709	73.60; 71.55; S12
B. Eibisch	(a) 112, 667	61.80 and 68.55; 73.60; 78.65; S5
R. Eibler	(a) 112, 175	75.50; S1.1; S1.2
M. El Hiti	(a) 116, 55	68.48 and 68.55; S1.3; S9.11
M.A. El Hiti	(a) 114, 185	73.60
	(a) 115, 185	73.60; S1.3
S.M. El-Mashri	(a) 113,K177	66.70; S12

M.K. El-Nimr	(a) 114, 377	77.20 and 77.40; S11.2
D. Elefant	(b) 151,K155	72.80; S1.61
K. Elk	(b) 151, 687	75.60
A.M. Elkorashy	(b) 152, 249	72.40; S8
V.M. Emeksuzyan	(a) 113, 315	61.80 and 73.20;.68.48; S5.11
Yu.S. Emelyanenko	(a) 111, 451	66.30; S7.11
L.M. Emiryan	(a) 116,K185	76.50; S8.16
V.V. Emtsev	(a) 116, K37	61.70; S5.11
S. Emura	(b) 153, 683	71.55; 78.50; 78.55; S9.11
R. Enderlein	(b) 153, 595	71.25 and 73.60; S5.11; S5.12
	(b) 156, 259	71.20
P. Enders	(b) 154, K5	66.30; S7.1
	(b) 156,K123	71.45; 71.38; S12.1
V.I. Engelko	(a) 115, 179	61.80 and 68.20; S1.61; S1.62
C. Engler	(b) 151, 555	68.20; 71.10; 73.20; S5.11
S. Enzo	(a) 115, 459	61.40 and 64.60; S1.1
J.-W. Erben	(a) 116, 735	61.80 and 68.55; 77.55; S9.12
Th. Erge	(a) 114, 705	76.60; 77.80; S12
R. Erikashvili	(a) 114, 757	78.20; S10
S. Erkoç	(b) 152, 447	63.10; 61.55; S1.2; S1.3; S2; S5
	(b) 152, K37	61.55; S4
	(b) 155, 461	63.10; 61.55; S1.2; S2; S5
	(b) 156,K105	68.45; S7.12
	(b) 156,K113	68.40
V.N. Ermakov	(b) 156, 333	71.45 and 74.20
M. Ermrich	(a) 116, K19	61.10
V.N. Erofeev	(a) 115, 199	64.70; 62.20; S9.16
R. Erre	(a) 113, 35	61.40; 64.75; S10.15
A.V. Ershov	(a) 116, 697	68.55 and 73.60; S5.11
C. Eschrich	(a) 113, 269	61.55 and 64.75; S1.2
U. Esposito	(b) 151,K177	75.10
P. Esteban-Puges	(b) 155, 571	71.55; 61.10; 78.55; S8.11
P.C. Euthymiou	(a) 114,K215	72.40; 78.50; S7.11
	(a) 116,K133	73.40 and 85; 61.80; S7.12
R.A. Evarestov	(b) 151, 275	75.10; 71.25; 71.50; S11
	(b) 152, 633	75.10; 71.25; 71.50; S11
P.J.S. Ewen	(a) 114, 587	78.65; 68.55; S8

R. Fabbri	(a) 115, 437	61.80; 61.10; S5.11
D.J. Fabian	(b) 152, 117	68.45 and 73.20; 71.35; S1
M. Faehnle	(a) 111, 617	75.80; S1.2
	(a) 112, 189	75.80; 75.50
	(a) 116, 819	75.80; S1.2
S. Fakioglu	(b) 151,K205	75.10
N.N. Faleev	(a) 113, 431	68.55; 61.10; S7.15
J.R. Faleiro Ferreira	(b) 155, 275	75.10
	(b) 156, 647	75.10
L.D. Falkovskaya	(b) 153, 271	72.10; 72.20
C. Falter	(b) 151, 503	63.20; S5.11
Y.D. Fan	(a) 116, K51	64.70; S1.1
C.S. Fang	(b) 154, 135	63.20; 71.36; 78.30; S7.15
Xiaoming Fang	(b) 152, K33	62.50 and 78.40; S8.15
P. Fantucci	(b) 153, 193	71.10; 71.25; S4
O.V. Farberovich	(b) 155, 179	71.25; 78.20; S10.1
	(b) 155, 215	71.25; 78.20; S8
	(b) 155, 491	71.20; 71.50; 79.60; S1.63

D. Farkas	(a) 111, 173	61.80 and 68.20; S1.1; S4
E. Farrés	(a) 111, 675	73.40 and 77.50; 73.60; 85; S10
	(a) 113, 83	68.55 and 73.40; 68.48; S10.1
	(a) 114, 167	68.55; S10.1
J.L. Farvacque	(b) 156, 717	78.20; 61.70; S7.12
Ya.V. Fattachov	(a) 112, 699	79.20; 64.70; S5.11; S5.12
J. Fayos	(b) 154, 439	61.50
A.B. Fazakas	(b) 154, 637	71.30; S12.1
A.A. Fedorov	(a) 115, 413	68.48; 61.80; S7.12; S7.15
D.L. Fedorov	(b) 154, 609	64.70 and 71.25; 71.35; 78.40; S8.15
Yu.M. Fedorov	(a) 115,K101	75.60; S10
N.V. Fedoseeva	(b) 155, 639	75.25; 75.30
S.N. Fedosov	(a) 114,K103	77.30; S12
	(a) 115, 293	77.80; 77.30; S12
V.M. Fedosyuk	(a) 111,K225	75.70; S1.2
	(a) 115, 279	75.60; 75.70; S1.2
V.K. Fedyanin	(b) 153, 207	71.38; S12
	(b) 155, K29	71.38; S12.1
	(b) 155, 481	68.20 and 68.45; 73.20
A. Felimban	(a) 115, K27	73.40; S8.11; S8.13
B.-X. Feng	(a) 114, 451	61.70; 78.70; S1.2
Duan Feng	(a) 116, 231	61.70; 77.80; S10
F. Fenske	(a) 115, 497	68.55 and 79.20; S1.2
I.D. Feranchuk	(b) 155, 169	71.38
J.R. Ferguson	(b) 154, K59	72.20 and 73.60
F. Fermi	(a) 115, 325	61.70 and 78.60; S9.12
W. Fernengel	(a) 115, 547	75.60; 75.25; S1.1; S1.2; S1.61; S1.62
T. Feudel	(a) 116, 571	61.70; 66.30; S5.11
N.M. Ficza	(a) 113,K147	61.70 and 85; 68.55; S5.11; S10.1
H. Fiedler	(b) 153,K119	68.55 and 73.60; 72.40; 78.65; S5.11
W. Figueiredo	(b) 153, K73	75.10
G.M. Filippov	(b) 151, K29	71.45; 71.50; 72.10
I.M. Filippov	(a) 113, 307	61.70 and 72.20; 61.80; S5.11
A.B. Filonov	(b) 155,K127	71.25 and 74.70; S10.15
V.I. Fineberg	(a) 113, 215	73.40 and 85; S5.11
V.A. Finkel	(b) 151,K161	74.70; S10.15
	(a) 113, 375	62.20 and 64.60; 75.30; S1.4
V.M. Finskaya	(a) 111, K89	74.70; S10.15
L. Firlej	(b) 154, 333	64.70 and 76.30; 72.20; S12.1
F. Firszt	(a) 111,K113	71.55 and 78.60; S8.12
D.W. Fischer	(b) 154, 11	61.70 and 71.55; 72.40; 78.30; 78.50; 78.55; S7.12
F. Fischer	(b) 154, 789	78.50; 78.55; S9.12
S.I. Fisher	(b) 155, 169	71.38
A.Ya. Fishman	(b) 156, 597	75.30; 76.60; S11.2
E. Flach	(b) 155, 317	78.40; 71.25; 71.35; S8.15
S. Flach	(b) 152, 311	75.30 and 75.50; 75.40
	(b) 155, K9	63.20 and 63.75; S10
R. Flagmeyer	(a) 111, 541	73.40 and 85; 78.55; S7.12
	(a) 112, 391	61.50 and 61.80; S5.11; S5.12; S7.11; S7.12
	(a) 113,K211	68.55] S7.15
	(a) 115, 413	68.48; 61.80; S7.12; S7.15
L.I. Flaks	(a) 113, 431	68.55; 61.10; S7.15

B. Flietner	(b) 154, 219	71.10
A.G. Fokin	(b) 153, 89	62.30
	(b) 154, 367	77.20 and 78.90
R. Follath	(b) 151, 703	76.80; S9.11
V.M. Fomin	(b) 156, 225	71.38 and 73.20
G.H. Fong	(a) 114, 273	74.70 and 78.70; S10.15
Guangyao Fong	(b) 155, 647	75.50; S11.2
A. Forkl	(a) 113, 181	75.80; S1.1; S1.2
	(a) 114, 685	75.60; S1.1; S1.4
	(a) 116, 803	75.60; S1.1
J. Fousek	(a) 112, 625	77.30 and 77.80; S12
A. Fousková	(a) 115, 445	61.70 and 78.20; S12
Z. Frait	(b) 154, 363	75.50 and 76.50; S1.1; S1.2
D. Fraitová	(b) 154, 363	75.50 and 76.50; S1.1; S1.2
	(b) 155, 281	76.50
R. Francini	(b) 151, 347	78.20; S11
C. de Francisco	(a) 112, K43	75.60; S11.2
	(a) 113, 541	75.60; S10.1
T.S. Frangulian	(b) 156, K23	61.70; S9.11
T. Frauenheim	(a) 112, 667	61.80 and 68.55; 73.60; 78.65; S5
D. Freude	(a) 113, 159	72.20 and 74.70; 76.60; S10.15
K. Friedland	(b) 156, 461	73.40; S7.12; S7.15
H.-G. Fritsche	(b) 154, 603	68.40; 62.20; S9.11
H.P. Fritzer	(b) 153, 307	74.70; 75.30; 78.30; S10.15
B. Fritzsch	(b) 156, 65	66.30
R. Fritzsch	(b) 156, 65	66.30
V.V. Frolov	(b) 153, 659	71.55; 61.70; S5.11
I. Frombold-Treu	(b) 153, K109	65.50
R. Fuchs	(a) 111, K83	72.30
I.Ya. Fugol	(b) 153, 539	63.90; 71.45; 78.55; S5.2
W. Fuhs	(a) 114, 413	72.40 and 73.40; 71.55; S5.11
T. Fujihara	(a) 113, K245	75.50; S11.2
S. Fujita	(b) 156, K17	61.70 and 66.30
T. Fukuda	(a) 112, 207	78.30; S11.1
Y. Fukuda	(a) 114, K245	78.60 and 79.75; S10
	(a) 115, K251	78.60 and 79.75; S9.16
M. Fukuhara	(b) 156, 583	74.20
T. Fukumori	(a) 113, K19	61.10 and 61.70; S7.12
K. Fukushima	(b) 153, K141	74.20; S10.15
V.N. Fulga	(a) 114, 259	71.55; 78.55; S8.16
P.C.W. Fung	(a) 116, K199	62.20 and 77.80; S12
J. Furthmueller	(a) 116, 819	75.80; S1.2
K. Futagami	(a) 113, K19	61.10 and 61.70; S7.12
M. Gaafar	(a) 111, K259	78.20; S12
L. Gabla	(a) 112, 811	61.80 and 78.55; S10
R.G. Gabrielyan	(b) 151, 655	76.80
G.V. Gadiyak	(b) 156, 243	68.25 and 68.40; 73.20; S5.11
K. Gaertner	(a) 112, 733	68.55; 61.80; S1.61
	(a) 112, 747	68.55 and 79.20; 64.60; S1.61
P.I. Gaiduk	(a) 112, 703	61.80 and 64.75; S5.11
G. Galambos	(a) 111, 173	61.80 and 68.20; S1.1; S4
R.R. Gałązka	(b) 154, 389	78.20; 61.60; S8.15
A. Galdikas	(a) 114, K211	72.20 and 72.80; S8.16
J. Galibert	(b) 153, 653	63.20 and 71.70; 72.20; S7.13

R.R. Galimov	(b) 153, 443	61.70
	(b) 154, 43	61.70
J.C. Gallardo	(b) 153, 179	71.25; 78.20
M.V. Galustashvili	(a) 114, 99	61.70; S9.11
M.Ya. Gamarnik	(a) 113, K45	64.70; S8.11
	(b) 156, K1	61.10 and 68.90; S1.3
B. Gambin	(b) 151, 513	62.20
N. Ganev	(a) 115, K13	62.20; 68.20; S1.1
V.V. Ganin	(a) 115,K195	74.70; 78.30; 79.20; 79.60
V.A. Ganshin	(a) 114, 457	61.70; 78.20; S11.1
Jin-Liang Gao	(b) 154,K137	71.38
M. Gapanov	(a) 114,K241	78.60 and 79.75; 61.80; S9.11
S.V. Gaponenko	(b) 156, 449	71.35 and 78.65; S8.12
V.I. Garasim	(a) 111, K99	72.80 and 75.50; S8
G. Garayagdev	(a) 116, K91	62.65 and 72.40; S8.15
J. García-Solé	(b) 153,K187	78.55; S9.11
R.K. Gartia	(a) 114, 407	78.60
N.M. Gasanly	(b) 153, K89	63.20 and 78.30; S8.16
	(b) 153, 727	78.30; 77.80; S8.16
M. Gasgnier	(a) 114, 11	68.55 and 73.40; 75.70; 78.65; S1.4; S1.63; S10
F.M. Gashimzade	(b) 152, 329	78.20; S8
	(b) 155,K161	78.30
V.I. Gatalskaya	(a) 111, K89	74.70; S10.15
	(a) 111,K209	74.70; S10.15
	(a) 113,K235	61.80 and 74.70; S10.15
	(a) 115,K205	74.70; S10.15
G.V. Gatalskii	(a) 113,K235	61.80 and 74.70; S10.15
B. Gather	(a) 116,K195	77.20; 77.80; S11.1
K. Gatterer	(b) 153, 307	74.70; 75.30; 78.30; S10.15
G.M. Gavrilenko	(b) 155, 481	68.20 and 68.45; 73.20
V.I. Gavrilenko	(a) 112, 805	61.70; 78.20; S5.11
	(b) 155, 723	78.20; 71.25; S5.11
V. Gavryushin	(b) 151, 721	71.55 and 78.50; S8.12; S10.1
	(b) 156, 697	78.40; 71.25; S11.1
G. Gawlik	(a) 112, 343	61.80; 61.70; S1.1
J. Gawlik	(a) 112, 347	61.80 and 68.20; S1.1
T. Gbehi	(b) 152, 415	61.70; 76.30; 78.50; 78.60; S11
H. Gdanitz	(a) 112, K75	66.30; S5.11
N. Gečiauskaitė	(b) 154, 341	76.80; 61.55; S1.63
B. Gegenheimer	(b) 155,K123	72.80 and 74.70; 72.20; S10.15
I.I. Gegusin	(b) 151, 581	71.25 and 78.70; S9.11
W. Gehlhoff	(b) 155, 597	71.55; 76.30; S7.11
H.D. Geiler	(a) 112, 727	68.55
V. Geist	(a) 112, 79	61.80; 61.10; S5.11; S5.12; S7.1
R.S. Gekht	(b) 155, 639	75.25; 75.30
P.V. Gel	(b) 152, K55	78.70; S1.61
P.V. Geld	(a) 114, 315	65.70 and 75.40; S1.61
V.D. Gelikonova	(a) 115,K205	74.70; S10.15
A. Gemperle	(a) 112, 335	61.80 and 68.20; S1
D. Genzow	(a) 114, 621	68.55; 78.55; S8
O. Georgieva	(a) 116, 213	61.60 and 61.70; S8.16
A.N. Georgobiani	(a) 111, 301	78.20 and 78.55; 78.60; S8.11
N. Georgoulas	(a) 116, 725	73.60 and 78.65; S6
K.H. Georgy	(a) 112, K91	62.20; S1.2
X. Gerbaux	(a) 115, 587	78.30; 77.80; S11

E. Gerdes	(a) 114, 387	77.20 and 77.40; S10.15
H. Gerecke	(b) 154, 565	63.20 and 68.30; 73.40; S7.12
	(b) 156, 151	63.20; 68.30; S7.12; S7.15
E. Gerlach	(b) 151, 241	72.20; 61.70; 71.45
A.I. German	(b) 152, 593	72.20
K. Germanova	(a) 113,K231	72.80; S7.12
K.P. Ghatak	(b) 151,K135	71.25 and 73.40; S8.15
	(b) 153,K209	73.40 and 85
	(b) 154, K29	62.20; S7
	(b) 154, K35	62.20; S8.11
	(b) 154,K121	73.40; S7.15
	(b) 154,K147	72.20
	(b) 155, K23	71.25; S4
A. Ghazi	(a) 116, K47	63.70; S9.11
C. Ghezzi	(b) 152, 101	73.40; 71.35; 72.40; 78.55; S7.15
G. Ghibaudo	(a) 113, 223	73.40 and 85
	(b) 153,K155	72.10
A. Ghosal	(b) 151,K135	71.25 and 73.40; S8.15
D.K. Ghosh	(a) 114, 381	78.20; 85; S7.15
K. Ghosh	(a) 111, K83	72.30
P.S. Ghosh	(a) 114, 381	78.20; 85; S7.15
S. Ghosh	(b) 152, 625	72.50; 71.25; S7.13
	(b) 152, 691	78.20; 77.60; S7.13
	(b) 156, 705	78.35; S7.13
A. Ghoshal	(b) 153,K209	73.40 and 85
	(b) 154,K121	73.40; S7.15
	(b) 154,K147	72.20
	(b) 155, K23	71.25; S4
E.V. Gilan	(a) 112, 305	61.80 and 73.40; S8.13
P. Gilliot	(b) 153, 403	78.20; S9
C. Giovanella	(b) 154, 273	74.50; 64.70; S10.15
E. Giovenale	(b) 151, 453	61.70; 78.50; 78.55; S9.11
A.A. Gippius	(a) 114, 265	61.60 and 72.80; 75.50; S1.4
D. Girginoudi	(a) 116, 725	73.60 and 78.65; S6
W. Girirat	(b) 152, K33	62.50 and 78.40; S8.15
G. Giro	(b) 153, 281	72.40 and 73.60; S12
V.V. Gladkii	(a) 115, 569	77.80; S9.16
C. Gladun	(b) 151,K155	72.80; S1.61
S.J. Gładysz	(b) 153,K171	75.40
G. Gładyszewski	(a) 112, 753	61.55 and 61.80; 68.48; S4
	(a) 115,K207	68.55 and 74.70; S10.15
P. Glaremin	(b) 156, 195	64.60; 78.20; 78.55; S11.1
V.B. Glazman	(b) 153, 659	71.55; 61.70; S5.11
G.A. Gledhill	(b) 151, 469	61.70; 78.50; S7.12
K.D. Glinchuk	(a) 112, K65	61.70; 71.55; S5.11
	(a) 113, 497	71.55 and 78.55; 61.80; S7.12
G.F. Glinskii	(b) 155, 501	71.36
	(b) 155, 513	71.36
I. Głowacki	(a) 114,K107	78.60; S12
L. Głusiec	(a) 112, 835	66.30; S1.1
	(a) 112, 839	62.20; 66.30; S1.1
Yu.P. Gnatenko	(b) 153, 667	71.35 and 71.55; 72.40; 78.55; S8
N.K. Gobran	(b) 155, 453	62.80; S1.1; S1.2; S4
M. Godlewski	(a) 114, 127	61.70; 76.30; S8.11
I. Godmanis	(a) 111, 335	78.60; S10.1
O. Goede	(b) 151, 175	71.35 and 71.55; 76.90; 78.55; S8.11; S8.15

O. Goede	(b) 151, 311	76.30; 75.30; S8.15
	(b) 154, 405	78.20; 75.30; S8.15
P. Goernert	(a) 115, 229	68.55 and 78.20; S11.2
G. Goetz	(a) 111, 59	68.48; 61.16; 61.80; S7.12
	(a) 111, K13	61.80 and 68.48; S1.2
	(a) 112, 289	61.80; 78.50; S7.11; S7.12
	(a) 112, 727	68.55
	(a) 112, 733	68.55; 61.80; S1.61
	(a) 112, 747	68.55 and 79.20; 64.60; S1.61
I.M. Golban	(b) 152, 225	72.20; S5
L.M. Goldenberg	(b) 154, 333	64.70 and 76.30; 72.20; S12.1
R. Goldhahn	(b) 154, 769	78.20; S7.12
M. Goldmann	(b) 152, 463	64.60; 68.45; S5
T. Goldmann	(a) 114, 387	77.20 and 77.40; S10.15
E. Goldys	(b) 154, 397	78.50; S7.11
O.A. Golikova	(a) 113, K1	61.40 and 72.20; S5
	(b) 155, 475	71.20; 73.60; 78.65; S5.11
I.Yu. Goliney	(b) 156, 211	78.65; S7.12; S7.15
E.M. Gololobov	(a) 114, K57	74.70; S10.15
	(a) 115, K79	61.70 and 74.70; S10.15
V.N. Golovin	(b) 151, K161	74.70; S10.15
M.F. Golovko	(b) 155, 549	71.55 and 73.40
Yu.I. Golovko	(a) 114, 559	64.70 and 68.55; 77.80; S10.15
V.P. Goltsev	(a) 111, K13	61.80 and 68.48; S1.2
A. Goltzené	(b) 156, 71	61.70; 76.90
M. Golz	(b) 151, 383	78.65
M.J.M. Gomes	(b) 151, 747	71.35 and 78.40; S9
A.V. Gomonnai	(b) 154, K197	61.80 and 78.30; S7.11
V.P. Gomozov	(a) 115, 207	68.55; 73.60; S10
Yu.G. Goncharov	(b) 153, 529	64.60 and 78.30; S8.16
R. Gonczarek	(b) 154, K157	74.20
J.L. Gondar	(b) 156, 259	71.20
Changde Gong	(b) 151, K77	75.40
	(b) 154, K63	75.10
J. Gong	(a) 111, K53	61.70 and 72.70; S7.12
Min Gong	(a) 111, K49	71.55; S5.11
F. Gonzáles	(a) 112, K53	78.60; S10.1
J. González	(a) 113, 187	75.50; S1.1; S1.2
J. Gonzalez Estevez	(a) 115, K233	75.50 and 75.80; S1
E.S.R. Gopal	(a) 114, 525	64.70; 61.40; 65.70; 72.80; S8
S.I. Gorbachuk	(a) 114, 305	76.30; S10.15
I.S. Gorban	(a) 115, 555	78.60; 71.55; S7
L.V. Gorchak	(a) 112, 305	61.80 and 73.40; S8.13
V.S. Gorelik	(b) 155, 113	63.20 and 68.30; 78.30; S8
S.D. Gorodetskii	(a) 115, K95	75.50; S1.1; S1.2
B.P. Gorshunov	(a) 116, K129	78.20; 74.70; S12
P. Górski	(b) 155, K157	78.20; S11.1
S.B. Goryachev	(b) 153, 443	61.70
	(b) 154, 43	61.70
B.N. Goshchitskii	(a) 112, 161	61.40 and 75.50; S11.2
T. Goto	(a) 111, K1	61.60; S1.61
V. Gottschalch	(a) 112, 519	61.70 and 68.55; S7.11; S7.12; S7.15
	(a) 113, 71	61.70 and 68.55; S7.11; S7.15
	(a) 114, 419	73.40 and 85; 78.60; S7.11; S7.15
F. Gouda	(a) 111, K265	61.25 and 77.20; S12

R.D. Gould	(a)	115, 237	68.55 and 73.60; S8.13
C. Gourdon	(b)	153, 641	71.35 and 78.55; S8.15
D. Gourier	(b)	152, 415	61.70; 76.30; 78.50; 78.60; S11
G. Govindaraj	(a)	113, 113	72.15 and 73.60
G.A. Govor	(a)	113, 403	64.70; 75.30; S1.2
R.P. Goyal	(b)	155, K81	62.20 and 78.20; S9.12
V.K. Goyal	(a)	116, 425	73.20 and 85; 61.70; 72.40
D.B. Goyer	(b)	154,K197	61.80 and 78.30; S7.11
A.V. Grachev	(b)	156, 403	61.70
A. Graja	(b)	154, 333	64.70 and 76.30; 72.20; S12.1
	(b)	155, 147	71.45; 63.20; S12.1
D. Grambole	(a)	112, 765	61.40 and 61.80; S12
U.M. Grassano	(b)	151, 347	78.20; S11
	(b)	151, 453	61.70; 78.50; 78.55; S9.11
V.I. Grebennikov	(b)	151, 623	74.40
S. Grebinskii	(a)	114,K211	72.20 and 72.80; S8.16
L.G. Grechko	(b)	153, 633	71.35 and 78.20
T. Greday	(a)	116, 521	61.55; 61.14; 64.70; S1.2
R.C. Greenhow	(b)	154, K47	71.10
I. Gregora	(a)	111, 345	78.55; 71.55; 78.30; S7.12
A.M. Grekhov	(b)	155, 475	71.20; 73.60; 78.65; S5.11
L. Grenacs	(b)	153, 721	76.60; 71.50; S1; S1.3; S4
V.P. Gribkovskii	(b)	156, 449	71.35 and 78.65; S8.12
V.I. Gridnev	(b)	152, 61	61.80; S5.11
J. Griesche	(a)	114, 621	68.55; 78.55; S8
J. Grigas	(a)	113, 597	77.20; 77.80; S11.1
I.I. Grigorchak	(a)	112, K61	61.70; 72.80; 78.50; S8
N.I. Grigorchuk	(b)	153, 633	71.35 and 78.20
A.A. Grigorenko	(b)	155, 215	71.25; 78.20; S8
B.L. Grigorenko	(b)	155, 179	71.25; 78.20; S10.1
V.G. Grigoriev	(b)	153, 65	61.80 and 79.20
E. Grilli	(a)	115, 599	78.55; 71.55; S8.16
M.L.N. Grillo	(a)	111, 625	76.30; S9.11
P. Grima	(a)	111, 405	61.60; 64.75; S8.16
G.A. Grinvalds	(a)	116,K125	61.80 and 78.40; 78.55; S11.2
J. Grňo	(a)	112, 327	61.70; 72.40; S5.11
R. Groene	(a)	113,K143	76.80 and 78.70; S1.1
R. Groessinger	(a)	112, 175	75.50; S1.1; S1.2
	(a)	115,K219	75.50; 75.30; S1.1
R. Groetzschel	(a)	112, 661	61.70 and 61.80; S5.11
	(b)	155, 77	63.20; 78.30; S7.13; S8.11; S9
	(a)	116, 735	61.80 and 68.55; 77.55; S9.12
S. Grosswig	(a)	115, 369	61.10; S5.11
H. Gruber	(b)	153, 307	74.70; 75.30; 78.30; S10.15
A.N. Gruzintsev	(a)	111, 301	78.20 and 78.55; 78.60; S8.11
V.G. Grybyk	(a)	116, 287	64.60; 62.20; 78.20; S12
M. Gryziński	(a)	112, 757	61.80 and 68.20
B. Gržeta	(a)	111, 417	61.55; 62.20; 64.70; S4
Bing-Lin Gu	(b)	153, 139	64.60; 64.75; S1.1; S1.2
V.A. Gubanov	(b)	151, 407	71.25; S1.61; S1.62
V.A. Gubanova	(b)	156, K77	76.30; 61.80; S9.11
P. Guenter	(b)	151, K1	66.30; 64.70; S12
M. Guittard	(a)	113, K5	61.16 and 61.60; S8
I.R. Gulakov	(a)	115,K113	72.80 and 78.60; S8.16
P.G. Guletskii	(b)	154, 623	71.25; 75.50; S1.1; S1.2
Z. Gumienny	(a)	111,K249	78.55; 71.55; S7.1
H.-E. Gumlich	(b)	153, 391	78.55; 71.70; S8.15

H.-E. Gumlich	(b) 154, 405	78.20; 75.30; S8.15
	(b) 155, 317	78.40; 71.25; 71.35; S8.15
	(b) 156, 561	71.35 and 71.70; 78.55; S8.15
Sheng-Li Guo	(b) 152, 577	71.70 and 76.30; S9.16
	(b) 153, K45	71.55 and 78.50; S11.1
	(b) 153,K167	71.70 and 76.30; S10.1
	(b) 154, K67	76.30; S9; S9.12
L.B. Guoba	(a) 111, 507	68.48; 79.20; S7.12
B.R.K. Gupta	(b) 155, K81	62.20 and 78.20; S9.12
S.K. Gupta	(a) 115, 569	77.80; S9.16
A.G. Gurevich	(a) 116,K185	76.50; S8.16
V.N. Gurin	(a) 114, 265	61.60 and 72.80; 75.50; S1.4
A.F. Gurov	(a) 115, K19	64.70; S4
V.A. Gurtov	(a) 116, 769	72.20; 71.55; 85; S5.11
A.G. Guseinov	(a) 113,K207	65.40; S8.16
A.M. Guseinov	(a) 116,K173	72.40; S8
D.A. Guseinov	(a) 115, K5	65.40 and 65.50; S8.16
N.G. Guseinov	(a) 111, K17	61.70; 72.80; 75.50; S8
	(a) 113,K207	65.40; S8.16
S.G. Guseinov	(b) 151, 759	78.20; S8.16
Ya.Yu. Guseinov	(a) 115, K53	72.80; S10.1
	(a) 116, 305	73.40; 72.20; S10
A.I. Gusev	(b) 151, 211	74.70; S1.61
	(a) 111, 443	61.70; S1.61
	(a) 113, 353	65.40; S1.61
	(b) 154, 453	61.50; 61.12; S1.61
	(b) 156, 11	61.60 and 75.30; 76.60; S1.61; S1.62; S10
N.A. Gushchina	(b) 153, 161	73.40; 71.25; S7.15
E.V. Gushchinskaya	(a) 111, 451	66.30; S7.11
N. Guskos	(b) 152, K9	74.70 and 76.30; S10.15
	(b) 153, 243	71.70; 76.30; S9.12
A.K. Gutakovskii	(a) 115, 413	68.48; 61.80; S7.12; S7.15
J. Gutiérrez	(a) 111, 279	75.80; S1.1; S1.2
M.Yu. Gutkin	(a) 113, 337	61.70 and 68.48
G.L. Gutsev	(b) 153, 659	71.55; 61.70; S5.11
	(b) 156, 319	71.55; S5.11
M. Guymont	(a) 113, K5	61.16 and 61.60; S8
M. Guzzi	(a) 115, 599	78.55; 71.55; S8.16
J. Gyulai	(a) 112, 237	61.80; 68.55; S5.11
	(a) 112, 721	73.40; 68.55; S5.11
	(a) 115, 75	61.80; 73.40; S5.11
	(a) 116, 561	66.30; 61.70; S5.11
M. Haase	(b) 153, 225	71.45; S12
V. Hadjicontis	(b) 156, 393	66.30; S1.2
A. Hadni	(a) 115, 587	78.30; 77.80; S11
H. Haefke	(a) 116, 39	66.30 and 68.48; S9.11
H. Haefner	(a) 114, 419	73.40 and 85; 78.60; S7.11; S7.15
J. Haertwig	(a) 115, 369	61.10; S5.11
H. Hagenschulte	(b) 154, 71	66.30; S1.3
T. Hai	(b) 156, 641	75.30 and 75.70
F. Haider	(a) 111, 71	61.16; 61.70; S1.2; S1.3
A. Hairie	(a) 113, 489	71.25; S7.16
R. Hajduk	(a) 113, 295	61.80; S5.11
E. Hajto	(a) 114, 587	78.65; 68.55; S8

33*

N.C. Halder	(a) 112, 589	73.60; S10.1
	(a) 113, 511	73.60; S10
T. Halicioglu	(b) 156,K105	68.45; S7.12
	(b) 156,K113	68.40
C. Hamann	(a) 114, 551	68.55; S1.3; S12
	(a) 116, 729	68.20 and 68.55; S12
J.F. Hamet	(a) 114, 467	61.80; S5.11
K. Hammer	(a) 112, 667	61.80 and 68.55; 73.60; 78.65; S5
Guosheng Han	(a) 114,K189	74.70; S10.15
K. Handrich	(b) 152, 377	61.40; 63.50
F. Hanna	(a) 115,K149	64.70; 77.80; S11.1
T. Hara	(a) 113, 459	73.30 and 73.60; 73.20; S1.61; S7.12
Ch. Hardalov	(a) 113,K231	72.80; S7.12
T.M. Haridasan	(b) 153, 131	65.50; 63.90; S9.12
	(b) 153, 579	63.20; 61.70; S8.13
	(b) 154, 55	66.30; S8.13
V.S. Haroutyunyan	(a) 111, 41	61.10
H. Hartmann	(a) 114, 127	61.70; 76.30; S8.11
U. Hartmann	(b) 151, 289	75.60
	(a) 115, 285	75.60
M. Hartmanova	(a) 116, 457	62.20
T. Hartung	(a) 112, 769	61.70 and 61.80; S12
F.S. Hasoon	(a) 112,K139	61.60 and 78.30; S7.15; s8.15
G.A. Hassan	(a) 116, K43	62.20; S1.3
H.H. Hassan	(a) 116, K43	62.20; S1.3
K. Hatakeyama	(a) 115,K215	75.50; S1.63
Z. Hatzopoulos	(a) 114, K41	61.80 and 72.40; S7.12; S7.15
G. Hauck	(a) 116, 545	61.30; 64.70; S12
Lanying He	(a) 116, 267	64.70 and 74.70; 61.10; S10.15
T. He	(a) 115, 539	75.60; 76.80; S10
Yizhen He	(a) 115, 467	61.40 and 64.70; S1.1; S1.2
Z.H. He	(a) 115, K71	74.70 and 79.60; S10.15
P.C. Healy	(b) 152, 519	71.45 and 79.60; 74.10; S11
G. Heber	(b) 151, 693	75.25 and 75.50
	(b) 152, 303	75.25 and 75.50
M. Hecker	(a) 115,K159	62.20
K.H. Heckner	(a) 112, 395	61.80 and 68.55; 73.40; S10.15
V. Heera	(a) 114, 599	68.48 and 68.55; S5.11
E. Hegenbarth	(a) 111,K165	65.40; S7
	(a) 114, K67	66.70 and 74.70; S10.15
	(b) 154, K93	61.40
	(b) 156, K97	66.70; S10.15
K. Heilig	(a) 114, 579	73.20 and 73.40; S5.11; S10
A. Heilmann	(a) 114, 551	68.55; S1.3; S12
R. Heilmann	(a) 114, 419	73.40 and 85; 78.60; S7.11; S7.15
	(a) 115, 257	71.55; 72.80; 78.55; S7.15
W. Heimbrodt	(b) 151, 311	76.30; 75.30; S8.15
	(b) 154, 405	78.20; 75.30; S8.15
E. Heiner	(b) 151, 645	72.10; 71.10; 78.20
	(b) 153, 295	72.10; 72.20
A. Heinrich	(b) 151,K155	72.80; S1.61
J. Heinrichs	(b) 151, 231	72.10
F. Heinricht	(a) 116, 145	61.16; 68.20
C. Heiser	(a) 112, 765	61.40 and 61.80; S12

B. Hellermann	(a) 116,K195	77.20; 77.80; S11.1
E.K. Henner	(b) 151, 663	76.30
W. Henrion	(b) 151, 375	78.65; S1.61
E. Hensel	(a) 112, 533	68.48 and 68.55; S1.1; S7
W. Hergert	(a) 111,K253	78.60
R. Hergt	(a) 111, 583	74.60 and 74.70; S10.15
H. Hermann	(a) 116, K19	61.10
E.B. Hermida	(b) 151, 71	62.20; S4; S11
A. Hernado	(a) 113, 187	75.50; S1.1; S1.2
L. Hernandez	(a) 113, 677	73.30 and 73.40; S7.11; S7.15
J. Herold	(a) 111, 59	68.48; 61.16; 61.80; S7.12
F. Herrmann	(a) 112, 765	61.40 and 61.80; S12
K.-H. Herrmann	(a) 116, 113	61.14
K.H. Herrmann	(a) 114,K249	72.40 and 78.30; S8.15
R. Herrmann	(b) 152, K5	72.20 and 74.70; S10.15
	(b) 152, 481	73.40; S7.13
	(b) 153, K63	74.70; S10.15
	(a) 113, 159	72.20 and 74.70; 76.60; S10.15
	(a) 116, 793	72.20 and 74.20; S10.15
P. Hertel	(a) 114, 765	78.20; 61.70; S10.15
D. Hesse	(a) 116, 67	68.48 and 68.55; S1.61
	(a) 116, 81	68.48 and 68.55; S1.61
H. Hesse	(a) 116,K195	77.20; 77.80; S11.1
Th. Heumann	(b) 154, 71	66.30; S1.3
J. Heydenreich	(a) 116, 81	68.48 and 68.55; S1.61
	(a) 116, 123	61.16; 61.70
R. Heyner	(a) 114, 609	68.20; 79.20; S1.2
M. Hidaka	(a) 115, 149	64.70; S9.16
T.D. Hien	(a) 114, 361	75.30 and 75.80; S1.1; S1.2; S1.4
Than Duc Hien	(a) 111, 591	75.30; 75.25; S1.1; S1.4
Nguyen Van Hieu	(b) 151,K171	74.20
T. Higashi	(a) 113,K245	75.50; S11.2
J.F. Higgs	(b) 155, 427	63.20; S5.2
P.G. Hill	(a) 114, 587	78.65; 68.55; S8
R. Hillebrand	(a) 116, 81	68.48 and 68.55; S1.61
	(a) 116, 123	61.16; 61.70
C.P. Hilton	(b) 151, 175	71.35 and 71.55; 76.90; 78.55; S8.11; S8.15
Y. Hirose	(a) 114, 693	72.15 and 75.50; 61.55; S1.2
R. Hirrle	(a) 111, 315	78.55 and 78.60; 71.55; S11.2
	(a) 111, 629	78.55 and 78.60; 71.55; S11.2
D. Hirsch	(b) 152, 505	71.20 and 79.60; S7
R. Hirte	(a) 116, 179	61.40; 61.16; 62.20; S12
K. Hisatake	(a) 113,K245	75.50; S11.2
J. Hlávka	(a) 112, 327	61.70; 72.40; S5.11
D. Hoehl	(a) 112, 419	61.70 and 61.80; 61.14; 68.48; S5.11
	(a) 112, 715	61.80 and 68.55; 68.48; S5.11
M. Hoehne	(b) 156, 325	71.55 and 76.30; S5.11
R. Hoehne	(a) 114, K93	75.30 and 75.50; S11.2
B. Hoenerlage	(b) 151, 747	71.35 and 78.40; S9
S. Hoenerlage	(b) 153, 403	78.20; S9
M. Hoericke	(a) 113,K211	68.55; S7.15
W. Hoerig	(a) 116,K119	61.70 and 78.50; S7.12
S.R. Hoffman	(b) 155, K33	71.20; 72.15; S1.2; S1.3
H. Hoffmann	(b) 153, 391	78.55; 71.70; S8.15
S. Hoffmann	(b) 153, 391	78.55; 71.70; S8.15

A. Hofmann	(a) 111, 243	75.50; S1.1; S1.2
C.A. Hogarth	(a) 113, 483	68.55 and 76.30; S10.15
	(a) 114, 215	78.65; 68.55; S10.15
	(a) 114, 225	68.55 and 76.30; S10.15
	(a) 116, 703	73.60; 73.40
W. Hohenau	(a) 111, 335	78.60; S10.1
B. Hohler	(a) 111, 89	66.30; 72.15; S1.2
E.K. Hollmann	(a) 114, K21	79.20; S1.2
L. Holtmann	(a) 113, K89	72.40; S11.1
V. Holý	(b) 151, 23	61.10 and 61.70; S5.11
	(b) 155, 339	61.10 and 61.70; S5.11
T. Holz	(b) 156, 259	71.20
D. Hommel	(a) 114, 127	61.70; 76.30; S8.11
J. Horák	(a) 113, 615	78.30; S8.15
	(a) 115, 87	61.70; 72.20; S8
Y. Horikago	(b) 155, K5	62.20; 63.20; S1.3
J. Horn	(a) 113,K103	61.70 and 75.50; S11.2
A. Horodeński	(a) 112, 757	61.80 and 68.20
	(a) 112, 821	79.90
G. Horváth	(a) 111,K185	73.60
A.A. Hosseini	(a) 113, 57	64.75; 61.80; S1.1
M. Hrabčák	(a) 115, K83	75.80; 61.80; S1.1
L. Hrivnák	(a) 116, K73	71.25; S7.12; S7.15
A.Z. Hrynkiewicz	(a) 113, 295	61.80; S5.11
A. Hu	(a) 112, 99	63.20; 78.35; S1.2
An Hu	(b) 155, 137	68.30; 78.30; S1.2; S1.3
Jifan Hu	(a) 113,K127	75.30; S1.1; S1.4
Shu-Hui Hu	(a) 116, K7	61.40 and 78.70; 74.70; S10.15
Xiaoyu Hu	(b) 154, 61	61.70; 62.20; S4
	(b) 155, 53	61.70; 62.20; S4
Jinqi Huang	(a) 114, 651	61.60 and 74.70; S10.15
L.J. Huang	(a) 116, K51	64.70; S1.1
Qingzhen Huang	(a) 114,K127	78.45; 78.30; 78.55; S11.1
Tzu-Yu Huang	(b) 152, 335	78.20; S1.61
Xin-Fan Huang	(a) 113, 453	68.48 and 73.40; S5.11
Yichuan Huang	(a) 114,K127	78.45; 78.30; 78.55; S11.1
O. Hudák	(b) 154, 743	75.40; S11.1
	(b) 155, 257	75.30 and 75.50; S11.1
F.K. Huebner	(a) 116, 389	75.60
I. Huettel	(a) 111, 367	72.70 and 73.40; 78.60; S7.15
J. Humlíček	(a) 111, 655	78.30; 73.40; S7.12
	(b) 155, 723	78.20; 71.25; S5.11
B. Hunger	(a) 112, 721	73.40; 68.55; S5.11
A. Hupfer	(b) 152, 505	71.20 and 79.60; S7
S.A. Hussein	(a) 114,K205	72.80; S8
S. Hyodo	(a) 111, 223	72.20; 61.70; S8.11
G. Iadonisi	(b) 153, 611	71.35 and 71.38
A. Ibarra	(a) 112, K53	78.60; S10.1
T.D. Ibragimov	(b) 155, 113	63.20 and 68.30; 78.30; S8
E.M. Ibragimova	(a) 115,K255	61.80 and 78.60; S10.1; S12
A. Ibrahim	(a) 115,K149	64.70; 77.80; S11.1
R. Ichikawa	(a) 113, 459	73.30 and 73.60; 73.20; S1.61; S7.12
S. Ichikawa	(a) 115,K215	75.50; S1.63
S. Ichinose	(b) 151, 95	63.20 and 71.45

S. Ichinose	(b) 155, 467	64.10 and 71.45; S12
J.O.A. Idiodi	(b) 153, 489	63.10; 63.20
	(b) 153, 501	63.10; 63.20
A.N. Igumentsev	(a) 112, K71	66.30; S11.2
T. Ikari	(a) 115, K43	71.55; S8
R. Ikeda	(a) 114,K143	66.30 and 76.60; S12
Y. Ikeda	(a) 116, 633	62.40; S1.2
K. Ikezaki	(a) 113, K67	73.60; S12
V.P. Ikonnikov	(b) 151, 269	75.30; 76.80; S11.2
B. Iliescu	(b) 152, 51	61.70; S9.11
I. Iliev	(a) 115,K243	78.20; S8.15
M.N. Iliev	(b) 152, K29	78.30; S11.1
A.I. Ilinikh	(b) 154, 679	74.20 and 74.50
R. Ilkhamov	(a) 112, 743	61.80 and 68.48; S1.61
R.A. Ilkhamov	(a) 112, 385	61.80; 66.30; S5.11
	(a) 112, 781	61.70 and 61.80; S5.11
L. Ilkov	(a) 111, 145	64.80; 72.20; 75.50; S11.2
M.A. Illarramendi	(a) 113, 187	75.50; S1.1; S1.2
R.M. Imamov	(a) 113,K153	61.10
	(a) 115,K133	61.10; 68.20; S5.11
	(a) 116, 197	61.10; 78.70; S7.12
T. Imura	(a) 115,K261	78.30; S8.15
J. Iñiguez	(a) 113, 541	75.60; S10.1
J.I. Iniguez	(a) 112, K43	75.60; S11.2
K. Inoue	(a) 113, K67	73.60; S12
M. Inoue	(b) 151, 441	61.60; S8.16
A. Ioan	(b) 152, 51	61.70; S9.11
S.P. Ionov	(b) 155, 397	64.60; 71.25; S8.15
	(b) 155,K137	74.20
K. Irmscher	(a) 116, 755	71.55
T.I. Isaeva	(a) 115,K101	75.60; S10
M. Ishiguro	(b) 153, 683	71.55; 78.50; 78.55; S9.11
K. Ishii	(a) 112, 207	78.30; S11.1
M. Ishii	(b) 152, K63	78.30; 63.20; S1.63
	(b) 155, K13	62.20; 63.20; 78.30; S1.63
	(b) 156,K167	62.20; 63.20; 78.30; S1.63
T. Ishikawa	(a) 115, 383	61.10; S7.12
B.B. Ismail	(a) 115, 237	68.55 and 73.60; S8.13
T.G. Ismailov	(b) 155,K161	78.30
H. Isotalo	(b) 154, 305	72.20; 72.60; 76.30; S12
A.G. Ivanchin	(b) 155, K1	61.70
R.D. Ivanchuk	(a) 111, K99	72.80 and 75.50; S8
V.A. Ivanov	(a) 112,K111	72.40; S8
V.I. Ivanov	(b) 151, 259	75.30 and 75.50; 76.50; S11.2
Z.G. Ivanov	(a) 113, 439	73.40 and 74.50; S7.12
T.I. Ivanova	(b) 152, 431	61.50 and 64.60
V.A. Ivanshin	(b) 151, K53	76.30 and 77.80; S11.1
I.N. Ivleva	(a) 115,K195	74.70; 78.30; 79.20; 79.60
T. Iwashita	(b) 152, 289	75.10 and 75.30
	(b) 154, 747	75.10; 75.30
E.N. Izakovich	(a) 115,K195	74.70; 78.30; 79.20; 79.60
	(a) 116, K13	61.50; 74.70; S10.15
E.M. Jaboori	(a) 113, 579	78.40; S10.15
	(a) 116,K111	78.65; S10.1
A. Jacobi	(a) 113,K135	78.20; S8.15
	(a) 116, 793	72.20 and 74.20; S10.15

B. Jacobs	(a) 115, 257	71.55; 72.80; 78.55; S7.15
K. Jacobs	(a) 111, 541	73.40 and 85; 78.55; S7.12
E. Jaeger	(a) 116,K219	72.40 and 85; S8.11; S8.12
H.U. Jaeger	(a) 116, 571	61.70; 66.30; S5.11
J. Jagielski	(a) 112, 343	61.80; 61.70; S1.1
L. Jahn	(a) 116,K179	75.70; 75.60; S4
M. Jain	(a) 114, K29	73.60 and 78.65; S8.16
Mukesh Jain	(a) 115,K181	73.60 and 78.65; S8.16
V.K. Jain	(a) 111, 625	76.30; S9.11
R. Jakubas	(a) 111, K27	64.70; 77.20; S12
	(a) 114,K111	78.20; S12
	(a) 115, K17	64.70 and 65.70; S12
W. Jakubowski	(a) 115, 81	66.30; S10
I. Jamnický	(a) 116, K77	68.25 and 71.55; 72.50; S5.11; S10
R. Janes	(b) 152, 583	71.50 and 76.30; 71.38; 76.40; S9
U. Jank	(a) 114,K249	72.40 and 78.30; S8.15
K. Jarašiūnas	(a) 112, 375	71.55 and 78.20; S5.11
	(a) 114,K183	73.40; 68.55; S7.12; S9.15
D. Jaworska	(a) 112, 385	61.80; 66.30; S5.11
	(a) 112, 793	61.80; 78.20; S7.12
M. Jayachandran	(a) 113,K217	73.30; 85; S8.11
M.K. Jayaraj	(a) 114,K233	78.60; S8.11
	(a) 115, K63	72.80; S10.1
C.K. Jayasankar	(b) 155, 221	71.70; S11.1
	(b) 155, 559	71.70; S9.16
J. Jędrkowiak	(a) 112, 357	61.80; 68.20; S1.1
Li Jen Ho	(a) 112, 781	61.70 and 61.80; S5.11
B. Jenichen	(a) 113,K211	68.55; S7.15
T.A. Jenkins	(a) 113, K71	72.15; S1.2
M. Jergel	(a) 111, 597	75.80; S1.1; S1.2
H. Jex	(b) 151, 703	76.80; S9.11
M. Jeżewski	(a) 116, 745	71.55; S7.11
A. Jezierski	(b) 153, 185	71.20; S1.2
K. Jezierski	(a) 112,K135	78.40; S7
Jiaben Ji	(a) 114,K189	74.70; S10.15
M.R. Ji	(a) 115, K71	74.70 and 79.60; S10.15
Chun-Sheng Jia	(a) 116, 371	74.70; S10.15
Wei Jia	(a) 114,K171	68.60 and 76.30; S6
B.-H. Jiang	(a) 114, 451	61.70; 78.70; S1.2
Jianyi Jiang	(a) 115, K75	74.70; 62.40; S10.15
Shan Jiang	(b) 152, K33	62.50 and 78.40; S8.15
Yi-Jian Jiang	(b) 156, 145	63.20 and 78.30; 71.36
I. Jiménez	(b) 154, 439	61.50
M. Jiménez de Castro	(a) 112, K53	78.60; S10.1
Changqing Jin	(a) 116, 267	64.70 and 74.70; 61.10; S10.15
Hanmin Jin	(b) 151, 249	76.80; S1.1; S1.4
Jirong Jin	(a) 114,K189	74.70; S10.15
Mingzhi Jin	(b) 151, 249	76.80; S1.1; S1.4
N.Y. Jin	(a) 116, 91	61.70; 61.16; S1.2; S1.3
Pei-Wan Jin	(b) 154,K137	71.38
Xiang-feng Jin	(a) 116, 709	73.60; 71.55; S12
Xin Jin	(a) 114,K189	74.70; S10.15
M. Jirsa	(a) 113, 573	75.70 and 76.50; 75.60; S11.2
R.N. Joarder	(b) 156, 181	65.50; 65.20; S2
I.P. Jones	(a) 113, 57	64.75; 61.80; S1.1
L. Jonikas	(a) 112, 375	71.55 and 78.20; S5.11

J.F.J. Jordan	(b) 155, 89	63.20; S10
S.K. Joshi	(a) 113, 603	78.70; S12
K. Jóźwikowski	(a) 111, 559	72.20; S8.15
A. Juarez	(b) 156, K89	71.55 and 78.55; 73.40; S7.12
Ch. Jung	(b) 155, 317	78.40; 71.25; 71.35; S8.15
Z. Juránek	(a) 111, 603	75.50; S1.1
M. Jurczyk	(a) 112,K121	75.50; S8.16
	(a) 114,K219	75.50; S1.1
	(a) 115,K229	75.50; S1.2
H. Käämbre	(a) 114,K241	78.60 and 79.75; 61.80; S9.11
	(a) 115,K109	79.75; S9.11
N.M. Kabachnik	(b) 155, 289	71.45 and 78.70; 79.20
A. Kabasi	(b) 154, 281	72.20 and 73.40; S7.12
A.V. Kabyshev	(a) 116, 221	61.40; 72.80; 77.20; S1.62
	(a) 116, K69	61.70 and 68.20; S6; S7
G.A. Kachurin	(a) 113,K165	61.70; S5.11
J. Kaczmarek	(a) 112, 245	61.80 and 68.20; S1.1
T. Kaczor	(a) 113, 199	61.70 and 78.50; 71.55; S10.15
H.-Matsuo Kagaya	(b) 153, K17	62.20; 63.20; S7.1
	(b) 155, K5	62.20; 63.20; S1.3
H. Kahnt	(a) 111, K7	66.30
S.L. Kakani	(b) 155, 241	74.10 and 75.40; S10
N.G. Kakazei	(a) 114, 305	76.30; S10.15
J. Kalinowski	(b) 153, 281	72.40 and 73.60; S12
G.S. Kalsi	(b) 151, 283	75.10 and 75.40
J. Kalus	(b) 154, 153	63.20; S12
	(b) 155, 89	63.20; S10
F.M. Kamarly	(a) 111, K17	61.70; 72.80; 75.50; S8
Y. Kameda	(a) 112, 467	61.40; 61.12; S8.16
J.V. Kameneckas	(a) 111, 507	68.48; 79.20; S7.12
B. Kamieńska	(a) 112, 839	62.20; 66.30; S1.1
B. Kamieńska-Stefaniuk	(a) 112, 339	61.80 and 68.20; S1.1
A.A. Kaminskii	(a) 112, 197	78.20 and 78.45; 78.55; S10
	(a) 112, K49	78.20; S10
	(a) 113, 193	63.20 and 78.55; 71.55; S10.15
	(a) 113, 585	78.30; 71.70; S11.2
	(a) 113,K257	78.45; S9.16; S11.2
	(a) 114, 713	78.45 and 78.55; 71.55; 78.50; S10
G. Kampfrath	(a) 114, 551	68.55; S1.3; S12
V.M. Kamyshov	(a) 111,K171	64.70; S11.2
T. Kanadani	(a) 112, K15	64.75; S4
	(a) 113,K199	62.20 and 64.75; S4
	(a) 114, K17	64.75 and 72.15; S4
	(a) 114,K197	72.15; S4
	(a) 115,K147	62.20; S4
	(a) 116,K153	61.70; S4
B.D. Kandilarov	(b) 152, 163	71.25
T. Kaneko	(b) 156, 49	61.80; 71.10
T. Kaneyoshi	(b) 151, 675	75.30 and 75.50
Z.C. Kang	(a) 113, 23	61.60 and 74.70; 61.16; S10
	(a) 114, 73	61.60; 61.16; S10
	(a) 114, 441	61.60; 61.16; 61.70; S10
Y. Kanie	(b) 154, 195	73.20 and 73.40

M. Kaniewska	(a) 116, 745	71.55; S7.11
J. Kaniewski	(a) 116, 745	71.55; S7.11
M. Kanis	(b) 151, 311	76.30; 75.30; S8.15
T. Kanomata	(a) 111, K1	61.60; S1.61
Yu.O. Kanter	(a) 115, 413	68.48; 61.80; S7.12; S7.15
M.M. Kantor	(a) 112, 335	61.80 and 68.20; S1
S.B. Kantorow	(a) 112, 695	61.70; 79.20; S5.11
V.G. Kapinos	(b) 155, 373	61.70
V.V. Kaplin	(b) 152, 61	61.80; S5.11
	(a) 112,K131	79.20
M.G. Kaplunov	(a) 116,K129	78.20; 74.70; S12
D. Kapor	(b) 156,K143	75.10
D.V. Kapor	(b) 155, K77	75.10
A. Kara	(a) 112, 89	62.20; 61.70; S8.11
J. Karaba	(b) 155,K143	75.10
S.A. Karandashov	(a) 115,K117	78.45 and 78.55; S7.12; S7.15
A.A. Karanovich	(a) 111, 257	61.70; 76.30; S5.11
S. Karbanov	(a) 111, 145	64.80; 72.20; 75.50; S11.2
A.G. Karoza	(a) 111,K233	75.70 and 78.20; S11.2
	(b) 152, 657	78.30; S11.2
V.N. Karpechev	(b) 156, 663	78.55; 71.55; S10.1
A.L. Karpei	(a) 111,K209	74.70; S10.15
B.V. Karpenko	(b) 153, 271	72.10; 72.20
V.V. Karyagin	(b) 153, 263	72.20
	(b) 156, 235	71.10 and 73.40; 72.20; S8.15
A.G. Kasanskii	(a) 112, K23	72.40 and 73.60; S5.11
C. Kaschner	(a) 112, 747	68.55 and 79.20; 64.60; S1.61
P. Kashyap	(a) 114, K29	73.60 and 78.65; S8.16
Preeti Kashyap	(a) 115,K181	73.60 and 78.65; S8.16
V.A. Kasiyan	(b) 154, 287	72.20; 71.30; 71.55; S8.12
	(b) 154, 691	72.20; 71.30; S8.12
V.Kh. Kasiyanenko	(b) 152, K55	78.70; S1.61
I. Kasko	(a) 112, 747	68.55 and 79.20; 64.60; S1.61
G. Kasner	(b) 151, 521	63.50
T.K. Kasumov	(a) 111,K205	72.40; 63.20; S8.16
	(a) 113,K241	72.40; S8.16
O.I. Kasyutich	(a) 115, 279	75.60; 75.70; S1.2
G.I. Kataev	(a) 114, K79	75.30; S1.4
V.N. Katerinchuk	(a) 115, K35	73.40 and 78.60; S8; S10.1
S. Katircioğlu	(b) 152, K37	61.55; S4
M. Kato	(a) 113, K67	73.60; S12
A.A. Katsnelson	(b) 155, 179	71.25; 78.20; S10.1
E.Z. Katsnelson	(a) 111,K233	75.70 and 78.20; S11.2
	(b) 152, 657	78.30; S11.2
A. Kawabuchi	(a) 115,K261	78.30; S8.15
T. Kawakubo	(b) 153, K93	71.55; 76.30; 78.50; S7.11
E.A. Kayafas	(a) 112, 123	72.20 and 77.50; S12
L.A. Kazakevich	(a) 113, 307	61.70 and 72.20; 61.80; S5.11
M.M. Kazanin	(a) 113, 143	71.55; 72.20; 76.70; S5
Kh.R. Kazdaev	(a) 115, 179	61.80 and 68.20; S1.61; S1.62
M. Kazimierski	(a) 114, K49	74.70; S1.4
A. Kazlauskas	(b) 151, 721	71.55 and 78.50; S8.12; S10.1
	(b) 156, 697	78.40; 71.25; S11.1
T.S. Kê	(a) 113, 393	62.40; S4
	(a) 113,K195	62.40; S4
	(a) 115, 119	62.40; S4
V. Keith	(a) 114, K71	75.60; S8.15
P.N. Keivan	(b) 154, 513	66.30; S5.12
S. Kemmler-Sack	(a) 111, 315	78.55 and 78.60; 71.55; S11.2

S. Kemmler-Sack	(a) 111, 629	78.55 and 78.60; 71.55; S11.2
	(b) 156, 339	74.70; 71.30; 72.20; S10.15
R. Kern	(a) 116, 23	68.20; 61.70; 62.20; S9.11
K. Kerpelin	(a) 113, 637	78.20; 64.60; 77.80; S10
E. Kessler	(a) 116, K61	68.45; S1.3; S4
F.R. Kessler	(a) 111, 639	78.20; 78.50; S5.11
	(b) 154,K191	78.30 and 78.40; 71.35; S5.11; S7.11
	(b) 156, K45	71.35; 72.20; 72.40; S5.12
P. Ketolainen	(b) 156, K81	61.70 and 78.50; S9.11
P.K. Khabibullaev	(b) 152, 395	61.70; 71.35; 71.38; S9.11
S.Kh. Khafizov	(a) 114, 713	78.45 and 78.55; 71.55; 78.50; S10
I.B. Khaibullin	(a) 112, 699	79.20; 64.70; S5.11; S5.12
M.M. Khalifa	(a) 113,K177	66.70; S12
M.A. Khan	(a) 113,K189	62.20; S1.2
M.N. Khan	(a) 114, 285	72.20 and 78.30; S10.15
Q.H. Khan	(a) 116, K47	63.70; S9.11
W.I. Khan	(a) 112, 221	73.40 and 85; S7.12
G.D. Khattak	(a) 114, K71	75.60; S8.15
V.N. Khelashvili	(b) 156, K67	76.60
B. Khelifa	(b) 155, 191	71.10 and 78.70; S5.11; S5.12
	(b) 156, 497	71.20 and 71.60; S5.11; S7.12
M.L. Khidekel	(a) 115,K195	74.70; 78.30; 79.20; 79.60
	(a) 116, K13	61.50; 74.70; S10.15
T.A. Khimich	(b) 155, K63	76.60; S10.1
V.I. Khitko	(a) 112, 331	61.80 and 73.40; S7.12
V.G. Khlabutin	(b) 152, 61	61.80; S5.11
V.V. Khodasevich	(a) 111, K13	61.80 and 68.48; S1.2
I.I. Khodos	(a) 114, 113	61.70; S7.12
A.F. Khokhlov	(a) 116, 697	68.55 and 73.60; S5.11
E.V. Kholopov	(b) 155,K131	74.20; S10.15
A. Kholov	(a) 114, 713	78.45 and 78.55; 71.55; 78.50; S10
V.A. Khonik	(a) 116, 255	62.40; S1.2
T. Khristov	(b) 152, 163	71.25
V.I. Khrupa	(a) 116,K141	61.10; 61.70; S5.11
A.V. Khrushchak	(b) 152, K55	78.70; S1.61
B.I. Khudik	(b) 151, 111	73.60 and 78.65
	(b) 153, 167	78.65
V.M. Khulugurov	(b) 155, 359	61.70 and 71.55; S9.11
Paramdeep Khurana	(a) 115, K57	72.80; S8
S.S. Khusmurodov	(a) 112, 781	61.70 and 61.80; S5.11
F.A. Khwaja	(b) 156, 111	64.10; 64.60; S9.11
H. Kido	(a) 111,K105	72.80 and 75.50; S10.15
R. Kiemel	(b) 156, 339	74.70; 71.30; 72.20; S10.15
J. Kierul	(a) 114, 545	72.15 and 73.60; S4
M. Kiknadze	(b) 151, K49	76.70; S9.12
	(b) 152, K13	76.70; S9.12
H.K. Kim	(b) 153,K183	78.55; S8
H.S. Kim	(a) 111, 47	61.16 and 61.70; S1.2
	(a) 115, 51	61.14; 61.70
M. Kinoshita	(b) 152, 533	61.50 and 71.45
G.A. Kiosse	(a) 114, K63	74.70; S10.15
H.R. Kirchmayr	(a) 112, 175	75.50; S1.1; S1.2
	(a) 115,K219	75.50; 75.30; S1.1
V.A. Kirikov	(a) 115, 569	77.80; S9.16
F.-G. Kirscht	(a) 111,K195	71.55; S5.11
Y. Kirsh	(a) 113, 631	78.60; S11

Y. Kirsh	(a) 114, 739	78.60; S10.1
É. Kisdi-Koszó	(a) 111, 603	75.50; S1.1
A.A. Kiselev	(b) 156, K57	74.20
V.G. Kiselev	(b) 152, 667	78.20
L.F. Kiss	(a) 114, 685	75.60; S1.1; S1.4
S. Kistinger	(b) 151, 241	72.20; 61.70; 71.45
M. Kisza	(a) 112, 761	61.40 and 61.80; S12
K. Kiszczak	(a) 112, 815	79.70
A. Kitade	(a) 115, 587	78.30; 77.80; S11
T. Kitano	(a) 115, 383	61.10; S7.12
H. Kitayama	(a) 115,K261	78.30; S8.15
A.V. Kityk	(a) 116, 287	64.60; 62.20; 78.20; S12
A.E. Kiv	(b) 156,K101	68.45; S5.11
M. Kiwaki	(b) 156,K163	61.70 and 76.30; S10.1
T. Kiyosawa	(a) 111, K21	64.70 and 64.75; S8.15
R. Klabes	(a) 113,K171	61.80; 66.30; 79.20; S8
J. Klamut	(a) 116, 367	74.70; S10.15
M. Klaua	(a) 116, 47	68.55; 68.48; S1.3
Yu.I. Klechin	(a) 116, 627	62.20; 61.40; S1.1
W. Klein	(a) 114, 419	73.40 and 85; 78.60; S7.11; S7.15
P. Kleinert	(b) 155, K99	71.38
	(b) 156, K41	71.38
C.A. Kleint	(a) 115, 165	64.70 and 77.80; 64.80; S10.15
J. Klejst	(a) 113, K37	62.65 and 66.70; S7.12
M. Klenner	(b) 151, 503	63.20; S5.11
Z. Kleszczewski	(b) 153, 573	62.65 and 78.70; 62.60
S.V. Kletskii	(a) 115, K31	68.55; S8.15
P.V. Klevtsov	(a) 114, 713	78.45 and 78.55; 71.55; 78.50; S10
J. Kliava	(a) 114, 305	76.30; S10.15
H. Kliem	(b) 155, 309	77.40
A. Klimakow	(a) 115,K135	61.70 and 66.30; S8.15
S.O. Klimonskii	(a) 113,K123	75.50; S8
A.E. Klimov	(a) 114, 161	73.60; 68.55; S8.15
A.A. Klochikhin	(b) 151, 319	78.30
M. Klopov	(b) 151,K119	63.20; S11.1
H.A. Klose	(a) 112, 395	61.80 and 68.55; 73.40; S10.15
M. Kłosek	(a) 111, 127	61.40 and 64.70; 62.40; S1.1; S1.2
G. Kluge	(a) 113, K55	68.55; 78.65; S8
	(a) 113,K171	61.80; 66.30; 79.20; S8
Th. Klupsch	(a) 113, 561	75.60
	(a) 114, 327	75.60; S11.2
A.M. Klushin	(a) 115,K205	74.70; S10.15
N.I. Klyui	(a) 112, 805	61.70; 78.20; S5.11
	(b) 155, 723	78.20; 71.25; S5.11
A.A. Klyukanov	(b) 155, 295	71.45 and 78.55; S8.12
K. Kneipp	(a) 114,K255	78.30; S1
U. Knoop	(b) 151, 693	75.25 and 75.50
	(b) 152, 303	75.25 and 75.50
P. Knothe	(a) 112, 533	68.48 and 68.55; S1.1; S7
	(a) 113, 321	61.70; 72.20; S5.11
T. Ko	(b) 154, 85	64.70; 61.10; 62.20; 72.15; S1.2
T. Kobayashi	(a) 116, 513	61.50 and 61.60; S9
T. Kobayasi	(b) 155, 191	71.10 and 78.70; S5.11; S5.12

T. Kobayasi	(b) 156, 497	71.20 and 71.60; S5.11; S7.12
V.F. Kobelev	(a) 113, K41	64.75; S7.11
A.P. Kobzev	(a) 112, 385	61.80; 66.30; S5.11
A.N. Kocharian	(b) 156, 633	64.75 and 75.30; S1.4
J. Kodeš	(a) 116, K85	72.20; S5.11
G.W. Koebrugge	(a) 112, 35	61.40 and 64.70; 62.20; 72.15; S1.1; S1.2
R. Koegler	(a) 113, 321	61.70; 72.20; S5.11
	(a) 113,K185	61.70; 72.80; S5.11
E. Koenigsberger	(b) 151, 101	61.50 and 64.75; 65.50; S9.11
N. Koguchi	(a) 111, K21	64.70 and 64.75; S8.15
I.P. Kogutyuk	(b) 154, 513	66.30; S5.12
M. Kohyama	(b) 152, 533	61.50 and 71.45
Z.G. Koinov	(b) 155, 501	71.36
	(b) 155, 513	71.36
T. Kojima	(a) 111, 233	72.60; 72.20; S1.61
K. Kokko	(b) 151, 139	71.10; S2
	(b) 153, 235	71.20; 65.40; 76.60; S1.3
	(b) 156,K119	71.20
V.B. Kokshenev	(b) 156, 119	63.10 and 65.40; 65.70; S5.2
E. Kolbig	(a) 114, 579	73.20 and 73.40; S5.11; S10
M. Kolenda	(a) 112, K57	61.50 and 68.55; S1.61
J. Kollár	(b) 151,K101	61.50
P. Kollár	(a) 111, 603	75.50; S1.1
N.P. Kolmakova	(a) 115, K87	65.70 and 75.80; S11.2
A.V. Kolobov	(a) 115,K139	66.30; S8
A.B. Kolpachev	(b) 156, 309	71.20; S1.61
F.F. Komarov	(a) 112, 273	61.80; S5.11
	(a) 112, 323	61.80; S5.11
	(a) 112, 703	61.80 and 64.75; S5.11
	(a) 115, 427	61.80; 61.70; 78.60; S5; S7.1
L.I. Komarov	(b) 155, 169	71.38
R.V. Konakova	(a) 113, 215	73.40 and 85; S5.11
A.V. Kondrachuk	(a) 112, 411	61.80; 68.20; S5.11
E.A. Kondrashkina	(a) 114, K7	61.10; 68.20; S5.12
Q.P. Kong	(a) 113, 345	62.20; 61.70
S.G. Konnikov	(a) 113, 431	68.55; 61.10; S7.15
L.A. Konopko	(b) 155, K45	74.70; S10.15
G.G. Konoplyanik	(a) 115, 427	61.80; 61.70; 78.60; S5; S7.1
V.V. Konotop	(a) 112, 579	72.20; 61.70
P. Konsin	(b) 156, K37	62.50 and 63.75; 63.70; 77.80; S8
S.A. Konstantinov	(a) 112, 335	61.80 and 68.20; S1
A.V. Konstantinovich	(b) 154,K151	72.80; S8
Ch.V. Kopetskii	(a) 114, 191	64.80 and 68.55; 79.20; S1.61
S. Kopta	(a) 113, 295	61.80; S5.11
T. Kopte	(a) 114, K67	66.70 and 74.70; S10.15
	(b) 156, K97	66.70; S10.15
V.G. Koptev	(a) 112, 197	78.20 and 78.45; 78.55; S10
S.L. Korableva	(b) 152, 191	71.70; 76.30; 78.50; S9.16
D.N. Korbutyak	(a) 114, 635	66.30 and 73.40; 78.65; S7.12
S.Z. Korczak	(b) 153, 361	76.30; S8.15
W. Korczak	(b) 153, 361	76.30; S8.15
N.N. Koren	(a) 112,K111	72.40; S8
N.S. Korets	(a) 115, 555	78.60; 71.55; S7
	(a) 115, 561	78.60; S7
Yu.N Korkishko	(a) 114, 457	61.70; 78.20; S11.1
K. Kornarzyński	(a) 112, 815	79.70
W. Korneta	(b) 154, K1	61.50

L.S. Kornienko	(b) 156, 663	78.55; 71.55; S10.1
V.M. Korol	(a) 113, 277	61.70; 71.55; S5.11
Yu.M. Korolev	(a) 116, K13	61.50; 74.70; S10.15
V.F. Korotov	(a) 112, 331	61.80 and 73.40; S7.12
F.P. Korshunov	(a) 113,K235	61.80 and 74.70; S10.15
M.M. Korsukova	(a) 114, 265	61.60 and 72.80; 75.50; S1.4
S.V. Kortov	(a) 114, 315	65.70 and 75.40; S1.61
E.R. Korutcheva	(b) 155,K119	74.20; S10.15
N.A. Korynevskii	(b) 153, 583	64.60; 77.80
	(b) 154, 519	64.60; 77.80
G.F. Korznikova	(a) 112, 137	75.50; 61.40; S1.1
A.E. Koshelev	(b) 155, K45	74.70; S10.15
Yu.Yu. Kostetskii	(b) 152, 191	71.70; 76.30; 78.50; S9.16
H. Kostial	(b) 156, 461	73.40; S7.12; S7.15
P. Koštial	(a) 116, K77	68.25 and 71.55; 72.50; S5.11; S10
B. Kosturek	(a) 114,K111	78.20; S12
V.Ya. Kostyuchenko	(b) 152, K59	78.20; 71.55; S7.13
H.D. Koswig	(a) 116, 545	61.30; 64.70; S12
L.A. Kosykh	(a) 113,K111	75.50; S12.1
Kh.V. Kotandjian	(b) 151, 655	76.80
D. Kothari	(a) 112, 131	72.20; S11.2
J. Kotlinski	(a) 112, 415	61.80; 71.55; S5.11
H. Kotoku	(b) 153, K17	62.20; 63.20; S7.1
E.V. Kotov	(a) 112, 323	61.80; S5.11
S.S. Kotsur	(b) 155, 697	78.20
X.C. Kou	(a) 112, 175	75.50; S1.1; S1.2
	(a) 115,K219	75.50; 75.30; S1.1
S.D. Kouimtzi	(b) 153, 675	61.80 and 71.55; S5.11; S5.12; S7.1
C.D. Kourkoutas	(a) 114,K215	72.40; 78.50; S7.11
G.A. Kourouklis	(b) 152, 89	64.60; 63.20; 78.30; S9.12
G.S. Kousik	(b) 154, 713	72.10 and 72.70; S5.11
J. Kováč	(a) 114, 679	61.80 and 75.30; 75.40; 75.60; S1.1; S1.2
	(a) 115, K83	75.80; 61.80; S1.1
D. Kovacheva	(a) 116, 213	61.60 and 61.70; S8.16
L. Kovács	(a) 113, K75	61.70 and 72.40; 78.50; S11.1
V.V. Kovalchuk	(b) 156,K101	68.45; S5.11
A.F. Kovalenko	(b) 155, 549	71.55 and 73.40
Z.D. Kovalyuk	(a) 112, K61	61.70; 72.80; 78.50; S8
	(b) 153, 667	71.35 and 71.55; 72.40; 78.55; S8
	(a) 114, 293	72.80; 71.55; S8.16
	(b) 154,K151	72.80; S8
	(a) 115, K35	73.40 and 78.60; S8; S10.1
	(b) 155, 717	78.20; S8.12
V.A. Kovarskii	(b) 151, K35	74.20
	(b) 155, 535	71.55; 63.20
	(b) 156,K157	64.70 and 78.50
S.V. Koveshnikov	(a) 111, 81	61.80 and 71.55; 73.40; S5.11
N.M. Kovtun	(b) 156, 597	75.30; 76.60; S11.2
A. Kowalczyk	(a) 114, 355	75.30; 61.55; S1.1; S1.2
A. Kozanecki	(a) 112, 777	61.80; 78.55; S7.11; S7.15
V.V. Kozel	(a) 112, 437	79.20; 75.50; S1.2
E.I. Kozhachenko	(a) 113,K265	78.55; S8.16
Z. Kozioł	(a) 113, 151	74.70; S10
A.B. Kozirev	(a) 114, K21	79.20; S1.2
A.I. Kozlov	(b) 154, 377	71.25 and 78.40; 71.35; S7

A.V. Kozlov	(b) 155, 215	71.25; 78.20; S8
G.V. Kozlov	(b) 153, 529	64.60 and 78.30; S8.16
V. Kozlovskii	(a) 111,K215	61.80 and 72.80; S7.12
I.D. Kozmik	(a) 112, K61	61.70; 72.80; 78.50; S8
W. Kraak	(b) 152, 481	73.40; S7.13
	(b) 153, K63	74.70; S10.15
	(a) 116, 793	72.20 and 74.20; S10.15
W.D. Kraeft	(b) 153,K109	65.50
E. Kraetzig	(a) 113, K75	61.70 and 72.40; 78.50; S11.1
	(a) 114, 765	78.20; 61.70; S10.15
	(a) 116,K195	77.20; 77.80; S11.1
T. Krajewski	(a) 111, 271	76.30; 61.80; 66.70; S11
A. Krapf	(b) 152, K5	72.20 and 74.70; S10.15
A.B. Krasilnikov	(a) 113, 421	68.55; 61.16; S5.11
I. Kraus	(a) 115, K13	62.20; 68.20; S1.1
E. Krause	(a) 113, K79	72.40; S8.11; S8.12
	(a) 116,K219	72.40 and 85; S8.11; S8.12
M. Krause	(b) 155, 317	78.40; 71.25; 71.35; S8.15
U. Krause	(b) 151, 479	61.70
E. Krautz	(b) 153, 307	74.70; 75.30; 78.30; S10.15
I.N. Kravtsova	(b) 151, 159	71.55; 76.30; S5.11; S5.12
K. Kreher	(a) 111, 541	73.40 and 85; 78.55; S7.12
J. Kreissl	(b) 155, 597	71.55; 76.30; S7.11
	(a) 116, 755	71.55
A. Krell	(a) 111, 109	61.70 and 62.20; S10.15
J. Krempaský	(b) 153,K147	72.20; S7.13
H. Kresse	(a) 111,K265	61.25 and 77.20; S12
J. Krier	(a) 116, 681	62.20 and 68.25; S1.2
R.M. Krishna	(b) 151, 615	71.10 and 76.30; S11
B.S. Krishnamurthy	(b) 156, 253	72.10 and 72.40; S7.12
N. Kristianpoller	(a) 113, 631	78.60; S11
N. Kristoffel	(b) 151,K119	63.20; S11.1
V.G. Krivenko	(b) 155, K63	76.60; S10.1
O.P. Krivoruchko	(a) 113, 359	64.75; 76.80; S10
A.S. Krochuk	(b) 154, K9	61.70; 78.50; S9.16
E. Kroener	(b) 151, 513	62.20
M. Kroetzsch	(a) 116, 537	61.50; 74.70; S10.15
M. Krohn	(a) 116, 23	68.20; 61.70; 62.20; S9.11
F. Królicki	(a) 112,K135	78.40; S7
H. Kronmueller	(a) 111, 89	66.30; 72.15; S1.2
	(a) 111, 243	75.50; S1.1; S1.2
	(a) 111, 611	75.60; 75.50; S1.1; S1.2
	(a) 112, 181	75.60; 75.50; S1.2
	(a) 112, 189	75.80; 75.50
	(b) 153, 367	75.60; 75.70
	(a) 113, 169	75.60; 75.50; S1.2
	(a) 113, 181	75.80; S1.1; S1.2
	(a) 113, 383	61.40 and 64.60; 75.60; S1.2
	(a) 114, 685	75.60; S1.1; S1.4
	(a) 115, 547	75.60; 75.25; S1.1; S1.2; S1.61; S1.62
	(a) 116, 803	75.60; S1.1
K. Krop	(a) 114, 361	75.30 and 75.80; S1.1; S1.2; S1.4
A. Krost	(b) 153, 391	78.55; 71.70; S8.15
S.P. Kruchinin	(b) 156, 333	71.45 and 74.20
E. Krueger	(b) 156, 345	74.20
A.E. Krumins	(a) 113, 637	78.20; 64.60; 77.80; S10
M.Ya. Krupotkin	(a) 113, 375	62.20 and 64.60; 75.30; S1.4

M. Krupski	(a) 116, 657	64.60; S9.16
K.R. Krylov	(a) 115, 267	74.70; 72.20; S10.15
J. Krynicki	(a) 112, 301	61.80 and 79.20; S5.11
	(a) 112, 405	61.70; S5.11
Ts.A. Kryskov	(a) 115, 555	78.60; 71.55; S7
A.Yu. Kryukov	(a) 115, K47	72.80; S12
J. Kuběna	(b) 151, 23	61.10 and 61.70; S5.11
	(b) 155, 339	61.10 and 61.70; S5.11
V. Kubertavičius	(b) 151, 721	71.55 and 78.50; S8.12; S10.1
N. Kubicki	(b) 153, K63	74.70; S10.15
T. Kubozoe	(a) 111, K201	71.55 and 79.75; S9.11
W. Kucharczyk	(b) 155, K157	78.20; S11.1
A.E. Kuchma	(b) 155, 525	71.25; 79.60; S9.11
W. Kuczyński	(a) 112, 617	77.80 and 78.20; 61.30; S12
R.V. Kudryavtseva	(a) 116, 697	68.55 and 73.60; S5.11
R. Kuduk	(a) 112, 799	61.70; 78.60; 79.20
G. Kuehn	(b) 152, K47	71.20; 79.60; S8.16
G. Kuehnel	(a) 112, K35	72.80; S7.12
R. Kuentzler	(a) 116, 399	75.40; 65.40; 71.20; S1.2
P. Kuivalainen	(b) 154, 305	72.20; 72.60; 76.30; S12
V.N. Kukin	(a) 113, 251	61.16 and 61.70
E. Kukleva	(a) 116, 545	61.30; 64.70; S12
A.B. Kuklov	(b) 155, 407	62.40
A.I. Kulak	(a) 111, 193	68.45 and 72.40; S10.1
V.A. Kulbachinskii	(b) 151, 185	72.60 and 74.90; S5
B.S. Kulbuzhev	(b) 153, 517	64.60 and 78.30; S8.16
	(b) 153, 529	64.60 and 78.30; S8.16
J. Kulda	(b) 153, 435	61.12
	(b) 156, 41	61.12
M. Kulik	(a) 112, 785	61.80; 78.20; S7.12
	(a) 112, 789	61.80 and 68.55; 78.20; S7.12
	(a) 112, 793	61.80; 78.20; S7.12
	(a) 112, 799	61.70; 78.60; 79.20
V.S. Kulikauskas	(a) 113, 285	61.80; 72.20; 79.20; S8.15
N.I. Kulikŏv	(b) 152, 685	78.20; S1
K. Kult	(a) 112, 335	61.80 and 68.20; S1
O.A. Kulyasova	(a) 115, K23	68.55 and 79.20; S5.11; S10.1
L.L. Kulyuk	(a) 113, K227	71.55 and 72.40; 76.30; 78.55; S7.11
A. Kumar	(a) 115, 307	77.40; S6
	(a) 115, K105	61.40 and 78.70; S4
D. Kumar	(a) 116, 425	73.20 and 85; 61.70; 72.40
Devendra Kumar	(a) 116, K81	61.50 and 72.20; S10.15
M. Kumar	(b) 156, 131	62.20; S9.15
	(b) 156, 171	62.20 and 64.60; S8
R. Navil Kumar	(a) 112, K113	72.80 and 77.20; 64.70; S11.1
Ravi Kumar	(a) 115, K125	68.55 and 85; 73.40; S5.11
S. Kumar	(b) 151, 225	72.10 and 72.15
V. Kumar	(a) 114, 659	72.40; 71.55; S7.12
	(b) 155, 209	71.45; S9; S10
V.S. Kishan Kumar	(b) 155, 679	61.70 and 78.50; 78.55; S11.1
Vikram Kumar	(a) 113, K261	78.55; 61.70; 71.55; S5.11
K. Kumazaki	(b) 151, 353	78.30; S8.15
	(b) 153, 751	78.30; S8.15
	(b) 156, 371	71.25 and 78.20; 77.20; S8.15
M. Kumeda	(b) 156, K163	61.70 and 76.30; S10.1
J. Kunsch	(a) 113, K79	72.40; S8.11; S8.12
	(a) 116, K219	72.40 and 85; S8.11; S8.12
A.B. Kunz	(b) 156, 545	68.45; S10.1

M.H. Kuok	(a) 114,K223	76.80; S11
G.G. Kurbanova	(b) 155, 113	63.20 and 68.30; 78.30; S8
A.V. Kurdyumov	(a) 116, K1	61.50 and 64.70; S1.62
J. Kuriata	(b) 152, K9	74.70 and 76.30; S10.15
I.N. Kurkin	(b) 151, K53	76.30 and 77.80; S11.1
	(b) 152, 191	71.70; 76.30; 78.50; S9.16
L.A. Kurochkin	(a) 111, K89	74.70; S10.15
	(a) 111,K209	74.70; S10.15
	(a) 115,K205	74.70; S10.15
N.D. Kushch	(a) 116,K129	78.20; 74.70; S12
S.S. Kushwaha	(b) 154, 553	63.20; 78.30; S7.1; S8.1
J.P. Kuska	(b) 151, 479	61.70
S.B. Kustov	(a) 116, 645	62.20 and 62.40
V.A. Kutasov	(b) 154, 669	72.10; S8
V.V. Kutasov	(a) 113, 143	71.55; 72.20; 76.70; S5
V.L. Kutuzov	(a) 112, 361	62.20; 61.80; S1.1
Yu.B. Kuzma	(a) 114, 265	61.60 and 72.80; 75.50; S1.4
J. Kuzmík	(a) 115,K121	78.65 and 85; S7.15
A.V. Kuznetsov	(b) 153, 271	72.10; 72.20
P.I. Kuznetsov	(b) 156, 449	71.35 and 78.65; S8.12
V.D. Kuznetsov	(a) 113,K123	75.50; S8
V.I. Kuznetsov	(a) 113, 359	64.75; 76.80; S10
V.N. Kuznetsov	(a) 113, 285	61.80; 72.20; 79.20; S8.15
T. Kuzuba	(b) 155, K13	62.20; 63.20; 78.30; S1.63
	(b) 156,K167	62.20; 63.20; 78.30; S1.63
J. Kvasnica	(b) 151, 39	66.30
J. Laakso	(b) 154, 305	72.20; 72.60; 76.30; S12
E. Tremolet de Lacheisserie	(a) 115,K233	75.50 and 75.80; S1
H.K. Lachowicz	(a) 113, 181	75.80; S1.1; S1.2
R. Lacroix	(b) 154,K143	71.70; 62.20; S10.1
F. Lagnel	(b) 151, 531	73.60; S10.15
G. Lahajnar	(b) 151, K1	66.30; 64.70; S12
W. Lai	(b) 154, 229	71.38; 73.40
R. Laiho	(b) 153,K159	75.30 and 75.50; S9.16
	(a) 115, 575	77.60; 62.20; S11.1
	(a) 116, 729	68.20 and 68.55; S12
V.D. Lakhno	(b) 152, 79	63.20; 71.38; S9.11
S.V.J. Lakshman	(b) 151, 599	71.55 and 78.50; S11
	(b) 151, 615	71.10 and 76.30; S11
	(b) 153, 257	71.70; 76.30; 78.50; S11
M.V.S. Lakshmi	(a) 111, 667	73.40 and 85; S5.11
W.R.L. Lambrecht	(b) 151, 565	71.10; S5.11; S7.1; S8.1
C. Landron	(a) 113, 35	61.40; 64.75; S10.15
J. van Landuyt	(a) 112, 451	61.16 and 64.60; S11.1
H. Lange	(b) 151, 375	78.65; S1.61
	(a) 115, 497	68.55 and 79.20; S1.2
	(a) 115,K243	78.20; S8.15
J. Langner	(a) 112, 757	61.80 and 68.20
V.B. Lapin	(b) 154, K43	71.20 and 71.25; S1.2
S. Lapinskas	(b) 156, 93	64.60 and 74.20
M.A. Lasheen	(a) 111,K259	78.20; S12
G.V. Lashkarev	(b) 151, 633	72.10; S8.15
L.N. Lashkevich	(a) 114, 635	66.30 and 73.40; 78.65; S7.12
A. Latuszynski	(a) 112, 815	79.70
A.V. Latyshev	(a) 113, 421	68.55; 61.16; S5.11
P. Lavallard	(b) 153, 641	71.35 and 78.55; S8.15

A.A. Lavrentev	(b) 155, 185	71.38; S12.1
A.V. Lazarenko	(a) 115, 179	61.80 and 68.20; S1.61; S1.62
Than Binh Le	(a) 116, 357	71.55; S7.15
A.B. Lebedev	(a) 116, 645	62.20 and 62.40
N.Yu. Lebedev	(b) 153, K97	61.80; S1.3
J. Lecomte-Beckers	(a) 116, 521	61.55; 61.14; 64.70; S1.2
J. Ledzion	(a) 114, 545	72.15 and 73.60; S4
Sanboh Lee	(a) 113, 365	61.70 and 62.20
Yinggang Lee	(b) 153, 501	63.10; 63.20
A. Lehmann	(b) 153, K85	78.30 and 78.65
G. Lehmann	(b) 151, 463	61.70; 76.30; S10.1
	(b) 155, K59	76.30; 71.55; S9.16
J. Lehmann	(b) 153, 721	76.60; 71.50; S1; S1.3; S4
L. Lehmann	(a) 111, 541	73.40 and 85; 78.55; S7.12
P. Lehuede	(a) 113, 35	61.40; 64.75; S10.15
L. Lejček	(a) 115, 445	61.70 and 78.20; S12
J. Lekki	(a) 113, 295	61.80; S5.11
M. Lemaitre-Blaise	(a) 115, 523	71.70; 78.50; 78.55; S10.15
F. Lemke	(b) 151,K151	72.10; S8
W. Lengauer	(a) 112, K81	63.20; S1.62
R. Lenhard	(a) 114, 475	61.70 and 71.55; 73.40; S5.11
R. Lenk	(b) 154, 733	72.10
	(b) 155, 627	72.10
K. Lenkeit	(b) 153, K97	61.80; S1.3
	(a) 115, 413	68.48; 61.80; S7.12; S7.15
Yu.N. Leonovich	(a) 111,K209	74.70; S10.15
P.I. Leontev	(a) 113,K117	75.40; S4
G.V. Leplyanin	(a) 113,K111	75.50; S12.1
C. Leroux	(a) 116, 529	61.55; 61.12; 76.80; S1.1
A.K. Letko	(a) 111,K209	74.70; S10.15
Th. Leuthner	(a) 116, 113	61.14
C. Levade	(a) 112, 89	62.20; 61.70; S8.11
M. Levalois	(a) 111, 211	71.20; S7.16
	(a) 113, 489	71.25; S7.16
Th. Leventouri	(b) 152, K9	74.70 and 76.30; S10.15
R.Z. Levitin	(a) 115, K87	65.70 and 75.80; S11.2
R. Levy	(b) 151, 747	71.35 and 78.40; S9
	(b) 153, 403	78.20; S9
B.M.S. Lewerentz	(a) 114,K147	61.80 and 73.30; S5.11
C. Lexcellent	(a) 116, 689	62.20; 66.30; S1.2
B.R. Li	(a) 114, 273	74.70 and 78.70; S10.15
	(a) 115, K67	64.70 and 72.80; S10
D.Y. Li	(b) 154, 85	64.70; 61.10; 62.20; 72.15; S1.2
H.-D. Li	(a) 113, 11	61.80 and 64.60; 61.55; S1.62
H.D. Li	(a) 116, K51	64.70; S1.1
J.Y. Li	(a) 115,K219	75.50; 75.30; S1.1
Kaiyang Li	(a) 116, 203	61.40; 75.60; 76.80; S1.1
Qiguang Li	(b) 152, K33	62.50 and 78.40; S8.15
S.Q. Li	(a) 114, 273	74.70 and 78.70; S10.15
X.H. Li	(a) 114, 273	74.70 and 78.70; S10.15
X.Z. Li	(a) 115, K91	76.80; S1.1
Xiaoyun Li	(a) 111,K123	78.70; S11.1
Y. Li	(a) 113, 345	62.20; 61.70
Yuan Li	(b) 153, 215	71.10; 71.25; S8.1
Z.Y. Li	(b) 156, 641	75.30 and 75.70
Jingkui Liang	(a) 114,K127	78.45; 78.30; 78.55; S11.1
	(a) 114, 651	61.60 and 74.70; S10.15
Kejun Liao	(a) 113,K223	72.40 and 73.60; S5.11; S6

Li-Ji Liao	(b) 156, 145	63.20 and 78.30; 71.36
A.V. Liapzev	(b) 156, K57	74.20
L.V. Liashok	(a) 115, 207	68.55; 73.60; S10
M. Libra	(a) 116, K65	73.60; S10.1
H. Lichte	(a) 116, 113	61.14
O. Lichtenberger	(a) 116, 47	68.55; 68.48; S1.3
C.-M. Liegener	(b) 156, 441	79.20; S9
J. Liese	(a) 114, 207	68.55; S8
G. Liidja	(b) 151, 29	61.70; S9.11
	(b) 152, 563	71.55 and 78.55; S9.11
Yu.V. Lilenko	(a) 113, 285	61.80; 72.20; 79.20; S8.15
C.-T. Lin	(b) 155, 77	63.20; 78.30; S7.13; S8.11; S9
M.X. Lin	(a) 115, 539	75.60; 76.80; S10
Mingxi Lin	(a) 113,K127	75.30; S1.1; S1.4
Y. Lin	(b) 155,K109	71.20; 76.60; S5.2
P.J. Lin-Chung	(b) 153, 215	71.10; 71.25; S8.1
U. Lindner	(b) 155, 263	75.10; 75.30
J.J. Liou	(a) 112, 651	73.40 and 85; S7.12
	(a) 113,K267	73.40 and 85
E. Lipinska	(a) 112, 347	61.80 and 68.20; S1.1
V.I. Lipnitskii	(a) 115,K113	72.80 and 78.60; S8.16
A. Lipowski	(b) 153, 315	75.30; S9
B. Lippold	(a) 115, 257	71.55; 72.80; 78.55; S7.15
	(a) 116, 537	61.50; 74.70; S10.15
N.M. Litovchenko	(a) 112, K65	61.70; 71.55; S5.11
V.G. Litovchenko	(a) 112, 805	61.70; 78.20; S5.11
	(b) 155, 723	78.20; 71.25; S5.11
E.A. Little	(a) 112, 55	61.80; 75.50; S1.1
S.A. Litvinenko	(a) 116, 615	61.70 and 64.75; S10.1
L.A. Litvinov	(a) 116,K125	61.80 and 78.40; 78.55; S11.2
V.L. Litvinov	(b) 154, 507	61.80
V.V. Litvinov	(a) 115, K9	61.70 and 71.55; S5.12
B.X. Liu	(a) 113, 11	61.80 and 64.60; 61.55; S1.62
	(a) 116, K51	64.70; S1.1
C.-L. Liu	(a) 114, 451	61.70; 78.70; S1.2
Gao-Chao Liu	(b) 153,K167	71.70 and 76.30; S10.1
Hongjian Liu	(a) 116, 267	64.70 and 74.70; 61.10; S10.15
Jie Liu	(b) 153,K131	71.45; 71.38; S12.1
Ke-Qin Liu	(b) 153,K167	71.70 and 76.30; S10.1
Shengyan Liu	(a) 116, 621	62.40; 61.70; S1.4; S3
Tongming Liu	(a) 111, K67	72.80; 72.40; 78.40; S12.1
Wei Liu	(a) 114, 177	68.25; S4
Zhongming Liu	(a) 116,K199	62.20 and 77.80; S12
J. Liwo	(a) 115,K185	73.25; S10.15
K. Loebe	(a) 111,K129	78.60; S8.11
H. Loeffler	(a) 111, 417	61.55; 62.20; 64.70; S4
	(a) 113, 269	61.55 and 64.75; S1.2
B. Loegel	(a) 116, 777	74.70; 75.30; S10.15
F.J. Lohmeier	(b) 151, 347	78.20; S11
	(b) 154, 789	78.50; 78.55; S9.12
U. Lohse	(a) 115,K171	73.40; 85; S7.12
C.D. Lokhande	(a) 111, 17	68.55; 73.40; 85; S8
N.R. Lokhande	(b) 153, K35	71.25 and 78.70; S8; S10.1
A.A. Lomov	(a) 115,K133	61.10; 68.20; S5.11
A.T. Lonchakov	(b) 154, 703	71.30 and 72.20; S5.12
C.A. Londos	(a) 113, 503	71.55; S5.11
Xiang Long	(a) 116, 267	64.70 and 74.70; 61.10; S10.15

V.V. Lopatin	(a) 116, 221	61.40; 72.80; 77.20; S1.62
	(a) 116, K69	61.70 and 68.20; S6; S7
F.J. López	(b) 151, 331	78.55. and 78.60; S9.16
	(b) 153,K199	78.50; S9.16
J. López Gondar	(b) 151, 127	71.25; S7.12; S7.15
F. López-Aguilar	(b) 152, 543	71.20 and 71.25; 71.50; S1.4
A. López-Echarri	(b) 154, 143	64.60; 65.40; S9.16
M.A. López-Quintela	(a) 111, 611	75.60; 75.50; S1.1; S1.2
E.R. Lopez-Tellez	(b) 154,K171	71.55 and 78.50; S9.11
P. Lošták	(a) 113, 615	78.30; S8.15
	(a) 115, 87	61.70; 72.20; S8
V.B. Lototskii	(a) 111, K99	72.80 and 75.50; S8
K.A.K. Lott	(a) 113, 483	68.55 and 76.30; S10.15
	(a) 114, 225	68.55 and 76.30; S10.15
P.P. Lottici	(a) 111, 411	61.60; 64.60; S8.15
	(b) 152, 39	61.60 and 78.70; S8.16
	(b) 154, 461	61.60; 78.70; S8.16
	(b) 154,K117	63.20; 78.30; S8.16
A. Lovas	(a) 114, 679	61.80 and 75.30; 75.40; 75.60; S1.1; S1.2
	(a) 115, K83	75.80; 61.80; S1.1
V.Z. Lozovskii	(b) 151, 111	73.60 and 78.65
	(b) 153, 167	78.65
Ke-Qin Lu	(b) 152, 577	71.70 and 76.30; S9.16
	(b) 153, K45	71.55 and 78.50; S11.1
	(b) 154, K67	76.30; S9; S9.12
Mu Lu	(a) 114,K189	74.70; S10.15
Tianchang Lu	(a) 111,K123	78.70; S11.1
Wei Lu	(b) 152, K33	62.50 and 78.40; S8.15
V.S. Lubimov	(b) 155,K137	74.20
W. Ludwig	(b) 151, 503	63.20; S5.11
Juh-Tzeng Lue	(b) 152, 335	78.20; S1.61
P.F. Lugakov	(a) 113, 307	61.70 and 72.20; 61.80; S5.11
P. Lukáš	(b) 153, 435	61.12
	(b) 156, 41	61.12
I. Lukeš	(a) 111, 655	78.30; 73.40; S7.12
K. Lukierska-Walasek	(b) 153, 561	63.75 and 64.60; 75.50
A.I. Lukomskii	(a) 111,K109	78.40 and 78.60; S8.16
	(a) 115,K113	72.80 and 78.60; S8.16
	(a) 115,K237	78.60; S7.1
L.N. Lukyanova	(b) 154, 669	72.10; S8
V.K. Lukyanyuk	(a) 114, 293	72.80; 71.55; S8.16
	(b) 155, 717	78.20; S8.12
J. Lunarški	(a) 112, 365	62.20; 61.80; S1.1
H.L. Luo	(a) 115, 539	75.60; 76.80; S10
Horng-Jye Luo	(b) 154, 713	72.10 and 72.70; S5.11
Le Luo	(a) 116, K7	61.40 and 78.70; 74.70; S10.15
Zundu Luo	(a) 114,K127	78.45; 78.30; 78.55; S11.1
Nguyen Hoang Luong	(a) 111, 591	75.30; 75.25; S1.1; S1.4
D. Lupu	(b) 155, 65	66.30; 71.38; 78.20; S1.2; S1.4; S3
V.S. Lutovinov	(b) 155, K63	76.60; S10.1
F. Luty	(b) 154, 167	64.60; S2
	(b) 154, 181	64.60; S9
W. Łużny	(a) 116, K25	61.12; S4
I.I. Lyapilin	(b) 153, 263	72.20
	(b) 156, 235	71.10 and 73.40; 72.20; S8.15
V.I. Lygin	(b) 156, 205	68.20; S10.1

V.S. Lysenko	(a) 112, K9	61.70; 71.55; 73.40; S5.11
	(a) 113, 653	73.40 and 85; 76.30; S5.11; S10.1
	(a) 115, 75	61.80; 73.40; S5.11
V.M. Lyubin	(a) 115,K139	66.30; S8
R.N. Lyubovskaya	(b) 154, 333	64.70 and 76.30; 72.20; S12.1
A.S. Lyutovich	(a) 114,K163	68.55; 61.80; S5.11
D. Machajdík	(a) 112, 781	61.70 and 61.80; S5.11
	(a) 114, 567	68.55 and 74.70; S10.15
M.N. Machavariani	(b) 154, 97	65.50
M. Machmer	(a) 113, K27	64.60; 66.30
W. Maciejewski	(b) 151,K191	75.70; 76.50
J. Maćkowiak	(b) 154, 115	61.20 and 64.60
M. Macucci	(b) 152, 601	72.10 and 72.20
D. Mączka	(a) 112, 695	61.70; 79.20; S5.11
	(a) 112, 799	61.70; 78.60; 79.20
	(a) 112, 815	79.70
R.S. Madatov	(a) 114, K45	72.80 and 78.40; S5.11; S5.12
R. Madelon	(a) 113, 489	71.25; S7.16
Y. Madhavan	(b) 153, 579	63.20; 61.70; S8.13
	(b) 154, 55	66.30; S8.13
K.N. Madhusoodanan	(a) 114, 525	64.70; 61.40; 65.70; 72.80; S8
N. Madi	(a) 112, 343	61.80; 61.70; S1.1
K. Madl	(a) 116, 561	66.30; 61.70; S5.11
K. Maeda	(a) 111, 223	72.20; 61.70; S8.11
	(a) 113,K245	75.50; S11.2
F. Maes	(b) 155, K55	76.30; 61.70; S9.11
L. Magafas	(a) 111,K175	73.60; S8.11
M.V. Magnitskaya	(b) 152, 685	78.20; S1
W. Magnus	(b) 153, K31	71.25
J. Maier	(a) 112, 115	73.60; S9
S.N. Maiti	(b) 156, 83	64.75; 61.10; S9.11
E. Majková	(b) 153,K147	72.20; S7.13
B. Majkusiak	(a) 116, 313	68.55; S10
	(a) 116, 833	73.40 and 85
K. Makariunas	(b) 154, 341	76.80; 61.55; S1.63
V.N. Makarov	(a) 112, 707	61.80; 64.70; S1.61
K. Maki	(a) 116, 803	75.60; S1.1
L.A. Makovetskaya	(a) 111,K109	78.40 and 78.60; S8.16
	(a) 115,K113	72.80 and 78.60; S8.16
S.A. Maksimenko	(b) 151, 17	61.10
V.M. Maksimenko	(b) 151, 159	71.55; 76.30; S5.11; S5.12
S.K. Maksimov	(a) 113, K23	61.16 and 61.70
	(a) 113, 251	61.16 and 61.70
I.D. Makuta	(a) 111, 193	68.45 and 72.40; S10.1
M.J. Małachowski	(a) 113, 467	72.40 and 73.60; S8.15
L. Malacký	(a) 115,K121	78.65 and 85; S7.15
J. Małecki	(a) 112, 617	77.80 and 78.20; 61.30; S12
J. Malek	(b) 153, 207	71.38; S12
V.G. Malesh	(b) 154,K197	61.80 and 78.30; S7.11
L. Malicskó	(a) 116, 55	68.48 and 68.55; S1.3; S9.11
O.K. Malinovskaya	(a) 111, 201	71.55 and 73.30; S8.12
I.E. Malinovskii	(b) 156, 449	71.35 and 78.65; S8.12
C. Malinowska-Adamska	(b) 151, 53	65.50; S1.2; S1.3; S2; S4
E. Maliszewski	(b) 154, 543	63.20; 62.20; S1.3

B. Mallik	(a) 111, K61	72.80; 73.25; S12
A. Mallouky	(a) 111, 181	73.60; S8
A.U. Malsagov	(b) 153, 517	64.60 and 78.30; S8.16
S.A. Malyshev	(a) 111, 451	66.30; S7.11
M.M. Malyush	(a) 115, 279	75.60; 75.70; S1.2
V.I. Malyutin	(a) 114, K75	75.70; S1.2
B.R. Mamatkulov	(a) 115,K167	73.40; S5.12
R. Mamatkulov	(a) 115,K167	73.40; S5.12
A.M. Mamedov	(b) 151, K87	78.40; 71.25; S11.1
F.I. Mamedov	(a) 111,K205	72.40; 63.20; S8.16
	(a) 113,K241	72.40; S8.16
V.S. Mamedov	(a) 114, K45	72.80 and 78.40; S5.11; S5.12
B.P. Mamko	(b) 152, K55	78.70; S1.61
A.P. Mamontov	(a) 113, 285	61.80; 72.20; 79.20; S8.15
R. Mamy	(b) 156,K109	68.55 and 79.60; S8
L.A. Manakova	(b) 155, 397	64.60; 71.25; S8.15
A. Manar	(b) 151, 747	71.35 and 78.40; S9
M.O. Manaresh	(b) 154, 11	61.70 and 71.55; 72.40; 78.30; 78.50; 78.55; S7.12
V.A. Manasson	(a) 115, K35	73.40 and 78.60; S8; S10.1
C. Mande	(b) 153, K35	71.25 and 78.70; S8; S10.1
M. Manfredi	(a) 112, 215	78.20; 71.55; S9.11
R. Mańka	(b) 155, 231	74.20; 71.38; S10
E. Mann	(b) 153, 465	61.70; 62.20; 62.40
C. Manolikas	(a) 111, 491	64.70; S1.63
	(a) 115, 399	61.14 and 64.70; S1.63
K. Mansikka	(b) 153, 235	71.20; 65.40; 76.60; S1.3
Bing Mao	(a) 116, 709	73.60; 71.55; S12
Ming Mao	(b) 155, 647	75.50; S11.2
	(a) 115, 467	61.40 and 64.70; S1.1; S1.2
P. Mao	(a) 115, K91	76.80; S1.1
R. Marasas	(b) 151, 133	71.35
I.G. Marchenko	(a) 115, K39	61.80 and 73.40; S5.11
I.Z. Marchuk	(a) 112, K61	61.70; 72.80; 78.50; S8
S. Marcinkevičius	(a) 115,K211	72.40; S7.16
M.J. Marcinkowski	(b) 152, 9	61.70
P. di Marco	(b) 153, 281	72.40 and 73.60; S12
H. Margolin	(b) 154, 61	61.70; 62.20; S4
	(b) 155, 53	61.70; 62.20; S4
D.F. Mariani	(a) 112, K53	78.60; S10.1
V.P. Markevich	(a) 111,K149	61.70; S5.11
P. Markoš	(b) 151, 571	71.38
M. Markowski	(b) 156, 105	64.10; 64.60
T.A. Marsagishvili	(b) 154, 97	65.50
J. Martan	(a) 112, 357	61.80; 68.20; S1.1
C. Marti	(b) 152, 463	64.60; 68.45; S5
A. Martín	(b) 151, 331	78.55. and 78.60; S9.16
	(b) 153,K199	78.50; S9.16
D. Martin	(b) 152, 101	73.40; 71.35; 72.40; 78.55; S7.15
P.J. Martin	(b) 152, 475	61.40 and 68.55; 78.70; S5; S5.11
Ph. Martin	(a) 114, K71	75.60; S8.15
G. Martinek	(a) 114, 685	75.60; S1.1; S1.4
J. Martins S. Filho	(b) 154,K179	75.80 and 78.20
L.F. Martynova	(a) 114, K87	76.60; S1.2
G. Marx	(a) 114, 609	68.20; 79.20; S1.2
P. Mary	(a) 114, 467	61.80; S5.11
J. Masalski	(a) 112, 357	61.80; 68.20; S1.1

A.V. Maslov	(a) 113,K153	61.10
I.A. Maslyanitsin	(a) 112, K49	78.20; S10
A. Masoero	(a) 116, 813	75.60
C.P. Massolo	(a) 116, 503	61.60; 61.70; 76.80; S10.1
C.D. Mateescu	(b) 156, 411	61.70; 66.30; 78.50; S9.11
E.L. Mathé	(a) 112, 511	61.70; S5.11
T. Mathew	(a) 116, 405	77.20 and 77.40; S11.1
M. Matlak	(b) 151, 203	74.70
	(b) 151,K187	75.50; S10.1
I. Matsubara	(a) 113,K245	75.50; S11.2
J. Matsui	(a) 115, 383	61.10; S7.12
F. de Matteis	(b) 151, 453	61.70; 78.50; 78.55; S9.11
R. Mattheis	(a) 116, 67	68.48 and 68.55; S1.61
	(a) 116, 81	68.48 and 68.55; S1.61
J.A.D. Matthew	(b) 154, K47	71.10
H. Matthies	(b) 155, K59	76.30; 71.55; S9.16
P. Matthys	(b) 155, K55	76.30; 61.70; S9.11
B.A. Matveev	(a) 115,K117	78.45 and 78.55; S7.12; S7.15
G.A. Matveev	(b) 154, 703	71.30 and 72.20; S5.12
E.E. Matyas	(a) 112,K111	72.40; S8
Z.A. Matysina	(b) 156, 419	64.60 and 65.40; S1.1; S1.2
P.W. Mauriz	(b) 151,K109	63.20 and 76.80
J. Mayer	(a) 116, K25	61.12; S4
P.S. Mazumdar	(a) 114, 407	78.60
N. Mazumder	(a) 116, 415	73.40 and 85; S5.11; S7.12
P. Mazzetti	(a) 116, 813	75.60
R.B. Mc Lellan	(a) 116,K157	65.50; S1.2; S1.4
D.R. McKenzie	(b) 152, 475	61.40 and 68.55; 78.70; S5; S5.11
M. Mebarki	(a) 115, 173	73.40; 73.20; S7.13; S7.15
N.I. Medvedeva	(b) 151, 407	71.25; S1.61; S1.62
G.A. Medvedkin	(a) 111, 289	78.20; S1.63; S8.16
	(b) 151, 711	78.20; S8.16
	(a) 115, 11	72.40; 78.20; S7.16
A. Meetsma	(a) 114, 91	61.60; S1.63
A. Mehdaoui	(a) 116, 777	74.70; 75.30; S10.15
S.K. Mehta	(a) 115, 135	64.75; 72.20; S8.15
D. Meimaris	(a) 115, 483	64.70 and 72.80; 61.40; S7.16
K. Meinel	(a) 116, 47	68.55; 68.48; S1.3
A.O. Mekhrabov	(b) 156, 419	64.60 and 65.40; S1.1; S1.2
E. Mekhtiev	(b) 151, 759	78.20; S8.16
L.A. Meleshchenko	(b) 152, 657	78.30; S11.2
M.A. Melikian	(b) 156, 633	64.75 and 75.30; S1.4
A.I. Melker	(b) 153, K5	61.70; 61.80; S1; S1.1
	(a) 116, 595	62.20
N.N. Melnik	(b) 153, 727	78.30; 77.80; S8.16
A.A. Melnikov	(a) 115, 427	61.80; 61.70; 78.60; S5; S7.1
Yu.S. Melnikova	(a) 113, 215	73.40 and 85; S5.11
A.N. Men	(a) 111,K171	64.70; S11.2
	(a) 116, 603	64.10
A. Menelle	(a) 116, 529	61.55; 61.12; 76.80; S1.1
Y. Mera	(a) 111, 223	72.20; 61.70; S8.11
T. Merceron	(a) 112, 601	75.30 and 75.50; S11.2
M. Merisalo	(a) 112, K81	63.20; S1.62
J.C. Merle	(b) 153, 403	78.20; S9
J. Mertsching	(b) 152, 551	61.65 and 71.25; S12.1
M. Metodieva	(b) 152, 163	71.25
A. Meyer	(b) 154, 469	61.50; S2
K. de Meyer	(b) 153, K31	71.25

M.M. Mezdrogina	(b)	155,K113	71.55; S5.11
V.A. Mezrin	(a)	114, 145	64.60; 61.70; S9.12; S10.1
J.L. Miane	(a)	115, 107	64.70; 61.14; 77.20; S11
V.P. Michailov	(a)	113,K257	78.45; S9.16; S11.2
R. Michalec	(a)	111,K133	61.12
	(a)	112, 29	61.12 and 62.65; S7.13
	(a)	113,K157	61.12
J. Michalski	(b)	152, K17	75.30
B. Michel	(a)	111,K137	61.10 and 68.20; S1.1
D. Michel	(a)	114, 705	76.60; 77.80; S12
R. Micnas	(b)	156,K165	75.10
G. Micocci	(a)	114, 253	71.55; S8
J. Mielnicki	(b)	154, 583	73.60
	(b)	156, 355	75.10 and 75.30
M. Miglierini	(a)	113,K143	76.80 and 78.70; S1.1
M. Mihálik	(a)	114, 679	61.80 and 75.30; 75.40; 75.60; S1.1; S1.2
	(a)	115, K83	75.80; 61.80; S1.1
E. Mihóková	(b)	154,K183	78.55; 66.90
A.I. Mikhailin	(a)	116, 595	62.20
V.V. Mikhailin	(b)	155, 215	71.25; 78.20; S8
I.P. Mikhailovskii	(a)	116,K105	76.30 and 78.55; S6
A.A. Mikhalenko	(b)	155, 359	61.70 and 71.55; S9.11
V.M. Mikheev	(b)	156, 577	72.10; 72.20; S8.15
V. Mikolaj	(a)	112,K119	72.70; 85; S5.11
P. Mikołajczak	(a)	112; 739	61.55 and 61.80; 68.48; S1.2
	(a)	112, 753	61.55 and 61.80; 68.48; S4
M. Militzer	(a)	113, 337	61.70 and 68.48
B.V. Mill	(a)	112, 197	78.20 and 78.45; 78.55; S10
M.A. Milyaev	(b)	155, 419	64.60; 75.30; S1.61
J. Mimkes	(a)	112, K5	66.30; S5.11
Ye Ming	(a)	113, K75	61.70 and 72.40; 78.50; S11.1
L.Q. Minh	(a)	115,K163	68.55 and 78.65; S12
R. Mirianashvili	(b)	151, K49	76.70; S9.12
	(b)	152, K13	76.70; S9.12
N.A. Mironova	(a)	116,K125	61.80 and 78.40; 78.55; S11.2
C.N. Mishra	(b)	155, K81	62.20 and 78.20; S9.12
S. Mishra	(a)	114,K115	77.20 and 77.40; 77.70; S11
S.K. Mishra	(b)	154, 127	63.20; S5.2
L.E. Misiak	(b)	154, 249	71.70 and 76.30; S9.16
J. Misiewicz	(a)	111,K249	78.55; 71.55; S7.1
	(a)	112,K135	78.40; S7
S.K. Misra	(b)	154, 249	71.70 and 76.30; S9.16
	(b)	154, 259	71.70; 76.30; 78.50; S12
E.M. Misyuryov	(a)	112, 699	79.20; 64.70; S5.11; S5.12
W.C. Mitchel	(b)	154, 11	61.70 and 71.55; 72.40; 78.30; 78.50; 78.55; S7.12
B. Mitra	(b)	153,K209	73.40 and 85
	(b)	154, K35	62.20; S8.11
	(b)	154,K121	73.40; S7.15
	(b)	154,K147	72.20
	(b)	155, K23	71.25; S4
G. Mitroaica	(b)	152, 51	61.70; S9.11
V.Ya. Mitrofanov	(b)	156, 597	75.30; 76.60; S11.2
A.I. Mitsek	(a)	113, 529	75.60 and 75.70; S1.1; S1.2; S11.2
	(b)	156, 507	71.45; 72.15; 75.80; S1.2; S1.4; S1.5
A. Mittas	(a)	116, 725	73.60 and 78.65; S6

J. Mittenbacher	(a) 112, 825	61.80 and 79.70; 61.16; S1.3; S4
Z. Mitura	(a) 115,K207	68.55 and 74.70; S10.15
D.B. Mizandrontsev	(b) 153, K5	61.70; 61.80; S1; S1.1
B. Mizera	(a) 112,K105	73.40; 85; S7.12
R. Mizeris	(a) 113, 597	77.20; 77.80; S11.1
H. Mizubayashi	(a) 116, 583	61.80 and 62.40; S1
	(a) 116, 633	62.40; S1.2
G. Mizutani	(b) 151, 591	71.35 and 78.20; S9
	(b) 152, 489	71.35 and 78.20; S9
Y. Mlik	(a) 115, 107	64.70; 61.14; 77.20; S11
N. Mliki	(a) 116, 669	64.70; S10
M. Mocker	(b) 151,K151	72.10; S8
P. Modrak	(b) 152, 203	71.50; S1.2
G. Moellenstedt	(a) 116, 13	61.14
K.-P. Moellmann	(a) 114,K249	72.40 and 78.30; S8.15
D.N. Mogilyanskii	(a) 115, 199	64.70; 62.20; S9.16
F.M. Mohammad	(b) 152, 145	71.25 and 78.70; S1
S. Noor Mohammad	(b) 156, 287	71.25; 71.55; S5.11; S5.12
M.A. Mohammed	(a) 115, 213	73.60; S10.1
Y.N. Mohapatra	(a) 114, 659	72.40; 71.55; S7.12
S.V. Moharil	(a) 111,K119	78.60; S12
P. Mohazzabi	(b) 153, 59	61.50
M.M. Moiseenko	(b) 151, 121	68.20 and 72.15
K. Mojejko-Kotlinska	(a) 112, 415	61.80; 71.55; S5.11
S.M. Mokhnyak	(b) 156, 377	78.20
O.M. Mokry	(a) 116, 287	64.60; 62.20; 78.20; S12
E.E. Mola	(b) 155, K93	73.30; S4
A. Molak	(b) 155, 231	74.20; 71.38; S10
S.L. Molodtsov	(b) 155, 525	71.25; 79.60; S9.11
J. Monecke	(b) 152, 123	71.10; 71.20; 71.55
	(b) 153, 623	71.10; 63.10
	(b) 154, 805	77.20; 64.80
	(b) 155, 437	68.30 and 78.30; 63.10
	(b) 155,K103	71.10; 71.55
M. Moneta	(a) 112, 773	61.80; S5.11
	(b) 154, K13	61.80
O. Monnereau	(a) 113, 23	61.60 and 74.70; 61.16; S10
	(a) 114, 73	61.60; 61.16; S10
	(a) 114, 441	61.60; 61.16; 61.70; S10
D. Monselesan	(b) 155, 349	61.30 and 62.10
E.N. Moos	(a) 111, 99	61.70; 68.25
A. Morawiec	(b) 154, 535	62.20; S1; S10
T. Morawska-Kowal	(a) 116, 367	74.70; S10.15
P. Morfouli	(a) 111, 529	73.60; 73.40; S5.11; S10.1
D.V. Morgan	(a) 115, 213	73.60; S10.1
D. Morin	(a) 113, 23	61.60 and 74.70; 61.16; S10
N. Morita	(a) 112, 207	78.30; S11.1
Yu.N. Morokov	(b) 156, 243	68.25 and 68.40; 73.20; S5.11
J.W. Moroń	(a) 111, 127	61.40 and 64.70; 62.40; S1.1; S1.2
I.O. Morozova	(b) 156, 663	78.55; 71.55; S10.1
T.V. Morozova	(a) 114, 457	61.70; 78.20; S11.1
M. Morsli	(b) 155, 609	72.20; 61.60; 71.20; S10
N. Moser	(a) 111, 243	75.50; S1.1; S1.2
	(a) 112, 181	75.60; 75.50; S1.2
	(a) 113, 181	75.80; S1.1; S1.2
V.V. Moshchalkov	(a) 114, 265	61.60 and 72.80; 75.50; S1.4
Yu. N. Moskvich	(a) 111, 477	64.70; 76.60; S11.1

Yu.N. Moskvich	(b) 156, 615	76.60; 66.30; 72.20; S9.16
V.M. Mostovoi	(a) 111,K145	61.70; S7.13
S. Motiejūnas	(b) 154, 341	76.80; 61.55; S1.63
E.V. Mozdor	(a) 113, 215	73.40 and 85; S5.11
P.E. Mozol	(b) 153, 667	71.35 and 71.55; 72.40; 78.55; S8
B. Mróz	(a) 115, 575	77.60; 62.20; S11.1
A. Mrwa	(a) 116, 729	68.20 and 68.55; S12
H. Mucha	(a) 114, K37	61.80 and 72.80; 85; S8.15
J. Mucha	(a) 112, K57	61.50 and 68.55; S1.61
R. Muehle	(a) 112, 769	61.70 and 61.80; S12
	(a) 112, 825	61.80 and 79.70; 61.16; S1.3; S4
I. Muehling	(b) 151, 545	61.40 and 71.20; S5
E. Mueller	(b) 155, K17	68.40; 64.10
H.-U. Mueller	(a) 116, 793	72.20 and 74.20; S10.15
E. Mugeński	(b) 151, 47	61.70; 76.30; 78.50; S9.11
	(a) 111,K249	78.55; 71.55; S7.1
N.N. Mukhamadeva	(a) 112, K97	66.30; S3
E.Kh. Mukhamedzhanov	(a) 113,K153	61.10
	(a) 116, 197	61.10; 78.70; S7.12
C.D. Mukherjee	(b) 154, K55	72.15 and 74.70; S10.15
R.K. Mukherjee	(b) 155, 541	71.70; 78.20; S9.16
S. Mukherjee	(b) 154,K107	62.65 and 65.70; 63.20
V.M. Mukhortov	(a) 114, 559	64.70 and 68.55; 77.80; S10.15
B.G. Mulimani	(b) 156, 253	72.10 and 72.40; S7.12
Kh.Ya. Mulyukov	(a) 112, 137	75.50; 61.40; S1.1
J.M. Muñoz	(a) 112, K43	75.60; S11.2
J. E. Muñoz Santiuste	(b) 153,K187	78.55; S9.11
S.P. Muntyan	(a) 114, K63	74.70; S10.15
F.M. Muntyanu	(a) 114, K63	74.70; S10.15
J. Murakowski	(b) 156, 105	64.10; 64.60
V.A. Murashov	(b) 155, K63	76.60; S10.1
M.N. Muraya	(a) 111,K209	74.70; S10.15
L.I. Murin	(a) 111,K149	61.70; S5.11
P.C. Murley	(b) 151,K105	61.80; S5.11
E. Murray	(a) 116, 399	75.40; 65.40; 71.20; S1.2
R. Murray	(b) 151, 469	61.70; 78.50; S7.12
Y.Satyanarayana Murthy	(a) 115,K175	73.60; S8.16
Y.V.G.S. Murti	(b) 151, 445	61.70; 71.55; 78.50; S9.11
B.S.R. Murty	(b) 154, 297	61.60 and 72.20; 75.50; S11
D.S.R. Murty	(b) 154, K85	78.70; S1.2
R.Sh. Musalimov	(a) 115, 451	61.70; S4
P.L. Muthal	(a) 111,K119	78.60; S12
G.S. Myakenkaya	(b) 153, 659	71.55; 61.70; S5.11
	(b) 156, 319	71.55; S5.11
S. Myhra	(b) 152, 519	71.45 and 79.60; 74.10; S11
R.S. Nadirzade	(b) 155,K161	78.30
V.A. Nadolinnyi	(a) 116,K105	76.30 and 78.55; S6
N. Nafari	(b) 152, 111	71.10 and 73.40
E.L. Nagaev	(b) 155, K71	71.55 and 75.30
L.C. Nagarajan	(a) 113, 511	73.60; S10
N. Nagasawa	(b) 151, 591	71.35 and 78.20; S9
	(b) 152, 489	71.35 and 78.20; S9
A.T. Nagat	(a) 114,K205	72.80; S8

V. Nagirnii	(b) 152, 563	71.55 and 78.55; S9.11
A. Nagornyi	(a) 115, 325	61.70 and 78.60; S9.12
P. Nagy	(b) 151, 571	71.38
B.Srinivasulu Naidu	(a) 115,K175	73.60; S8.16
D. Nakamura	(a) 114, 157	64.70 and 77.80; S12
	(a) 114,K143	66.30 and 76.60; S12
Y. Nakamura	(a) 114, 693	72.15 and 75.50; 61.55; S1.2
H. Nakashima	(a) 115, K43	71.55; S8
H. Nakayama	(a) 112, 207	78.30; S11.1
K. Nandakumar	(a) 114, 525	64.70; 61.40; 65.70; 72.80; S8
P.S. Nandhra	(b) 151, 469	61.70; 78.50; S7.12
S.H. Naqvi	(b) 156, 111	64.10; 64.60; S9.11
T.V. Narasaiah	(a) 113, 597	77.20; 77.80; S11.1
G.L. Narendra	(b) 153, 257	71.70; 76.30; 78.50; S11
F.S. Nasredinov	(b) 155,K113	71.55; S5.11
Y. Nasu	(b) 151, K81	78.60; S9.11
K. Natarajan	(a) 111,K269	73.40 and 85; S5.11
	(a) 115,K265	73.40 and 85; 61.70; S5.11
B.A. Natig	(b) 153, K89	63.20 and 78.30; S8.16
	(a) 114,K119	77.80; 61.80; S8.16
I. Natkaniec	(a) 114, 497	64.70; 61.12; S9
	(a) 116, K25	61.12; S4
G.A. Naumovets	(a) 112, K9	61.70; 71.55; 73.40; S5.11
K. Navaneethakrishnan	(b) 154, 643	71.55; 71.38; S5.11; S7.11
W. Nawrocik	(a) 114, 497	64.70; 61.12; S9
B.A. Nayafov	(a) 114, K45	72.80 and 78.40; S5.11; S5.12
B.B. Nayak	(b) 153, 153	71.20 and 73.60; 78.70; S8.15
A.B. Nazarenko	(b) 151,K167	74.30
I.V. Nazarenko-Baryakhtar	(b) 153, 167	78.65
H.N. Nazareno	(b) 153, 179	71.25; 78.20
A.N. Nazarov	(a) 112, K9	61.70; 71.55; 73.40; S5.11
	(a) 113, 653	73.40 and 85; 76.30; S5.11; S10.1
	(a) 115, 75	61.80; 73.40; S5.11
E. Nebauer	(a) 115,K171	73.40; 85; S7.12
D.D. Nedeoglo	(b) 154, 287	72.20; 71.30; 71.55; S8.12
	(b) 154, 691	72.20; 71.30; S8.12
C. Neelmeijer	(a) 112, 765	61.40 and 61.80; S12
	(a) 116,K147	61.80 and 78.30; 63.20; S7.11
H. Negishi	(b) 151, 441	61.60; S8.16
I.G. Neizvestnyi	(a) 114, 161	73.60; 68.55; S8.15
V. Nekvasil	(b) 153, K69	71.70 and 75.60; S1.2
V.V. Nemoshkalenko	(b) 152, K55	78.70; S1.61
A. Neogi	(b) 152, 691	78.20; 77.60; S7.13
	(b) 156, 705	78.35; S7.13
A.I. Nepomnyachikh	(b) 155, 365	61.70; 76.30; 78.40; S9.11
K.E. van Ness	(a) 116,K213	78.70; S12
A.I. Nesterov	(b) 156, 403	61.70
G. Neumann	(b) 155, 385	61.70
H. Neumann	(b) 152, K47	71.20; 79.60; S8.16
	(a) 112,K139	61.60 and 78.30; S7.15; S8.15
	(a) 115, 257	71.55; 72.80; 78.55; S7.15
	(a) 116,K119	61.70 and 78.50; S7.12
	(a) 116,K147	61.80 and 78.30; 63.20; S7.11
P. Nevermann	(a) 112, 229	73.40 and 78.60; 68.55; 78.55; S7.15
R.C. Newman	(b) 151, 469	61.70; 78.50; S7.12
Yongmin Ni	(a) 114, 651	61.60 and 74.70; S10.15

J.E. Nicholls	(b) 151, 175	71.35 and 71.55; 76.90; 78.55; S8.11; S8.15
H.Q. Nie	(b) 156, 363	75.10 and 75.30
S. Nieber	(b) 153, 367	75.60; 75.70
J. Nieder	(b) 156, K45	71.35; 72.20; 72.40; S5.12
O.V. Nielsen	(a) 111, 279	75.80; S1.1; S1.2
R. Nies	(a) 111, 639	78.20; 78.50; S5.11
A.N. Nigam	(a) 115,K105	61.40 and 78.70; S4
R.R. Nigmatullin	(b) 153, 49	61.40
G. Nihoul	(a) 114, 73	61.60; 61.16; S10
I.Ya. Nikiforov	(b) 156, 309	71.20; S1.61
K.G. Nikiforov	(a) 116,K185	76.50; S8.16
E.S. Nikitin	(a) 114, 559	64.70 and 68.55; 77.80; S10.15
S.A. Nikitin	(a) 113,K117	75.40; S4
	(a) 114, K99	75.30; S1.4
S.Yu. Nikitin	(b) 155,K127	71.25 and 74.70; S10.15
S.K. Nikogosyan	(a) 113, 43	61.80; 66.70; 72.20; S7.11
L.I. Nikolaev	(b) 152, K55	78.70; S1.61
V.V. Nikolaev	(a) 116,K141	61.10; 61.70; S5.11
A.Ya. Nikolaich	(a) 113, K41	64.75; S7.11
V.I. Nikolaichik	(a) 114, 113	61.70; S7.12
P.K. Nikolyuk	(b) 152, K55	78.70; S1.61
V.K. Nikulin	(b) 153, 161	73.40; 71.25; S7.15
M. Nimis	(b) 152, 67	61.70; 78.50; S9.11
	(b) 156, K5	61.70; S9.11
N. Nishimaki	(b) 151, K81	78.60; S9.11
S.V. Nistor	(b) 156, 411	61.70; 66.30; 78.50; S9.11
G.P. Nizhnikova	(b) 155, 491	71.20; 71.50; 79.60; S1.63
S. Nizioł	(a) 116, K25	61.12; S4
J.S. Nkoma	(b) 153, 383	68.20 and 78.55; S7.11
D.O. Northwood	(a) 115, 125	62.20; S4
J. Notholt	(a) 114, 207	68.55; S8
G. Nouet	(a) 114, 467	61.80; S5.11
A. Novak	(a) 114, 531	64.70; 78.30; S11.1
P. Novák	(b) 155, 669	76.60; 75.25; S11.2
A. Novikov	(a) 112, 733	68.55; 61.80; S1.61
A.P. Novikov	(a) 112, 273	61.80; S5.11
	(a) 112, 323	61.80; S5.11
D.V. Novikov	(a) 114, K7	61.10; 68.20; S5.12
	(a) 115,K133	61.10; 68.20; S5.11
V.N. Novikov	(a) 115, 475	61.40 and 64.70; S1.1; S1.2
R. Novotný	(a) 113, 615	78.30; S8.15
S. Nowak	(a) 113, 199	61.70 and 78.50; 71.55; S10.15
J.L. Nowiński	(a) 115, 81	66.30; S10
W. Nowy-Wiechuła	(b) 151, 47	61.70; 76.30; 78.50; S9.11
Yu.G. Nurullaev	(a) 116,K173	72.40; S8
Kho Khyu Nyan	(a) 114, 265	61.60 and 72.80; 75.50; S1.4
K.P. Oboth	(b) 154, 789	78.50; 78.55; S9.12
L. Ochertianova	(a) 114,K211	72.20 and 72.80; S8.16
H.L. Oczkowski	(a) 111,K113	71.55 and 78.60; S8.12
G. Oelgart	(a) 111, 541	73.40 and 85; 78.55; S7.12
	(a) 112, 229	73.40 and 78.60; 68.55; 78.55; S7.15
	(a) 114, 419	73.40 and 85; 78.60; S7.11; S7.15

G. Oelgart	(a) 115, 257	71.55; 72.80; 78.55; S7.15
T. Oerd	(b) 156, K37	62.50 and 63.75; 63.70; 77.80; S8
J.-E. Oesterholm	(b) 154, 305	72.20; 72.60; 76.30; S12
S.G. Ogloblin	(b) 151, 319	78.30
S. Ohara	(b) 151, 441	61.60; S8.16
T. Ohgaku	(a) 111, 165	62.20; 62.40; S9.11
H. Ohkubo	(a) 115, K43	71.55; S8
M. Ohta	(a) 112, K15	64.75; S4
S. Ohta	(a) 115,K215	75.50; S1.63
E. Ojala	(b) 153, 235	71.20; 65.40; 76.60; S1.3
M. Okada	(b) 153, K93	71.55; 76.30; 78.50; S7.11
Yu.G. Okladnoy	(b) 152, 217	72.20; 75.30
B.L. Oksengendler	(a) 114,K163	68.55; 61.80; S5.11
S. Okuda	(a) 116, 583	61.80 and 62.40; S1
	(a) 116, 633	62.40; S1.2
W. Okulski	(b) 153,K115	71.25 and 73.60; S8
	(b) 155, 709	71.25 and 78.30
M. Okumura	(a) 116, 583	61.80 and 62.40; S1
J. Olejniczak	(a) 116, 367	74.70; S10.15
A. Olés	(a) 116, K29	61.12; S1.61
E. Oleszkiewicz	(a) 112, 761	61.40 and 61.80; S12
L.E. Oliveira	(b) 151, 605	71.55; S5.12
M.T. Olm	(b) 152, 583	71.50 and 76.30; 71.38; 76.40; S9
M. Olszowy	(a) 113, 609	77.30; 61.80; S12
V.V. Onoochin	(b) 153, K25	71.38
O.R. Onufriv	(b) 154, K9	61.70; 78.50; S9.16
Z. Onyszkiewicz	(b) 153, 315	75.30; S9
	(b) 156, 105	64.10; 64.60
	(b) 156, K71	75.70
A. Opanowicz	(a) 116, 343	71.55 and 78.60
V.A. Oparin	(a) 113, K41	64.75; S7.11
K. Orazov	(a) 116, K91	62.65 and 72.40; S8.15
S.M. Orel	(a) 116, 295	62.30
S. Orfanakos	(a) 114, 491	61.70; 77.30
U. Orhun	(b) 154, 249	71.70 and 76.30; S9.16
V.F. Orlenko	(a) 116,K119	61.70 and 78.50; S7.12
L.K. Orlik	(a) 115,K155	64.70; S7
V.N. Orlov	(a) 115, K87	65.70 and 75.80; S11.2
Y. Osaka	(a) 115,K261	78.30; S8.15
Z. Osetek	(b) 152, 267	75.70 and 76.80
G.M. Osetinskij	(a) 112, 781	61.70 and 61.80; S5.11
Yu.N. Osetskii	(b) 155, 373	61.70
V.A. Osipov	(b) 153, 207	71.38; S12
	(b) 155, K29	71.38; S12.1
M.G. Ostapenko	(b) 152, 239	72.80; 71.55; S12.1
N.F. Ostrovskaya	(a) 116, K1	61.50 and 64.70; S1.62
V.E. Osukhovskii	(a) 114, K75	75.70; S1.2
J. Oswald	(a) 111, 345	78.55; 71.55; 78.30; S7.12
A.I. Otko	(a) 116,K191	76.30; 64.70; S11.1
T. Otmianowski	(a) 112, 339	61.80 and 68.20; S1.1
H. Otop	(a) 116,K115	77.80 and 79.75; S12
V.V. Otroshok	(b) 155, 365	61.70; 76.30; 78.40; S9.11
E.V. Ottenberg	(a) 112, 335	61.80 and 68.20; S1
G. Otto	(a) 112, 391	61.50 and 61.80; S5.11; S5.12; S7.11; S7.12
J. Otto	(b) 151, 363	78.30 and 78.50; S9.11
V.V. Ovcharov	(a) 113, K23	61.16 and 61.70

V.V. Ovcharov	(a) 113,K149	61.14
S.G. Ovchinnikov	(b) 156, 403	61.70
I.A. Ovidko	(b) 151, K7	61.70
T. Ovsenik	(b) 151, K1	66.30; 64.70; S12
M.Yu. Ovsyannikov	(a) 112, 361	62.20; 61.80; S1.1
A.E. Owen	(a) 114, 587	78.65; 68.55; S8
R.P. Ozerov	(b) 155, 201	71.45; 78.70
	(a) 115, 515	71.20; 61.55; S10
M. Ožvoldová	(a) 115, 315	78.40 and 78.60; S10.15
G. Paasch	(b) 151, 145	71.25 and 73.40; S5.11; S5.12; S7.1
G. Pacchioni	(b) 153, 193	71.10; 71.25; S4
J. Pačes	(a) 113, 573	75.70 and 76.50; 75.60; S11.2
A.K. Pachauri	(b) 155, 73	63.90; 77.20; S9.11
A. Padmaja	(b) 152, 455	62.20; S1.2; S3
H. Pagnia	(a) 111, 387	73.40 and 85; S1.3; S6; S10
K.F. Pai	(b) 154, 135	63.20; 71.36; 78.30; S7.15
A. Pajączkowska	(a) 113, 151	74.70; S10
	(b) 155,K123	72.80 and 74.70; 72.20; S10.15
Z. Pająk	(a) 114, 497	64.70; 61.12; S9
M. Pajěk	(a) 112, 781	61.70 and 61.80; S5.11
U. Pal	(a) 111, 515	68.55; S8.12
	(a) 114, 721	78.65; S8.13
G.V. Palchik	(a) 115, K9	61.70 and 71.55; S5.12
P. Paliwal	(b) 156, 171	62.20 and 64.60; S8
R.K. Paliwal	(b) 155, 241	74.10 and 75.40; S10
M. Palpacuer	(a) 113, 623	78.30; 78.65; S12
H.O. Pamuk	(b) 156,K113	68.40
G. Pananakakis	(a) 111, 529	73.60; 73.40; S5.11; S10.1
J. Pancír	(a) 115, 87	61.70; 72.20; S8
B.M. Pande	(a) 114,K159	62.20; S1.1
K.B. Pandeya	(a) 113, 603	78.70; S12
V.E. Panin	(b) 153, K11	61.70; 62.20; S4
R.S. Pank	(a) 116, 703	73.60; 73.40
D. Panknin	(a) 113, 331	61.70; S7.12
V.V. Pankov	(a) 112, K71	66.30; S11.2
V.I. Panov	(a) 116, K33	61.70; S5.11
A. Paola	(b) 155, K93	73.30; S4
G.J. Papaioannou	(a) 114,K215	72.40; 78.50; S7.11
	(a) 116,K133	73.40 and 85; 61.80; S7.12
K. Paprocki	(a) 112, 301	61.80 and 79.20; S5.11
	(a) 112, 743	61.80 and 68.48; S1.61
C. Paracchini	(a) 115, 325	61.70 and 78.60; S9.12
S.M. Paraskevas	(b) 152, K9	74.70 and 76.30; S10.15
M.-P. Pardo	(a) 113, K5	61.16 and 61.60; S8
R. Pareja	(a) 111, 463	61.70; 78.70; S5.11
A. Parisini	(b) 152, 101	73.40; 71.35; 72.40; 78.55; S7.15
H. L. Park	(b) 153,K183	78.55; S8
J.S. Park	(a) 116, 709	73.60; 71.55; S12
Om Parkash	(a) 116, K81	61.50 and 72.20; S10.15
V. Parkhutik	(a) 114, K33	73.40; S4; S10.1
J. Partyka	(a) 112, 815	79.70
L.Ya. Pasenko	(a) 116,K185	76.50; S8.16
R. Pasianot	(a) 111, 173	61.80 and 68.20; S1.1; S4

B. Pasquier	(a) 114, 531	64.70; 78.30; S11.1
G. Pastor	(b) 154, 439	61.50
M. Pastori	(a) 115, 599	78.55; 71.55; S8.16
J. Pastrňák	(a) 111, 345	78.55; 71.55; 78.30; S7.12
S.N. Pastushonok	(a) 115,K225	75.50; S10.15
A.P. Pathak	(b) 153, 455	61.80 and 78.70; S5; S5.11; S5.12
Manjunatha Pattabi	(a) 114,K179	73.60; 68.55; S1.3
S.M. Pattalwar	(b) 151,K115	65.40; S4
J.S. Patuschenko	(a) 116,K179	75.70; 75.60; S4
P. Paufler	(a) 112, 519	61.70 and 68.55; S7.11; S7.12; S7.15
	(a) 113, 71	61.70 and 68.55; S7.11; S7.15
W. Paulus	(a) 114, 91	61.60; S1.63
E. Paumier	(a) 113, 489	71.25; S7.16
M. Pavel	(a) 115, 445	61.70 and 78.20; S12
E. Pavlidou	(b) 153, K51	72.20; S8
D.A. Pavlov	(a) 116, 697	68.55 and 73.60; S5.11
S.T. Pavlov	(b) 155,K161	78.30
J. Pavlovský	(a) 112, K1	61.60; 76.80; S1.1
S.H. Pawar	(a) 111, 17	68.55; 73.40; 85; S8
R. Pawellek	(a) 111, 617	75.80; S1.2
	(a) 112, 189	75.80; 75.50
Cz. Pawlaczyk	(a) 113, 447	73.40 and 77.40; S11
M. Pawłowska	(b) 154, K89	78.55; S9.16
A. Pawłowski	(a) 113, 447	73.40 and 77.40; S11
B. Pawłowski	(a) 112, 773	61.80; S5.11
K.-E. Peiponen	(b) 156, K81	61.70 and 78.50; S9.11
A. Pejączkowska	(b) 152, K1	72.20 and 74.70; S10.15
K. Pekała	(b) 152, K1	72.20 and 74.70; S10.15
M. Pekała	(b) 152, K1	72.20 and 74.70; S10.15
	(b) 155,K123	72.80 and 74.70; 72.20; S10.15
B. Pellegrini	(b) 152, 601	72.10 and 72.20
C. Pelosi	(a) 113, 677	73.30 and 73.40; S7.11; S7.15
N.F. Perelman	(a) 113,K227	71.55 and 72.40; 76.30; 78.55; S7.11
P. Perrier	(b) 153, 653	63.20 and 71.70; 72.20; S7.13
V.V. Peshev	(a) 114,K139	61.80; S7.11
M. Pessa	(a) 116, 443	68.55; S7.11; S7.12
J. Petersson	(a) 114, 705	76.60; 77.80; S12
M. Petit	(a) 111, 155	64.60; S10.1
A.G. Petrosyan	(a) 113, 193	63.20 and 78.55; 71.55; S10.15
A.S. Petrov	(a) 113, 285	61.80; 72.20; 79.20; S8.15
E.G. Petrov	(b) 152, 239	72.80; 71.55; S12.1
K. Petrov	(a) 116, 213	61.60 and 61.70; S8.16
P.P. Petrov	(b) 153, 115	64.60; 61.60; S8.15
V. Petrov	(b) 152, K29	78.30; S11.1
V.V. Petrov	(a) 111,K141	61.70 and 78.50; S5.11
	(a) 112, 561	61.70 and 71.55; 72.20; 78.30; S5.11
Yu.A. Petrov	(a) 111,K171	64.70; S11.2
M.L. Petrovskii	(a) 111, K89	74.70; S10.15
	(a) 111,K209	74.70; S10.15
H. Pfeiffer	(a) 113, 159	72.20 and 74.70; 76.60; S10.15
S. Phanjoubam	(a) 112, 131	72.20; S11.2
J. Philip	(a) 114, 525	64.70; 61.40; 65.70; 72.80; S8

F. Phillipp	(a) 116, 91	61.70; 61.16; S1.2; S1.3
C. Picard	(b) 151, 531	73.60; S10.15
S. Pick	(b) 156, 243	68.25 and 68.40; 73.20; S5.11
R. Pickenhain	(b) 154, 219	71.10
J. Piechota	(a) 113, 151	74.70; S10
J. Piekoszewski	(a) 112, 757	61.80 and 68.20
L. Pierre	(a) 113, 23	61.60 and 74.70; 61.16; S10
V. Pierron-Bohnes	(a) 116, 529	61.55; 61.12; 76.80; S1.1
U. Pietsch	(a) 113,K203	65.70; S7.15
	(a) 113,K211	68.55; S7.15
M. Piłat	(b) 153, K21	71.25 and 73.60
I.V. Pilipovich	(b) 152, 347	78.70; 71.25; S5.11
Ya. Pilkevich	(a) 111,K215	61.80 and 72.80; S7.12
A.N. Pilyankevich	(a) 116, K1	61.50 and 64.70; S1.62
Lin-Rui Ping	(b) 153,K167	71.70 and 76.30; S10.1
S.L. Pinjare	(a) 113,K261	78.55; 61.70; 71.55; S5.11
S.M. Pintus	(a) 115, 413	68.48; 61.80; S7.12; S7.15
J. Piotrowski	(a) 113, 467	72.40 and 73.60; S8.15
	(a) 114, K37	61.80 and 72.80; 85; S8.15
E. Pippel	(a) 116, 165	61.70; 61.16; S1.61; S10
J. Piprek	(a) 112, 551	66.30 and 73.40; S10.1
A.S.T. Pires	(b) 155, K67	75.40
R. Pirlot	(b) 153, 721	76.60; 71.50; S1; S1.3; S4
T.V. Pishkova	(b) 155, 535	71.55; 63.20
D.I. Piskunov	(a) 113, 251	61.16 and 61.70
P. Piskunowicz	(a) 114, 505	64.70 and 65.70; S11.1
P. Pissis	(a) 111, 325	77.30
R. Pitis	(b) 154, 637	71.30; S12.1
V.S. Pivovar	(a) 112, 331	61.80 and 73.40; S7.12
I. Placencia	(a) 111, 675	73.40 and 77.50; 73.60; 85; S10
	(a) 113, 83	68.55 and 73.40; 68.48; S10.1
	(a) 114, 167	68.55; S10.1
S.K. Plachkova	(a) 111, 469	61.70 and 66.70; S8.16
	(a) 116, 213	61.60 and 61.70; S8.16
N.M. Plakida	(b) 155, K9	63.20 and 63.75; S10
P.A. Platonov	(b) 155, 373	61.70
A.N. Plevako	(a) 111,K209	74.70; S10.15
N. Plugaru	(a) 113,K253	61.55 and 75.50; S1.1
C. Pochrybniak	(a) 112, 757	61.80 and 68.20
A. Podgorski	(a) 112, 343	61.80; 61.70; S1.1
E.A. Podlipko	(a) 112, 323	61.80; S5.11
V.G. Podoprigora	(b) 155, 117	68.30; S12
A. Poeppl	(a) 115, 247	71.55; 76.30; S10.1
A. Poggi	(a) 113, K61	73.25; S5.11
	(a) 114, 247	71.55; 72.20; 78.30; S5.11
A.D. Pogrebnyak	(a) 112, 361	62.20; 61.80; S1.1
	(a) 114, 481	61.70; 61.80; 78.70; S7.12
B. Pohoryles	(a) 116, 349	61.70 and 71.55; 72.40; S5.12
E.P. Pokatilov	(b) 156, 225	71.38 and 73.20
J. Pokluda	(b) 151, 85	62.20
G.N. Polisskii	(a) 116, K91	62.65 and 72.40; S8.15
I.A. Politov	(a) 114, 99	61.70; S9.11
T. Poloczek	(a) 111, 127	61.40 and 64.70; 62.40; S1.1; S1.2
M. Połomska	(b) 154,K103	64.70; 77.20; S11
I.S. Polovtsev	(a) 114, 233	61.70 and 71.55; 76.60; S5.11
A.M. Polubotko	(b) 156, 677	78.30
L.P. Poluchankina	(a) 111,K209	74.70; S10.15

A.M. Polyakov	(b) 156, 615	76.60; 66.30; 72.20; S9.16
M.E. Polyakov	(b) 153, 479	61.70
T.V. Polyakova	(a) 111,K209	74.70; S10.15
Yu.I. Polygalov	(b) 154, 377	71.25 and 78.40; 71.35; S7
G. Pompe	(a) 111,K165	65.40; S7
W. Pompe	(a) 111, 109	61.70 and 62.20; S10.15
M.J. Ponnambalam	(b) 153, 79	61.70 and 71.55; 78.20
K. Ponnuraju	(a) 113, 477	78.65; 68.55; S8.15
A.I. Ponomarev	(a) 115, 267	74.70; 72.20; S10.15
B.K. Ponomarev	(a) 111,K237	75.60; S1.1
	(a) 112,K127	75.60; S1.1
V. Pop	(a) 113,K253	61.55 and 75.50; S1.1
A. Popitsch	(b) 153, 307	74.70; 75.30; 78.30; S10.15
Yu.A. Popkov	(b) 154, 353	75.30 and 75.50
A.S. Poplavnoi	(b) 154, 377	71.25 and 78.40; 71.35; S7
A.P. Popov	(b) 155, 185	71.38; S12.1
V.B. Popov	(a) 112, K9	61.70; 71.55; 73.40; S5.11
V.P. Popov	(a) 113,K165	61.70; S5.11
	(a) 115, 413	68.48; 61.80; S7.12; S7.15
S. Popović	(a) 111, 417	61.55; 62.20; 64.70; S4
R. Poprawski	(a) 113, K31	64.70; 65.40; S11
A. Poradzisz	(a) 112, 431	79.20; S9.11
P. Porcher	(a) 115, 523	71.70; 78.50; 78.55; S10.15
M. Porte	(a) 112, 601	75.30 and 75.50; S11.2
W. Possart	(a) 114,K175	68.48 and 79.60; S4; S12
M. Posselt	(a) 112, 263	61.80
T. Possner	(a) 111, K7	66.30
Z. Postawa	(a) 112, 431	79.20; S9.11
L. Potocký	(a) 111, 603	75.50; S1.1
B. Poumellec	(b) 151, 531	73.60; S10.15
J. Pouzet	(a) 111, 181	73.60; S8
F. Povolo	(b) 151, 71	62.20; S4; S11
A.A. Povzner	(a) 114, 315	65.70 and 75.40; S1.61
S.K. Poznyak	(a) 111, 193	68.45 and 72.40; S10.1
J. Pozo	(b) 153,K163	75.10
I. Pozsgai	(a) 114, 635	66.30 and 73.40; 78.65; S7.12
R.B. Prabhu	(b) 151, K45	72.15; S1.4
	(b) 151,K141	71.20
A.K. Pradhan	(b) 152, K69	78.30; 63.20; S11.1
L. Pranevičius	(a) 114, K25	61.80 and 68.48; S5.11
	(a) 116,K161	66.30 and 68.48; S7.11
C. Prasad	(b) 154, 201	71.25; S7.1
Ch.Durga Prasad	(a) 116, K81	61.50 and 72.20; S10.15
G. Prasad	(a) 115, 135	64.75; 72.20; S8.15
G.M.Prasad	(b) 155, 209	71.45; S9; S10
P.S.R. Prasad	(a) 116, 275	64.70 and 78.30; S12
P.S.S. Prasad	(b) 152, 171	61.80 and 71.55; 78.50; S11.1
A. Pratap	(b) 154, K23	63.50; S2
Chr. Pratsch	(b) 154, 315	73.40
U. Preppernau	(b) 152, K5	72.20 and 74.70; S10.15
I.I. Prikhodko	(a) 111, K13	61.80 and 68.48; S1.2
M.T. Primatarowa	(b) 156,K125	71.55
A.D. Prins	(b) 151, 759	78.20; S8.16
V.D. Prochukhan	(b) 151, 711	78.20; S8.16
J. Prokeš	(a) 114, 643	71.55; S5.11
A.V. Prokhorovich	(a) 113, 497	71.55 and 78.55; 61.80; S7.12
V.K. Prokopenko	(a) 113,K107	75.30 and 76.60; S11.1
	(b) 156, 597	75.30; 76.60; S11.2
V.S. Prosolovich	(a) 112, 561	61.70 and 71.55; 72.20; 78.30; S5.11

M.T. Protasova	(a) 113, 359	64.75; 76.80; S10
N. Pruss	(b) 153, K63	74.70; S10.15
	(a) 113, 159	72.20 and 74.70; 76.60; S10.15
	(a) 116, 793	72.20 and 74.20; S10.15
N.A. Prytkova	(a) 114, K57	74.70; S10.15
	(a) 115, K79	61.70 and 74.70; S10.15
S.G. Psakhie	(b) 153, K11	61.70; 62.20; S4
V.Yu. Ptitsin	(a) 112, K65	61.70; 71.55; S5.11
K.K. Pukhov	(a) 113, 193	63.20 and 78.55; 71.55; S10.15
G. Pukinskas	(b) 153, 741	78.20; S8
	(b) 154, 389	78.20; 61.60; S8.15
V.N. Pushkar	(a) 113, 529	75.60 and 75.70; S1.1; S1.2; S11.2
A.L. Pushkarchuk	(b) 155,K127	71.25 and 74.70; S10.15
A.A. Puzanov	(b) 153, 65	61.80 and 79.20
G. Puzonas	(b) 156, 697	78.40; 71.25; S11.1
T.M. Pyatkova	(a) 111, 499	61.70 and 68.55; S1.61
	(a) 116, 331	68.55; 61.70; S1.61; S5.11
H. Pykacz	(a) 115, K17	64.70 and 65.70; S12
N.B. Pyshnaya	(a) 115,K191	71.55; S7.11
K. Pyszniak	(a) 112, 799	61.70; 78.60; 79.20
Z. Pytel	(b) 154, K1	61.50
Xiaoding Qi	(a) 114,K127	78.45; 78.30: 78.55; S11.1
Renyuan Qian	(a) 111, K67	72.80; 72,40; 78.40; S12.1
	(a) 116, 709	73.60; 71.55; S12
Y.T. Qian	(a) 115, K71	74.70 and 79.60; S10.15
Yitai Qian	(b) 154, K51	74.70; S10.15
Minwang Qiu	(a) 114,K127	78.45; 78.30; 78.55; S11.1
Doan Nhat Quang	(b) 154, 237	71.25 and 72.20; S7.13
S. Quereshi	(a) 114, 299	72.80; S12
M. Quintero	(a) 111, 405	61.60; 64.75; S8.16
	(a) 115, 157	64.70; S8.15
G. Raatz	(a) 113, 131	68.55; 68.48; S1.2; S10.1
L.M. Rabkin	(b) 153, 517	64.60 and 78.30; S8.16
G. Račiukaitis	(b) 151, 721	71.55 and 78.50; S8.12; S10.1
	(b) 156, 697	78.40; 71.25; S11.1
S.I. Radautsan	(a) 113,K227	71.55 and 72.40; 76.30; 78.55; S7.11
	(a) 113,K265	78.55; S8.16
	(a) 114, 259	71.55; 78.55; S8.16
	(a) 115,K191	71.55; S7.11
	(a) 116,K185	76.50; S8.16
M.G. Radchenko	(a) 114, 559	64.70 and 68.55; 77.80; S10.15
S. Radhakrishna	(b) 152, 171	61.80 and 71.55; 78.50; S11.1
B. Radjenovic	(b) 156, 487	71.20; 71.30; S5.11
D. Rafaja	(a) 113,K181	66.30
S. Lakshmi Raghavan	(a) 114, 749	78.20; 61.70; S9.11
S.M. Mujibur Rahman	(b) 156, 181	65.50; 65.20; S2
P.A. Raikerus	(a) 116, 769	72.20; 71.55; 85; S5.11
N. Raj	(b) 152, 135	71.36; 71.45; 75.10

B. Rajchel	(a) 113, 295	61.80; S5.11
O.S. Rajora	(a) 114, 199	68.60; 68.55; S1.2
A.E. Rakhshani	(a) 113, 123	68.55; 78.65; S10
P.N. Ram	(b) 156, 137	63.20 and 71.55; 63.50; S9.11
R.K. Ram	(b) 154, 553	63.20; 78.30; S7.1; S8.1
K. Ramachandran	(b) 153, 579	63.20; 61.70; S8.13
	(b) 154, 55	66.30; S8.13
	(b) 154,K113	63.20; S8.15
	(b) 155, K85	66.30; S8.13
C.H. Ramakrishna	(b) 156, 253	72.10 and 72.40; S7.12
M.V. Ramana	(a) 116,K205	78.60; S10.15
N. Ramesh	(a) 114, 749	78.20; 61.70; S9.11
K. Ramkumar	(a) 111, 667	73.40 and 85; S5.11
	(a) 111,K269	73.40 and 85; S5.11
	(a) 115,K265	73.40 and 85; 61.70; S5.11
R. Ranganathan	(b) 154, K55	72.15 and 74.70; S10.15
M. Rani	(b) 154, K23	63.50; S2
A.V. Krishna Rao	(a) 114,K115	77.20 and 77.40; 77.70; S11
B. Srinivasa Rao	(b) 156, K27	62.20; S9.11
C. Nageswara Rao	(b) 154, K85	78.70; S1.2
J.L. Rao	(b) 151, 599	71.55 and 78.50; S11
	(b) 151, 615	71.10 and 76.30; S11
	(b) 153, 257	71.70; 76.30; 78.50; S11
K.V. Rao	(a) 114,K115	77.20 and 77.40; 77.70; S11
	(a) 115, 301	77.20 and 77.40; S11.1
R. Ramji Rao	(b) 152, 455	62.20; S1.2; S3
R.V. Gopala Rao	(b) 153, 421	61.25 and 61.50; S9.11
	(b) 154, 427	61.25; S1.3; S2
	(a) 115, 357	61.25; S9.12
S. Rao	(a) 111, 173	61.80 and 68.20; S1.1; S4
V.V. Rao	(a) 114, 721	78.65; S8.13
T.V. Rapinchuk	(a) 115,K237	78.60; S7.1
I.M. Rarenko	(b) 154, 513	66.30; S5.12
P. Rashev	(a) 116, 521	61.55; 61.14; 64.70; S1.2
R. Rataj	(b) 155,K123	72.80 and 74.70; 72.20; S10.15
B.K. Rath	(b) 156, 137	63.20 and 71.55; 63.50; S9.11
R.P.S. Rathore	(b) 156, 77	63.20; 62.20; S3
A.M. Ratner	(b) 153, 539	63.90; 71.45; 78.55; S5.2
B. Rau	(a) 115,K247	78.65; S5
A.K. Ray	(a) 116, 703	73.60; 73.40
A.K. Raychaudhuri	(b) 154, K55	72.15 and 74.70; S10.15
N.A. Razik	(a) 116, 383	75.60; S1.1
M.S.K. Razmi	(b) 156, 111	64.10; 64.60; S9.11
C. Razzetti	(a) 111, 411	61.60; 64.60; S8.15
	(b) 152, 39	61.60 and 78.70; S8.16
	(b) 154, 461	61.60; 78.70; S8.16
A. Razzitte	(b) 155, K93	73.30; S4
K. Recker	(b) 155, K59	76.30; 71.55; S9.16
M.R. Reddy	(b) 151, 599	71.55 and 78.50; S11
P.Jayarama Reddy	(a) 115,K175	73.60; S8.16
R. Reetz	(a) 111,K129	78.60; S8.11
L. Reggiani	(b) 152, 601	72.10 and 72.20
F.M. Reicha	(a) 116, 55	68.48 and 68.55; S1.3; S9.11
D. Reichardt	(b) 156, 471	68.20 and 73.20; S5; S5.11; S5.12
M. Reiche	(a) 114, 247	71.55; 72.20; 78.30; S5.11
J. Reichel	(a) 111,K195	71.55; S5.11
M.F. Reid	(b) 155, 559	71.70; S9.16

J. Reineke	(a) 114, 579	73.20 and 73.40; S5.11; S10
T. Reininger	(a) 111, 243	75.50; S1.1; S1.2
H. Reis	(a) 115, 497	68.55 and 79.20; S1.2
Ch. Rellin	(b) 152, K5	72.20 and 74.70; S10.15
I.A. Remizov	(b) 155, 117	68.30; S12
G.E. Remnev	(a) 112, 361	62.20; 61.80; S1.1
A.A. Rempel	(b) 151, 211	74.70; S1.61
	(a) 113, 353	65.40; S1.61
	(b) 154, 453	61.50; 61.12; S1.61
F. Remy	(a) 114, 73	61.60; 61.16; S10
	(a) 114, 441	61.60; 61.16; 61.70; S10
P. Renaudin	(a) 112, 601	75.30 and 75.50; S11.2
M. Renteria	(a) 116, 503	61.60; 61.70; 76.80; S10.1
S. Rentsch	(a) 114, 757	78.20; S10
F.G. Requejo	(a) 116, 503	61.60; 61,70; 76.80; S10.1
J. Resch	(b) 152, 377	61.40; 63.50
V.Ya. Reznichenko	(b) 152, 675	78.20; S8.11
I.M. Reznik	(b) 155, 201	71.45; 78.70
H. Rhan	(a) 112, 519	61.70 and 68.55; S7.11; S7.12; S7.15
	(a) 113, 71	61.70 and 68.55; S7.11; S7.15
	(a) 113,K211	68.55; S7.15
B. Rheinlaender	(a) 112, 229	73.40 and 78.60; 68.55; 78.55; S7.15
	(a) 114, 419	73.40 and 85; 78.60; S7.11; S7.15
	(a) 114,K237	78.45; S7.15
G.Yu. Riabukhin	(b) 156, 225	71.38 and 73.20
D. Ribarov	(a) 115,K243	78.20; S8.15
F.S. Richardson	(b) 155, 221	71.70; S11.1
	(b) 155, 559	71.70; S9.16
E. Richter	(a) 112, 533	68.48 and 68.55; S1.1; S7
H.H. Richter	(a) 112, 721	73.40; 68.55; S5.11
J. Richter	(b) 153, 711	75.10; 75.30
	(b) 154, K79	75.50
R. Richter	(a) 114, 765	78.20; 61.70; S10.15
V. Riede	(a) 112,K139	61.60 and 78.30; S7.15; s8.15
	(a) 116,K147	61.80 and 78.30; 63.20; S7.11
D.V. Riekstinya	(a) 116,K125	61.80 and 78.40; 78.55; S11.2
R. Riera	(b) 152, 357	71.35 and 78.30; S9
R. Riesenberg	(a) 112, K85	62.20; 61.80; 75.60; S12
G. Riontino	(a) 114, 511	64.75; 72.15; S1.2; S4
D. Ristau	(a) 115, 191	78.65; 77.55; S9; S10
Lj. Ristovski	(b) 153,K125	71.35
J. Rivas	(a) 111, 611	75.60; 75.50; S1.1; S1.2
	(a) 113, 541	75.60; S10.1
A. Rizzo	(a) 114, 253	71.55; S8
M. Robino	(b) 153, 403	78.20; S9
V.E. Rodionov	(a) 116,K169	68.55; S6
T.V. Rodzina	(a) 115,K143	61.70; S7.13
H. Roeppischer	(b) 154, 769	78.20; S7.12
M. Roesler	(a) 114, 327	75.60; S11.2
	(b) 156, K85	79.20; S4
	(b) 156, 515	65.40 and 71.45
A. Rogalski	(a) 111, 559	72.20; S8.15
	(a) 113, 467	72.40 and 73.60; S8.15
M. Rolland	(a) 113, 623	78.30; 78.65; S12
Z.A. Romanenko	(a) 111,K209	74.70; S10.15
A.E. Romanov	(a) 113, 337	61.70 and 68.48

A.E. Romanov	(a) 115,K159	62.20
I.G. Romanov	(a) 112, 361	62.20; 61.80; S1.1
S.N. Romanov	(b) 153, K5	61.70; 61.80; S1; S1.1
B.N. Romanyuk	(a) 112, 805	61.70; 78.20; S5.11
A.A. Romanyukha	(b) 151, K59	76.30; S12
	(b) 151, K65	76.30; S12
A.M. Roshchupkin	(b) 151, 121	68.20 and 72.15
O.S. Roshupkina	(a) 115,K195	74.70; 78.30; 79.20; 79.60
W. Rosiński	(a) 112, 245	61.80 and 68.20; S1.1
A. Rost	(a) 115, 165	64.70 and 77.80; 64.80; S10.15
A.H. Rostomyan	(a) 116, 483	61.10 and 78.70
	(a) 116, 493	61.10 and 78.70
A.M. Rostomyan	(a) 116, 483	61.10 and 78.70
	(a) 116, 493	61.10 and 78.70
P. Roth	(b) 154, K93	61.40
S. Roth	(a) 112, 667	61.80 and 68.55; 73.60; 78.65; S5
T. Rother	(b) 153,K109	65.50
M. Roushdy	(b) 155, 453	62.80; S1.1; S1.2; S4
D. Roy	(b) 156, 83	64.75; 61.10; S9.11
S.K. Roy	(a) 116, 415	73.40 and 85; S5.11; S7.12
S.A. Różański	(a) 112, 617	77.80 and 78.20; 61.30; S12
E.I. Rozum	(b) 152, 61	61.80; S5.11
	(a) 112,K131	79.20
A. Rozycki	(a) 114, 531	64.70; 78.30; S11.1
S.V. Rubanov	(a) 115, 413	68.48; 61.80; S7.12, S7.15
V.A. Rubtsov	(a) 115,K155	64.70; S7
V.Yu. Rud	(a) 113, 207	72.40 and 73.40; 85; S7.16
Yu.V. Rud	(a) 111, 289	78.20; S1.63; S8.16
	(b) 151, 711	78.20; S8.16
	(a) 113, 207	72.40 and 73.40; 85; S7.16
	(a) 115, 11	72.40; 78.20; S7.16
T.E. Rudenko	(a) 113, 653	73.40 and 85; 76.30; S5.11; S10.1
V.N. Rudko	(b) 156, 211	78.65; S7.12; S7.15
A.S. Rudnev	(a) 114, 481	61.70; 61.80; 78.70; S7.12
P. Rudolph	(a) 112, 569	71.55; 64.75; 78.50; 78.55; S8.13
W. Rudolph	(a) 112, 765	61.40 and 61.80; S12
J. Rudzińska-Girulska	(b) 156, 303	71.25; 78.70; S1.2; S1.3
H. Ruecker	(b) 153, 595	71.25 and 73.60; S5.11; S5.12
I. Ruiz-Larrea	(b) 154, 143	64.60; 65.40; S9.16
C. Ruiz-Mejia	(b) 154,K171	71.55 and 78.50; S9.11
R.A. Rupp	(a) 113, 637	78.20; 64.60; 77.80; S10
H. Ruppersberg	(a) 116, 681	62.20 and 68.25; S1.2
E.A. Ryabenko	(b) 156, 205	68.20; S10.1
A.P. Ryabov	(b) 153, 727	78.30; 77.80; S8.16
T.N. Ryabtseva	(a) 116, 255	62.40; S1.2
A.I. Ryazanov	(b) 154, 497	61.70
Yu.I. Rybalko	(b) 153, 539	63.90; 71.45; 78.55; S5.2
A.O. Rybaltovskii	(b) 156, 663	78.55; 71.55; S10.1
H. Rzewuski	(a) 112, 405	61.70; S5.11
F.A. Saadalah	(a) 111, 137	62.20; S4
V.A. Saakyan	(a) 113, 43	61.80; 66.70; 72.20; S7.11
J. Sabín del Valle	(b) 151, 127	71.25; S7.12; S7.15

R.Kh. Sabirov	(b) 151,K201	76.20
K. Sadaiyandi	(b) 154,K113	63.20; S8.15
	(b) 155, K85	66.30; S8.13
L. Sadlowski	(b) 152, K9	74.70 and 76.30; S10.15
O.L. Sadovskaya	(a) 112, K19	64.70; 66.70; S5.11
M.S. Sadygov	(a) 111, K73	72.40 and 73.40; S8.16
O.M. Sadykhov	(a) 116,K173	72.40; S8
E.K. Sadykov	(b) 156, 605	76.80; 76.30
S. Saha	(a) 111, 515	68.55; S8.12
	(a) 114, 721	78.65; S8.13
M. Sahay	(b) 154, 201	71.25; S7.1
S.N. Sahu	(a) 111,K179	68.55; 78.65; S4; S5; S7.11
A.G. Saif	(a) 111, 573	74.60
	(b) 152, 261	74.60
	(b) 152, 617	74.20; 74.70; S10.15
J.S. Saif	(b) 151, 495	66.70
R. Sakai	(a) 113,K131	75.70; S1.1
A. Sakakibara	(a) 112, K15	64.75; S4
	(a) 113,K199	62.20 and 64.75; S4
	(a) 114, K17	64.75 and 72.15; S4
	(a) 114,K197	72.15; S4
	(a) 116,K153	61.70; S4
A. Sakalas	(a) 111, K43	73.30; S7.12
S.I. Sakhaeva	(b) 152, 191	71.70; 76.30; 78.50; S9.16
V.P. Sakhnenko	(b) 152, 431	61.50 and 64.60
C. Salá	(b) 153, K31	71.25
A.P. Salalas	(a) 111, 507	68.48; 79.20; S7.12
A. Salazar	(a) 113, 187	75.50; S1.1; S1.2
M. Ben Salem	(a) 116, 669	64.70; S10
W. Salewski	(a) 114,K255	78.30; S1
I.Kh. Salikhov	(b) 152, 191	71.70; 76.30; 78.50; S9.16
V.N. Salimgareeva	(a) 113,K111	75.50; S12.1
E.A. Salkov	(b) 153, 667	71.35 and 71.55; 72.40; 78.55; S8
A. Salokatve	(a) 116, 443	68.55; S7.11; S7.12
V.N. Salomatov	(b) 155, 359	61.70 and 71.55; S9.11
L.K. Samanta	(a) 114, 381	78.20; 85; S7.15
B.K. Samantaray	(a) 111, 515	68.55; S8.12
T. Sanadze	(b) 151, K49	76.70; S9.12
	(b) 152, K13	76.70; S9.12
A.M. Sandauer	(a) 111,K219	72.20; S8
P. Šandera	(b) 151, 85	62.20
L.M. Sandratskii	(b) 154, 623	71.25; 75.50; S1.1; S1.2
	(b) 155,K105	71.25; 78.20; S1.4
R. Sandrik	(a) 112, 781	61.70 and 61.80; S5.11
N.S. Sannikova	(a) 113,K111	75.50; S12.1
S.P. Sanyal	(b) 156, K27	62.20; S9.11
V.B. Sapre	(b) 153, K35	71.25 and 78.70; S8; S10.1
V.V. Saraikin	(a) 114, 457	61.70; 78.20; S11.1
Z.K. Saralidze	(a) 114, 99	61.70; S9.11
O.G. Sarbey	(b) 156, 461	73.40; S7.12; S7.15
A.L. Sarce	(b) 154, 105	62.20; 61.80; 65.50
R.M. Sardarly	(a) 114,K119	77.80; 61.80; S8.16
V. Sargunas	(a) 116,K161	66.30 and 68.48; S7.11
V.R. Sargūnas	(a) 111, 507	68.48; 79.20; S7.12
C.K. Sarkar	(b) 156, 253	72.10 and 72.40; S7.12
	(b) 156,K145	76.30; 72.20; S7.15
S.E. Sarkisov	(a) 113, 193	63.20 and 78.55; 71.55; S10.15

S.E. Sarkisov	(a) 113, 585	78.30; 71.70; S11.2
H.N.K. Sarma	(a) 112,K101	73.60; S4
E.F. Sarmento	(b) 151, 675	75.30 and 75.50
G.S. Sastry	(a) 116,K205	78.60; S10.15
M.S. Murali Sastry	(a) 111, K37	68.55; 73.60; S1.3; S4
	(a) 114,K179	73.60; 68.55; S1.3
S.B.S. Sastry	(b) 155, 679	61.70 and 78.50; 78.55; S11.1
H. Sato	(b) 153,K141	74.20; S10.15
K. Sato	(a) 115, 61	61.16; S1.2; S4
T. Satow	(a) 112, 467	61.40; 61.12; S8.16
A. Satpathy	(b) 153, 421	61.25 and 61.50; S9.11
	(b) 154, 427	61.25; S1.3; S2
S. Satpathy	(b) 153, 455	61.80 and 78.70; S5; S5.11; S5.12
M.R. Sattarov	(a) 114, K79	75.30; S1.4
M. Satyam	(a) 111,K269	73.40 and 85; S5.11
	(a) 114,K259	73.40 and 85; S5.11
	(a) 115,K265	73.40 and 85; 61.70; S5.11
G.K. Savchuk	(b) 156, 687	78.70; S5.11
E.J. Savino	(a) 111, 173	61.80 and 68.20; S1.1; S4
	(b) 154, 105	62.20; 61.80; 65.50
A.F. Savitskii	(b) 151, 269	75.30; 76.80; S11.2
	(b) 152, 639	75.50 and 76.80; S11.2
	(b) 152, K51	75.30; S11.2
	(b) 155, 249	75.30; S11.2
P.I. Savitskii	(b) 154,K151	72.80; S8
M.M. Savosta	(b) 155, 669	76.60; 75.25; S11.2
V.V. Savyak	(a) 112, 11	72.20
E.L. Saw	(a) 114,K223	76.80; S11
H.C. Saxena	(b) 155, 73	63.90; 77.20; S9.11
K. Saxena	(a) 111, 309	77.80; S12
	(a) 114, 369	77.30 and 77.80; S12
N.S. Saxena	(b) 154, K23	63.50; S2
A. Scacco	(b) 151, 453	61.70; 78.50; 78.55; S9.11
L.M. Scarfone	(b) 155, K33	71.20; 72.15; S1.2; S1.3
G. Schaarschmidt	(a) 115,K247	78.65; S5
U. Schade	(b) 154, K5	66.30; S7.1
H. Schaefer	(a) 113, 551	75.60; S1.1
A. Schaper	(a) 116, 179	61.40; 61.16; 62.20; S12
W. Scharff	(a) 112, 667	61.80 and 68.55; 73.60; 78.65; S5
K. Scheerschmidt	(a) 116, 123	61.16; 61.70
A. Schenk	(b) 154, 815	72.10 and 73.40; 77.50; S8.15
M. Schenk	(a) 115,K135	61.70 and 66.30; S8.15
V. Schiller	(a) 112, 721	73.40; 68.55; S5.11
A. Schindler	(a) 112, 391	61.50 and 61.80; S5.11; S5.12; S7.11; S7.12
	(a) 116,K147	61.80 and 78.30; 63.20; S7.11
J. Schleifer	(b) 154, 153	63.20; S12
J. Schloesser	(b) 153, 773	71.25 and 78.20; S8.11
K. Schmalz	(a) 111,K195	71.55; S5.11
	(a) 112, K75	66.30; S5.11
	(a) 113, K9	66.30; S5.11
	(a) 113, 315	61.80 and 73.20; 68.48; S5.11
	(a) 114, 233	61.70 and 71.55; 76.60; S5.11
	(a) 116, K37	61.70; S5.11
U. Schmelzer	(b) 154, 153	63.20; S12
	(b) 155, 89	63.20; S10
H. Schmidt	(a) 114, 621	68.55; 78.55; S8

K. Schmidt	(a) 113, K56	68.55; 78.65; S8
H.A. Schneider	(b) 154,K175	78.30; 72.20; S7.1
W. Schnelle	(a) 115, 505	73.60; S4
	(a) 116, 337	73.60; 72.10; S4
	(a) 116, 715	73.60; 72.10; S4
Ch. Schnittler	(b) 151,K123	71.25 and 73.90
	(b) 152, 153	71.10; 71.20; 71.25; 71.50
	(b) 153, K39	71.10; 71.50
G. Schramm	(a) 116, K97	78.55; S5.11
	(a) 116,K209	78.55; S5.11
B. Schreckenbach	(a) 116, 735	61.80 and 68.55; 77.55; S9.12
J. Schreiber	(b) 152, 311	75.30 and 75.50; 75.40
W.H. Schreiner	(a) 112, 105	68.48 and 68.55; S1.1; S4
H. Schrunner	(b) 151, 101	61.50 and 64.75; 65.50; S9.11
A. Schubert	(a) 111,K137	61.10 and 68.20; S1.1
J.-W. Schuenemann	(a) 113, 403	64.70; 75.30; S1.2
W. Schueppel	(a) 114, 327	75.60; S11.2
E. Schulz	(a) 111, 361	78.70; 71.25; S10
	(a) 116, 179	61.40; 61.16; 62.20; S12
R. Schulz	(a) 114,K237	78.45; S7.15
S. Schulze	(b) 152, 505	71.20 and 79.60; S7
W. Schumacher	(b) 155, 263	75.10; 75.30
B. Schumann	(a) 114,K183	73.40; 68.55; S7.12; S9.15
J. Schumann	(b) 151,K155	72.80; S1.61
M. Schvoerer	(a) 115, 335	78.60; 61.70; S9.12
C. Schwab	(b) 156, 71	61.70; 76.90
R. Schwabe	(a) 111, 541	73.40 and 85; 78.55; S7.12
G. Schwarz	(a) 114,K237	78.45; S7.15
E. Schwarzmann	(a) 111, K93	74.70 and 76.30; S10.15
U. Schwertmann	(a) 114, K11	61.10 and 64.70; 61.60; S10.1
C.B. Scruby	(a) 112, 55	61.80; 75.50; S1.1
A. Seeger	(b) 153, 465	61.70; 62.20; 62.40
	(a) 116, 91	61.70; 61.16; S1.2; S1.3
H.K. Sehgal	(a) 114, K29	73.60 and 78.65; S8.16
	(a) 115,K181	73.60 and 78.65; S8.16
P. Seidel	(a) 115, 273	74.70; 74.50
Yu.M. Seidov	(b) 153,K203	78.30; 71.70; 75.30; S1.62
P. Sekar	(b) 153, 131	65.50; 63.90; S9.12
P. Selbmann	(b) 151,K151	72.10; S8
R. Sellmann	(a) 112, K5	66.30; S5.11
O.V. Selyugin	(b) 155, 155	71.38
H.-Ch. Semmelhack	(a) 116, 537	61.50; 74.70; S10.15
K. Sen	(a) 116, 425	73.20 and 85; 61.70; 72.40
P.K. Sen	(b) 151, 739	78.35; S7.13
S. Sengupta	(b) 154,K107	62.65 and 65.70; 63.20
	(b) 156, 83	64.75; 61.10; S9.11
E.A. Senokosov	(b) 155, 295	71.45 and 78.55; S8.12
A.D. Serazetdinov	(b) 156, 205	68.20; S10.1
P.V. Serba	(a) 116, 555	61.80
P.P. Seregin	(b) 155,K113	71.55; S5.11
G. Serfőző	(a) 115, 75	61.80; 73.40; S5.11
M. Serginov	(a) 113, 207	72.40 and 73.40; 85; S7.16
J. Serna	(a) 115, 389	61.40 and 61.70; 78.70; S12
M. Servidori	(a) 115, 437	61.80; 61.10; S5.11
M. Setvak	(a) 112, 689	61.70 and 61.80
V.F. Shabanov	(b) 155, 117	68.30; S12
V.G. Shadrov	(a) 111,K225	75.70; S1.2
S.S. Shaimeev	(a) 116, K33	61.70; S5.11
S. Shakhdgan	(b) 156, 697	78.40; 71.25; S11.1

V.I. Shakhovtsov	(b) 151, 159	71.55; 76.30; S5.11; S5.12
A.D. Shaligram	(a) 111, 457	61.70; 72.20; S5.11
A.M. Shamah	(a) 115,K149	64.70; 77.80; S11.1
V.N. Shambalev	(a) 112, K71	66.30; S11.2
H.A. Shamuratov	(a) 116,K169	68.55; S6
B.D. Shanina	(b) 151, 159	71.55; 76.30; S5.11; S5.12
J. Shanker	(b) 153, 123	62.20; S9.12
	(b) 156, 131	62.20; S9.15
B.Ya. Shapiro	(b) 154, 679	74.20 and 74.50
E.F. Shapovalova	(a) 112, 155	75.50; S10.15
	(a) 113,K249	75.60; S10
N. Sharif	(a) 112,K139	61.60 and 78.30; S7.15; s8.15
A.K. Sharma	(b) 155, 73	63.90; 77.20; S9.11
B.B. Sharma	(a) 115, 135	64.75; 72.20; S8.15
B.K. Sharma	(b) 152, 145	71.25 and 78.70; S1
K.C. Sharma	(b) 151, 225	72.10 and 72.15
R.G. Sharma	(a) 116, 377	72.15; S1.1; S1.2
R.K. Sharma	(a) 115, 135	64.75; 72.20; S8.15
S.B. Sharma	(b) 156, 171	62.20 and 64.60; S8
S.K. Sharma	(a) 115,K125	68.55 and 85; 73.40; S5.11
K.V. Shastov	(a) 113, 285	61.80; 72.20; 79.20; S8.15
P.P. Shatskii	(b) 152, K23	75.60 and 75.80
S.S. Sheinin	(a) 111, 47	61.16 and 61.70; S1.2
	(a) 115, 51	61.14; 61.70
R.I. Shekhmametev	(b) 155, K51	74.20
V.Sh. Shekhtman	(a) 115, 199	64.70; 62.20; S9.16
A.U. Sheleg	(a) 111, 485	62.20 and 64.60; S7
M.U. Sheleg	(a) 115, 279	75.60; 75.70; S1.2
A.L. Shelekhov	(b) 152, 685	78.20; S1
L.E. Shelimova	(a) 111, 469	61.70 and 66.70; S8.16
A.A. Shemyakov	(a) 113,K107	75.30 and 76.60; S11.1
	(b) 156, 597	75.30; 76.60; S11.2
D.X. Shen	(a) 113, K15	61.80; S5.11
Hui Ming Shen	(a) 113, K85	62.20 and 74.70; S10.1
Huimin Shen	(a) 116,K199	62.20 and 77.80; S12
V.A. Shenderovskii	(a) 111, K99	72.80 and 75.50; S8
Yubao Sheng	(a) 114,K189	74.70; S10.15
T.D. Shermergor	(b) 153, 89	62.30
	(b) 154, 367	77.20 and 78.90
V. Shershulskii	(a) 114, K33	73.40; S4; S10.1
V.P. Shestov	(a) 112, 437	79.20; 75.50; S1.2
S.Y. Shete	(b) 151,K115	65.40; S4
Songqing Shi	(a) 111, K67	72.80; 72.40; 78.40; S12.1
Y. Shi	(a) 113, K15	61.80; S5.11
Yan Shi	(b) 151, 249	76.80; S1.1; S1.4
K. Shibata	(b) 152, K63	78.30; 63.20; S1.63
S. Shigetomi	(a) 115, K43	71.55; S8
V.D. Shigorin	(a) 112, K49	78.20; S10
M.Sh. Shikhsaidov	(a) 114, 113	61.70; S7.12
Y. Shimada	(a) 113,K131	75.70; S1.1
N.M. Shimanskaya	(a) 114, K57	74.70; S10.15
	(a) 115, K79	61.70 and 74.70; S10.15
S. Shimanuki	(b) 151, K81	78.60; S9.11
T. Shimizu	(b) 156,K163	61.70 and 76.30; S10.1
T. Shiozawa	(a) 116, 513	61.50 and 61.60; S9
V.B. Shipilo	(a) 115,K237	78.60; S7.1
K.E. Shirinov	(b) 153, K89	63.20 and 78.30; S8.16
M.M. Shirinov	(b) 151, 759	78.20; S8.16
	(b) 152, 317	64.60 and 78.30; S8.16

D.M. Shirokov	(a) 112, 781	61.70 and 61.80; S5.11
V.B. Shirokov	(b) 153, 529	64.60 and 78.30; S8.16
E.M. Shishonok	(a) 115,K237	78.60; S7.1
Sham-Tsong Shiue	(a) 113, 365	61.70 and 62.20
Yu.A. Shiyanov	(a) 114, 191	64.80 and 68.55; 79.20; S1.61
A.P. Shkadarevich	(a) 112, 197	78.20 and 78.45; 78.55; S10
L.V. Shmeleva	(b) 155, 117	68.30; S12
M.I. Shmiglyuk	(b) 154, 209	71.35; 78.20; S10.1
Nguyen Hong Shon	(b) 154, 661	72.50; 72.55
S. Shoval	(a) 113, 631	78.60; S11
B.D. Shrivastava	(a) 113, 603	78.70; S12
K.N. Shrivastava	(b) 153, 547	63.90; 76.80; 78.30
V.R. Shtirz	(a) 112, 161	61.40 and 75.50; S11.2
A. Shukla	(b) 154, 553	63.20; 78.30; S7.1; S8.1
R.K. Shukla	(a) 115,K105	61.40 and 78.70; S4
Yu.M. Shulga	(a) 115,K195	74.70; 78.30; 79.20; 79.60
V.N. Shumskii	(a) 114, 161	73.60; 68.55; S8.15
Yu.N. Shvachko	(b) 151, K59	76.30; S12
	(b) 151, K65	76.30; S12
A.M. Shvaika	(b) 156, 377	78.20
P. Siciliano	(a) 114, 253	71.55; S8
A.S. Sidorenko	(b) 155, K45	74.70; S10.15
Yu.Yu. Sidorin	(b) 156, K1	61.10 and 68.90; S1.3
D. Siebert	(a) 111, K93	74.70 and 76.30; S10.15
W. Siegel	(a) 112, K35	72.80; S7.12
J. Sielanko	(a) 112, 315	61.80 and 79.20; S5; S5.11
	(a) 112, 319	61.80; 79.20; S5.11
	(a) 112, 839	62.20; 66.30; S1.1
A. Siemko	(a) 113, 181	75.80; S1.1; S1.2
	(a) 114, K83	75.30; S1.1
J. Siemon	(a) 114, 621	68.55; 78.55; S8
K. Sierański	(a) 111, K57	72.80; S7.11
J. Sietsma	(a) 112, 35	61.40 and 64.70; 62.20; 72.15; S1.1; S1.2
A.J. Sievers	(b) 151, K97	63.20 and 78.50; S9.11
R. Signerski	(b) 153, 281	72.40 and 73.60; S12
J. Šikula	(a) 111, 367	72.70 and 73.40; 78.60; S7.15
A. Šileika	(b) 154, 389	78.20; 61.60; S8.15
P. Silfsten	(b) 156, K81	61.70 and 78.50; S9.11
J. Simanowski	(a) 114, 387	77.20 and 77.40; S10.15
A.V. Simashkevich	(a) 112, 305	61.80 and 73.40; S8.13
	(b) 154, 287	72.20; 71.30; 71.55; S8.12
	(b) 154, 691	72.20; 71.30; S8.12
A.V. Siminel	(a) 113,K227	71.55 and 72.40; 76.30; 78.55; S7.11
O. Simmich	(a) 113, K27	64.60; 66.30
	(a) 113, 269	61.55 and 64.75; S1.2
A.Yu. Simonov	(b) 155, K45	74.70; S10.15
Z. Šimša	(a) 115, 229	68.55 and 78.20; S11.2
A. Singh	(a) 111,K191	68.55 and 73.60; S1.61
H. Singh	(b) 152, 145	71.25 and 78.70; S1
M. Singh	(b) 154, 589	71.25 and 72.10; 78.20
	(b) 154,K131	71.60
N. Singh	(b) 154, K19	63.20; S2
	(b) 156, K33	63.20; S1.1
N. Dhiren Singh	(a) 112,K101	73.60; S4
R. Singh	(b) 153, 123	62.20; S9.12
	(b) 156, 131	62.20; S9.15
S.J. Singh	(a) 114, 407	78.60

S.K. Singh	(b) 153, 153	71.20 and 73.60; 78.70; S8.15
S.N. Singh	(a) 115,K125	68.55 and 85; 73.40; S5.11
S.P. Singh	(b) 154, K19	63.20; S2
T.N. Singh	(b) 154, 127	63.20; S5.2
V.K. Singh	(a) 115, 135	64.75; 72.20; S8.15
J. Sitek	(a) 113,K143	76.80 and 78.70; S1.1
F.F. Sizov	(a) 113, 107	73.40; S8.15
L.I. Skatkov	(a) 115, 207	68.55; 73.60; S10
E.M. Skok	(b) 152, K59	78.20; 71.55; S7.13
R. Skomski	(b) 153, 711	75.10; 75.30
	(b) 156,K149	75.60 and 75.90
M.Ya. Skorokhod	(a) 111,K145	61.70; S7.13
	(a) 116,K141	61.10; 61.70; S5.11
W. Skorupa	(a) 112, 661	61.70 and 61.80; S5.11
	(a) 114, 135	61.80; S5.11
T. Skrabka	(a) 112,K105	73.40; 85; S7.12
M. Skrinjar	(b) 156,K143	75.10
M.J. Škrinjar	(b) 155, K77	75.10
A.V. Skripov	(b) 151, K59	76.30; S12
	(b) 151, K65	76.30; S12
N.A. Skubenko	(b) 153, 667	71.35 and 71.55; 72.40; 78.55; S8
P.A. Skubenko	(b) 153, 667	71.35 and 71.55; 72.40; 78.55; S8
L.N. Skuja	(a) 114, 731	78.55; S10.1
A.I. Skvortsov	(b) 156, 605	76.80; 76.30
V.N. Skvortsova	(a) 116,K125	61.80 and 78.40; 78.55; S11.2
M.J. Słaby	(a) 112,K105	73.40; 85; S7.12
J. Slama	(a) 115, 533	75.60; S1.1; S1.2
A.M. Slupskii	(a) 112,K131	79.20
E.M. Slyusarenko	(b) 154, 475	66.30
S. van Smaalen	(a) 114, 91	61.60; S1.63
A. Smal	(a) 115,K207	68.55 and 74.70; S10.15
V. Smatko	(a) 114, 567	68.55 and 74.70; S10.15
Z. Smetana	(a) 111, 285	75.50; S1.2; S1.4
	(b) 153, K69	71.70 and 75.60; S1.2
	(a) 114,K229	75.40; S1.4
A.S. Smirnov	(a) 116,K125	61.80 and 78.40; 78.55; S11.2
V.P. Smirnov	(b) 151, 275	75.10; 71.25; 71.50; S11
	(b) 152, 633	75.10; 71.25; 71.50; S11
I.S. Smirnova	(a) 116, 237	61.70; 74.70; S10.15
W.M.A. Smit	(b) 156, K93	78.55; 78.50; S11.1
	(b) 156, 537	71.70; 78.55; S8.11
G.B. Smith	(b) 152, 475	61.40 and 68.55; 78.70; S5; S5.11
I.O. Smith	(a) 115, 125	62.20; S4
P.V. Smith	(b) 154,K127	71.10; 62.20; S1.61
G.W. Smith II	(b) 156, 583	74.20
E.M. Smolarenko	(a) 115,K155	64.70; S7
M.A. Smondyrev	(b) 155, 155	71.38
S.V. Smorodinov	(a) 114,K139	61.80; S7.11
L. Smrčka	(a) 111, K77	74.70; S10.15
O. Smrčková	(a) 111, K77	74.70; S10.15
F. Smutný	(b) 154, 777	78.30; 77.80; S9.16
	(b) 154,K103	64.70; 77.20; S11
I.I. Snighireva	(a) 114, 113	61.70; S7.12
L. Sobczyk	(a) 111, K27	64.70; 77.20; S12
E. Sobeslavsky	(a) 114, 135	61.80; S5.11
M. Abdus Sobhan	(b) 156, 287	71.25; 71.55; S5.11; S5.12

J. Sobhanadri	(a) 113, 477	78.65; 68.55; S8.15
V.V. Sobolev	(b) 154, 377	71.25 and 78.40; 71.35; S7
H. Sobotta	(a) 112,K139	61.60 and 78.30; S7.15; s8.15
	(a) 116,K147	61.80 and 78.30; 63.20; S7.11
N.V. Sochinskii	(a) 115, K31	68.55; S8.15
K. Sokalski	(b) 152, 275	75.10 and 75.30
V. Sokol	(a) 114, K33	73.40; S4; S10.1
O.B. Sokolov	(b) 151, 623	74.40
V.I. Sokolov	(a) 116, 615	61.70 and 64.75; S10.1
V.O. Sokolov	(b) 156, 663	78.55; 71.55; S10.1
Zh.D. Sokolovskaya	(a) 115,K195	74.70; 78.30; 79.20; 79.60
	(a) 116, K13	61.50; 74.70; S10.15
V.S. Soldatov	(b) 154, 475	66.30
I. Soletta	(a) 115, 459	61.40 and 64.60; S1.1
V.S. Solovev	(a) 112, 703	61.80 and 64.75; S5.11
T. Soma	(b) 153, K17	62.20; 63.20; S7.1
	(b) 155, K5	62.20; 63.20; S1.3
G. Somé	(a) 116, 689	62.20; 66.30; S1.2
F. Somma	(b) 151, 453	61.70; 78.50; 78.55; S9.11
H. Sommer	(b) 152, K47	71.20; 79.60; S8.16
	(a) 112, 533	68.48 and 68.55; S1.1; S7
R. Sommerfeldt	(a) 113, K75	61.70 and 72.40; 78.50; S11.1
K. Somogyi	(a) 114, 635	66.30 and 73.40; 78.65; S7.12
T. Soovik	(b) 152, 563	71.55 and 78.55; S9.11
L.E. Soshnikov	(a) 111, 485	62.20 and 64.60; S7
J. Sosnowski	(b) 154, 543	63.20; 62.20; S1.3
W. Soszka	(a) 112, 425	61.80; 68.20; 79.20; S1
N. Sotnik	(a) 111, 387	73.40 and 85; S1.3; S6; S10
M. Souiri	(b) 156, 71	61.70; 76.90
E.N. Sovyak	(b) 155, 549	71.55 and 73.40
M. Sowa	(a) 112, 835	66.30; S1.1
	(a) 112, 839	62.20; 66.30; S1.1
R. Spagnolo	(a) 114, 519	65.70 and 66.70; 72.80; S5.11
H.N. Spector	(b) 151, 133	71.35
V.N. Spector	(a) 116, K13	61.50; 74.70; S10.15
W. Sperber	(a) 113, K99	75.60
	(a) 116, 389	75.60
V.N. Spiridonov	(a) 113, 315	61.80 and 73.20; 68.48; S5.11
A.V. Spitsyn	(a) 111, 301	78.20 and 78.55; 78.60; S8.11
B. Sreedhar	(b) 153, 257	71.70; 76.30; 78.50; S11
K. Srinivasan	(a) 114, 749	78.20; 61.70; S5.11
C.M. Srivastava	(a) 111,K229	75.80; 76.50; S1.1
K.K. Srivastava	(a) 115, K57	72.80; S8
K.S. Srivastava	(b) 155, 127	68.30; 71.45; S7.12
S.K. Srivastava	(a) 116, 837	73.40 and 85
J.L. Staehli	(b) 152, 101	73.40; 71.35; 72.40; 78.55; S7.15
A. Stahl	(b) 153, 773	71.25 and 78.20; S8.11
M. Stahl	(b) 154, 815	72.10 and 73.40; 77.50; S8.15
I.M. Starchenko	(a) 111,K209	74.70; S10.15
P.M. Starik	(a) 111, K99	72.80 and 75.50; S8
N.V. Starostin	(b) 155, K51	74.20
Z. Stary	(a) 115, 87	61.70; 72.20; S8
I.V. Stasyuk	(b) 155, 697	78.20
	(b) 156, 377	78.20
V. Stavrov	(a) 114,K131	61.70; 79.20; S5.11
O. Stefan,Jr.	(a) 114, 643	71.55; S5.11
P. Stefański	(b) 151,K195	75.30; S1.1
	(a) 114, 355	75.30; 61.55; S1.1; S1.2

P. Stefanski	(b) 156, 657	75.30; S1.1; S1.2; S1.4
P. Stegger	(b) 151, 463	61.70; 76.30; S10.1
J. van der Stel	(a) 112, 35	61.40 and 64.70; 62.20; 72.15; S1.1; S1.2
V.F. Stelmakh	(a) 112, 381	66.30; 61.80; 62.20; S5.11
	(a) 112, 695	61.70; 79.20; S5.11
	(a) 115, 427	61.80; 61.70; 78.60; S5; S7.1
S.I. Stenin	(a) 113, 421	68.55; 61.16; S5.11
H. Stenzel	(a) 116, 39	66.30 and 68.48; S9.11
O. Stenzel	(a) 112, 667	61.80 and 68.55; 73.60; 78.65; S5
	(a) 115,K247	78.65; S5
Yu.S. Stepanchuk	(a) 116, 295	62.30
A. Stepanescu	(a) 116, 813	75.60
V. Stepankevicius	(b) 153,K193	78.55; 71.35; 78.30; S7.12; S7.15
S.A. Stepanov	(a) 114, K7	61.10; 68.20; S5.12
V.N. Stepanov	(a) 115,K255	61.80 and 78.60; S10.1; S12
V.S. Stepanyuk	(b) 155, 179	71.25; 78.20; S10.1
	(b) 155, 215	71.25; 78.20; S8
D. Stephan	(a) 112, 79	61.80; 61.10; S5.11; S5.12; S7.1
J.D. Stephenson	(a) 115, 97	61.70; 65.70; S5.11; S5.12
J. Stępień-Damm	(a) 116, 367	74.70; S10.15
Yu.L. Steshenko	(b) 153, 539	63.90; 71.45; 78.55; S5.2
H. Stettin	(a) 111,K265	61.25 and 77.20; S12
J. Steuer	(a) 114, 665	74.10 and 74.50; S1
A.D. Stevens	(b) 154, 483	61.70; 76.30; S9
W.M. Stobbs	(a) 115, 61	61.16; S1.2; S4
U. Stoepel	(a) 115, 165	64.70 and 77.80; 64.80; S10.15
Gh. Stoicescu	(b) 156, 411	61.70; 66.30; 78.50; S9.11
V.F. Stolarenko	(b) 156, K23	61.70; S9.11
A. Stolovich	(b) 151, 29	61.70; S9.11
V.V. Stolyarov	(a) 112, 137	75.50; 61.40; S1.1
W. Stolzmann	(b) 153,K109	65.50
	(b) 156, 515	65.40 and 71.45
M.I. Strashnikova	(b) 152, 675	78.20; S8.11
P. Streitenberger	(b) 152, 111	71.10 and 73.40
W. Stręk	(b) 154, K89	78.55; S9.16
E.A. Streltsov	(a) 111, 193	68.45 and 72.40; S10.1
V.A. Streltsov	(a) 115, 515	71.20; 61.55; S10
E.K. Stribuk	(a) 111, K89	74.70; S10.15
	(a) 111,K209	74.70; S10.15
	(a) 113,K235	61.80 and 74.70; S10.15
	(a) 115,K205	74.70; S10.15
G. Strinati	(b) 153, 611	71.35 and 71.38
M. Stróżak	(a) 112, 739	61.55 and 61.80; 68.48; S1.2
E.E. Strumban	(a) 113,K227	71.55 and 72.40; 76.30; 78.55; S7.11
H. Stubb	(b) 154, 305	72.20; 72.60; 76.30; S12
N.V. Stuchinskaya	(a) 115, 561	78.60; S7
S.A. Studenikin	(b) 152, K59	78.20; 71.55; S7.13
N.M. Stus	(a) 115,K117	78.45 and 78.55; S7.12; S7.15
U. Stutenbaeumer	(b) 156, 561	71.35 and 71.70; 78.55; S8.15
Gang Su	(b) 156, K51	74.20
Wenhui Su	(a) 116, 267	64.70 and 74.70; 61.10; S10.15
M. Subotowicz	(a) 112, 353	61.80; 68.20; S1.1

M. Subotowicz	(a) 112, 415	61.80; 71.55; S5.11
	(a) 112, 739	61.55 and 61.80; 68.48; S1.2
	(a) 112, 753	61.55 and 61.80; 68.48; S4
	(b) 153, 361	76.30; S8.15
	(a) 115,K207	68.55 and 74.70; S10.15
R. Subrahmanyam	(b) 151, 445	61.70; 71.55; 78.50; S9.11
M. Suchańska	(a) 112, 437	79.20; 75.50; S1.2
A.V. Sudarkin	(a) 113,K149	61.14
P. Sueptitz	(a) 113, K55	68.55; 78.65; S8
	(a) 113,K171	61.80; 66.30; 79.20; S8
A. Suga	(a) 113, 459	73.30 and 73.60; 73.20; S1.61; S7.12
H. Sugier	(a) 113, K49	68.45; S10.1
	(a) 116,K165	61.80 and 68.45; S10.1
M. Suhrke	(b) 154, 315	73.40
A.A. Sukhovskii	(a) 111, 477	64.70; 76.60; S11.1
	(b) 156, 615	76.60; 66.30; 72.20; S9.16
S.V. Sukhvalo	(a) 114, 573	73.60; S1.63
B. Sukumar	(b) 154, 643	71.55; 71.38; S5.11; S7.11
A.M. Suleimanov	(b) 153,K203	78.30; 71.70; 75.30; S1.62
V.B. Sulimov	(b) 156, 663	78.55; 71.55; S10.1
C. Sulkowski	(a) 114, K53	74.70; S1.62
Cz. Sulkowski	(a) 116, 367	74.70; S10.15
N.S. Sullivan	(b) 155,K109	71.20; 76.60; S5.2
K. Sumiyama	(a) 114, 693	72.15 and 75.50; 61.55; S1.2
K. Sun	(a) 115, 539	75.60; 76.80; S10
Shifang Sun	(b) 154, K51	74.70; S10.15
X.K. Sun	(a) 112, 175	75.50; S1.1; S1.2
	(a) 115,K219	75.50; 75.30; S1.1
Yupin Sun	(a) 115, K75	74.70; 62.40; S10.15
J. Suñé	(a) 111, 675	73.40 and 77.50; 73.60; 85; S10
	(a) 113, 83	68.55 and 73.40; 68.48; S10.1
	(a) 114, 167	68.55; S10.1
Yu.R. Suprun-Belevich	(a) 112, 381	66.30; 61.80; 62.20; S5.11
Yu.P. Surov	(a) 111,K243	78.70; S11
	(a) 114, 481	61.70; 61.80; 78.70; S7.12
K.D. Sushkevich	(a) 112, 305	61.80 and 73.40; S8.13
E. Susi	(a) 113, K61	73.25; S5.11
	(a) 114, 247	71.55; 72.20; 78.30; S5.11
L.G. Suslina	(b) 154, 609	64.70 and 71.25; 71.35; 78.40; S8.15
A. Sütő	(b) 151,K101	61.50
A.V. Suvorov	(a) 112, 707	61.80; 64.70; S1.61
S. Suzuki	(a) 113, 51	61.70 and 62.40; S1
	(a) 113,K245	75.50; S11.2
P. Švec	(a) 115, 143	64.70; 66.70; S8
B.G. Svensson	(a) 114,K147	61.80 and 73.30; S5.11
P. Svoboda	(a) 111, 285	75.50; S1.2; S1.4
	(b) 153, K69	71.70 and 75.60; S1.2
	(a) 114,K229	75.40; S1.4
S. Swarup	(a) 115,K105	61.40 and 78.70; S4
K. Swiatek	(a) 114, 127	61.70; 76.30; S8.11
H.K. Sy	(b) 156, 455	71.45 and 73.20
D. Sýkorová	(a) 111, K77	74.70; S10.15
M.C.R. Symons	(b) 152, 583	71.50 and 76.30; 71.38; 76.40; S9
	(b) 154, 483	61.70; 76.30; S9
V.N. Syromyatnikov	(b) 155, 419	64.60; 75.30; S1.61

O.I. Syrotynskii	(a) 116,K129	78.20; 74.70; S12
I.A. Szabó	(b) 155, 385	61.70
L. Szaro	(a) 113, 97	73.20 and 73.30; 72.40
A. Szász	(b) 152, 117	68.45 and 73.20; 71.35; S1
	(b) 155, 179	71.25; 78.20; S10.1
	(b) 155, 215	71.25; 78.20; S8
J. Szatkowski	(a) 111, K57	72.80; S7.11
A. Szlaferek	(a) 114, 355	75.30; 61.55; S1.1; S1.2
	(b) 156, 657	75.30; S1.1; S1.2; S1.4
J. Szymaszek	(a) 114, K49	74.70; S1.4
H. Szymçzak	(a) 113, 151	74.70; S10
M. Szymonski	(a) 112, 431	79.20; S9.11
W. Szyszko	(a) 112, 315	61.80 and 79.20; S5; S5.11
	(a) 112, 835	66.30; S1.1
	(a) 112, 839	62.20; 66.30; S1.1
A. Szytuła	(a) 116, K29	61.12; S1.61
V. Tabakova	(a) 116,K101	78.55; S9.11
K. Tabata	(a) 111,K105	72.80 and 75.50; S10.15
S. Tabor	(b) 156,K165	75.10
M. Taege	(a) 116, 179	61.40; 61.16; 62.20; S12
M.M. Tagiev	(b) 152, 317	64.60 and 78.30; S8.16
E.V. Tagirov	(b) 152, 329	78.20; S8
Y. Tagishi	(a) 116, 583	61.80 and 62.40; S1
M.M. Tagyev	(b) 151, 759	78.20; S8.16
Luu Tuan Tai	(a) 111, 591	75.30; 75.25; S1.1; S1.4
M. Taibi	(a) 115, 523	71.70; 78.50; 78.55; S10.15
M.A. Tairov	(a) 111, 289	78.20; S1.63; S8.16
	(b) 151, 711	78.20; S8.16
	(a) 113, 207	72.40 and 73.40; 85; S7.16
	(a) 115, 11	72.40; 78.20; S7.16
W. Takahara	(a) 116, 787	74.70; 61.80; S10.15
S. Takahashi	(a) 111, K21	64.70 and 64.75; S8.15
A. Takase	(a) 115,K215	75.50; S1.63
N. Takeuchi	(a) 111, 165	62.20; 62.40; S9.11
	(a) 114,K245	78.60 and 79.75; S10
	(a) 115,K251	78.60 and 79.75; S9.16
M.G. Takwale	(a) 111, 457	61.70; 72.20; S5.11
G.N. Talalakin	(a) 115,K117	78.45 and 78.55; S7.12; S7.15
V.M. Talanov	(a) 115, K1	61.60; S11.2
	(a) 116, 603	64.10
D. Talapatra	(b) 155, 541	71.70; 78.20; S9.16
G. Tamulaitis	(b) 153,K193	78.55; 71.35; 78.30; S7.12; S7.15
S. Tamulevičius	(a) 114, K25	61.80 and 68.48; S5.11
T. Tanabe	(a) 114,K143	66.30 and 76.60; S12
K.A. Tanaka	(a) 116, 787	74.70; 61.80; S10.15
A. Tandon	(b) 155, 127	68.30; 71.45; S7.12
C.Q. Tang	(a) 113,K161	61.70 and 78.70; S1.2
S.Y. Tang	(b) 154, 589	71.25 and 72.10; 78.20
	(b) 154,K131	71.60
Y.S. Tang	(b) 154,K187	78.20 and 78.65
	(b) 156,K153	78.55; S5.11
Yi Tang	(a) 114,K189	74.70; S10.15
K. Taniguchi	(b) 154, 195	73.20 and 73.40
E.A. Taraban	(a) 113, 359	64.75; 76.80; S10
E. Taranko	(b) 155, 481	68.20 and 68.45; 73.20

R. Taranko	(b) 155, 481	68.20 and 68.45; 73.20
S.A. Taraskin	(a) 113, K31	64.70; 65.40; S11
P. Tarkowski	(a) 112, 353	61.80; 68.20; S1.1
T. Tarnóczi	(a) 114, K83	75.30; S1.1
E. Tarnowska	(a) 112, 385	61.80; 66.30; S5.11
L. Tarricone	(b) 152, 101	73.40; 71.35; 72.40; 78.55; S7.15
J. Tatarkiewicz	(b) 153, 11	61.80 and 78.20; 78.30; 78.50; 78.55; S5.11; S7.12
	(b) 154, K99	61.70 and 63.20; S5.11
	(b) 155, 77	63.20; 78.30; S7.13; S8.11; S9
M. Taut	(b) 153, 225	71.45; S12
A. Tawfik	(a) 114, 377	77.20 and 77.40; S11.2
M. Taya	(a) 114, 157	64.70 and 77.80; S12
D.A. Tayurskii	(b) 152, 645	76.20; 76.30; 76.60; 76.70
H. Teichler	(b) 154, 649	71.55; S5.11
M.J. Tello	(b) 154, 143	64.60; 65.40; S9.16
Mingkang Teng	(a) 111,K123	78.70; S11.1
M.A. Teplov	(b) 152, 191	71.70; 76.30; 78.50; S9.16
S.V. Teplov	(a) 112, 437	79.20; 75.50; S1.2
A. Tepore	(a) 114, 253	71.55; S8
P. Terreni	(b) 152, 601	72.10 and 72.20
L. Terziev	(a) 116, 521	61.55; 61.14; 64.70; S1.2
L. Tesař	(a) 114, 475	61.70 and 71.55; 73.40; S5.11
K.P. Thakur	(b) 155, 445	64.60; S1.5
A. Thanailakis	(a) 116, 725	73.60 and 78.65; S6
N.D. Theodore	(a) 114, 105	61.70; S7.12
J. Thery	(b) 152, 415	61.70; 76.30; 78.50; 78.60; S11
M. Theuerkauf	(b) 156, 431	63.50
W. Thielemann	(a) 112, 229	73.40 and 78.60; 68.55; 78.55; S7.15
A. Thomas	(a) 113,K171	61.80; 66.30; 79.20; S8
	(a) 116, 735	61.80 and 68.55; 77.55; S9.12
H. Thomas	(a) 112, K29	71.55; 85; S7.12
J.G. Thompson	(b) 152, 519	71.45 and 79.60; 74.10; S11
F. Thrum	(a) 111, 551	72.15 and 73.60; S1.2; S10.1
Nguyen Phu Thuy	(a) 111, 591	75.30; 75.25; S1.1; S1.4
Decheng Tian	(a) 114,K135	61.70 and 78.70; 64.70; S4
	(a) 115, K91	76.80; S1.1
	(a) 116, 203	61.40; 75.60; 76.80; S1.1
	(a) 116, 621	62.40; 61.70; S1.4; S3
C. Tien	(a) 112, 99	63.20; 78.35; S1.2
I.M. Tiginyanu	(a) 111, 301	78.20 and 78.55; 78.60; S8.11
	(a) 113,K265	78.55; S8.16
	(a) 114, 259	71.55; 78.55; S8.16
	(a) 115,K191	71.55; S7.11
S.A. Tijs	(a) 113,K165	61.70; S5.11
G.F. Tikhinskii	(b) 151,K161	74.70; S10.15
B. Tillack	(a) 112, 721	73.40; 68.55; S5.11
W.A. Tiller	(b) 156,K105	68.45; S7.12
	(b) 156,K113	68.40
D.R. Tilley	(b) 152, 135	71.36; 71.45; 75.10
I.N. Timchenko	(b) 154, 287	72.20; 71.30; 71.55; S8.12
	(b) 154, 691	72.20; 71.30; S8.12
S.D. Tiron	(b) 154, 209	71.35; 78.20; S10.1
A.M. Tishin	(a) 113,K117	75.40; S4
	(a) 114, K79	75.30; S1.4

A.M. Tishin	(a) 114, K99	75.30; S1.4
A.S. Titi	(a) 114, 83	61.16 and 61.70; S1.2
K. Tittelbach	(a) 116, K37	61.70; S5.11
K. Tittelbach-Helmrich	(a) 113, K9	66.30; S5.11
S.S.K. Titus	(a) 114, 525	64.70; 61.40; 65.70; 72.80; S8
A.N. Tiwari	(a) 111,K229	75.80; 76.50; S1.1
D. Tjapkin	(b) 156, 487	71.20; 71.30; S5.11
A.S. Tkachenko	(a) 112, K9	61.70; 71.55; 73.40; S5.11
	(a) 113, 653	73.40 and 85; 76.30; S5.11; S10.1
N.V. Tkachenko	(b) 154, 353	75.30 and 75.50
V.V. Tkachev	(a) 115, 427	61.80; 61.70; 78.60; S5; S7.1
N.T. Toan	(a) 115,K163	68.55 and 78.65; S12
N. Toichev	(b) 156, K77	76.30; 61.80; S9.11
V.V. Tokarev	(a) 111, 499	61.70 and 68.55; S1.61
	(a) 116, 331	68.55; 61.70; S1.61; S5.11
Y. Tokumitsu	(a) 115,K261	78.30; S8.15
H. Tollet	(a) 114, 567	68.55 and 74.70; S10.15
V.A. Tolokonnikov	(a) 111, 201	71.55 and 73.30; S8.12
A. Tomas	(a) 113, K5	61.16 and 61.60; S8
D. Tomáš	(a) 113, K95	75.60; S11.2
I. Tomáš	(a) 113, K95	75.60; S11.2
M. Tomášek	(b) 156, 243	68.25 and 68.40; 73.20; S5.11
P.E. Tomaszewski	(a) 111, K27	64.70; 77.20; S12
	(a) 112, K13	64.70 and 77.80; 77.70; S11.1
Zh.M. Tomilo	(a) 114, K57	74.70; S10.15
	(a) 115, K79	61.70 and 74.70; S10.15
A. Tomita	(a) 114,K245	78.60 and 79.75; S10
	(a) 115,K251	78.60 and 79.75; S9.16
J.W. Tomm	(a) 114, 621	68.55; 78.55; S8
K. Tompa	(b) 153, 721	76.60; 71.50; S1; S1.3; S4
A.G. Toneian	(b) 156, 633	64.75 and 75.30; S1.4
V. Topa	(b) 152, 51	61.70; S9.11
V.I. Torgashev	(b) 153, 517	64.60 and 78.30; S8.16
	(b) 153, 529	64.60 and 78.30; S8.16
	(b) 154, 777	78.30; 77.80; S9.16
E.E. Tornau	(b) 156, 93	64.60 and 74.20
M. Torres	(b) 154, 439	61.50
A.L. Tóth	(a) 113,K147	61.70 and 85; 68.55; S5.11; S10.1
	(a) 116,K137	61.16 and 72.20
L. Tóth	(b) 155,K143	75.10
M. Toulemonde	(a) 114, 467	61.80; S5.11
R. Tovar	(a) 111, 405	61.60; 64.75; S8.16
	(a) 115, 157	64.70; S8.15
M.V. Tovarnitskii	(a) 114, 293	72.80; 71.55; S8.16
J.E. Townsend	(a) 114, 739	78.60; S10.1
P.D. Townsend	(a) 114, 739	78.60; S10.1
I. Trabjerg	(b) 154, K89	78.55; S9.16
C. Trallero-Giner	(b) 152, 357	71.35 and 78.30; S9
M. Trapp	(a) 113, 315	61.80 and 73.20; 68.48; S5.11
V.Yu. Trashchakov	(a) 115,K201	74.70 and 78.70; S10.15
H.-R. Trebin	(b) 155, 349	61.30 and 62.10
A.B. Treshchalov	(b) 154, 763	78.40; 71.35; S9.11
D. Triantis	(a) 114, 491	61.70; 77.30
Ch.G. Tricalinos	(b) 153, K1	61.80; S4
S.B. Trickey	(b) 155, K93	73.30; S4
Ch. Trikalinos	(b) 152, K9	74.70 and 76.30; S10.15

J. Troeltzsch	(a) 115,K139	66.30; S8
O.A. Troitskii	(b) 151, 121	68.20 and 72.15
E. Trojnar	(a) 116, 367	74.70; S10.15
M. Trott	(b) 151,K123	71.25 and 73.90
	(b) 152, 153	71.10; 71.20; 71.25; 71.50
	(b) 153, K39	71.10; 71.50
S. Trott	(b) 151,K123	71.25 and 73.90
	(b) 152, 481	73.40; S7.13
I.O. Troyanchuk	(a) 112, 155	75.50; S10.15
	(a) 113,K107	75.30 and 76.60; S11.1
	(a) 113,K249	75.60; S10
	(a) 115,K225	75.50; S10.15
J. Trpcevska	(a) 115, K13	62.20; 68.20; S1.1
V.M. Trukhan	(a) 115,K155	64.70; S7
Truong Ba Ha	(b) 155, 685	78.70
U. Trutschel	(b) 151, 383	78.65
J. Trzcianka	(a) 112, 793	61.80; 78.20; S7.12
	(a) 112, 835	66.30; S1.1
K. Trzebiatowski	(a) 115,K185	73.25; S10.15
T. Tsang	(a) 114, K1	61.30; S12
L. T. Tsankov	(a) 114, 341	76.80
W.S. Tse	(b) 154, 135	63.20; 71.36; 78.30; S7.15
M.S. Tseluevskii	(a) 115, K79	61.70 and 74.70; S10.15
N.K. Tsenev	(a) 115, 451	61.70; S4
I.M. Tsidilkovskii	(b) 154, 703	71.30 and 72.20; S5.12
	(a) 115, 267	74.70; 72.20; S10.15
V.I. Tsidilkovskii	(a) 115, 267	74.70; 72.20; S10.15
V.G. Tsirelson	(b) 155, 201	71.45; 78.70
	(a) 115, 515	71.20; 61.55; S10
T. Tsuboi	(b) 151, K91	63.20 and 78.40; S9.12
	(b) 153,K159	75.30 and 75.50; S9.16
	(b) 156, K81	61.70 and 78.50; S9.11
S. Tsunekawa	(a) 112, 207	78.30; S11.1
J.W. Tucker	(b) 156,K139	75.10
A.I. Tugushi	(b) 155,K153	76.60
	(b) 156, K67	76.60
C. Tuijn	(b) 155, 107	65.50; 64.60
M. Tuleta	(a) 112, 811	61.80 and 78.55; S10
V.E. Tupitsyn	(b) 154, 377	71.25 and 78.40; 71.35; S7
M. Tu[illegible]aturi	(a) 114, 519	65.70 and 66.70; 72.80; S5.11
M. D. Turner	(b) 153, 59	61.50
A. Turos	(a) 112, 253	61.70; S1.1; S1.2; S1.3; S1.4
	(a) 112, 343	61.80; 61.70; S1.1
U.S. Tursunov	(b) 155,K113	71.55; S5.11
M. Turtenwald	(a) 115, 273	74.70; 74.50
D.M. Turtsevich	(a) 114, K57	74.70; S10.15
	(a) 115, K79	61.70 and 74.70; S10.15
I.E. Tyschenko	(a) 113,K165	61.70; S5.11
A.G. Tyurin	(a) 115, K47	72.80; S12
Yu.N. Tzarev	(b) 153, 161	73.40; 71.25; S7.15
Do Ngoc Uan	(a) 114, 551	68.55; S1.3; S12
K. Uematsu	(a) 113,K245	75.50; S11.2
O. Uemura	(a) 112, 467	61.40; 61.12; S8.16
V.V. Uglov	(a) 111, K13	61.80 and 68.48; S1.2
N.A. Ukhin	(b) 154, 507	61.80
J. Ulański	(a) 114,K107	78.60; S12

S. Ulbricht	(a) 116, 571	61.70; 66.30; S5.11
J. Ulner	(b) 151,K181	75.70; 75.80
	(b) 153, 331	75.10 and 75.30
W. Ulrici	(b) 155, 597	71.55; 76.30; S7.11
Y. Umakoshi	(a) 116, 787	74.70; 61.80; S10.15
C. Umbach	(b) 156, 371	71.25 and 78.20; 77.20; S8.15
K. Unger	(b) 154, 219	71.10
W. Unger	(a) 114,K175	68.48 and 79.60; S4; S12
A. Unkroth	(a) 112, K85	62.20; 61.80; 75.60; S12
A. Urbaniak-Kucharczyk	(b) 152, 267	75.70 and 76.80
	(b) 154, 327	75.30 and 75.70
	(b) 154, 583	73.60
E. Urbanova	(a) 113, 615	78.30; S8.15
V.I. Urenev	(a) 115, K9	61.70 and 71.55; S5.12
A.R. Urmanov	(b) 153, 65	61.80 and 79.20
V.V. Ursakii	(a) 115,K191	71.55; S7.11
G.I. Urushadze	(b) 156, K63	72.20 and 73.40
N. Uryû	(b) 152, 289	75.10 and 75.30
	(b) 154, 747	75.10; 75.30
A.P. Ushakova	(a) 114, 113	61.70; S7.12
V.D. Usikov	(a) 113, 667	73.40; 72.40
E.V. Ustelemova	(b) 155, 419	64.60; 75.30; S1.61
V.V. Ustinov	(b) 151, K59	76.30; S12
	(b) 151, K65	76.30; S12
E. Utkina	(a) 114, K33	73.40; S4; S10.1
J. Utzig	(b) 153, 377	76.80; S1.2; S5.12
A.M. Vainrub	(b) 155, 147	71.45; 63.20; S12.1
J. Vaitkus	(a) 112, 375	71.55 and 78.20; S5.11
P. Vajda	(b) 154, 321	72.15; 75.50; S1.4
O. Valassisdes	(b) 153, K51	72.20; S8
V. Valcheev	(a) 113,K231	72.80; S7.12
R.Z. Valiev	(a) 112, 137	75.50; 61.40; S1.1
	(a) 115, 451	61.70; S4
S.A. Valiev	(a) 113, 653	73.40 and 85; 76.30; S5.11; S10.1
C.P.G. Vallabhan	(a) 112,K113	72.80 and 77.20; 64.70; S11.1
	(a) 114,K233	78.60; S8.11
	(a) 115, K63	72.80; S10.1
M. Vallade	(a) 111, 155	64.60; S10.1
D.M. Vanderwalker	(a) 111, 119	61.70 and 62.50; S1.2
	(a) 112, 73	61.70; 61.16; S1.2
P. Vaněk	(b) 154, 777	78.30; 77.80; S9.16
A.V. Vannikov	(a) 115, K47	72.80; S12
R.A. Vardanyan	(b) 156, 527	61.70 and 71.55; 72.90; S8.1
J. Varghese	(a) 113, 123	68.55; 78.65; S10
J. Varrio	(a) 116, 443	68.55; S7.11; S7.12
V.S. Varychenko	(a) 115, 427	61.80; 61.70; 78.60; S5; S7.1
M.A.Z. Vasconcellos	(a) 112, 105	68.48 and 68.55; S1.1; S4
P. Vašek	(a) 111, K77	74.70; S10.15
A.V. Vasilev	(a) 116, K33	61.70; S5.11
V.V. Vasilev	(a) 116,K105	76.30 and 78.55; S6
M.A. Vasiliev	(a) 115, K95	75.50; S1.1; S1.2
O.I. Vasin	(a) 114, 161	73.60; 68.55; S8.15
P. Vašina	(a) 111, 367	72.70 and 73.40; 78.60; S7.15
A.B. Vassilikou-Dova	(b) 155, K59	76.30; 71.55; S9.16
L. Vatskichev	(a) 112, K57	61.50 and 68.55; S1.61

M. Vatskicheva	(a) 112, K57	61.50 and 68.55; S1.61
J. Vávra	(a) 111,K133	61.12
	(a) 112, 29	61.12 and 62.65; S7.13
K. Vavřina	(a) 116, K85	72.20; S5.11
P.R. Vaya	(a) 113, 477	78.65; 68.55; S8.15
M. Vazquez	(a) 113, 187	75.50; S1.1; S1.2
	(a) 115, 547	75.60; 75.25; S1.1; S1.2; S1.61; S1.62
R.V. Vedrinskii	(b) 151, 581	71.25 and 78.70; S9.11
H.B. van Veldhuizen	(a) 112, 35	61.40 and 64.70; 62.20; 72.15; S1.1; S1.2
I. Veltruský	(b) 155, 669	76.60; 75.25; S11.2
V.K. Venkatesan	(a) 113,K217	73.30; 85; S8.11
G. Vértesy	(a) 113, 573	75.70 and 76.50; 75.60; S11.2
M. Verwerft	(a) 112, 451	61.16 and 64.60; S11.1
S.G. Veselko	(b) 156, 527	61.70 and 71.55; 72.90; S8.1
J.L. Vicente	(b) 155, K93	73.30; S4
Yu.A. Vidadi	(a) 115, K53	72.80; S10.1
Nguyen Ai Viet	(b) 151,K171	74.20
D. Vignaud	(b) 156, 717	78.20; 61.70; S7.12
J. Vignolo	(a) 112, K53	78.60; S10.1
L. Viña	(b) 156, 371	71.25 and 78.20; 77.20; S8.15
A.V. Vinogradov	(a) 113,K149	61.14
E.A. Vinogradov	(b) 153, 727	78.30; 77.80; S8.16
S.L. Vinogradov	(a) 111, 201	71.55 and 73.30; S8.12
A.V. Vinokurov	(b) 152, 191	71.70; 76.30; 78.50; S9.16
H. Vinzelberg	(b) 151,K155	72.80; S1.61
F. Viras	(a) 115, 483	64.70 and 72.80; 61.40; S7.16
R. Visocekas	(b) 152, 415	61.70; 76.30; 78.50; 78.60; S11
A.A. Vitkauskas	(a) 111, 507	68.48; 79.20; S7.12
D. Vivien	(b) 152, 415	61.70; 76.30; 78.50; 78.60; S11
D. Vladikova	(a) 111, 145	64.80; 72.20; 75.50; S11.2
A.F. Vladimirov	(a) 111, 99	61.70; 68.25
V.I. Vladimirov	(a) 113, 337	61.70 and 68.48
G. Vlasak	(a) 111, 597	75.80; S1.1; S1.2
S.I. Vlaskina	(a) 116,K169	68.55; S6
K.B. Vlasov	(b) 155, 419	64.60; 75.30; S1.61
C.M. van Vliet	(b) 154, 713	72.10 and 72.70; S5.11
O.G. Vlokh	(a) 116, 287	64.60; 62.20; 78.20; S12
G. Voelkel	(b) 151, K53	76.30 and 77.80; S11.1
	(a) 115, 247	71.55; 76.30; S10.1
W. Vogelsberger	(b) 155, K17	68.40; 64.10
A.V. Voitsekhovskii	(a) 113, 285	61.80; 72.20; 79.20; S8.15
A.A. Volkov	(b) 153, 529	64.60 and 78.30; S8.16
V.A. Volpyas	(a) 114, K21	79.20; S1.2
V. Vorlíček	(a) 111, 345	78.55; 71.55; 78.30; S7.12
	(a) 111, 655	78.30; 73.40; S7.12
V.V. Vorobev	(a) 113, 375	62.20 and 64.60; 75.30; S1.4
A. Vorobeva	(a) 114, K33	73.40; S4; S10.1
S.A. Vorobiev	(b) 152, 61	61.80; S5.11
	(a) 112,K131	79.20
Yu.D. Vorobiev	(a) 114, K75	75.70; S1.2
S.P. Voronyuk	(b) 155, 717	78.20; S8.12
N. Vouroutzis	(a) 111, 491	64.70; S1.63
	(a) 115, 399	61.14 and 64.70; S1.63
N.V. Vugman	(a) 111, 625	76.30; S9.11
H. Wabnitz	(a) 114, 757	78.20; S10

H. Wada	(b) 152, K63	78.30; 63.20; S1.63
A.H. Wafik	(a) 111, 265	75.60; S1.2
	(a) 116, 383	75.60; S1.1
G. Wagner	(a) 112, 519	61.70 and 68.55; S7.11; S7.12; S7.15
	(a) 113, 71	61.70 and 68.55; S7.11; S7.15
H. Wakao	(a) 113,K245	75.50; S11.2
K. Walasek	(b) 153, 561	63.75 and 64.60; 75.50
H. Waldmann	(b) 153, 391	78.55; 71.70; S8.15
F. Wallrafen	(b) 155, K59	76.30; 71.55; S9.16
F. Walz	(a) 111, 611	75.60; 75.50; S1.1; S1.2
B.-Y. Wang	(a) 114, 451	61.70; 78.70; S1.2
C.H. Wang	(a) 113, K15	61.80; S5.11
Chunzheng Wang	(b) 154, 259	71.70; 76.30; 78.50; S12
En-Ge Wang	(b) 155, 587	71.55; 71.20; 73.20; S7.12
Guanghou Wang	(a) 111,K123	78.70; S11.1
Huai-Yu Wang	(b) 155, 587	71.55; 71.20; 73.20; S7.12
Hui-su Wang	(b) 151, 167	71.70; S8.1
L. Wang	(b) 153, 343	71.25 and 75.10
Qiang Wang	(b) 153, 139	64.60; 64.75; S1.1; S1.2
Renhui Wang	(b) 152, 33	61.14; 61.70; S1.2; S4
S.J. Wang	(a) 114, 273	74.70 and 78.70; S10.15
S.X. Wang	(a) 115, K71	74.70 and 79.60; S10.15
Shao-Ling Wang	(a) 114,K135	61.70 and 78.70; 64.70; S4
T.-M. Wang	(a) 114, 451	61.70; 78.70; S1.2
Wanlu Wang	(a) 113,K223	72.40 and 73.60; S5.11; S6
Wen-cai Wang	(a) 112, 541	61.40 and 68.55; S1.1; S5.12
Xiao-Gang Wang	(a) 116, K7	61.40 and 78.70; 74.70; S10.15
Ye Ning Wang	(a) 113, K85	62.20 and 74.70; S10.1
Yening Wang	(a) 116,K199	62.20 and 77.80; S12
Yifeng Wang	(a) 116, 267	64.70 and 74.70; 61.10; S10.15
Yong-Liang Wang	(b) 154, 649	71.55; S5.11
Yuansheng Wang	(a) 115, 467	61.40 and 64.70; S1.1; S1.2
Z. Wang	(a) 114, 273	74.70 and 78.70; S10.15
Zhenxi Wang	(a) 113,K127	75.30; S1.1; S1.4
Zhongguang Wang	(a) 114, 177	68.25; S4
Zidan Wang	(b) 151, K77	75.40
	(b) 154, K63	75.10
B.M. Wanklyn	(a) 115, 149	64.70; S9.16
S. Waplak	(a) 111, 271	76.30; 61.80; 66.70; S11
J.B. Ward	(a) 114,K223	76.80; S11
M.S. Wartak	(b) 152, K43	63.70
J. Wąsicki	(a) 114, 497	64.70; 61.12; S9
H. Wasiewicz	(b) 153, 361	76.30; S8.15
Y. Watanabe	(a) 111,K201	71.55 and 79.75; S9.11
B. Weber	(a) 112, 733	68.55; 61.80; S1.61
G. Weber	(b) 151, 605	71.55; S5.12
H.-J. Weber	(b) 156, 195	64.60; 78.20; 78.55; S11.1
R. Wedell	(b) 151, 479	61.70
G.Z. Wei	(b) 156, 363	75.10 and 75.30
J.P. von der Weid	(b) 153, 757	71.35 and 78.60; S9.11
	(b) 153, 765	71.35 and 78.60; S9.15
N. Weiden	(a) 113, 163	76.60; S9.16
H. Weinert	(b) 153,K193	78.55; 71.35; 78.30; S7.12; S7.15
	(a) 114,K183	73.40; 68.55; S7.12; S9.15
A. Weiss	(a) 113, 163	76.60; S9.16

H. Welling	(a) 115, 191	78.65; 77.55; S9; S10
E. Wendler	(a) 112, 289	61.80; 78.50; S7.11; S7.12
G. Wendrock	(a) 111, 417	61.55; 62.20; 64.70; S4
H. Wenzel	(a) 111, 377	73.40 and 85; S7.11; S7.15
P. Werner	(a) 112, 715	61.80 and 68.55; 68.48; S5.11
	(a) 116, 81	68.48 and 68.55; S1.61
Z. Werner	(a) 112, 757	61.80 and 68.20
J. Wernisch	(a) 114, 629	79.20; S1; S5; S5.11
W. Wesch	(a) 111, 59	68.48; 61.16; 61.80; S7.12
	(a) 112, 289	61.80; 78.50; S7.11; S7.12
P. Wesołowski	(a) 115, 81	66.30; S10
J.M. Wesselinowa	(b) 151,K209	77.80
	(b) 156,K129	74.20
R.N. West	(b) 155, 191	71.10 and 78.70; S5.11; S5.12
	(b) 156, 497	71.20 and 71.60; S5.11; S7.12
A. Whittaker	(a) 113,K267	73.40 and 85
G. Wiatrowski	(b) 154, 583	73.60
	(b) 156, 355	75.10 and 75.30
J. Wiechuła	(b) 151, 47	61.70; 76.30; 78.50; S9.11
H. Wieczorek	(a) 114, 413	72.40 and 73.40; 71.55; S5.11
G.A. Wiegers	(a) 114, 91	61.60; S1.63
J. Wiehl	(a) 111, 315	78.55 and 78.60; 71.55; S11.2
	(a) 111, 629	78.55 and 78.60; 71.55; S11.2
A. Wierzbicki	(b) 156, K71	75.70
E. Wieser	(a) 113, 321	61.70; 72.20; S5.11
H. Will	(a) 113, K27	64.60; 66.30
M.R. Willis	(a) 114, 299	72.80; S12
W. Windsch	(a) 114, 705	76.60; 77.80; S12
W. Wischert	(a) 111, 315	78.55 and 78.60; 71.55; S11.2
	(a) 111, 629	78.55 and 78.60; 71.55; S11.2
	(b) 156, 339	74.70; 71.30; 72.20; S10.15
H. Witte	(a) 112, K35	72.80; S7.12
W. Witthuhn	(b) 154, K99	61.70 and 63.20; S5.11
	(b) 155, 77	63.20; 78.30; S7.13; S8.11; S9
A. Witzmann	(a) 112, 747	68.55 and 79.20; 64.60; S1.61
J.J. Wnuk	(a) 114, K49	74.70; S1.4
R. Wojnecki	(b) 152, 203	71.50; S1.2
Z. Wokulski	(a) 114, K53	74.70; S1.62
J. Wolak	(b) 154, 333	64.70 and 76.30; 72.20; S12.1
T. Wołejko	(a) 114, 505	64.70 and 65.70; S11.1
J. Wolf	(a) 113,K203	65.70; S7.15
D. Wolff	(a) 112, 727	68.55
E. Wolska	(a) 114, K11	61.10 and 64.70; 61.60; S10.1
J. Woltersdorf	(a) 113, 131	68.55; 68.48; S1.2; S10.1
	(a) 116, 165	61.70; 61.16; S1.61; S10
J.C. Woolley	(a) 111, 405	61.60; 64.75; S8.16
	(a) 115, 157	64.70; S8.15
P. Wróbel	(b) 153,K177	74.20 and 75.10
A. Wrzeciono	(b) 151,K195	75.30; S1.1
	(a) 114, 355	75.30; 61.55; S1.1; S1.2
	(b) 156, 657	75.30; S1.1; S1.2; S1.4
F.M. Wu	(a) 113, K15	61.80; S5.11
Ping-Feng Wu	(b) 152, 577	71.70 and 76.30; S9.16
X.F. Wu	(b) 154, 85	64.70; 61.10; 62.20; 72.15; S1.2
Xiang Wu	(b) 152, 499	71.25; 73.40
H.-J. Wuensche	(a) 111, 377	73.40 and 85; S7.11; S7.15
	(b) 154, 815	72.10 and 73.40; 77.50; S8.15

M. Wurlitzer	(a) 116, 537	61.50; 74.70; S10.15
Hua Xia	(a) 112, 99	63.20; 78.35; S1.2
	(b) 155, 137	68.30; 78.30; S1.2; S1.3
Jiansheng Xia	(b) 154, K51	74.70; S10.15
Yuebo Xia	(a) 114, 177	68.25; S4
Jun-kai Xie	(b) 151, 167	71.70; S8.1
L.M. Xie	(b) 153, 107	63.20 and 78.30; S5.11
Sishen Xie	(a) 114, 651	61.60 and 74.70; S10.15
Rufeng Xu	(a) 113,K127	75.30; S1.1; S1.4
Z.-Y. Xu	(a) 114, 451	61.70; 78.70; S1.2
D.A. Yablonskii	(b) 152, K23	75.60 and 75.80
M.R. Yakheev	(b) 151, 337	78.30; 77.80; S8.16
	(b) 153, 727	78.30; 77.80; S8.16
A.I. Yakimov	(a) 113, 519	72.20; 72.60; S5.11
E.B. Yakimov	(a) 111, 81	61.80 and 71.55; 73.40; S5.11
V.N. Yakimovich	(a) 115,K155	64.70; S7
R.A. Yakshibaev	(a) 111, 431	61.80 and 66.90; S8.15
	(a) 112, K97	66.30; S3
G.G. Yakushcheva	(b) 156, 449	71.35 and 78.65; S8.12
M. Yamada	(a) 112, K15	64.75; S4
K. Yamakawa	(a) 111, 89	66.30; 72.15; S1.2
H. Yamamoto	(b) 154, 195	73.20 and 73.40
R. Yamamoto	(b) 152, 533	61.50 and 71.45
T. Yamane	(a) 116, 787	74.70; 61.80; S10.15
N. Yamashita	(b) 156, 537	71.70; 78.55; S8.11
V.A. Yanchuk	(b) 156,K101	68.45; S5.11
Chuan-zhang Yang	(b) 153, 323	75.10
F.M. Yang	(a) 115,K219	75.50; 75.30; S1.1
Jin Yang	(b) 153,K131	71.45; 71.38; S12.1
Jiping Yang	(a) 114, 757	78.20; S10
S.H. Yang	(a) 111, K53	61.70 and 72.70; S7.12
Yan-Jun Yang	(a) 116, 371	74.70; S10.15
Zhengju Yang	(b) 154, 167	64.60; S2
	(b) 154, 181	64.60; S9
A.Yu. Yangirov	(a) 111, K17	61.70; 72.80; 75.50; S8
B. Yangui	(a) 116, 669	64.70; S10
P.H. Yannakopoulos	(a) 114, K41	61.80 and 72.40; S7.12; S7.15
Xixian Yao	(a) 114,K189	74.70; S10.15
C.T. Yap	(a) 114,K223	76.80; S11
S. Yariv	(a) 113, 631	78.60; S11
N.A. Yarykin	(a) 111, 81	61.80 and 71.55; 73.40; S5.11
G.Yu. Yashin	(b) 153, K25	71.38
S.M. Yavadova	(a) 115, K5	65.40 and 65.50; S8.16
Huaqing Yin	(a) 115, K75	74.70; 62.40; S10.15
L. Yin	(a) 114, K1	61.30; S12
P. Yli-Lahti	(b) 154, 305	72.20; 72.60; 76.30; S12
Zhi-pu You	(a) 111, K49	71.55; S5.11
Huifang Young	(a) 116,K199	62.20 and 77.80; S12
W.H. Young	(b) 154, 469	61.50; S2
F.N. Yousif	(b) 154, K47	71.10
Moheb H. Youssef	(b) 155, 453	62.80; S1.1; S1.2; S4
T.H. Youssef	(a) 111, 137	62.20; S4
Y.A. Youssef	(a) 113, 123	68.55; 78.65; S10

Gong Yu	(b) 152, 73	61.40 and 68.30; 78.30; S5.11
I. Yu	(b) 153,K183	78.55; S8
Zhu-Xin Yu	(a) 114,K135	61.70 and 78.70; 64.70; S4
J.S. Yuan	(a) 113,K267	73.40 and 85
Lanping Yue	(a) 111,K155	62.40; S1
I.R. Yukhnovskii	(b) 153, 583	64.60; 77.80
	(b) 154, 519	64.60; 77.80
A.D. Yumatoff	(b) 151, 581	71.25 and 78.70; S9.11
V.A. Yunkin	(a) 111, 81	61.80 and 71.55; 73.40; S5.11
E.N. Yurchenko	(a) 113, 359	64.75; 76.80; S10
V.A. Yurchenko	(a) 112, 331	61.80 and 73.40; S7.12
M.A. Yurishchev	(b) 153, 703	75.10; S12.1
Yu.V. Yushkievich	(a) 112, 815	79.70
Yu.I. Yuzyk	(b) 153, 517	64.60 and 78.30; S8.16
	(b) 153, 529	64.60 and 78.30; S8.16
Yu.I. Yuzyuk	(b) 154, 777	78.30; 77.80; S9.16
K. Yvon	(a) 115, 515	71.20; 61.55; S10
V.A. Zablotskii	(a) 112, 145	75.60
V.N. Zabolotskii	(a) 111, 431	61.80 and 66.90; S8.15
H.-J. Zacharias	(a) 113,K103	61.70 and 75.50; S11.2
A.A. Zahab	(a) 115, 491	68.55 and 73.60; S8
R.I. Zainullina	(b) 155, 419	64.60; 75.30; S1.61
A.G. Zaitsev	(a) 114, K21	79.20; S1.2
A.M. Zaitsev	(a) 115, 427	61.80; 61.70; 78.60; S5; S7.1
I.N. Zakharchenko	(a) 114, 559	64.70 and 68.55; 77.80; S10.15
L. Zakharenko	(a) 111,K215	61.80 and 72.80; S7.12
L.F. Zakharenkov	(a) 113, K41	64.75; S7.11
C. Zaldo	(a) 114, 397	78.50; 78.55; S10
A.V. Zalesskii	(b) 155, K63	76.60; S10.1
M. Załužny	(b) 153, K21	71.25 and 73.60
	(b) 153,K115	71.25 and 73.60; S8
	(b) 155, 709	71.25 and 78.30
H. Zangar	(a) 115, 107	64.70; 61.14; 77.20; S11
A. Zani	(a) 115, 437	61.80; 61.10; S5.11
X. Zaoui	(b) 156,K109	68.55 and 79.60; S8
M.B. Zapart	(a) 116,K191	76.30; 64.70; S11.1
W. Zapart	(a) 116,K191	76.30; 64.70; S11.1
R.G. Zaporozhchenko	(b) 152, 347	78.70; 71.25; S5.11
V.A. Zaporozhchenko	(b) 152, 347	78.70; 71.25; S5.11
G. Zardas	(a) 114,K215	72.40; 78.50; S7.11
G.E. Zardas	(a) 114, K41	61.80 and 72.40; S7.12; S7.15
	(a) 116,K133	73.40 and 85; 61.80; S7.12
I.M. Zaritskii	(a) 113, 653	73.40 and 85; 76.30; S5.11; S10.1
	(a) 115, 75	61.80; 73.40; S5.11
E.V. Zarochentsev	(a) 116, 295	62.30
D.I. Zarovskii	(a) 111, 499	61.70 and 68.55; S1.61
A.V. Zastavnoy	(a) 113, 277	61.70; 71.55; S5.11
E.A. Zavadskii	(a) 112, 145	75.60
M. Závětová	(a) 111, 655	78.30; 73.40; S7.12
G.S. Zavt	(b) 155, 147	71.45; 63.20; S12.1
D.M. Zayachuk	(a) 111, K99	72.80 and 75.50; S8
M.H. Zayan	(a) 116, 383	75.60; S1.1
N.S. Zayats	(a) 113, 497	71.55 and 78.55; 61.80; S7.12
S. Zazubovich	(b) 152, 563	71.55 and 78.55; S9.11

K. Ždánský	(a) 112, 639	73.40 and 85; S7.15
	(a) 116, 357	71.55; S7.15
A.P. Zdebskii	(a) 116, K91	62.65 and 72 40; S8.15
A. Zehe	(a) 115, 223	68.20; 68.55; 79.20; S5.11
	(b) 156, 65	66.30
	(b) 156, K89	71.55 and 78.55; 73.40; S7.12
	(a) 116, 735	61.80 and 68.55; 77.55; S9.12
T. Zemčík	(a) 112, K1	61.60; 76.80; S1.1
D. Zenke	(a) 116, 179	61.40; 61.16; 62.20; S12
A. Zentgraf	(a) 112, 747	68.55 and 79.20; 64.60; S1.61
A. Zentko	(a) 114, 679	61.80 and 75.30; 75.40; 75.60; S1.1; S1.2
Yu.E. Zerov	(b) 151, 61	63.10 and 63.50
J.Th. Zettler	(a) 112, 395	61.80 and 68.55; 73.40; S10.15
Cai-Guo Zhan	(a) 114,K135	61.70 and 78.70; 64.70; S4
De-Gang Zhang	(b) 153,K131	71.45; 71.38; S12.1
Fangqing Zhang	(b) 152, 73	61.40 and 68.30; 78.30; S5.11
	(b) 154, K39	61.70 and 68.60; S5.12
	(a) 114,K171	68.60 and 76.30; S6
H.T. Zhang	(b) 153, 465	61.70; 62.20; 62.40
Hong Zhang	(a) 114,K127	78.45; 78.30; 78.55; S11.1
	(a) 116, K7	61.40 and 78.70; 74.70; S10.15
J.L. Zhang	(a) 115, K67	64.70 and 72.80; S10
K.Y. Zhang	(b) 156, 363	75.10 and 75.30
Li-Yuan Zhang	(b) 155, 587	71.55; 71.20; 73.20; S7.12
Peng-Xiang Zhang	(b) 156, 145	63.20 and 78.30; 71.36
Q.R. Zhang	(a) 115, K71	74.70 and 79.60; S10.15
Qirui Zhang	(b) 154, K51	74.70; S10.15
Shougong Zhang	(a) 113,K127	75.30; S1.1; S1.4
Shouye Zhang	(b) 155, 647	75.50; S11.2
W. Zhang	(a) 112, 99	63.20; 78.35; S1.2
X.K. Zhang	(a) 112, 99	63.20; 78.35; S1.2
Xiao-Wen Zhang	(b) 153, 139	64.60; 64.75; S1.1; S1.2
Xing-Kui Zhang	(b) 155, 137	68.30; 78.30; S1.2; S1.3
Yitong Zhang	(a) 114,K189	74.70; S10.15
Yuling Zhang	(a) 114, 651	61.60 and 74.70; S10.15
M.Zh. Zhanibekov	(b) 156, K77	76.30; 61.80; S9.11
Min-Guang Zhao	(b) 153, 249	71.70; 76.30; S1.2
R.W. Zhao	(a) 115,K219	75.50; 75.30; S1.1
Sang-bo Zhao	(b) 151, 167	71.70; S8.1
	(b) 152, 181	71.70; 78.50; S10.1
T.S. Zhao	(a) 112, 175	75.50; S1.1; S1.2
Y.G. Zhao	(b) 153, 107	63.20 and 78.30; S5.11
O.A. Zharov	(a) 113,K149	61.14
N.V. Zhavoronkov	(a) 113,K257	78.45; S9.16; S11.2
N.A. Zheltova	(a) 115,K195	74.70; 78.30; 79.20; 79.60
	(a) 116, K13	61.50; 74.70; S10.15
Fanchun Zheng	(a) 115, K75	74.70; 62.40; S10.15
Keqin Zheng	(a) 116, 621	62.40; 61.70; S1.4; S3
Siming Zheng	(b) 155, 653	75.10
Wen-Chen Zheng	(b) 151, K11	61.70; 62.20; 76.30; S11.1
	(b) 151, K17	61.70 and 71.70; 76.30; S10.1
	(b) 151,K145	71.70; 76.30; S9.12; S10; S10.1
	(b) 153,K103	65.70; 71.70; 76.30; S11.1
	(b) 154, K73	76.30; 71.70; S9.15
	(b) 154,K167	71.70 and 76.30; S11.1

Wen-Chen Zheng	(b) 155,K149	71.70 and 76.30; S9.16
N.N. Zholonko	(b) 156, 119	63.10 and 65.40; 65.70; S5.2
Jia-lin Zhong	(b) 153, 323	75.10
Kang-Wei Zhou	(b) 152, 181	71.70; 78.50; S10.1
Shu-qin Zhou	(a) 116, 709	73.60; 71.55; S12
X. Zhou	(a) 113, 11	61.80 and 64.60; 61.55; S1.62
Z.Y. Zhou	(a) 115, 149	64.70; S9.16
G.I. Zhovnir	(a) 115, K31	68.55; S8.15
A.W. Zhu	(a) 113, 393	62.40; S4
	(a) 113,K195	62.40; S4
Daoben Zhu	(a) 111, K67	72.80; 72.40; 78.40; S12.1
J.J. Zhu	(a) 115, K91	76.80; S1.1
Meifang Zhu	(a) 111, 67	
S.N. Zhu	(a) 112, 99	63.20; 78.35; S1.2
S. Zhukauskas	(a) 111, K43	73.30; S7.12
A.P. Zhukov	(a) 111,K237	75.60; S1.1
	(a) 112,K127	75.60; S1.1
V.P. Zhukov	(b) 151, 407	71.25; S1.61; S1.62
A.F. Zhuravlev	(b) 151, 111	73.60 and 78.65
T.S. Zhuravleva	(a) 115, K47	72.80; S12
S.P. Zhvavyi	(a) 112, K19	64.70; 66.70; S5.11
M. Zielecki	(a) 112, 365	62.20; 61.80; S1.1
J. Zielinski	(b) 151, 203	74.70
O. Ziemann	(a) 114,K237	78.45; S7.15
P. Ziesche	(b) 152, 111	71.10 and 73.40
Z. Zikmund	(a) 112, 625	77.30 and 77.80; S12
M.B. Zikry	(a) 112, K91	62.20; S1.2
L.A. Zilberman	(a) 111,K145	61.70; S7.13
L.G. Zimin	(b) 156, 449	71.35 and 78.65; S8.12
H. Zimmermann	(a) 112, 569	71.55; 64.75; 78.50; 78.55; S8.13
M. Zimpel	(b) 155, 581	71.45 and 71.70
J. Zinkiewicz	(a) 112, 835	66.30; S1.1
V.S. Zinovev	(a) 116, K57	65.70 and 68.25; 64.30; S5.11; S10.1
J. Żmija	(a) 113, 199	61.70 and 78.50; 71.55; S10.15
J.-P. Zoellner	(b) 151, 145	71.25 and 73.40; S5.11; S5.12; S7.1
I. Zólomy	(a) 111, 371	73.40 and 85
	(a) 115,K129	73.40 and 85; S5.11
	(a) 116,K223	73.40 and 85; S5.11
I.V. Zolotukhin	(a) 116, 255	62.40; S1.2
N.V. Zotova	(a) 115,K117	78.45 and 78.55; S7.12; S7.15
E.M. Zub	(a) 114, 573	73.60; S1.63
G. Zubauskas	(a) 116,K161	66.30 and 68.48; S7.11
H. Zuber	(b) 156, 561	71.35 and 71.70; 78.55; S8.15
V.E. Zubkus	(b) 156, 93	64.60 and 74.20
J. Żuk	(a) 112, 789	61.80 and 68.55; 78.20; S7.12
	(a) 112, 799	61.70; 78.60; 79.20
A. Žukauskas	(b) 153,K193	78.55; 71.35; 78.30; S7.12; S7.15
K. Zukowska	(a) 112, 785	61.80; 78.20; S7.12
P.W. Żukowski	(a) 112, 695	61.70; 79.20; S5.11
J. Zukrowski	(a) 114, 361	75.30 and 75.80; S1.1; S1.2; S1.4
I. Zupančič	(b) 151, K1	66.30; 64.70; S12
I.P. Zvyagin	(b) 152, 231	72.10; 72.20; 72.40
	(b) 152, 231	72.10; 72.20; 72.40

Substance Classification*)

S1. Transition metals and their alloys and compounds (for oxides, see S10)
- S1.1. Fe
- S1.2. Other 3d metals
- S1.3. Noble metals
- S1.4. Lanthanides
- S1.5. Actinides
- S1.61. Carbides, silicides, germanides
- S1.62. Nitrides, phosphides, arsenides
- S1.63. Sulfides, selenides, tellurides

S2. Alkali metals and their alloys and compounds (for alkali halides, see S9.11)

S3. Alkaline earth metals and their alloys and compounds (for alkaline earth halides, see S9.12)

S4. Other metals and their alloys

S5. Nonmetallic elements
- S5.11. Silicon
- S5.12. Germanium
- S5.2 Solidified gases

S6. Group IV compounds (for transition metal compounds, see S1.61)

S7. Group V compounds (for transition metal compounds, see S1.62)
- S7.1. III–V compounds
- S7.11. Phosphides
- S7.12. Arsenides
- S7.13. Antimonides
- S7.15. Mixed crystals
- S7.16. Higher compounds

S8. Chalcogenides (for transition metal compounds, see S1.63)
- S8.1. II–VI compounds
- S8.11. Sulfides
- S8.12. Selenides
- S8.13. Tellurides
- S8.15. Mixed crystals
- S8.16. Higher compounds

S9. Halides
- S9.11. Alkali halides
- S9.12. Alkaline earth halides
- S9.15. Mixed crystals
- S9.16. Higher compounds

S10. Oxides
- S10.1. Simple oxides
- S10.15. Mixed crystals

S11. Other higher compounds
- S11.1. Salts of oxy-acids
- S11.2. Spinels, garnets, ferrites

S12. Organic compounds
- S12.1. Quasi-one-dimensional systems

For comments on the use of the classification scheme see p. 13.

---

*) Outside the ICSTI Classification for Physics.

Manuscripts and letters for physica status solidi (b) — basic research and physica status solidi (a) — applied research

should be sent to

Leipziger Straße 3–4, P. B. 1233, 1086 Berlin

or

Kunst und Wissen, Erich Bieber OHG, Postfach 330104, D-1000 Berlin 33.

The object of "physica status solidi (b)" and "physica status solidi (a)" is to publish rapidly important and topical results concerning fundamental research in the field of solid state physics and aspects of application and experimental techniques of fundamental solid state research, respectively. The issues of "physica status solidi (b)" and "physica status solidi (a)" will be published on the 1st and the 16th day of each month, respectively. Either series will contain the following categories:

1. Review articles written by arrangement with the editors. They will be remunerated.
2. Original papers selected from those which have not been submitted for publication elsewhere, after being reviewed. (The latest date for acceptance of contributions is the 8th and the 23rd day of each month for "physica status solidi (b)" and "physica status solidi (a)", respectively, 50 days before the delivery of the journal.)
3. Short notes published in the off-set part of the journal. They are, if necessary after being reviewed, selected by editorial staff. (The latest date for acceptance of manuscripts for this part is the 12th and the 27th day of each month for "physica status solidi (b)" and "physica status solidi (a)", respectively, 16 days before the delivery of the journal.)

Editorial Information: General

Instructions for preparing of figures are given in the second issue of each volume

Manuscripts should be submitted preferentially in English. Besides, papers in German, French, and Russian are admitted. The final decision on the incorporation of the papers in the series (b) or (a) is taken by the editorial staff after consultation of the referee. Papers on liquids are only published in physica status solidi if the problems treated are also relevant for the solid state. Authors are asked to observe the following rules:

1. Manuscripts have to be written as consisely as possible. At most, 20 and 4 pages (28 lines per page; abstracts, tables, references, figures, and captions of figures included, where 3 figures are counted as about one page) are admitted for original papers and short notes, respectively. All manuscripts should be typewritten on one side only. Two copies are required. Badly arranged manuscripts are returned.
2. The author's name, initials, and institution should be given.
3. One abstract of about ten lines in English and one in German, French, or Russian are required for original papers.
4. Because of lack of space conclusions (if any) must not have the character of a summary (i.e. substantially a repetition of the contents of the abstract).
5. Formulae should not be written too small and not with pencil. Separate lines for formulae are desirable. Indistinct writing will lead to a return of the manuscript.
6. Drawings should be enclosed on separate sheets of paper, not exceeding the maximum and minimum sizes of 20×30 and 10×15 cm^2, respectively. Photographs should be submitted, on white paper, only for review articles and original papers. Duplicates are required. Detailed instructions for preparing of figures are given on cover 4 of the second issue of each volume.
7. Captions for the figures should be submitted on separate sheets of paper, and special instructions with regard to the size of illustrations should be given on these.
8. References should be numbered in order of appearance and should be given, on a separate sheet, in the form used in physica status solidi.
9. SI-units should be used.
10. Each paper must contain a proposal for the subject classification according to the classification scheme appearing in the January issues of each annual set.

The authors will be sent galley proofs of review articles and original papers. Proof reading of short notes and pre-printed titles will be done by the editorial staff.

For each review article, original paper, and short note a total of 50 reprints is available free of charge. Requests for additional reprints (against payment) should be given together with the submitted manuscripts. **The deadline for orders of additional reprints is the 15th and the 30th of each month** for "physica status solidi (b)" and "physica status solidi (a)", respectively, two weeks before the delivery of the journal.

FSC
www.fsc.org
MIX
Papier | Fördert gute Waldnutzung
FSC® C083411

Kolibri 360 GmbH
Ferdinand-Jühlke-Straße 7
99095 Erfurt, Deutschland
produktsicherheit@kolibri360.de